Applied Mathematical Sciences

EDITORS

Fritz John
Courant Institute of Mathematical Sciences
New York University
New York, NY 10012

J.E. Marsden
Department of Mathematics
University of California
Berkeley, CA 94720

Lawrence Sirovich
Division of Applied Mathematics
Brown University
Providence, RI 02912

ADVISORS

M. Ghil New York University
J.K. Hale Brown University
J. Keller Stanford University
K. Kirchgässner Universität Stuttgart
B. Matkowsky Northwestern University
J.T. Stuart Imperial College
A. Weinstein University of California

EDITORIAL STATEMENT

The mathematization of all sciences, the fading of traditional scientific boundaries, the impact of computer technology, the growing importance of mathematical-computer modelling and the necessity of scientific planning all create the need both in education and research for books that are introductory to and abreast of these developments.

The purpose of this series is to provide such books, suitable for the user of mathematics, the mathematician interested in applications, and the student scientist. In particular, this series will provide an outlet for material less formally presented and more anticipatory of needs than finished texts or monographs, yet of immediate interest because of the novelty of its treatment of an application or of mathematics being applied or lying close to applications.

The aim of the series is, through rapid publication in an attractive but inexpensive format, to make material of current interest widely accessible. This implies the absence of excessive generality and abstraction, and unrealistic idealization, but with quality of exposition as a goal.

Many of the books will originate out of and will stimulate the development of new undergraduate and graduate courses in the applications of mathematics. Some of the books will present introductions to new areas of research, new applications and act as signposts for new directions in the mathematical sciences. This series will often serve as an intermediate stage of the publication of material which, through exposure here, will be further developed and refined. These will appear in conventional format and in hard cover.

MANUSCRIPTS

The Editors welcome all inquiries regarding the submission of manuscripts for the series. Final preparation of all manuscripts will take place in the editorial offices of the series in the Division of Applied Mathematics, Brown University, Providence, Rhode Island.

SPRINGER SCIENCE+BUSINESS MEDIA, LLC

Applied Mathematical Sciences | Volume 57

Applied Mathematical Sciences

1. John: **Partial Differential Equations**, 4th ed.
2. Sirovich: **Techniques of Asymptotic Analysis.**
3. Hale: **Theory of Functional Differential Equations**, 2nd ed.
4. Percus: **Combinatorial Methods.**
5. von Mises/Friedrichs: **Fluid Dynamics.**
6. Freiberger/Grenander: **A Short Course in Computational Probability and Statistics.**
7. Pipkin: **Lectures on Viscoelasticity Theory.**
9. Friedrichs: **Spectral Theory of Operators in Hilbert Space.**
11. Wolovich: **Linear Multivariable Systems.**
12. Berkovitz: **Optimal Control Theory.**
13. Bluman/Cole: **Similarity Methods for Differential Equations.**
14. Yoshizawa: **Stability Theory and the Existence of Periodic Solutions and Almost Periodic Solutions.**
15. Braun: **Differential Equations and Their Applications**, 3rd ed.
16. Lefschetz: **Applications of Algebraic Topology.**
17. Collatz/Wetterling: **Optimization Problems.**
18. Grenander: **Pattern Synthesis: Lectures in Pattern Theory, Vol I.**
20. Driver: **Ordinary and Delay Differential Equations.**
21. Courant/Friedrichs: **Supersonic Flow and Shock Waves.**
22. Rouche/Habets/Laloy: **Stability Theory by Liapunov's Direct Method.**
23. Lamperti: **Stochastic Processes: A Survey of the Mathematical Theory.**
24. Grenander: **Pattern Analysis: Lectures in Pattern Theory, Vol. II.**
25. Davies: **Integral Transforms and Their Applications**, 2nd ed.
26. Kushner/Clark: **Stochastic Approximation Methods for Constrained and Unconstrained Systems.**
27. de Boor: **A Practical Guide to Splines.**
28. Keilson: **Markov Chain Models—Rarity and Exponentiality.**
29. de Veubeke: **A Course in Elasticity.**
30. Sniatycki: **Geometric Quantization and Quantum Mechanics.**
31. Reid: **Sturmian Theory for Ordinary Differential Equations.**
32. Meis/Markowitz: **Numerical Solution of Partial Differential Equations.**
33. Grenander: **Regular Structures: Lectures in Pattern Theory, Vol. III.**
34. Kevorkian/Cole: **Perturbation Methods in Applied Mathematics.**
35. Carr: **Applications of Centre Manifold Theory.**
36. Bengtsson/Ghil/Källén: **Dynamic Meterology: Data Assimilation Methods.**
37. Saperstone: **Semidynamical Systems in Infinite Dimensional Spaces.**
38. Lichtenberg/Lieberman: **Regular and Stochastic Motion.**

(continued on inside back cover)

H.-J. Reinhardt

Analysis of Approximation Methods for Differential and Integral Equations

With 20 Figures

Springer Science+Business Media, LLC

H.-J. Reinhardt
Johann-Wolfgang-Goethe-Universität
Fachbereich Mathematik
6000 Frankfurt Main
Federal Republic of Germany

AMS Subject Classification: 41-02, 34A45, 35A35, 41A65, 45L05

ISBN 978-0-387-96214-6 ISBN 978-1-4612-1080-1 (eBook)
DOI 10.1007/978-1-4612-1080-1

Library of Congress Cataloging-in-Publication Data
Reinhardt, H.-J.
Analysis of approximation methods for differential and integral equations.
(Applied mathematical sciences; v. 57)
Includes bibliographies and index.
1. Approximation theory. 2. Differential equations. 3. Differential equations, Partial. 4. Integral equations. I. Title. II. Series: Applied mathematical sciences (Springer-Verlag New York Inc.): v. 57.
QA1.A647 vol. 57 510 s 85-16546
[QA297] [511'.4]

9 8 7 6 5 4 3 2 1

Preface

This book is primarily based on the research done by the Numerical Analysis Group at the Goethe-Universität in Frankfurt/Main, and on material presented in several graduate courses by the author between 1977 and 1981. It is hoped that the text will be useful for graduate students and for scientists interested in studying a fundamental theoretical analysis of numerical methods along with its application to the most diverse classes of differential and integral equations.

The text treats numerous methods for approximating solutions of three classes of problems: (elliptic) boundary-value problems, (hyperbolic and parabolic) initial value problems in partial differential equations, and integral equations of the second kind. The aim is to develop a unifying convergence theory, and thereby prove the convergence of, as well as provide error estimates for, the approximations generated by specific numerical methods. The schemes for numerically solving boundary-value problems are additionally divided into the two categories of finite-difference methods and of projection methods for approximating their variational formulations.

In accordance with our aims, we present in Part I approximation methods to each of the aforementioned classes of problems, state results concerning the solvability of the underlying approximate equations, and, for nonlinear problems, consider iterative procedures for their solution. Then, in Part II, we develop our underlying convergence theory for sequences of equations based on the concept of "discrete convergence". In Part III and IV, we reconsider the problem areas mentioned above and show, by means of our theory, the convergence of solutions obtained by specific methods when applied to a series of examples.

The convergence theory of approximation methods that we present in the text is applicable to a series of classes of both linear and nonlinear problems and will in many cases enable us to obtain two-sided error estimates. The methods we consider, for example, encompass finite element methods as well as finite-difference approximations for both ordinary and partial differential equations. Similarly, projection methods and methods based on quadrature formulas for numerically treating integral equations of the second kind can be analyzed with the techniques presented here. Moreover, the general convergence results can still be applied to other problems and other classes of approximation methods (to, say, initial value problems in ordinary differential equations, collocation methods for initial and boundary-value problems, etc.). The general convergence theory presented in the text was essentially developed by F. Stummel. Further developments and refinements were made by R. D. Grigorieff and his group (at the Technical University in Berlin) and by the author. At some places in the text, in particular in Part IV, unpublished results are contained in the presentation. It is appropriate, at this point, to mention the earlier contributions of Aubin (1967), Browder (1967), Céa (1964), Pereyra (1967), Petryshyn (1967b, 1968a), Stetter (1965a, 1965b, 1966), Vainikko (1967) in developing a convergence theory for approximation methods.

By necessity, we must limit the scope of the material presented in this book. The convergence theory, on the one hand, is developed in a very general setting, but, on the other, is restricted to problems where the approximating equations are expressed in terms of equicontinuously equidifferentiable mappings. These, of course, include linear problems. In the concrete applications, we mostly study problems with one spatial dimension. In higher dimensional problems, however (e.g. Poisson's equation or the two-dimensional heat equation), we consider only examples having rectangular spatial domains. We would like to mention that the approximation theory of finite elements can be treated by the analysis we develop in this book; but, due to the basic orientation of our presentation, we shall study finite element methods only in the context of specific examples. Moreover, there are numerous variants of the schemes considered in the text which will not be discussed because of lack of space. The concrete methods we consider serve to demonstrate the applicability of our general convergence theory as well as provide analytical techniques.

This book may also serve as a reference for a series of well-known and other numerical schemes for the problem classes considered. For practical purposes, the numerical methods can be chosen - and used - according to their stability and convergence properties provided in the text. It should be noted, however, that one and the same method may be stable, inversely stable and convergent or may not be stable, etc. depending on the norms underlying the analysis. For example, inverse stability of the well-known Crank-Nicolson-Galerkin method approximating the heat equation is explored three-fold in the text with the result that this method is conditionally stable with respect to the maximum norm, unconditionally stable in the sense of the von Neumann stability criterion, and unconditionally stable relative to suitable Sobolev norms which are even stronger than the maximum norms. Furthermore, there are schemes which produce converging approximations but only for restricted classes of problems; this phenomenon is expressed by the concept of stable convergence. We like to emphasize again that any stability or convergence statement is only relative to the underlying norms. The interested, more practically orientated reader is invited to study such phenomena by means of computational experiments which, for most of the examples considered in the book, can be performed on personal computers.

We now want to give some technical hints which should be noted in order to make the reading of the text easier. The book consists of thirteen chapters and is organized in four parts. Each chapter contains different sections and is preceeded by an introduction. The same notation will be used for labeling formulas and specifying conditions; we refer to, e.g., formula (60) in Chapter 4 simply by 4.(60). A different notation will be employed for denoting theorems, lemmas, important properties, and propositions, e.g., Theorem 5.9, Lemma 6.3, Property 7.6, Proposition 12.8, etc.. In the text and at the conclusion of each chapter, we cite references only by author and year of publication, and give additional works pertinent to the study but not specifically referred to in the text. The full reference complete with title of the cited work can be found in the bibliography following the final chapter. At a few places in the text, comments are made concerning extensions of the results we present but, in general, we do not give an extensive discussion of related literature. The reader who is not interested in all problem areas in the text should select the relevant chapters according to the following diagram:

Problems	Chap. in Part I	Chap.	
Boundary-value problems	1	8	
Variational equations	2	9	Part III
Integral equations	3	10	
Initial value problems	4	11-13	Part IV

In order to appreciate the convergence analysis in Chapters 8 to 13, it is, however, necessary to study - or, at least, to take notice of - the convergence theory developed in Part II (Chapters 5-7). We strongly recommend though that the reader has a basic knowledge of numerical analysis and functional analysis in order to gain the most benefit from this book.

The author would like to acknowledge his deep indebtedness to Professor F. Stummel who stimulated his interest in and from whom he learned about numerical analysis, first as a student and later as a collaborator. For various improvements, such as shorter proofs and better exposition at various places, the author is especially obliged to Professor R. D. Grigorieff who has been kind enough to read most of the manuscript. The author would like to express his appreciation to Professor

H. D. Victory, Jr., for his careful translation of - and, in some cases, suggestions for improving - the original German manuscript. For reading and discussing several chapters, the author is indebted to Professor K. H. Müller, Professor I. Sloan, and Privatdozent J. Lorenz, and to some of his students for reading parts of earlier versions of the manuscript. Special thanks are due to Mrs. H. Meßner for her prompt preparation of a preliminary version of this book, and to Mrs. Kate MacDougall for her careful typing of the final copy.

Contents

Part I
Presentation of Numerical Methods

In this preliminary part of the book, we introduce several problem areas and numerous suitable approximation methods. The problems we consider are

boundary-value problems in ordinary and partial differential equations;
integral equations of the second kind; and
initial value problems in partial differential equations.

For the first class of problems, we discuss two different techniques for deriving approximation methods, namely

finite-difference methods, and
projection methods for the associated variational formulation.

The equivalent variational formulation of boundary-value problems, rather than the classical differential equation formulation, serves as a basis for deriving approximations based on projection methods which include well-known Galerkin methods.

The integral equations we consider are approximated via

quadrature methods, and
projection methods.

For the third problem area, initial value problems in partial differential equations, the approximation methods may again be subdivided into two categories,

finite-difference methods, and
Galerkin methods.

The latter are further subdivided into continuous-time and discrete-time Galerkin methods. The similar nomenclature used for the approximation methods of both initial value problems and boundary-value problems indicate that techniques for deriving approximation schemes for the latter play a substantial role in obtaining approximation methods for the former. We consider linear and nonlinear examples in all problem areas and, correspondingly, derive linear and nonlinear methods.

We shall not be able, in general, to give an in-depth presentation of all the various numerical schemes appropriate for the problems we consider. Only at a few places in the text will it be feasible to consider the numerous variants of each respective method. Elsewhere in the text, however, we shall refer the interested reader to the cited literature for a thorough treatment of other methods related to those in the text.

An important goal of Part I is to acquaint the reader with well-known schemes for approximating problems arising in each of the three problem areas. Also, we formulate both the exact and approximate problems as operator equations which are basic for analyzing the stability and convergence properties in Parts III and IV of the text by the general convergence theory developed in Part II. Much of the discussion and formulas presented here will be referred to elsewhere in the text. On these grounds, we would strongly advise the reader to be familiar with the material in Part I.

The numerical methods we develop in this part will moreover be expressed in linear and nonlinear systems of equations which are suitable for numerical calculations; in addition, we shall cite results concerning their solvability.

Chapter 1
Finite-Difference Methods for Boundary-Value Problems

In this chapter, we approximate by means of finite-differences several prototype examples of boundary-value problems in both ordinary and partial differential equations. For each individual problem, we develop one or more finite-difference schemes and state some results on the solvability of the associated (linear or nonlinear) systems of equations. Then we formulate the original problem and its approximations as operator equations in suitable function spaces. In such a setting, we are then able to investigate accuracy properties of the finite-difference approximations themselves by analyzing the behavior of the truncation errors.

The methods we consider can be found in any standard numerical treatment of the class of problems presented in this chapter. A representative list of references is provided at the end of this chapter. From a perusal of the open literature, it becomes quite obvious that a great many distinct finite-difference approximations can be constructed for any specific problem. We make this rather clear for some of the examples presented in the chapter. In the next chapter, we shall see that we can generate still other approximations to our examples by means of so-called variational methods. A convergence analysis will be carried out in Chapter 8 for the finite-difference methods discussed here.

1.1 SAMPLE PROBLEMS

Before discussing in detail finite-difference approximations for various classes of differential equations, we present in this rather brief section a few prototype examples of linear, one- and two-dimensional boundary-value problems.

Example 1: An Inhomogeneous, One-Dimensional Boundary-Value Problem of Second Order

Differential Equation (D.E.): $-(pu')' + qu = f$ in $[a,b]$, $b > a$,

Boundary Conditions (B.C.): $u(a) = \alpha$, $u(b) = \beta$.

This differential equation models a diffusion process for one-dimensional regions, where p (> 0) is the diffusion coefficient; q (≥ 0) the absorption coefficient; and

f represents the source of the diffusing substance. We wish to find a solution u which has a differentiable "flow", $J = pu'$. The above differential equation is written in "self-adjoint" form. □

Example 2: Cantilevered Beam Problem (a linear, fourth-order, ordinary D.E.)

D.E.: $(pu'')'' + qu = f$ in $[0,\ell]$,
B.C.: $u(0) = u'(0) = u''(\ell) = (pu'')'(\ell) = 0$.

The solution $u \in C^4[0,\ell]$ represents the displacement of an elastic beam of length ℓ which is clamped at $x = 0$ and has a free end at $x = \ell$. The problem data p and q are assumed to be elements of $C^2[0,\ell]$ and $C[0,\ell]$, respectively, where p and q are related to the flexural rigidity and the compression modulus, resp.; f represents the load density. □

Example 3: Poisson's Equation

D.E.: $-\Delta u = f$ in G,
B.C.: $u|\partial G = g$.

Here, G is a bounded region in $\mathbb{R}^2$; ∂G its boundary; and $\Delta = \partial^2/\partial x_1^2 + \partial^2/\partial x_2^2$ the Laplacian. Poisson's equation describes, for example, the electrostatic field due to an external charge density f; the (steady-state) displacement of a perfectly elastic membrane due to an external force f, etc.. For $f = 0$, the D.E. is called *Laplace's equation* and describes the steady temperature distribution in a plate, the torsion of a beam, the flow of an incompressible ideal fluid through a two-dimensional channel, etc. □

The two-dimensional counterpart to Example 2 - a fourth order partial differential equation - is the plate bending problem. Examples 1 and 2 are special cases of an ordinary differential equation of order $2r$,

$$\sum_{j=0}^{r} (-1)^j [p_j(x)u^{(j)}]^{(j)} = f(x), \quad a \le x \le b,$$

with boundary conditions

$$\sum_{k=0}^{r-1} (\alpha_{jk}u^{(k)}(a) + \beta_{jk}u^{(k)}(b)) = \gamma_j, \quad j = 1,\dots,2r.$$

A general, *linear ordinary differential equation of second order* can be written as

$$u''(x) + p(x)u'(x) + q(x)u(x) = f(x), \quad a \le x \le b.$$

In general, associated separated boundary conditions are of the form

$$\alpha_0 u(a) - \alpha_1 u'(a) = \gamma_0, \quad \beta_0 u(b) + \beta_1 u'(b) = \gamma_1.$$

If $\alpha_1 = \beta_1 = 0$, then we have *Dirichlet boundary conditions* (or *boundary conditions of the first kind*); if $\alpha_0 = \beta_0 = 0$, we have *Neumann boundary conditions* or *boundary conditions of the second kind;* boundary conditions in the above general form are called *boundary conditions of the third kind*. There are still other types of boundary conditions, e.g., periodic ones.

Example 3 is a special case of a *general elliptic differential equation of second order* (in n dimensions),

$$\sum_{j,k=1}^{n} \frac{\partial}{\partial x_j} \left(a_{jk}(x)\frac{\partial u}{\partial x_k}\right) + a_0(x)u(x) = f(x), \qquad x \in G,$$

with *Dirichlet boundary conditions,*

$$u|\partial G = g.$$

In general, the boundary conditions can also include directional derivatives.

By using Example 1, we now show that we may as well consider Dirichlet boundary-value problems with homogeneous boundary conditions only. The same result is also true for general ordinary differential equations of order $2r$ and for general elliptic differential equations of second order. Indeed, if we solve the following boundary-value problem in lieu of Example 1,

$$-(pz')' + qz = f - \tilde{f} \quad \text{in} \quad [a,b],$$

with homogeneous boundary conditions, $z(a) = z(b) = 0$, where

$$\tilde{f}(x) = \frac{\alpha-\beta}{b-a} p'(x) + (qs)(x), \qquad s(x) = \alpha \frac{b-x}{b-a} + \beta \frac{x-a}{b-a},$$

then we easily obtain the solution of Example 1 by

$$u = z + s.$$

This is immediate, since $s(a) = \alpha$, $s(b) = \beta$, and

$$-(ps')' + qs = \tilde{f}.$$

1.2. Finite-Difference Methods for Linear, Second Order Ordinary Differential Equations

In this section, we consider general finite-difference schemes on uniform meshes for approximating boundary-value problems of linear, second-order ordinary differential equations. In a few special cases, we state results on the solvability of the difference equations and on the order of the truncation error. We shall not consider existence and uniqueness questions for the exact problem itself; for this, we refer the reader to the appropriate literature.

Let $C^r[a,b]$, $b > a$, be the linear space of all, continuous real- or complex-valued functions, with continuous derivatives up to order r on the open, bounded interval $(a,b) \subset \mathbb{R}$, having well-defined, one-sided limits at both a and b. We consider the following *general boundary-value problem for a linear, second order ordinary differential equation:* We seek a function $u \in C^2[a,b]$ which solves the ordinary differential equation,

$$u''(x) + p(x)u'(x) + q(x)u(x) = f(x), \quad a \leq x \leq b, \tag{1a}$$

subject to the boundary conditions

$$\alpha_0 u(a) - \alpha_1 u'(a) = \gamma_0, \quad \beta_0 u(b) + \beta_1 u'(b) = \gamma_1, \tag{1b}$$

where $p,q,f \in C[a,b]$ are given functions and $\alpha_0, \alpha_1, \beta_0, \beta_1, \gamma_0, \gamma_1 \in \mathbb{K}$ ($= \mathbb{R}$ or $\mathbb{C}$) are specified numbers. In order that the boundary-value problem makes sense, we first assume that

$$|\alpha_0| + |\alpha_1| > 0, \quad |\beta_0| + |\beta_1| > 0.$$

We refer to Coddington & Levinson (1955), Chapt. 7, Walter (1976), V., etc. for statements assuring the existence and uniqueness of solutions.

The differential operator occurring in (1a) is denoted by

$$(Lv)(x) \equiv v''(x) + p(x)v'(x) + q(x)v(x), \quad v \in C^2[a,b],$$

and the boundary conditions can be more succinctly expressed by defining

$$\ell_0(v) \equiv \alpha_0 v(a) - \alpha_1 v'(a), \quad \ell_1(v) \equiv \beta_0 v(b) + \beta_1 v'(b).$$

We use finite-difference methods to approximate the solution of (1a,b). In order to properly describe such methods, we let Λ be a sequence of positive *mesh sizes* (or *mesh widths*) h converging to zero. For every $h \in \Lambda$, let an equidistant mesh be defined by

$$I_h \equiv \{x \in \mathbb{R}: x = x_j \equiv a_h + jh, \quad j = 0,\dots,J_h\}$$

where $J_h \in \mathbb{N}$, $hJ_h = b_h - a_h$ with a_h and b_h given real numbers (approximating a and b, respectively) such that $b_h > a_h$. The following mesh

$$I_h' \equiv \{x \in \mathbb{R}: x = a_h + jh, \quad j = 1,\dots,J_h-1\}$$

will also be of use in our discussion. In the simplest case, $a_h = a$ and $b_h = b$ for all h. It is sometimes advantageous, however, to allow mesh points outside $[a,b]$, in order to better approximate the derivative terms in the boundary conditions. The vector space of all functions defined on I_h or I_h' with values in $\mathbb{K}$ is denoted as $C(I_h)$ or $C(I_h')$, respectively. We approximate the derivatives in (1a) by the following difference quotients,

$$D_h^+ v_h(x) = \frac{1}{h}(v_h(x+h) - v_h(x)), \quad x, x+h \in I_h,$$

$$D_h^- v_h(x) = \frac{1}{h}(v_h(x) - v_h(x-h)), \quad x, x-h \in I_h,$$

$$D_h v_h(x) = \frac{1}{2h}(v_h(x+h) - v_h(x-h)), \quad x \in I_h',$$

$$D_h^2 v_h(x) = \frac{1}{h^2}(v_h(x+h) - 2v_h(x) + v_h(x-h)), \quad x \in I_h', \quad v_h \in C(I_h).$$

In order of appearance, these difference quotients are labeled respectively as the *forward, backward, central difference quotient of first order* and the *central difference quotient of second order.* It is well known that, for sufficiently smooth functions, the first three approximate the first derivative with errors $O(h)$, $O(h)$ and $O(h^2)$, respectively, whereas the fourth difference quotient approximates the second derivative with accuracy $O(h^2)$. Here, $O(\cdot)$ represents the well-known *Landau symbol*, i.e., $\phi(h) = O(h^r)$ means that $|\phi(h)/h^r|$ is bounded as $h \to 0$.

A possible finite-difference approximation to the differential equation (1a) is then given by

$$(L_h u_h)(x) \equiv D_h^2 u_h(x) + p_h(x) D_h u_h(x) + q_h(x) u_h(x) = f_h(x), \quad x \in I_h', \tag{2a}$$

where p_h, q_h, and f_h are approximations of p, q, and f, respectively, which, in the simplest case, can be the restrictions of the given functions to the mesh points. The boundary conditions are approximated by

$$\begin{aligned} \ell_{0,h}(u_h) &\equiv \alpha_{0,h}(a_h) - \alpha_{1,h} D_h^+ u_h(a_h) = \gamma_{0,h}, \\ \ell_{1,h}(u_h) &\equiv \beta_{0,h} u_h(b_h) + \beta_{1,h} D_h^- u_h(b_h) = \gamma_{1,h}, \end{aligned} \tag{2b}$$

where the numbers $\alpha_{0,h}$, $\alpha_{1,h}$, $\beta_{0,h}$, $\beta_{1,h}$, $\gamma_{0,h}$, and $\gamma_{1,h} \in \mathbb{K}$ are approximations of α_0, α_1, β_0, β_1, γ_0 and γ_1, respectively. By use of the central difference quotients in (2a), we have approximated the first and second derivatives with the same order of accuracy. In several cases - for example, whenever p_h takes on very large values - it may be more advantageous to approximate the first derivative by forward or backward difference quotients.

We can clearly express L_h in the form

$$(L_h v_h)(x) = a_{-1,h}(x) v_h(x-h) + a_{0,h}(x) v_h(x) + a_{1,h}(x) v_h(x+h), \quad x \in I_h', \tag{3}$$

where $a_{i,h}$, $i = 0, \pm 1$, are the following functions defined on the mesh I_h',

$$\begin{aligned} a_{-1,h}(x) &= \frac{1}{h^2}(1 - \frac{h}{2} p_h(x)), \quad a_{0,h}(x) = -\frac{1}{h^2}(2 - h^2 q_h(x)), \\ a_{1,h}(x) &= \frac{1}{h^2}(1 + \frac{h}{2} p_h(x)). \end{aligned} \tag{4}$$

The above approximations lead to the following linear system of equations,

$$L_h u_h = f_h, \quad \ell_{0,h}(u_h) = \gamma_{0,h}, \quad \ell_{1,h}(u_h) = \gamma_{1,h},$$

for determining the mesh function $u_h \in C(I_h)$. For the case of Dirichlet boundary conditions, we can provide a first criterion for solvability. Indeed, let $\alpha_1 = \beta_1 = \alpha_{1,h} = \beta_{1,h} = 0$ and $\alpha_{0,h} \neq 0$, $\beta_{0,h} \neq 0$, and let

$$|a_{-1,h}| + |a_{1,h}| \leq |a_{0,h}|, \ |a_{1,h}| < |a_{0,h}| \quad \text{in } I_h'. \tag{5}$$

Then the associated $(J_h-1) \times (J_h-1)$ tridiagonal matrix is regular since it satisfies the *weak row sum criterion*. The latter means for an $n \times n$ matrix $A = (a_{ij})$ that (cf. Stummel & Hainer (1982), 8.2.2.)

$$\sum_{\substack{j \neq i \\ j=1}}^{n} |a_{ij}| \leq |a_{ii}|, \quad \sum_{j=i+1}^{n} |a_{ij}| < |a_{ii}|, \quad i = 1,\ldots,n .$$

For approximation (2a), condition (5) is always satisfied for sufficiently small h whenever p_h is uniformly bounded (with respect to h) and $q_h \leq 0$. Under these conditions, the associated matrix (multiplied by -1) is moreover an irreducibly diagonally dominant M-matrix. (For the definition of an irreducibly diagonally dominant matrix, we refer to Varga (1962), Sec. 1.4 and 1.5; we shall define the notion of an M-matrix in Section 1.4 before Theorem 1.1.)

We now express the boundary-value problem and its approximations as operator equations. We view the differential operator L as a mapping from $C^2[a,b]$ onto $C[a,b]$. The boundary conditions are included by defining an operator $A: E \to F$ by

$$Av \equiv (Lv, \ell_0(v), \ell_1(v)), \quad v \in E \equiv C^2[a,b], \quad F \equiv C[a,b] \times \mathbb{R}^2.$$

Solving the boundary-value problem (1a,b) is then tantamount to solving the equation

$$Au = w,$$

where $w \in F$ denotes an element $w \equiv (f, \gamma_0, \gamma_1)$. The finite-difference equations can be expressed in an analogous way. Indeed, L_h defines a mapping from $C(I_h)$ into $C(I_h')$. The boundary conditions are incorporated by defining

$$A_h v_h \equiv (L_h v_h, \ell_{0,h}(v_h), \ell_{1,h}(v_h)), \quad v_h \in C(I_h),$$

where $A_h: E_h \to F_h$ with $E_h \equiv C(I_h)$, $F_h \equiv C(I_h') \times \mathbb{R}^2$. Equations (2a,b) are then equivalent to

$$A_h u_h = w_h$$

where $w_h \equiv (f_h, \gamma_{0,h}, \gamma_{1,h}) \in F_h$.

The maximum norms are obvious ones to use for analyzing the above finite-difference methods,

$$||v||_E \equiv \max_{a\le x\le b} |v(x)|, \quad v \in E, \quad ||v_h||_{E_h} \equiv \max_{x\in I_h} |v_h(x)|, \quad v_h \in E_h, \tag{6}$$

$$||z||_F \equiv \max_{a\le x\le b} |g(x)| + |\xi_0| + |\xi_1|, \quad z = (g,\xi_0,\xi_1) \in F,$$

$$||z_h||_{F_h} \equiv \max_{x\in I_h'} |g_h(x)| + |\xi_0| + |\xi_1|, \quad z_h = (g_h,\xi_0,\xi_1) \in F_h. \tag{7}$$

We can also take account of the derivatives by setting

$$||v||_E = \max_{i=0,1,2} \max_{a\le x\le b} |v^{(i)}(x)|, \quad v \in E. \tag{8a}$$

In a corresponding manner, we can select a norm in E_h involving the maxima of the first and second difference quotients,

$$||v_h||_{E_h} = \max\left(\max_{x\in I_h} |v_h(x)|, \max_{x,x+h\in I_h} |D_h^+ v_h(x)|, \max_{x\in I_h'} |D_h^2 v_h(x)|\right), \quad v_h \in E_h. \tag{8b}$$

With regard to error estimates, it is, of course, desirable to show these with respect to the norms defined in (8). The choice of norms will essentially be determined by whatever methods are available for proving such estimates.

The sense in which the finite difference equations constitute an approximation of the boundary value problem can be made precise by the so-called *truncation errors*. These are the errors which occur when we apply the finite-difference operator L_h and the approximate boundary conditions $\ell_{0,h}$, $\ell_{1,h}$ to the solution u of the given boundary-value problem. More precisely, in the simple case where $a_h = a$, $b_h = b$, $h \in \Lambda$, the *truncation error* $\tau_h^{(0)}$ associated with the difference equation is given by

$$L_h u = Lu + \tau_h^{(0)} \tag{9a}$$

whereas the *truncation errors* $\tau_{0,h}$, $\tau_{1,h}$ associated with the approximate boundary conditions are specified by

$$\ell_{o,h}(u) = \ell_0(u) + \tau_{0,h}, \quad \ell_{1,h}(u) = \ell_1(u) + \tau_{1,h}. \tag{9b}$$

(Strictly speaking, we cannot actually apply $L_h, \ell_{0,h}$, and $\ell_{1,h}$ to u per se, since these are defined only for mesh functions. We have implicitly assumed that u has been replaced by its restriction $u|I_h$ to the mesh I_h. If mesh points lie outside $[a,b]$, then u must be extended in an appropriate manner (cf. Section 5.3.).) For the case $p_h = p|I_h'$ and $q_h = q|I_h'$, a Taylor series expansion shows that

$$\max_{x \in I_h'} |\tau_h^{(0)}(x)| = O(h^2) \quad (h \to 0)$$

for a solution u of (1a,b) in $C^4[a,b]$. For the boundary approximations (2b), we have the order of accuracy

$$|\tau_{h,0}| + |\tau_{h,1}| = O(h) \quad (h \to 0)$$

when $\alpha_{i,h} = \alpha_i$, $\beta_{i,h} = \beta_i$, $h \in \Lambda$. However, if we modify the approximations of the boundary condition by using central difference quotients of first order, then we can easily obtain an order of accuracy $O(h^2)$ also for the truncation errors associated with the boundary conditions. With Dirichlet boundary conditions $\alpha_{1,h} = \beta_{1,h} = 0$, $\alpha_{0,h} = \alpha_0$, $\beta_{0,h} = \beta_0$, we can easily see that the latter truncation errors will vanish.

The above definition of truncation errors agrees with the usual one in the literature in case the right-hand sides $w_h = (f_h, \gamma_{0,h}, \gamma_{1,h})$ of the approximating equations are given by $(f|I_h', \gamma_0, \gamma_1)$; then $\tau_h^{(0)} = L_h(u|I_h) - f_h$, $\tau_{i,h} = \ell_{i,h}(u|I_h) - \gamma_{i,h}$, $i = 0,1$.

As a final remark, we point out that Example 1 of Section 1.1 will also represent a special case of the boundary-value problems considered in (1a,b), if p is assumed continuously differentiable and the solution u lies in $C^2[a,b]$. In lieu of the so-called self-adjoint form presented in Section 1.1, we can express Example 1 by

$$-pu'' - p'u' + qu = f \quad \text{in} \quad [a,b], \quad u(a) = u(b) = 0.$$

If we, moreover, assume that $p \neq 0$ in $[a,b]$, and consider the simple case $a_h = a$, $b_h = b$, then we obtain the following tridiagonal system

$$(1 - \frac{h}{2}\tilde{p}_j)v_{j-1} - (2+h^2\tilde{q}_j)v_j + (1 + \frac{h}{2}\tilde{p}_j)v_{j+1} = -h^2\tilde{f}_j,$$
$$j = 1,\ldots,J-1, \quad v_0 = v_J = 0,$$

for determining $v_j = u_h(x_j)$, $j = 0,\ldots,J$ $(= J_h)$. Here, $\tilde{p}_j$, $\tilde{q}_j$, and $\tilde{f}_j$, $1 \le j \le J-1$, are given respectively by $\tilde{p}_j = p'(x_j)/p(x_j)$, $\tilde{q}_j = q(x_j)/p(x_j)$ and $\tilde{f}_j = f(x_j)/p(x_j)$, $1 \le j \le J-1$.

1.3. A FINITE-DIFFERENCE APPROXIMATION TO THE CANTILEVERED BEAM PROBLEM

We have already introduced the "Cantilevered Beam Problem" as Example 2 of Section 1.1. In this section, we provide a finite-difference approximation and examine the accuracy of the resulting truncation errors.

In the interval $[0,\ell]$, we define two equidistant meshes with mesh width h,

$$I_h \equiv \{x \in \mathbb{R}: x = jh, \quad j = -1,0,1,\dots,J_h+2\},$$

$$I_h' \equiv \{x \in \mathbb{R}: x = jh, \quad j = 1,\dots,J_h\},$$

where $J_h \in \mathbb{N}$ has the property that $J_h h = \ell$. Using the difference quotients defined in Section 1.2, we approximate the exact boundary-value problem,

$$\begin{aligned} &(pu'')'' + qu = f \quad \text{in} \quad [0,\ell], \\ &u(0) = \alpha^{(1)}, \quad u'(0) = \alpha^{(2)}, \quad u''(\ell) = \beta^{(1)}, \quad (pu'')'(\ell) = \beta^{(2)}, \end{aligned} \tag{10}$$

with given functions $p \in C^2[0,\ell]$, $q,f \in C[0,\ell]$, by the finite-difference equations

$$(L_h u_h)(x) \equiv D_h^2[p_h(D_h^2 u_h)](x) + q_h(x)u_h(x) = f_h(x), \quad x \in I_h', \tag{11a}$$

with boundary conditions

$$u_h(0) = \alpha_h^{(1)}, \quad D_h u_h(0) = \alpha_h^{(2)}, \quad D_h^2 u_h(\ell) = \beta_h^{(1)}, \quad D_h(p_h D_h^2 u_h)(\ell) = \beta_h^{(2)}. \tag{11b}$$

Here, p_h, q_h and f_h are approximations of p, q, and f, respectively, and $\alpha_h^{(i)}$, $\beta_h^{(i)}$, $i = 1,2$, are approximations of the right-hand sides in the boundary conditions which, for the problem in Section 1.2, are zero. An obvious way to define p_h, q_h and f_h is to restrict p, q, and f, respectively, to the mesh points, if we assume that p is also defined outside of $[0,\ell]$. For this case and when $\alpha_h^{(i)} = \alpha^{(i)}$, $\beta_h^{(i)} = \beta^{(i)}$, $i = 1,2$, we can express (11a) and (11b), respectively, as

$$\begin{aligned} &p_{j-1}v_{j-2} - 2(p_{j-1}+p_j)v_{j-1} + (p_{j-1}+4p_j+p_{j+1}+q_j h^4)v_j \\ &-2(p_j+p_{j+1})v_{j+1} + p_{j+1}v_{j+2} = h^4 f_j, \quad j = 1,\dots,J, \end{aligned} \tag{12a}$$

and

$$\begin{aligned} &v_0 = \alpha^{(1)}, \quad v_{-1} - v_1 = 2h\alpha^{(2)}, \quad v_{J-1} - 2v_J + v_{J+1} = h^2\beta^{(1)}, \\ &p_{J-1}(2v_{J-1}-v_{J-2}) + (p_{J+1}-p_{J-1})v_J - p_{J+1}(2v_{J+1}-v_{J+2}) = 2h^3\beta^{(2)}, \end{aligned} \tag{12b}$$

where $J = J_h$ and $v_j = u_h(x_j)$, $p_j = p(x_j)$, etc.

We should note that the finite-difference equations have been constructed such that only points from I_h occur in the formation of $L_h u_h$ and the approximations of the boundary conditions. The use of mesh points outside $[0,\ell]$ is to achieve as good as possible an approximation to the boundary conditions. We could have, for example, defined I_h without using the point $x_{-1} = -h$, and then approximated the boundary condition $u'(0) = 0$ by the forward difference quotient at the point $x = 0$. However, the order of accuracy would be $O(h)$, in contrast to the previous $O(h^2)$ for a sufficiently smooth solution u. Indeed, there is a large number of other possibilities for replacing derivatives by difference quotients - e.g. replacing the boundary condition $u'(0) = 0$ by a difference quotient of accuracy $O(h^2)$ without using the point $x = -h$ - however, investigation of such questions is beyond the immediate scope of the text.

The equations (12a,b) represent a linear, algebraic system of equations for the unknowns $u_h(x)$, $x \in I_h$, with as many equations as unknowns. The proof of the existence of a solution u_h would require a nontrivial investigation of the properties of the finite-difference equations.

In order to formulate problems (10) and (11) as operator equations, we select $E \equiv C^4[0,\ell]$, $F \equiv C[0,\ell] \times \mathbb{K}^4$ and define a mapping $A: E \to F$ by

$$Av \equiv (Lv, v(0), v'(0), v''(\ell), (pv'')'(\ell)), \quad v \in E.$$

Solving the original boundary-value problem (10) is then tantamount to solving $Au = w$, where $w \in F$ is the element $w \equiv (f, \alpha^{(1)}, \alpha^{(2)}, \beta^{(1)}, \beta^{(2)})$. A possible choice of norms in E and F is

$$||v||_E \equiv \max_{0 \le i \le 4} \left(\max_{0 \le x \le \ell} |v^{(i)}(x)| \right), \quad v \in E,$$

$$||z||_F \equiv \max\left(\max_{0 \le x \le \ell} |g(x)|, |\zeta_1|, |\zeta_2|, |\zeta_3|, |\zeta_4| \right),$$

$$z = (g, \zeta_1, \zeta_2, \zeta_3, \zeta_4) \in F.$$

In an analogous manner, we can consider the finite-difference equations in the space $E_h = C_h^4(I_h)$ of mesh functions defined on I_h with the norm

$$||v_h||_{E_h} \equiv \max_{0 \le i \le 4} \left(\max_{x \in I_h^{(i)}} |D_h^i v_h(x)| \right), \quad v_h \in E_h,$$

where $D_h^3 v_h$ and $D_h^4 v_h$ are difference quotients of third and fourth order, respectively, and the maxima are taken over appropriate meshes $I_h^{(i)} \subset I_h$. As the image space, we take $F_h = C(I_h') \times \mathbb{K}^4$ equipped with the maximum norm

$$||z_h||_{F_h} \equiv \max\left(\max_{x \in I_h'} |g_h(x)|, |\zeta_1|, |\zeta_2|, |\zeta_3|, |\zeta_4| \right),$$

$$z_h = (g_h, \zeta_1, \zeta_2, \zeta_3, \zeta_4) \in F_h.$$

The mapping $A_h: E_h \to F_h$ associated with (11) is then defined by

$$A_h v_h \equiv (L_h v_h, v_h(0), D_h v_h(0), D_h^2 v_h(\ell), D_h(p_h D_h^2 v_h)(\ell)), \quad v_h \in E_h.$$

In addition, if we let $w_h \in F_h$ be the element $w_h \equiv (f_h, \alpha_h^{(0)}, \alpha_h^{(1)}, \beta_h^{(0)}, \beta_h^{(1)})$, then we can express the finite-difference equations (11a,b) as $A_h u_h = w_h$.

The accuracy of the approximations to the boundary-value problem by finite-difference equations is described by the behavior of the truncation error, as in the case of the boundary-value problem in Section 1.2. This will be well-defined, provided that the solution u of (10) can be extended outside of $[0,\ell]$. If we write $\tau_h = (\tau_h^{(0)},\tau_h^{(1)},\tau_h^{(2)},\tau_h^{(3)},\tau_h^{(4)})$, then the *truncation error* associated with the difference equation has the form $\tau_h^{(0)} = L_h(u|I_h) - f|I_h'$ with L_h defined in (11a); the *truncation errors* associated with the approximate boundary conditions can be likewise expressed as

$$\tau_h^{(1)} = u(0) - \alpha^{(1)} (= 0),\ \tau_h^{(2)} = D_h u(0) - \alpha^{(2)},\ \tau_h^{(3)} = D_h^2 u(\ell) - \beta^{(1)},$$

$$\tau_h^{(4)} = D_h(p_h D_h^2 u)(\ell) - \beta^{(2)}.$$

If the solution u and the function p can be extended outside of $[0,\ell]$ as C^4- and C^2-functions, respectively, then

$$||\tau_h||_{F_h} \to 0 \quad (h \to 0).$$

Under additional regularity assumptions on u and p, the truncation errors converge to zero as $O(h^2)$.

1.4. FINITE-DIFFERENCE METHODS FOR A NONLINEAR BOUNDARY-VALUE PROBLEM

In this section, we consider a boundary-value problem for a nonlinear, ordinary differential equation of second order, which also includes Example 1 from Section 1.1 as a special case. By retaining the self-adjoint form, we shall see that we can obtain a finite-difference approximation to the linear part different from that presented at the end of Section 1.2. Results on the solvability of the nonlinear finite-difference equations are proved by monotonicity arguments. We present the associated Gauss-Seidel and Newton methods for iteratively solving the nonlinear system of equations. Concluding remarks again examine the order of the truncation errors.

We consider the following problem (cf. Ciarlet, Schultz, and Varga (1967)),

$$-(pu')'(x) + (qu)(x) = f(x,u(x)) \quad \text{in} \quad [0,1], \quad u(0) = \gamma_0,\ u(1) = \gamma_1, \tag{13}$$

and seek a solution $u \in C^2[0,1]$. Initially, we assume that both q and f are continuous, and that $p \in C^1[0,1]$. Other assumptions are specified in the text as they are needed.

With an equidistant mesh size $h > 0$, we define meshes in $[0,1]$ by

$$I_h \equiv \{x \in [0,1]:\ x = x_j \equiv jh,\ j = 0,\dots,J_h\},$$

$$I_h' \equiv \{x \in [0,1]:\ x = jh,\ j = 1,\dots,J_h-1\},$$

where $J_h \in \mathbb{N}$, $hJ_h = 1$. A well-known finite-difference approximation is given by (cf. Marchuk (1975), Sec. 2.1)

$$\begin{aligned}(L_h u_h)(x_j) \equiv -\frac{1}{h^2}\{p_{j+1/2}(u_h(x_{j+1})-u_h(x_j)) - p_{j-1/2}(u_h(x_j) - u_h(x_{j-1}))\}\\ + q(x_j)u_h(x_j) = f_h(x_j,u_h(x_j)),\quad 1 \le j \le J_h-1,\end{aligned} \tag{14}$$

with boundary conditions $u_h(0) = \gamma_{0,h}$, $u_h(1) = \gamma_{1,h}$. Here, $x_{j\pm1/2} = (x_j+x_{j\pm1})/2$, $p_{j\pm1/2} = p(x_{j\pm1/2})$, and f_h represents an approximation to f. Selecting $f_h(x,y) = f(x,y)$, $x \in I_h'$, $y \in \mathbb{R}$, and defining $v_j \equiv u_h(x_j)$, $q_j \equiv q(x_j)$, we can then write the system of equations (14) as

$$-p_{j-1/2}v_{j-1} + (p_{j-1/2} + p_{j+1/2} + q_jh^2)v_j - p_{j+1/2}v_{j+1} = h^2 f(x_j,v_j), \tag{15}$$
$$1 \le j \le J_h-1.$$

If we set $E \equiv C^2[0,1]$, $F \equiv C[0,1] \times \mathbb{K}^2$, and

$$A\colon v \in E \to (Lv - f(.,v),v(0),v(1)) \in F,$$

with $Lv \equiv -(pv')' + qv$, then we see that (13) is equivalent to

$$Au = w,$$

where $w \equiv (0,\gamma_0,\gamma_1)$ is the right-hand side. For the finite-difference approximation, we select $E_h \equiv C(I_h)$ and $F_h \equiv C(I_h) \times \mathbb{K}^2$. A mapping A_h is then defined by (14) as follows,

$$A_h\colon v_h \in E_h \to (L_hv_h - f_h(\cdot,v_h),v_h(0),v_h(1)) \in F_h.$$

The finite-difference equations in (14) can then be also expressed as

$$A_hu_h = w_h \quad (\equiv (0,\gamma_{0,h},\gamma_{1,h})).$$

The following theorem shows that the system (15) is uniquely solvable under appropriate assumptions on the functions p, q, and f. In order to prove this result, we shall need the concept of an M-matrix and of an isotone, diagonal mapping. A matrix $\mathbf{A} = (a_{ij})$ is called an *M-matrix* in case $\mathbf{A}$ is invertible, $a_{ij} \le 0$, $i \ne j$, and $\mathbf{A}^{-1} \ge 0$. The last requirement means that each component of $\mathbf{A}^{-1}y$ is nonnegative whenever y also has this property. (This is expressed as $y \ge 0 \Rightarrow \mathbf{A}^{-1}y \ge 0$.) A mapping $T\colon D \subset \mathbb{R}^n \to \mathbb{R}^m$ is called *isotone*, if $Tx \le Ty$ for $x \le y$, $x,y \in D$. Finally, a mapping $\Phi\colon D \subset \mathbb{R}^n \to \mathbb{R}^n$ is called *diagonal*, if $\Phi_j(v_1,\dots,v_n) = \Phi_j(v_j)$, $1 \le j \le n$. Using results from Ortega & Rheinboldt (1970), we can now show that (15) is uniquely solvable.

<u>Theorem 1.1</u>: Let $p(x) \geq c_0 > 0$, $q(x) \geq 0$, and f satisfy

$$f(x,y) - f(x,y') \leq 0, \quad x \in [0,1], \quad y \geq y', \quad y,y' \in \mathbb{R}. \tag{16}$$

Then the system (15) is uniquely solvable.

<u>Proof</u>: We write (15) as $\mathbf{A}_h v + \Phi_h(v) = w_h$, where $\mathbf{A}_h$ is a tridiagonal matrix defined by the left-hand side of (15), and $\Phi_h = (\Phi_{1h},\ldots,\Phi_{nh})$, $n \equiv J_h - 1$, is given by

$$\Phi_{jh}(v_1,\ldots,v_n) \equiv -h^2 f(x_j,v_j), \quad j = 1,\ldots,n.$$

Under the above assumptions, $\mathbf{A}_h$ is an M-matrix (cf. Ortega & Rheinboldt (1970), Sec. 2.4). The mapping Φ_h is continuous and diagonal. It is also isotone since due to (16) we have that

$$\Phi_{jh}(v) - \Phi_{jh}(u) = h^2[f(x_j,u_j) - f(x_j,v_j)] \leq 0, \quad j = 1,\ldots,n,$$

whenever $u_j \geq v_j$, $j = 1,\ldots,n$. By Theorem 13.5.6 in Ortega and Rheinboldt (1970), $\mathbf{A}_h + \Phi_h$ is a homeomorphism of $\mathbb{R}^n$ onto itself, and hence (15) is uniquely solvable. □

If f possesses a continuous partial derivative with respect to the second argument, then (16) is clearly equivalent to

$$f_y(x,y) \leq 0, \quad x \in [0,1], \quad y \in \mathbb{R}. \tag{17}$$

We now turn to investigating the iterative solution of system (15). The *Gauss-Seidel method* has the form

$$\begin{aligned} v_j^{(t+1)} &= (p_{j-1/2} + p_{j+1/2} + q_j h^2)^{-1}\{p_{j-1/2} v_{j-1}^{(t+1)} + p_{j+1/2} v_{j+1}^{(t)} \\ &\quad + h^2 f(x_j, v_j^{(t)})\}, \quad j = 1,\ldots,J_h - 1, \quad t = 0,1,2,\ldots . \end{aligned} \tag{18}$$

For the *Newton method*, we get after calculating the Jacobian matrix,

$$\begin{aligned} -p_{j-1/2} v_{j-1}^{(t+1)} &+ (p_{j-1/2} + p_{j+1/2} + q_j h^2) v_j^{(t+1)} - p_{j+1/2} v_{j+1}^{(t+1)} \\ &- h^2 f_y(x_j, v_j^{(t)}) v_j^{(t+1)} = h^2\{f(x_j, v_j^{(t)}) - f_y(x_j, v_j^{(t)}) v_j^{(t)}\}, \\ &\qquad j = 1,2,\ldots,J_h - 1, \quad t = 0,1,\ldots . \end{aligned} \tag{19}$$

We must thus solve a tridiagonal system of equations in each iteration step. Theorem 13.3.8 of Ortega & Rheinboldt (1970) furnishes the global convergence of Newton's method to the solutions of (15) provided that the mapping Φ_h, defined in the proof of Theorem 1.1, is continuously differentiable, isotone, and convex. This is the case when f has a continuous partial derivative with respect to the second argument, and when (17) is satisfied along with

$$f(x,y') - f(x,y) \leq f_y(x,y)(y'-y), \quad x \in [0,1], \quad y,y' \in \mathbb{R}. \tag{20}$$

The truncation error associated with the difference equation is given in the case of (15) by

$$\tau_h^{(0)}(x) = (L_h u)(x) - f(x,u(x)), \quad x \in I_h',$$

where u denotes the solution of (13). The truncation errors associated with the approximate boundary conditions vanish. If $p \in C^3[0,1]$ and the solution of (13) is in $C^4[0,1]$, then we see by a Taylor series expansion that

$$\max_{x \in I_h'} |\tau_h^{(0)}(x)| = O(h^2) \quad (h \to 0). \tag{21}$$

A three-point difference scheme which even has a degree of accuracy of $O(h^4)$ can be found in Babuska, Prager and Vitasek (1966), Sec. 4.3.3. This improvement is achieved by a suitable approximation of an integral identity due to Marchuk for the solution of self-adjoint second-order ordinary differential equations (cf. Marchuk (1975), Sect. 2.1).

1.5. FINITE-DIFFERENCE APPROXIMATIONS FOR TWO-DIMENSIONAL ELLIPTIC EQUATIONS

We begin by considering *Poisson's differential equation* with Dirichlet boundary conditions as a prototype example of an elliptic boundary-value problem:

$$-\Delta u = f \quad \text{in} \quad G, \quad u|\partial G = g, \tag{22}$$

where G is an open, connected, bounded subset of $\mathbb{R}^2$. For the derivation of a finite-difference approximation, we let a uniform grid in $\mathbb{R}^2$ be defined by

$$\mathbb{R}_h^2 \equiv \{x \in \mathbb{R}^2 : x = (m_1h_1, m_2h_2), \quad m_1, m_2 \text{ integers}\}$$

where h_1 and h_2 are positive mesh sizes in the x_1- and x_2-directions, respectively. With $\overline{G}$ denoting the closure of G, we let $\overline{G}_h \equiv \overline{G} \cap \mathbb{R}_h^2$ be the set of all grid points in $\overline{G}$. Further let

$$G_h \equiv \{x = (x_1,x_2) \in \mathbb{R}^2 : [x_1-h_1, x_1+h_1] \times [x_2-h_2, x_2+h_2] \subset \overline{G}\}$$

be the *interior* of $\overline{G}_h$ and $\partial G_h \equiv \overline{G}_h - G_h$ the *boundary* of $\overline{G}_h$. For rectangles, as in the example below, we shall choose h_1, h_2 in such a way that ∂G_h contains precisely the mesh points on the boundary ∂G.

The simplest finite-difference approximation of the Laplace operator is obtained by approximating the second partial derivatives by central difference quotients of second order,

$$\Delta_h v_h \equiv D^2_{1,h} v_h + D^2_{2,h} v_h, \quad v_h \in C(\overline{G}_h), \tag{23}$$

where

$$D^2_{j,h} v_h(x) \equiv (1/h_j^2)(v_h(x + h_j e_j) - 2v_h(x) + v_h(x - h_j e_j)), \quad j = 1,2,$$

and $e_j = (\delta_{j1}, \delta_{j2})$, $j = 1,2$, are the standard basis vectors in $\mathbb{R}^2$. Δ_h is called the *Laplacian difference operator* (or the *discrete Laplace operator*). Its *truncation error* can be represented as

$$\begin{aligned} \tau_h^{(0)}(x) &= \Delta_h u(x) - \Delta u(x) \\ &= \frac{1}{24}\left[h_1^2 \frac{\partial^4 u}{\partial x_1^4}(\xi_1, x_2) + h_2^2 \frac{\partial^4 u}{\partial x_2^4}(x_1, \xi_2)\right], \quad x = (x_1, x_2) \in G_h, \end{aligned} \tag{24}$$

with points $\xi_j \in (x_j - h_j, x_j + h_j)$, $j = 1,2$; we have implicitly assumed that $u \in C^4(G)$.

The boundary-value problem (22) can be approximated by

$$-\Delta_h u_h = f_h \quad \text{in} \quad G_h, \qquad u_h = g_h \quad \text{in} \quad \partial G_h. \tag{25}$$

Here, the right-hand side f and the boundary condition g are also approximated. In the simplest case, $f_h(x) = f(x)$, $x \in G_h$, $g_h(x) = g(x)$, $x \in \partial G_h$. The diagram at the right shows in the case $h_1 = h_2$ the neighboring points of $x \in G_h$ together with their respective weights, with which the function values appear in the difference approximation (25) (when multiplied by h^2).

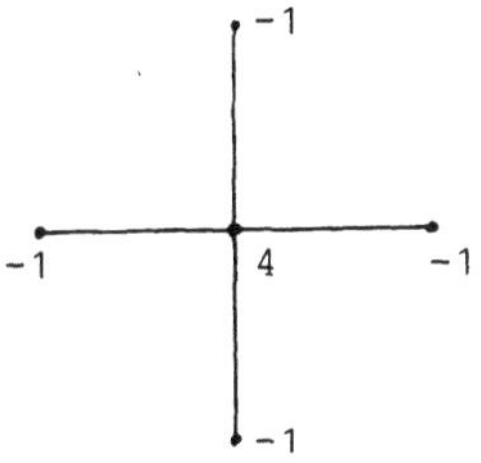

As an *example*, we choose a rectangular region G and a square mesh, i.e., $h = h_1 = h_2$. The mesh points of G_h may be enumerated in two different ways:

	3	6	9	12	15
	2	5	8	11	14
	1	4	7	10	13

	11	12	13	14	15
	6	7	8	9	10
	1	2	3	4	5

For each of the two cases presented, the associated systems of equations have a coefficient matrix consisting of tridiagonal block matrices. After multiplying by h^2, these become

$$\begin{pmatrix} A_3. & -I_3 & 0 & \cdots & 0 \\ -I_3 & A_3 & -I_3 & & \\ & -I_3 & A_3 & -I_3 & \\ & & -I_3 & A_3 & -I_3 \\ 0 & \cdots & 0 & -I_3 & A_3 \end{pmatrix}, \text{ respectively } \begin{pmatrix} A_5 & -I_5 & 0 \\ -I_5 & A_5 & -I_5 \\ 0 & -I_5 & A_5 \end{pmatrix}$$

where I_i denotes the $i \times i$-identity matrices, $i = 3,5$, and

$$A_3 = \begin{pmatrix} 4 & -1 & 0 \\ -1 & 4 & -1 \\ 0 & -1 & 4 \end{pmatrix}, \quad A_5 = \begin{pmatrix} 4 & -1 & 0 & . & 0 \\ -1 & 4 & -1 & & \\ 0 & -1 & 4 & -1 & \\ & & -1 & 4 & -1 \\ 0 & . & 0 & -1 & 4 \end{pmatrix}$$

It is obvious that the larger matrices above are irreducible diagonally dominant M-matrices, and hence are invertible. Moreover, they are symmetric and banded with band widths of 7 and 11, respectively. The system of equations can be therefore solved in an efficient manner by Gaussian elimination for banded matrices. Also suitable, for very large systems of equations, are iterative methods, of which the Jacobi method, Gauss-Seidel-method, relaxation methods and also ADI schemes are well-known examples. The above form of the linear systems suggests, moreover, that block-iteration methods may be used (cf. the references cited at the end of this chapter).

In an entirely analogous manner, we can treat a general *elliptic differential operator of second order,*

$$(Lv)(x) \equiv -\sum_{j=1}^{2} \left\{A_j(x) \frac{\partial^2 v}{\partial x_j^2}(x) + B_j(x) \frac{\partial v}{\partial x_j}(x)\right\} + C(x)v(x), \quad x \in G, \tag{26}$$

where the coefficients A_j, B_j, C are assumed to be continuous in $\overline{G}$ and to satisfy

$$A_j(x) > 0, \quad j = 1,2, \quad C(x) \geq 0, \quad x \in \overline{G}.$$

The associated Dirichlet boundary-value problem

$$Lu = f \text{ in } G, \quad u = g \text{ in } \partial G$$

is approximated on the mesh $\overline{G}_h$ by

$$L_h u_h = f_h \quad \text{in} \quad G_h, \quad u_h = g_h \quad \text{in} \quad \partial G_h,$$

where the difference operator L_h is obtained by approximating the derivatives by central difference quotients of first or second order,

$$L_h \equiv - \sum_{j=1}^{2} (A_j D_{j,h}^2 + B_j D_{j,h}) + C. \tag{27}$$

The truncation errors associated with the difference equation again approach zero as the squares of the mesh sizes,

$$\tau_h^{(0)}(x) = O(h_1^2 + h_2^2) \quad (h_1, h_2 \to 0). \tag{28}$$

REFERENCES

Babuska, Prager & Vitasek (1966), Braun (1983), Ciarlet, Schultz & Varga (1967)*, Coddington & Levinson (1955), Collatz (1966), Courant & Hilbert (1966), Forsythe & Wasow (1960), Garabedian (1964), Gladwell & Wait (1979), Hartman (1964), van der Houwen (1968), Isaacson & Keller (1966), John (1967,1982), Keller (1968, 1976), Mitchell & Griffiths (1980), Ortega & Rheinboldt (1970), Smirnow (1964), Smith (1978), Stoer & Bulirsch (1978), Stummel & Hainer (1982), Törnig (1979), Varga (1962), Wachspress (1966), Walter (1976)

*Article

Chapter 2
Projection Methods for Variational Equations

In this chapter, we reformulate the examples presented in Section 1.1 as variational equations, and then derive methods for approximating their solutions. In Section 2.1, we establish the relation between a linear operator equation and the associated variational formulation whenever the underlying linear space is a prehilbert space and then derive variational formulations in detail for each of the sample problems in Section 1.1. In order to deduce, however, existence and uniqueness of solutions to our examples via the fundamental Lax-Milgram Lemma, we must require our underlying spaces to be Hilbert spaces. Consequently, in Section 2.2, we complete to a Hilbert space each of the prehilbert spaces associated with the respective problem and discuss the concept of a generalized (or weak) solution. We then approximate the solutions of our variational problems in Sections 2.3 and 2.4 by considering and solving each problem in a finite-dimensional subspace of the respective Hilbert spaces obtained in Section 2.2. This procedure can be viewed as a projection method. Among the many special types of projection methods in this chapter, Ritz-Galerkin methods are used for approximating solutions to the linear examples. In Section 2.5, we shall see that projection methods can just as well be used to approximate solutions to nonlinear problems. These methods result in a nonlinear system of equations which must be solved iteratively, e.g., by Newton methods. We conclude this chapter by presenting a projection method for the nonlinear boundary-value problem introduced in Section 2.2 and show how approximations can be constructed by a procedure analogous to the Ritz-Galerkin method.

At this point, we would like to remark that other variational formulations exist for the examples considered. In addition, the choice of the finite-dimensional subspaces - e.g., spaces of piecewise polynomial functions or the linear span of certain eigenfunctions as in the case of "spectral methods" - leads to a large number of possible approximations of one and the same boundary-value problem. Unfortunately, however, these interesting aspects cannot be pursued in the text.

2.1. BASIC PROPERTIES OF VARIATIONAL EQUATIONS AND SAMPLE PROBLEMS

In this section, we reformulate the examples of one- and two-dimensional boundary-value problems given in Section 1.1 as variational equations and discuss the various properties of the associated bilinear forms. The motivation and use of the term "variational formulation" will become rather transparent if we take into account the different ways of expressing a given linear operator equation. An essential tool for ensuring the existence and uniqueness of solutions to such problems is the Lax-Milgram Lemma or its generalization by Babuska and Aziz (1972). In order to apply these results, we must have the underlying spaces complete; such a requirement leads to the concept of a generalized (or weak) solution to each of our sample problems (see the revised examples in Section 2.2).

We begin this discussion by first exploring the relationship between a given linear operator equation in a (pre-) Hilbert space and an equation expressed by an associated bilinear form. We have, for every linear operator $A: D(A) \subset E \to F$ with domain of definition $D(A)$ and prehilbert spaces E and F (over $\mathbb{K} = \mathbb{R}$ or $\mathbb{K} = \mathbb{C}$), the *sesquilinear form*

$$a(u,v) \equiv (Au,v), \quad u \in D(A), \quad v \in F. \tag{1}$$

(In case $\mathbb{K} = \mathbb{R}$, (1) is also called a *bilinear form*). This form is *bounded* (or equivalently, *continuous*) - i.e., there is an $\alpha_1 \geq 0$ such that

$$|a(u,v)| \leq \alpha_1 ||u|| \; ||v||$$

for every $u \in D(A)$ and every $v \in F$ - if, and only if, A itself is bounded, in which case α_1 can be taken as $||A||$. In the following, we shall denote the norm and scalar products in both E and F by the common notation $||\cdot||$ and $(.,.)$, respectively. Conversely, whenever F is a Hilbert space, each bounded sesquilinear form $a(.,.): E \times F \to \mathbb{K}$ defines a bounded linear mapping $A: E \to F$ via

$$(Au,v) = a(u,v), \quad u \in E, \quad v \in F, \tag{2}$$

because of the Riesz Representation Theorem, since $a(u,.)$ is a bounded semilinear (or anti-linear or conjugate-linear) functional on F for every fixed $u \in E$.

We now give some terminology useful for stating our main ideas and results. A sesquilinear form $a(.,.)$ on $E \times E$, with E a prehilbert space, is called

symmetric $\Longleftrightarrow$ $a(u,v) = \overline{a(v,u)}$, $u,v \in E$;
positive semidefinite $\Longleftrightarrow$ a is symmetric and $a(u,u) \geq 0$, $u \in E$;
positive definite $\Longleftrightarrow$ a is symmetric and $a(u,u) > 0$, $0 \neq u \in E$,
elliptic $\Longleftrightarrow$ $\exists \alpha_0 > 0 \; \ni: |a(u,u)| \geq \alpha_0 ||u||^2$ for every $u \in E$.

Whenever the sesquilinear form is symmetric, ellipticity clearly implies positive definiteness. An operator $A: D(A) \subset E \to E$ is called *symmetric (positive semi-definite, positive definite)* whenever the respective property holds for the associated sesquilinear form $a(.,.): D(A) \times D(A) \to \mathbb{K}$. Furthermore, a mapping A is *densely defined* in E whenever its domain of definition is dense in E. We note that $D(A) \subset E$ is a linear subspace, which is itself a Hilbert space in case it is closed and E is complete.

We now turn to the sample problems from Section 1.1. Here, $E = F$. For each problem, we give $D(A)$ and E as well as the associated sesquilinear form and examine its symmetry, positivity, and ellipticity properties. For the sake of simplicity, we shall assume that all functions are real-valued, and we shall restrict our study to homogeneous Dirichlet boundary conditions (see the remark at the end of Section 1.1).

Example 1: (see Example 1 of Section 1.1):

$$Au \equiv -(pu')' + qu;$$

$$D(A) \equiv \{u \in C^2[a,b]: \ u(a) = u(b) = 0\};$$

$$(u,v)_0 \equiv \int_a^b u(x)v(x)dx.$$

Integrating by parts, we obtain the bilinear form

$$a(u,v) = (pu',v')_0 + (qu,v)_0, \quad u,v \in D(A).$$

In addition to the L^2-scalar product, we consider

$$(u,v)_1 \equiv (u,v)_0 + (u',v')_0, \quad u,v \in C^1[a,b],$$

and denote the appropriate norms by $||\cdot||_i$, $i = 0,1$. We can easily show the following result.

Lemma 2.1: If $p \in C^1[a,b]$, $q \in C[a,b]$, $p(x) \geq c_0 > 0$, and $q \geq 0$, $x \in [a,b]$, then A is symmetric and there is a constant $\alpha_0 > 0$ such that

$$a(u,u) \geq \alpha_0||u||_1^2, \quad u \in D(A). \tag{3}$$

Proof: The symmetry of A is clear. From the assumptions on p and q, we can easily conclude that

$$a(u,u) \geq c_0(u',u')_0 + (qu,u)_0 \geq c_0||u'||_0^2, \quad u \in D(A).$$

We now show that

$$||u||_0 \leq (b-a)||u'||_0 \tag{4}$$

for all $u \in C^1[a,b]$ with $u(a) = 0$. Using the preceding inequality, we see that (3) follows from (4) with $\alpha_0 = c_0(1 + (b-a)^2)^{-1}$. We now turn to proving (4). Since $u(a) = 0$, we have

$$u(x) = \int_a^x u'(s)ds, \quad a \leq x \leq b;$$

now using Hölder's Inequality, we see that

$$|u(x)|^2 \leq \left[\int_a^b |u'(s)|ds\right]^2 \leq \int_a^b ds \int_a^b |u'(s)|^2 ds, \quad a \leq x \leq b,$$

and hence

$$\int_a^b |u(x)|^2 dx \leq (b-a)^2 \int_a^b |u'(x)|^2 dx. \qquad \square$$

Inequality (3) in Lemma 2.1 shows that A is elliptic with respect to the scalar product $(.,.)_1$ and, in particular, is positive definite. $\square$

Example 2: (see Example 2 of Section 1.1):

$Au \equiv (pu'')'' + qu;$

$D(A) \equiv \{u \in C^4[0,\ell]: u(0) = u(0) = u''(\ell) = (pu'')'(\ell) = 0\};$

$a(u,v) = (pu'',v'')_0 + (qu,v)_0, \quad u,v \in D(A)$

(after integrating by parts twice);

$(u,v)_2 \equiv (u,v)_0 + (u',v')_0 + (u'',v'')_0, \quad u,v \in C^2[0,\ell].$

Lemma 2.2.: If $p \in C^2[0,\ell]$, $q \in C[0,\ell]$, $p(x) \geq c_0 > 0$ and $q(x) \geq 0$, $x \in [0,\ell]$, then A is symmetric and there is a constant $\alpha_0 > 0$ such that

$$a(u,u) \geq \alpha_0||u||_2^2, \quad u \in D(A). \tag{5}$$

Proof: The symmetry of A is trivial. By assumption,

$$a(u,u) = (pu'',u'')_0 + (qu,u)_0 \geq c_0||u''||_0^2, \quad u \in D(A).$$

We note that the proof of Lemma 2.1 has yielded

$$||u||_0 \leq \ell||u'||_0 \quad \text{for} \quad u \in C^1[0,\ell] \quad \text{with} \quad u(0) = 0.$$

Now applying this inequality with u' in place of u, we get

$$||u'||_0 \leq \ell||u''||_0 \quad \text{for} \quad u \in C^2[0,\ell] \quad \text{with} \quad u'(0) = 0.$$

Combining the above two inequalities, we have

$$||u||_2^2 \leq (1 + \ell^2 + \ell^4)||u''||_0^2 \leq (1+\ell^2)^2||u''||_0^2,$$

and thereby obtain the following inequality,

$$a(u,u) \geq c_0||u''||_0^2 \geq c_0(1+\ell^2)^{-2}||u||_2^2, \quad u \in D(A). \qquad \square$$

Example 3: (see Example 3 of Section 1.1):

$$Au \equiv -\Delta u;$$

$$D(A) \equiv \{u \in C^2(G) \cap C^1(\bar{G}): \Delta u \in L^2(G),\ u|\partial G = 0\}.$$

An essential tool for representing the associated bilinear form is *Green's formula*,

$$\int_G \Delta u\ v\ dx + \int_G \nabla u\cdot\nabla v\ dx = \int_{\partial G} \frac{\partial u}{\partial n}\ v\ ds$$

where the *gradient* is defined by $\nabla u = (\frac{\partial u}{\partial x_1}, \frac{\partial u}{\partial x_2})$ and $\frac{\partial u}{\partial n}$ is the (outer) *normal derivative*. The Green's formula enables us to represent the associated bilinear form as

$$a(u,v) = [u,v]_1 \equiv \int_G \nabla u.\nabla v\ dx, \quad u,v \in D(A).$$

This bilinear form is also labeled the *Dirichlet-Integral*. With the scalar product $(u,v)_1$ defined as $(u,v)_1 \equiv (u,v)_0 + [u,v]_1$, we obtain the following result.

Lemma 2.3: A is symmetric and satisfies the following inequality,

$$a(u,u) \geq \alpha_0||u||_1^2, \quad u \in D(A) \tag{6}$$

for some constant $\alpha_0 > 0$, where $||u||_1^2 \equiv ||u||_0^2 + |u|_1^2$, $|u|_1 \equiv [u,u]_1^{1/2}$.

Proof: Inequality (4) appearing in the proof of Lemma 2.1 is also valid in several dimensions,

$$||u||_0 \leq c|u|_1, \quad u \in D(A), \tag{7}$$

where the constant C depends on the region G. (This is the Poincaré-Friedrichs Inequality for functions in $D(A)$). From (7), we have then that

$$||u||_1^2 = ||u||_0^2 + |u|_1^2 \leq (1 + C^2)|u|_1^2 = (1 + C^2)a(u,u), \quad u \in D(A),$$

thereby proving (6). $\square$

The above examples are special cases of the following, general situation. Let $A: D(A) \subset E \to E$ be a differential operator, where E is a prehilbert space and $D(A)$, the domain of definition of the operator A, incorporates the homogeneous boundary conditions in its definition.

Problem: Let $w \in E$. We seek $u \in D(A)$ such that

$$Au = w. \tag{8}$$

In case A is densely defined, this problem is clearly equivalent to a formulation in terms of the associated bilinear form $a(.,.)$: $D(A) \times D(A) \to \mathbb{R}$ defined in (1).

Variational Formulation: Let $w \in E$. We seek a $u \in D(A)$ such that

$$a(u,v) = f(v) \quad \forall v \in D(A), \tag{9}$$

where the linear functional f is defined on E by $f(v) \equiv (w,v)$, $v \in E$.

The term "variational" will be justified in due course. Toward this end, we consider, in addition, the following

Minimization Problem: Let $w \in E$. We seek $u \in D(A)$ such that

$$J(u) = \inf_{v \in D(A)} J(v), \tag{10}$$

where the quadratic functional J is given by $J(v) \equiv \frac{1}{2} a(v,v) - f(v)$, $v \in D(A)$.

The next result points out the equivalence of each problem to the others.

Theorem 2.4: Let A be densely defined and positive semidefinite. Then $u \in D(A)$ is a solution of (8) (and (9)), if, and only if, u is a solution of (10). If A is, moreover, positive definite, then each of (8), (9), or (10) has at most one solution.

Proof: For $u,v \in D(A)$,

$$2(J(v)-J(u)) = (Av-w,v-u) + (v-u,Au-w). \tag{11}$$

If u is a solution of (9), then, because $(A(v-u),v-u) \geq 0$, $w = Au$ in (11) yields

$$2(J(v) - J(u)) \geq 0 \quad \forall v \in D(A),$$

i.e., u is a solution of (10). Conversely, if u is a solution of (10), then we substitute in (11) $v = u + tz$, $t > 0$, with $z \in D(A)$, to get

$$0 \leq 2(J(v) - J(u)) = t(Au-w,z) + t^2(Az,z) + t(z,Au-w).$$

Dividing by t and letting $t \to 0$, yields

$$0 \leq (Au-w,z) + (z,Au-w) = 2(Au-w,z), \quad z \in D(A).$$

Substituting $-z$ for z, we get

$$0 \geq 2(Au-w,z)$$

and hence

$$(Au-w,z) = 0, \quad z \in D(A),$$

thereby showing that u is a solution of (9). The uniqueness of the solution of (9) is trivial in case A is positive definite. □

In the next lemma, we determine the Fréchet-derivative of J and use it to show that J is convex. A functional $J: V \subset E \to \mathbb{R}$ is *Fréchet-differentiable* at $u \in E$ (and the linear mapping $J'(u)$ is denoted as the associated *Fréchet-derivative* at u), if there exists a linear functional $J'(u)$ with the property that, for each $\varepsilon > 0$, there is a $\delta > 0$ such that

$$u + v \in V \quad \text{and} \quad |J(u+v) - J(u) - J'(u)v| \leq \varepsilon ||v||$$

for all $v \in E$ with $||v|| \leq \delta$. The Fréchet-derivative is uniquely defined (cf. Dieudonné (1969), VIII.1). Since we consider no other derivative than the Fréchet-derivative, we shall omit the term "Fréchet" when no confusion can arise. V denotes a linear subspace of E which will at times be chosen to be $D(A)$.

Lemma 2.5: Let $a(.,.): V \times V \to \mathbb{R}$ be positive definite and bounded. Then the quadratic functional J has a bounded derivative at each $u \in V$ given by

$$J'(u)v = a(u,v) - f(v), \quad v \in V, \tag{12}$$

and is moreover strictly convex, i.e.,

$$J(\lambda u + (1-\lambda)v) < \lambda J(u) + (1-\lambda)J(v), \quad u \neq v \in V, \quad 0 < \lambda < 1,$$

or, equivalently,

$$J(v) > J(u) + J'(u)(v-u), \quad u \neq v \in V.$$

Proof: (i) $J(u+v) - J(u) - J'(u)v$

$$= \frac{1}{2} a(u+v,u+v) - f(u+v) - \frac{1}{2} a(u,u) + f(u) - a(u,v) + f(v)$$

$$= \frac{1}{2} a(v,v), \quad u,v \in V.$$

Because of the boundedness of $a(.,.)$, we have

$$|J(u+v) - J(u) - J'(u)v| \leq \frac{\alpha}{2} ||v||^2,$$

and thereby the definition of the Frechet-derivative is satisfied with $\delta = 2\varepsilon/\alpha$.

(ii) If we now substitute in (i) $v-u$ for v and use the positive definiteness, we get

$$J(v) - J(u) - J'(u)(v-u) = \frac{1}{2} a(v-u,v-u) > 0, \quad u \neq v. \quad □$$

The following lemma is a result from the theory of optimization and is cited without proof.

Lemma 2.6: A necessary condition for a differentiable functional J to have an extremum at $u_0 \in V$, is

$$J'(u_0)v = 0 \quad \forall v \in V. \tag{13}$$

If J is strictly convex, then (13) is both necessary and sufficient for u_0 to be a unique minimum of J. □

This lemma, however, states nothing about the existence of a solution. The Fréchet-derivative $J'(u)v$ is also called the *first variation* of J. Equation (13) is called the *variational equation* for this reason and agrees with (9) in case $V = D(A)$.

The existence (and uniqueness) question for not necessarily symmetric bilinear forms is settled by the well-known *Lax-Milgram-Lemma*.

Theorem 2.7: Let V be a Hilbert space and let $a(.,.): V \times V \to \mathbb{R}$ be a bounded elliptic bilinear form. Then, for every bounded linear functional f on V, the variational equation (9) has a unique solution $u \in V$ satisfying the inequality

$$\alpha_0 ||u|| \leq \sup_{0 \neq v \in V} |f(v)|/||v||. \quad \square \tag{14}$$

We began this section by considering a more general situation, namely mappings between two (pre)Hilbert spaces E and F. For a bounded linear mapping $A: E \to F$ where E and F are prehilbert spaces, the equation

$$Au = w, \quad w \in F, \tag{15}$$

is clearly equivalent to

$$a(u,v) = f(v), \quad v \in F, \tag{16}$$

for determining $u \in E$. In analogy to (9), we also denote the latter equation as the associated *variational equation*. Here $a(.,.)$ is given in (1) and the bounded linear functional f by $f(v) = (w,v)$, $w \in F$. We have the following generalization of the Lax-Milgram Lemma which is due to Babuska and Aziz (1972).

Theorem 2.8: Let $A: E \to F$ be a bounded linear mapping between Hilbert spaces E and F which has the following properties:

$$\mu_0 ||v|| \leq \sup_{0 \neq \psi \in F} |(Av,\psi)|/||\psi||, \quad v \in E, \tag{17a}$$

$$\sup_{v \in E} |(Av,\psi)| > 0 \quad \forall 0 \neq \psi \in F. \tag{17b}$$

Then equation (15) is uniquely solvable for every $w \in F$ and the solution $u \in E$ satisfies the estimate,

$$\mu_0 ||u|| \leq ||w||. \tag{18}$$

Proof: A is clearly injective because of (17a) and the inverse $A^{-1}: R(A) \to E$ is bounded by $1/\mu_0$. This fact, together with the completeness of E, implies that

$R(A)$ is closed (cf. Kantorovich & Akilov (1964), Ch. XII, Sec. 1.1). Because of the completeness of F and because of (17b), A is surjective; for otherwise we would have some $0 \neq w_0 \in F$ with $(Au,w_0) = 0$ for every $u \in E$. This would contradict (17b). □

2.2. SAMPLE PROBLEMS (REVISED)

We now apply the Lax-Milgram Lemma to Examples 1-3. We recognize at the outset that the domain of definition of the linear operator in each example is not complete in the respective norm appearing in (3), (5) and (6). The completion of each underlying domain space leads to the concept of a generalized (or weak) solution. In the concluding paragraphs of this section we shall derive an associated variational equation for the nonlinear example in Section 1.4 by a procedure similar to that used for linear problems.

Example 1 (Revised): Let V be the completion of $D(A)$ with respect to $||\cdot||_1$. Then $V \subset L^2(a,b)$, because a Cauchy sequence in the norm $||\cdot||_1$ is also a Cauchy sequence in $L^2(a,b)$ and, thus, has a limit in $L^2(a,b)$. Further, $V \subset W^{1,2}(a,b) \equiv \{v \in L^2(a,b)$: their generalized derivatives $v' \in L^2(a,b)\}$. A function $v \in L^2(a,b)$ possesses a *generalized derivative* in $L^2(a,b)$ whenever there is a sequence $v_k \in C^1[a,b]$ converging to v in $L^2(a,b)$ for which the derivatives $\{v_k'\}$ form a Cauchy sequence in $L^2(a,b)$. The limit $v' = \lim_{k\to\infty} v_k'$ is called the *generalized derivative* (or *strong* L^2*-derivative*) of v and clearly lies in $L^2(a,b)$. Moreover, by integration by parts (and by passing to the limit), we see that

$$\int_a^b v'\phi dx = -\int_a^b v\phi' dx, \quad \phi \in C_0^\infty(a,b),$$

where $C_0^\infty(a,b)$ denotes the set of infinitely differentiable functions with compact support in (a,b). If a function $v \in L^2(a,b)$ satisfies the last equation with a function from $L^2(a,b)$ denoted by v', then this is already sufficient for v' being the generalized derivative of v (cf. Agmon (1965)).

The space V is moreover dense in $L^2(a,b)$, since $C_0^\infty(a,b)$ is dense in $L^2(a,b)$. Thus, the inclusions

$$C_0^\infty(a,b) \subset D(A) \subset V \subset W^{1,2}(a,b) \subset L^2(a,b)$$

hold. We remark here that V is also equal to the completion of $C_0^\infty(a,b)$ with respect to the $||\cdot||_1$-norm, which we customarily denote as $H_0^1(a,b)$.

The bounded bilinear form $a(u,v) = (pu',v')_0 + (qu,v)_0$ can be uniquely extended to V (since $D(A)$ is also dense in V). From now on, the notation

$a(.,.)$ will also denote the extended bilinear form. Because of continuity of the norm, we also have

$$|a(u,u)| \geq \alpha_0 ||u||_1^2, \quad u \in V. \tag{19}$$

(It is well known that the Poincaré-Friedrichs inequality for functions in $H_0^1(a,b)$ is valid with bound $(b-a)/\pi$ - note that in (4) only $u(a) = 0$ is required.) The extended bilinear form is also bounded with precisely the same bound as before, i.e.

$$|a(u,v)| \leq \alpha_1 ||u||_1 ||v||_1, \quad u,v \in V.$$

From the analysis in Section 2.1 (cf. (2)), we see that the extension of $a(.,.)$ to $V \times V$ defines a bounded, linear mapping $\hat{A}: V \to V$ via $(\hat{A}u,v)_1 = a(u,v)$, $u,v \in V$. Note that any inhomogeneous right-hand side $w \in L^2(a,b)$ defines a continuous linear functional $f(v) = (w,v)_0$ on $V = H_0^1(0,1)$. Hence, w can be identified with an element in $H^{-1}(0,1)$ (the dual of $H_0^1(0,1)$) and has its representation in V itself via

$$(\hat{w},v)_1 = (w,v)_0 \quad (= f(v)), \quad v \in V.$$

The variational equation (9) is therefore equivalent to $\hat{A}u = \hat{w}$, where the solution is sought in V. The Lax-Milgram Lemma now yields the following result for this example.

Theorem 2.9: Under the assumptions of Lemma 2.1, there is, for every $w \in L^2(a,b)$, a unique solution $u \in V$ of the variational equation

$$(pu',v')_0 + (qu,v)_0 = (w,v)_0, \quad v \in V. \tag{20}$$

Proof: The linear functional $f(v) = (w,v)_0$ is trivially bounded with respect to $||\cdot||_1$. The assertion thus follows from the Lax-Milgram Lemma. □

The solution of the variational equation (20) is not necessarily twice continuously differentiable. From the results in Section 2.1, we see that solving the variational equation is equivalent to finding the minimum of the associated quadratic functional in the larger space V. For this reason, we denote this solution as the *generalized* solution of the given differential equation. The solution of (20) is also labeled a *weak solution,* and the problem defined by (20) is termed a *weak form* of the given boundary problem. The following theorem shows that a sufficiently regular generalized solution is also a classical solution.

Theorem 2.10: Let $w \in C[a,b]$, and $u \in C^2[a,b]$. Then $u \in D(A)$ and is a solution of $Au = w$, if, and only if, $u \in V$ and is a solution of (20).

Proof: (i) By integration by parts, we can easily see that a classical solution will satisfy (20) for all $v \in D(A)$. By continuity, (20) holds for all $v \in V$.

(ii) For the converse, we first show that

$$|v(c)| \le C||v||_1, \quad c \in [a,b], \quad v \in C^1[a,b]. \tag{21}$$

We note that, for fixed $c \in [a,b]$,

$$v(x) = v(c) + \int_c^x v'(s)ds, \quad x \in [a,b],$$

and, using Hölder's Inequality, we see that

$$\begin{aligned}|v(c)|^2 &\le 2\left\{|v(x)|^2 + \left[\int_a^b |v'(s)|ds\right]^2\right\} \\ &\le 2\left\{|v(x)|^2 + (b-a)\int_a^b |v'(s)|^2 ds\right\}.\end{aligned}$$

Further, integration yields

$$\begin{aligned}(b-a)|v(c)|^2 &\le 2\left\{\int_a^b |v(x)|^2 dx + (b-a)^2\int_a^b |v'(x)|^2 dx\right\} \\ &\le 2\max(1,(b-a)^2)||v||_1^2.\end{aligned}$$

Inequality (21) then results with $C = \sqrt{2/(b-a)}\max(1,b-a)$.

Now, let $u \in V$ be a generalized solution which satisfies the regularity conditions of the theorem. Then there exist $u_k \in D(A)$, $k \in \mathbb{N}$, with $||u-u_k||_1 \to 0$ $(k \to \infty)$. The relation $u(a) = 0$ follows from (21) together with $u_k(a) = 0$ and $u \in C^1[a,b]$, since

$$|u(a)| = |u(a) - u_k(a)| \le c||u-u_k||_1 \to 0 \quad (k \to \infty).$$

Inequality (21) is also valid with b in place of c and hence $u(b) = 0$. By integrating by parts, we know that

$$(Au,v) = a(u,v) = (w,v)_0, \quad v \in C_0^\infty(a,b).$$

Since $C_0^\infty(a,b)$ is dense in $L^2(a,b)$, $Au = w$ almost everywhere in (a,b). Since, moreover, both Au and w are continuous, they are equal at all points of $[a,b]$. □

<u>Example 3 (Revised)</u>: In this case as in Example 1, $D(A)$ is not complete in the $||\cdot||_1$-norm. Let V denote its completion. The bilinear form $a(u,v) = [u,v]_1$ has a bounded extension to all of V and inequality (6) is thus valid for all $u \in V$. The Lax-Milgram Lemma can be applied to guarantee the existence and uniqueness of a (generalized) solution u of

$$[u,v]_1 = (w,v)_0, \quad v \in V,$$

for every $w \in L^2(G)$. This generalized solution is also the unique solution of the minimization problem (cf. (10))

$$J(u) = \inf_{v \in V} J(v). \quad \square$$

Example 2 (Revised): As in the other examples, the Lax-Milgram Lemma cannot be applied at the outset, since $D(A)$ is not complete in the $||\cdot||_2$-norm. If we let V denote its completion, then

$$C_0^\infty(0,\ell) \subset D(A) \subset V \subset L^2(0,\ell)$$

where the inclusions are dense with respect to the L^2-topology. The functions v in the completed space V preserve only the boundary conditions $v(0) = v'(0) = 0$ of functions in $D(A)$ but do not retain the boundary conditions at $x = \ell$ containing derivatives of second and third order. In order to see that the conditions at $x = 0$ are maintained by all functions $v \in V$, we let $v \in V$ and let $v_k \in D(A)$, $k \in \mathbb{N}$, such that $||v-v_k||_2 \to 0$ $(k \to \infty)$. Then one knows by estimates of the type of (21) that the v_k and v_k' converge uniformly to v and v', respectively. The limit functions are therefore continuous. In particular, at $x = 0$ we get

$$v(0) = \lim_{k\to\infty} v_k(0) = 0, \qquad v'(0) = \lim_{k\to\infty} v_k'(0) = 0.$$

To see that the boundary conditions at $x = \ell$ are lost by the completion, in general, similar arguments as in Strang & Fix (1973), Sec. 1.3, can be used. The boundary conditions $v(0) = v'(0) = 0$ are denoted as *forced* or *geometric boundary conditions*. The following theorem shows that a sufficiently smooth solution $u \in V$ of the variational equation

$$(pu'',v'')_0 + (qu,v)_0 = (w,v)_0, \quad v \in V,$$

will moreover satisfy the boundary conditions $u''(\ell) = (pu'')'(\ell) = 0$, which are labeled as *natural* or *dynamic boundary conditions*.

Theorem 2.11: Under the assumptions of Lemma 2.2, there is to every $w \in L^2(0,\ell)$ a unique generalized solution $u \in V$ of the variational equation associated with Example 2. For $w \in C[0,\ell]$ and $u \in C^4[0,\ell]$, $u \in D(A)$ and solves the differential equation $Au = w$, if, and only if, $u \in V$ is a solution of the corresponding variational problem.

Proof: As in the other examples, the first assertion of the theorem follows from the Lax-Milgram Lemma. To prove the second assertion, we note that the classical solution satisfies the variational equation for $v \in D(A)$ and hence for all $v \in V$ by continuity. Conversely, a generalized solution $u \in V$ satisfies the forced boundary conditions (see above). Using the hypothesized regularity of u and integration by parts, we obtain

$$(Au,v)_0 = (pu'')'v|_0^\ell - pu''v'|_0^\ell + (pu'',v'')_0 + (qu,v)_0 = (w,v)_0, \quad v \in C_0^\infty(0,\ell).$$

Because $C_0^\infty(0,\ell)$ is dense in $L^2(0,\ell)$, we see that $Au = w$ by continuity. With $v \in V$, the forced boundary conditions then yield

$$pu''v'|_{x=\ell} - (pu'')'v|_{x=\ell} = 0, \quad v \in V.$$

We select a function $v \in D(A)$ such that either

$$v(\ell) = 1, \quad v'(\ell) = 0 \quad \text{or} \quad v(\ell) = 0, \quad v'(\ell) = 1;$$

substituting either type of function into the preceding equality, we get

$$(pu'')'|_{x=\ell} = 0 \quad \text{and} \quad pu''|_{x=\ell} = 0.$$

Since $p > 0$, $u''(\ell) = 0$ results. □

Example 4: We consider the following nonlinear boundary-value problem from Section 1.4:

$$-(pu')'(x) + (qu)(x) = f(x,u(x)), \quad 0 \le x \le 1, \tag{22}$$

with homogeneous boundary conditions $u(0) = u(1) = 0$. We choose the same notation as in Example 1. Our assumptions are

$$p \in C^2[0,1], \quad q \in C[0,1], \quad p(x) \ge c_0 > 0, \quad q(x) \ge 0, \quad x \in [0,1],$$
$$f(.,.), f_y(.,.) \in C([0,1] \times \mathbb{R})$$

and

$$f_y(x,y) \le \lambda < \Lambda \equiv \inf_{0 \ne v \in V} a(v,v)/||v||_0^2, \quad x \in [0,1], \quad y \in \mathbb{R}, \tag{23}$$

where $f_y \equiv \partial f/\partial y$ and $||\cdot||_0$ denotes the L^2-norm. As in Example 1, let V denote the completion of $D(A)$ and $\hat{A}$, the operator defined on V by $a(.,.)$. The corresponding bilinear form is then symmetric, bounded, and elliptic on V. With the constant α_0 from (19), $\Lambda \ge \alpha_0$, since $a(v,v) \ge \alpha_0||v||_1^2 \ge \alpha_0||v||_0^2$, $v \in V$.

For the nonlinear variational problem corresponding to (22), we seek a $u \in V$ such that

$$(pu',v')_0 + (qu,v)_0 = \int_0^1 f(x,u(x))v(x)dx, \quad v \in V. \tag{24}$$

Because $D(A)$ is dense in V, the boundary-value problem (22) and the variational problem (24) are equivalent in the sense of the statements of Theorem 2.10.

In the following, we shall show that, as with linear problems, the variational problem (24) is equivalent to a minimization problem; and that solving the variational equation is tantamount to finding a zero of the first variation of the associated functional. The following theorem shows the equivalence to a minimization problem.

<u>Theorem 2.12</u>: Under the above assumptions, a solution u of the variational problem (24) is a strict minimum of the functional

$$J(v) \equiv \frac{1}{2} a(v,v) - \int_0^1 \int_0^{v(x)} f(x,\eta)\,d\eta\,dx \tag{25}$$

on V, and is unique. Conversely, a minimum of J is a solution of the variational equation (24).

<u>Proof</u>: We have (cf. (11)),

$$\begin{aligned} 2(J(v)-J(u)) &= (\hat{A}v,v-u)_1 + (v-u,\hat{A}u)_1 - 2\int_0^1\int_{u(x)}^{v(x)} f(x,\eta)\,d\eta\,dx, \\ &= (\hat{A}(v-u),v-u)_1 + 2(\hat{A}u,v-u)_1 - 2\int_0^1\int_{u(x)}^{v(x)} f(x,\eta)\,d\eta\,dx. \end{aligned}$$

Let u be a solution of (24). If we set $e = v-u$, then the previous equation yields

$$\begin{aligned} J(v) - J(u) &= \frac{1}{2}(\hat{A}e,e)_1 - \int_0^1\int_{u(x)}^{(u+e)(x)} f(x,\eta)\,d\eta\,dx + \int_0^1 f(x,u(x))e(x)\,dx \\ &= \frac{1}{2}(\hat{A}e,e)_1 - \int_0^1\int_{u(x)}^{(u+e)(x)} [f(x,\eta) - f(x,u(x))]\,d\eta\,dx. \end{aligned}$$

Because of (23), we obtain the estimate

$$\begin{aligned} \int_0^1\int_{u(x)}^{(u+e)(x)} [f(x,\eta) - f(x,u(x))]\,d\eta\,dx &\le \lambda\int_0^1\int_{u(x)}^{(u+e)(x)} [\eta-u(x)]\,d\eta\,dx \\ &= \frac{\lambda}{2}\int_0^1 e^2(x)\,dx. \end{aligned}$$

Since $\Lambda > \lambda$, we have finally that

$$J(v) - J(u) \ge \frac{1}{2}(\hat{A}e,e)_1 - \frac{\lambda}{2}\,||e||_0^2 \ge \frac{1}{2}(\Lambda-\lambda)\,||e||_0^2,$$

with the result that $J(v) > J(u)$ for all $v \neq u$, thereby showing that u is a strict minimum. We see then that u is unique.

Conversely, let u be a minimum of $J(\cdot)$. If we substitute $v = u + tz$, $t > 0$, $z \in V$, into the first equation of this proof, we get

$$0 \le 2(J(v)-J(u)) = 2t(\hat{A}u,z)_1 + t^2(\hat{A}z,z)_1 - 2\int_0^1\int_{u(x)}^{(u+tz)(x)} f(x,\eta)\,d\eta\,dx.$$

In this expression, we note that

$$\lim_{t\to 0} \frac{1}{t}\int_{u(x)}^{(u+tz)(x)} f(x,\eta)\,d\eta = f(x,u(x))z(x).$$

Dividing by t, and passing to the limit as $t \to 0$, we obtain

$$0 \leq (\hat{A}u,z)_1 - \int_0^1 f(x,u(x))z(x)dx, \qquad z \in V.$$

We get the reverse inequality with $-z$ in place of z, and hence the result that

$$a(u,z) = \int_0^1 f(x,u(x))z(x)dx, \quad z \in V. \qquad \square$$

As in Lemma 2.5, we now give the Fréchet-derivative and show the convexity of the functional $J(\cdot)$.

Lemma 2.13: Under the above assumptions, $J(\cdot)$ has at each $u \in V$ a bounded Fréchet-derivative given by

$$J'(u)v = a(u,v) - \int_0^1 f(x,u(x))v(x)dx, \tag{26}$$

and, moreover, $J(\cdot)$ is strictly convex.

Proof: (i) With $w = u+v$, $u,v \in V$, we have (cf. Proof of Thm. 2.12)

$$\int_0^1\int_0^{w(x)} f(x,\eta)d\eta dx - \int_0^1\int_0^{u(x)} f(x,\eta)d\eta dx - \int_0^1 f(x,u(x))v(x)dx$$
$$= \int_0^1\int_{u(x)}^{w(x)} [f(x,\eta) - f(x,u(x))]d\eta dx \leq \frac{\lambda}{2}\int_0^1 v^2(x)dx.$$

Therefore, the derivative of the mapping $W(u) \equiv \int_0^1\int_0^{u(x)} f(x,\eta)d\eta dx$, is given by

$$W'(u)v = \int_0^1 f(x,u(x))v(x)dx.$$

From Lemma 2.5, we see that the derivative of the first term of J, $a(v,v)$, is given by $a(u,v)$, and (26) easily follows.

(ii) As in the proof of Theorem 2.12, use of (25) yields

$$J(w) - J(u) - J'(u)(w-u)$$
$$= \frac{1}{2}(\hat{A}(w-u),w-u)_1 + (\hat{A}u,w-u)_1 - \int_0^1\int_{u(x)}^{w(x)} f(x,\eta)d\eta dx$$
$$- (\hat{A}u,w-u)_1 + \int_0^1 f(x,u(x))(w(x) - u(x))dx$$
$$= \frac{1}{2}(\hat{A}(w-u),w-u)_1 - \int_0^1\int_{u(x)}^{w(x)} [f(x,\eta) - f(x,u(x))]d\eta dx$$
$$\geq \frac{1}{2}(\Lambda-\lambda)||w-u||_0^2 > 0, \quad w \neq u, \quad u,w \in V. \qquad \square$$

Using Lemma 2.6, we can now see that satisfying the variational equation (24) is a necessary and sufficient criterion for $J(\cdot)$ to have a unique minimum. We cannot invoke the Lax-Milgram Lemma to deduce the existence of a solution of the nonlinear variational problem, but instead we must rely on other methods taken from the theory of ordinary differential equations. □

2.3. THE RITZ METHOD

Formulating a linear operator equation as a minimization problem in a Hilbert space allows us to use the Ritz method (also known as the Rayleigh-Ritz method) to obtain approximate solutions. This method, however, is applicable only for problems with positive semidefinite operators. We shall see in the following section, though, that the Ritz method is a special case of general Galerkin methods which can be used advantageously in the numerical treatment of general linear operator equations. For Example 1 of the preceding section, we shall apply the Ritz method with continuous, piecewise linear trial functions, and then shall derive the corresponding system of linear algebraic equations.

The derivation of the Ritz method relies essentially on Theorem 2.4 which shows the equivalence of the given operator equation (8) to the minimization problem (10). We thus begin with the problem of finding a solution $u \in D(A)$ of

$$Au = w,$$

where A is a linear, positive semidefinite operator mapping a dense subspace $D(A)$ of E into E and where $w \in E$. From Theorem 2.4, solving this operator equation is then equivalent to minimizing

$$J(v) = \frac{1}{2}(Av,v) - (w,v)$$

over $D(A)$. An obvious technique for approximating u is to minimize this functional on finite-dimensional subspaces of $D(A)$. As we shall see, this problem leads to solving a linear algebraic system.

Let E_n be a finite-dimensional subspace of $D(A)$. Then the *Ritz method* consits of finding a solution $u_n \in E_n$ to

$$J(u_n) = \inf_{v_n \in E_n} J(v_n). \tag{27}$$

As in Section 2.1, this minimization problem is also equivalent to an operator equation. In order to define the associated operator, we shall need the concept of a projection. For any closed subspace E_n of a Hilbert space E with scalar product $(.,.)$, an *orthogonal projection operator* $P_n: E \to E_n$ can be defined via

$$(P_n u,v) = (u,v), \quad u \in E, \quad v \in E_n. \tag{28}$$

The existence of a uniquely determined orthogonal projection $P_n u$ of $u \in E$ is a consequence of the Riesz Representation Theorem for Hilbert spaces. In general, however, we call a bounded linear mapping $P: N \to M$ between normed spaces N and $M \subset N$ a *projection operator* in case it is surjective and idempotent (i.e., $P^2 = P$). The following result shows the equivalence of the Ritz method to an operator equation.

Theorem 2.14: Suppose the mapping $A: D(A) \to E$ is positive semidefinite. Then $u_n \in E_n$ is a solution of (27) if, and only if,

$$A_n u_n = P_n w \tag{29}$$

where $A_n \equiv P_n A|E_n$.

Proof: By definition (28) of the orthogonal projection P_n, we see that

$$(Av_n,v_n) = (P_n Av_n,v_n) = (A_n v_n,v_n), \quad v_n \in E_n, \tag{30}$$

and that $(w,v_n) = (P_n w,v_n)$, $v_n \in E_n$. We thus obtain for the restriction of J to E_n,

$$J(v_n) = \frac{1}{2}(A_n v_n,v_n) - (P_n w,v_n), \quad v_n \in E_n.$$

From (30), we see that A_n is likewise positive semidefinite and symmetric on E_n, and we trivially note that A_n is densely defined in E_n. Application of Theorem 2.4 to A_n instead of A yields the assertion. □

In order to express (29) as a linear system of equations, let $\{\phi_1,\ldots,\phi_m\}$ be a basis of E_n. Then u_n can be represented as

$$u_n = \sum_{k=1}^{m} c_k\phi_k. \tag{31}$$

Equation (29) is equivalent to

$$0 = (P_n Au_n - P_n w,\phi_j) = (Au_n - w,\phi_j), \quad j = 1,\ldots,m. \tag{32}$$

Inserting u_n from (31) then yields the following system of equations equivalent to (29),

$$\sum_{k=1}^{m} (A\phi_k,\phi_j)c_k = (w,\phi_j), \quad j = 1,\ldots,m.$$

For Example 1 of Section 2.2, we now derive the system of equations for the Ritz method when the finite-dimensional subspace E_n is the space of continuous functions linear over each interval between the nodes $a = x_0 < x_1 < \ldots < x_{n+1} = b$

and vanishing at $x = a$ and $x = b$. More precisely, if we denote

$$I_k = (x_{k-1}, x_k),\quad h_k = x_k - x_{k-1},\quad k = 1,\dots,n+1,\quad h = \max_{1\le k\le n+1} h_k,$$

our trial functions in E_n have the form

$$u_n(x) = u_n(x_{k-1})\,\frac{x_k - x}{h_k} + u_n(x_k)\,\frac{x - x_{k-1}}{h_k},\quad x \in I_k,\quad k = 1,\dots,n+1.$$

With the *"roof functions"* as basis, i.e.,

$$\phi_k(x) = \begin{cases} 0, & a \le x \le x_{k-1}, \\ (x - x_{k-1})/h_k, & x_{k-1} \le x \le x_k, \\ (x_{k+1} - x)/h_{k+1}, & x_k \le x \le x_{k+1}, \\ 0, & x_{k+1} \le x \le b, \\ & k = 1,\dots,n, \end{cases}$$

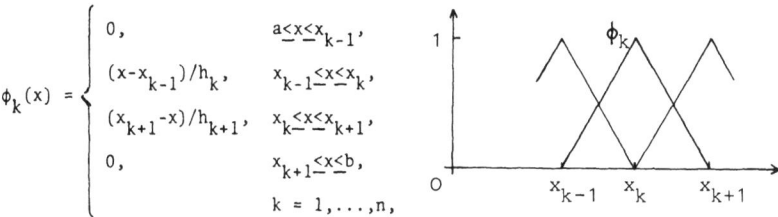

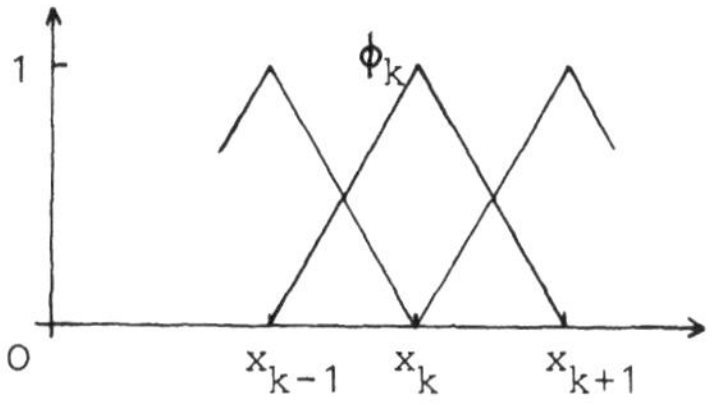

the unknowns c_k in (32) are precisely the values of the approximations u_n at the nodal points x_k, i.e., $c_k = u_n(x_k)$.

By abuse of notation, we also let $D(A)$ denote the completion of $\{u \in C^2[a,b]: u(a) = u(b) = 0\}$ with respect to the norm $||\cdot||_1$. Every function $u_n \in E_n$ possesses a generalized derivative in $L^2(a,b)$ given by

$$u_n'(x) = \frac{u_n(x_k) - u_n(x_{k-1})}{h_k},\quad x \in I_k,\ k = 1,\dots,n+1,$$

so that $E_n \subset D(A)$. This is because the above step function lies in $L^2(a,b)$ and satisfies the conditions for generalized derivatives (cf. Section 2.2). Hence $u_n \in D(A)$, for u_n can even be represented as a limit of a sequence $v^{(\nu)} \in C_0^\infty(a,b)$ with respect to the norm $||\cdot||_1$.

The system of equations (32) for the Ritz method has the following form for this particular example,

$$\sum_{k=1}^{n} [(p\phi_k', \phi_j')_0 + (q\phi_k, \phi_j)_0] u_n(x_k) = (w, \phi_j)_0,\quad j = 1,\dots,n. \tag{33a}$$

To determine the coefficients of this system, we calculate the occurring scalar products for the roof functions defined on a not necessarily equidistant mesh:

$$(p\phi_k',\phi_j')_0 = \begin{cases} -p_{j,j-1} \\ p_{j,j} \\ -p_{j,j+1} \\ 0 \end{cases}, \quad (q\phi_k,\phi_j)_0 = \begin{cases} q_{j,j-1}, & k = j-1, \\ q_{j,j}, & k = j, \\ q_{j,j+1}, & k = j+1, \\ 0 & |k-j| > 1, \end{cases}$$

with

$$p_{j,j-1} = \frac{1}{h_j^2}\int_{I_j} p\,dx, \quad p_{j,j+1} = \frac{1}{h_{j+1}^2}\int_{I_{j+1}} p\,dx, \quad p_{j,j} = p_{j,j-1} + p_{j,j+1},$$

$$q_{j,j-1} = \int_{I_j} q\phi_j\phi_{j-1}dx, \quad q_{j,j} = \int_{x_{j-1}}^{x_{j+1}} q\phi_j^2 dx, \quad q_{j,j+1} = \int_{I_{j+1}} q\phi_j\phi_{j+1}dx.$$

For brevity, we set $v_j = u_n(x_j)$, and (33a) then has the form of a finite-difference method:

$$(q_{j,j-1}-p_{j,j-1})v_{j-1} + (q_{j,j}+p_{j,j})v_j + (q_{j,j+1}-p_{j,j+1})v_{j+1} = (w,\phi_j)_0, \quad j = 1,\dots,n. \tag{33b}$$

For the special case where p and q are constant, (33) becomes

$$p\left(\frac{v_j-v_{j-1}}{h_j} - \frac{v_{j+1}-v_j}{h_{j+1}}\right) + \frac{q}{6}(h_jv_{j-1} + 2(h_j+h_{j+1})v_j + h_{j+1}v_{j+1}) = (w,\phi_j)_0, \quad j = 1,2,\dots,n. \tag{34}$$

If we now use Simpson's rule on each subinterval to approximate the integral on the right-hand side, then

$$(w,\phi_j)_0 = \int_{x_{j-1}}^{x_{j+1}} w\phi_j dx = \frac{1}{6}(2h_jw(x_{j-1/2}) + (h_j+h_{j+1})w(x_j) + 2h_{j+1}w(x_{j+1/2})) + O(h_j^5 + h_{j+1}^5).$$

For equidistant mesh widths $h_j = h$, $j = 1,\dots,n+1$, and constant functions p, q, (34) can be expressed in the following form,

$$-\frac{p}{h^2}(v_{j-1}-2v_j+v_{j+1}) + \frac{q}{6}(v_{j-1}+4v_j+v_{j+1}) = \frac{1}{h}(w,\phi_j)_0, \quad j = 1,\dots,n. \tag{35}$$

2.4. GALERKIN METHODS AND THE METHOD OF LEAST SQUARES

We have shown for the Ritz method that solving the minimization problem (27) is equivalent to solving the system of equations (32) if we assume that A is positive semidefinite. However, the approximations given in (29) are meaningful also if A is not necessarily positive semidefinite; this is even true for nonlinear A as we shall see in the following section. Equation (29) motivates the use of projection methods which we shall now proceed to establish for general normed space settings.

Let A be a linear mapping of a subspace $D(A)$ of a normed space E into a normed space F. Let E_n be a subspace of $D(A)$ and P_n a projection of F onto a subspace $F_n \subset F$. In this section, we assume that $\dim E_n = \dim F_n < \infty$. A *projection method* consists of determining a solution $u_n \in E_n$ to the following equation for given $w \in F$:

$$P_n A u_n = P_n w. \tag{36}$$

Equation (36) is then uniquely solvable for every $w \in F$ in case the restriction $P_n A|E_n$ is an injective mapping of E_n into F_n. We shall see that (36) can equivalently be expressed as a linear algebraic system of equations. Let $\{\psi_1,\ldots,\psi_m\}$ be a basis of F_n. For every $y \in F$, there are uniquely determined constants $\alpha_j = \alpha_j(y)$, $j = 1,\ldots,m$, such that

$$P_n y = \sum_{j=1}^{m} \alpha_j(y)\psi_j.$$

By specifying mappings Ψ_j, $j = 1,\ldots,m$, via

$$\Psi_j : y \in F \to \alpha_j(y) \in \mathbb{R}, \quad j = 1,\ldots,m,$$

we obtain linear functionals Ψ_j, $j = 1,\ldots,m$, in the dual space F^* of F which satisfy

$$\langle \psi_k, \Psi_j \rangle = \delta_{jk}, \quad j,k = 1,\ldots,m. \tag{37}$$

Equation (36) is thus equivalent to

$$\langle Au_n, \Psi_j \rangle = \langle w, \Psi_j \rangle, \quad j = 1,\ldots,m. \tag{38}$$

Relation (37) also shows that $\{\Psi_j,\ j = 1,\ldots,m\}$ forms a basis of F_n^*.

The desired solution $u_n \in E_n$ of (36) can be expressed as

$$u_n = \sum_{k=1}^{m} c_k \phi_k \tag{39}$$

where now $\{\phi_1,\ldots,\phi_m\}$ is a basis in E_n. As a result, we obtain the following linear algebraic system equivalent to (36),

$$\sum_{k=1}^{m} \langle A\phi_k, \Psi_j \rangle c_k = \langle w, \Psi_j \rangle, \quad j = 1,\ldots,m. \tag{40}$$

The system of equations in (40) is a linear system of m equations for m unknowns.

According to results in finite-dimensional linear algebra, we need only show uniqueness of a solution in order to guarantee both the existence and the uniqueness of a solution. The following theorem provides such a condition for unique solvability.

Theorem 2.15: Let $\dim E_n = \dim F_n < \infty$. Then (36) has a unique solution $u_n \in E_n$ for every $w \in F$ if, and only if, for every $v_n \in E_n$ the following relation holds:

$$\langle Av_n, \Psi\rangle = 0 \quad \forall\Psi \in F_n^* \Rightarrow v_n = 0. \tag{41}$$

Proof: Assuming (41), $v_n = \Sigma_k c_k\phi_k$ satisfies the homogeneous system of equations corresponding to (40). Thus, by (41), it has only the trivial solution and the inhomogeneous system (40) is always uniquely solvable. Conversely, let $v_n \in E_n$ with $\langle Av_n, \Psi\rangle = 0$ for every $\Psi \in F_n^*$. Then v_n is a solution of $P_nAv_n = 0$. Since the latter equation has only the trivial solution, $v_n = 0$. □

The existence of a positive number $c_n > 0$, such that

$$\sup_{0\neq\Psi\in F_n^*} \frac{|\langle Av_n, \Psi\rangle|}{||\Psi||} \geq c_n||v_n||, \quad v_n \in E_n, \tag{42}$$

is sufficient for (41) to hold. This is precisely condition (17a) of Theorem 2.8 which generalizes the Lax-Milgram Lemma if the finite-dimensional spaces E_n and F_n are taken as Hilbert spaces equipped with Euclidean norms. The second condition (17b) in Theorem 2.8 guaranteeing surjectivity is superfluous, if we assume $\dim E_n = \dim F_n < \infty$. We should point out here that our results up to now concerning solvability of projection methods have been obtained without assuming existence of A^{-1}.

In the following, we shall derive some particular, well-known methods by making adroit choices of the spaces F_n and the projections P_n. In these derivations, we assume $E = F$ (= H) is a Hilbert space, and, for conciseness, we set $V = D(A)$ ($\subset H$) for the domain of definition of A. Let P_n be the orthogonal projection of H onto a subspace $F_n \subset V$ and $\{\phi_1,\ldots,\phi_m\}$ a basis of $E_n \subset V$. We seek a solution $u_n \in E_n$ of the form (39). The various methods are earmarked by the choice of F_n when E_n is given.

1) The Petrov-Galerkin Method: The general method described in (36) for Hilbert space settings is often designated as the Petrov-Galerkin method. Any finite-dimensional subspace $F_n \subset H$ such that $\dim F_n = \dim E_n$ may be selected. Let $\{\psi_1,\ldots,\psi_m\}$ be a basis in F_n. We then obtain the following system of equations for determining c_k, $k = 1,\ldots,m$:

$$\sum_{k=1}^{m} (A\phi_k,\psi_j)c_k = (w,\psi_j), \quad j = 1,\ldots,m. \tag{43}$$

2) The Bubnov-Galerkin Method: Here we select $\psi_j = \phi_j$, $j = 1,\dots,m$, and thus obtain (32). If A is, moreover, positive semidefinite, then we get the Ritz (or Rayleigh-Ritz) method. For this reason, the Bubnov-Galerkin method is often labeled in this case as the *Ritz-Galerkin method*.

3) The Method of Moments: To describe this method, let M be an injective mapping of V into H. We select $\psi_j = M\phi_j$, $j = 1,\dots,m$, with the result that (36) becomes

$$\sum_{k=1}^{m} (A\phi_k, M\phi_j)c_k = (w, M\phi_j), \quad j = 1,\dots,m. \tag{44}$$

4) The Method of Least Squares: This arises as a special case of the method of moments when A is itself injective. In (44), we set $M = A$ and obtain the following system of equations

$$\sum_{k=1}^{m} (A\phi_k, A\phi_j)c_k = (w, A\phi_j), \quad j = 1,\dots,m. \tag{45}$$

If A is positive semidefinite, then (45) directly corresponds to the Ritz-Galerkin method with respect to the scalar product given by $[u,v] = (u,Av)$. The name of this method is motivated by the following proposition.

Theorem 2.16: Let A be injective. Then there is a uniquely determined solution $u_n \in [\phi_1,\dots,\phi_m]$ of (45). A solution $u_n \in E_n$ of (45) can be characterized by

$$||Au_n - w|| = \inf_{v_n \in E_n} ||Av_n - w||. \tag{46}$$

Proof: The element $u_n \in E_n$ solves (46) if Au_n constitutes the best approximation of w from the subspace $AE_n = [A\phi_1,\dots,A\phi_m]$. This is characterized in a well-known manner by

$$(Au_n - w, w_n) = 0, \quad w_n \in AE_n.$$

This equation is equivalent to (36) if P_n is the orthogonal projection onto AE_n. This orthogonal projection exists and is uniquely determined; and, since A is injective, u_n is also uniquely determined. □

We proceed to state several results which guarantee from properties of A itself the unique solvability of the system of equations for several of the above methods. For the Bubnov-Galerkin method, we see that the assumptions of the Lax-Milgram Lemma for A also ensure the validity of (42) with $\langle .,. \rangle = (.,.)$ and $c_n = \alpha_0$; hence equation (43) (with $\psi_j = \phi_j$) is uniquely solvable. For a positive definite A, the Ritz-Galerkin procedure leads to a system of linear equations with an associated positive definite matrix $A = (a_{jk})$, $a_{jk} \equiv (A\phi_k, \phi_j)$, and the unique solvability of the linear system is assured.

We now show what form the Ritz-Galerkin method has for Poisson's equation (cf. Example 3 in Section 2.1 and 2.2) if the underlying region G is subdivided into rectangles. For notational purposes, we set $x = x_1$ and $y = x_2$. We consider a closed region $\overline{G} \subset R^2$ of the form:

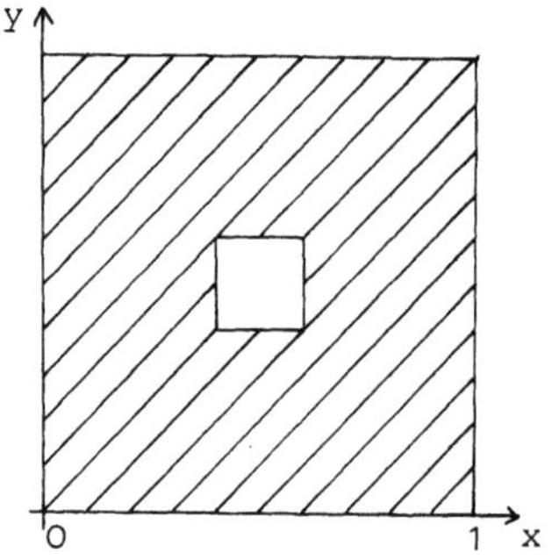

We subdivide the appropriate segments on the x- and y-axes into $(n+1)$ and $(m+1)$ equal subintervals, respectively, having nodes

$$x_k = kh_x, \quad k = 0,\ldots,n+1,$$

$$y_j = jh_y, \quad j = 0,\ldots,m+1,$$

with the mesh widths

$h_x = 1/(n+1)$ and

$h_y = 1/(m+1)$.

We let G_h denote the mesh points lying in G $(= \overline{G} - \partial G)$. $\overline{G}$ is thus subdivided into rectangles. With the functions

$$X_k(x) = \begin{cases} \frac{1}{h_x}(x-x_{k-1}), & x_{k-1} \leq x \leq x_k, \\ \frac{-1}{h_x}(x-x_{k+1}), & x_k \leq x \leq x_{k+1}, \\ 0 & , \text{ otherwise}, \end{cases}$$

$$Y_j(y) = \begin{cases} \frac{1}{h_y}(y-y_{j-1}), & y_{j-1} \leq y \leq y_j, \\ \frac{-1}{h_y}(y-y_{j+1}), & y_j \leq y \leq y_{j+1}, \\ 0 & , \text{ otherwise}, \end{cases}$$

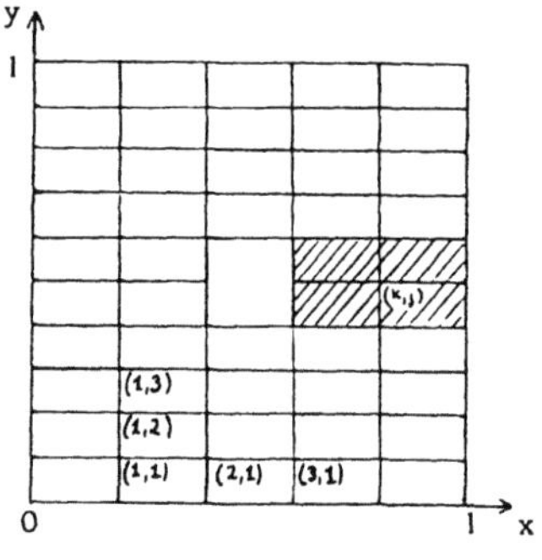

we define functions on $\overline{G}$ by

$$\phi_{k,j}(x,y) = X_k(x)Y_j(y), \quad (x_k,y_j) \in G_h.$$

As the finite-dimensional subspace $E_n = E_h$ of the Ritz-Galerkin method we now choose the span of the $\phi_{k,j}$; E_h is then a subspace of V. Every function $u_h \in E_h$ is thus expressible in the form

$$u_h(x,y) = \sum_{(x_p,y_q)\in G_h} v_{p,q}\phi_{p,q}(x,y)$$

where $v_{p,q} = u_h(x_p,y_q)$.

In order to approximate the solution of Poisson's equation via the Ritz-Galerkin method, we solve

$$\sum_{(x_p,y_q)\in G_h} a_{k,j}^{p,q} v_{p,q} = g_{k,j}, \quad (x_k,y_j) \in G_h, \tag{47a}$$

where

$$g_{k,j} = \iint_G w\phi_{k,j}\,dydx$$

and

$$a_{k,j}^{p,q} = \iint_G \left(\frac{\partial\phi_{p,q}}{\partial x}\frac{\partial\phi_{k,j}}{\partial x} + \frac{\partial\phi_{p,q}}{\partial y}\frac{\partial\phi_{k,j}}{\partial y}\right) dy\ dx$$

$$= \iint_G \left(Y_q Y_j \frac{dX_p}{dx}\frac{dX_k}{dx} + X_p X_k \frac{dY_q}{dy}\frac{dY_j}{dy}\right) dy\ dx.$$

Since $\phi_{k,j}$ is nonzero only on the four rectangles possessing (x_k,y_j) as a common corner point (cf. the previous illustration), then clearly

$$a_{k,j}^{p,q} = 0 \quad \text{for} \quad |p-k| > 1,\ |q-j| > 1.$$

For the other coefficients, we see that

$$a_{k,j}^{k,j} = \frac{4}{3h_x h_y}(h_x^2 + h_y^2), \quad a_{k,j}^{k,j-1} = a_{k,j}^{k,j+1} = \frac{-1}{3h_x h_y}(2h_x^2 - h_y^2),$$

$$a_{k,j}^{k-1,j} = a_{k,j}^{k+1,j} = \frac{-1}{3h_x h_y}(2h_y^2 - h_x^2),$$

$$a_{k,j}^{k-1,j-1} = a_{k,j}^{k+1,j-1} = a_{k,j}^{k-1,j+1} = a_{k,j}^{k+1,j+1} = \frac{-1}{6h_x h_y}(h_x^2 + h_y^2).$$

Whenever the mesh widths in the x- and y-directions are equal, $h = h_x = h_y$, the system of equations (47a) has the form

$$\begin{aligned}\frac{8}{3}v_{k,j} - \frac{1}{3}(v_{k-1,j} &+ v_{k+1,j} + v_{k-1,j-1} + v_{k,j-1} + v_{k+1,j-1} \\ &+ v_{k-1,j+1} + v_{k,j+1} + v_{k+1,j+1}) = g_{k,j}, \quad (x_k,y_j) \in G_h.\end{aligned} \tag{47b}$$

We now order the values $v_{k,j}$ in the following way,

$$V = (v_{11}, v_{21}, \ldots, v_{n1}, v_{12}, v_{22}, \ldots, v_{nm-1}, v_{1m}, \ldots, v_{nm}),$$

where only the indices of inner mesh points (from G_h) occur. Then (47b) can be written as a system of equations with block structure,

$$\begin{pmatrix} D_1 & -A_1 & 0 & \cdots & 0 \\ -A_2 & D_2 & -A_2 & & \\ & \ddots & \ddots & \ddots & \\ & & -A_{m-1} & D_{m-1} & -A_{m-1} \\ 0 & \cdots & 0 & -A_m & D_m \end{pmatrix} V = G, \tag{47c}$$

where $G = (g_{11}, g_{21}, \ldots, g_{n1}, g_{12}, \ldots, g_{nm})$, and

$$A_j = \frac{1}{3}\begin{pmatrix} 1 & 1 & 0 & \cdots & 0 \\ 1 & 1 & 1 & & \\ & \ddots & \ddots & \ddots & \\ & & 1 & 1 & 1 \\ 0 & \cdots & 0 & 1 & 1 \end{pmatrix}, \quad D_j = \frac{1}{3}\begin{pmatrix} 8 & -1 & 0 & \cdots & 0 \\ -1 & 8 & -1 & & \\ & \ddots & \ddots & \ddots & \\ & & -1 & 8 & -1 \\ 0 & \cdots & 0 & -1 & 8 \end{pmatrix},$$

and the number of rows and columns of A_j, D_j is equal to the number of mesh points (x_k, y_j) in G_h for fixed y_j. Whenever we solve such a system iteratively, we should, of course, take the block structure into account. Appropriate methods include Block-Jacobi methods, Block-Gauss-Seidel methods, and Block-Relaxation methods, and we refer the reader to Stoer & Bulirsch (1978) and Varga (1962) for a detailed discussion of these methods. Since we have here a sparse matrix, even direct methods can be effectively applied.

2.5. PROJECTION METHODS FOR NONLINEAR PROBLEMS

The fundamental equation (36) describing the projection method is also meaningful if A is a nonlinear operator. The solutions to the associated nonlinear (finite-dimensional) system of equations are determined iteratively, e.g., by Jacobi or Gauss-Seidel methods or by Newton's method. If we carry out only one step of Newton's method for each member of an increasing sequence of finite-dimensional subspaces, then we have the "projective Newton's method". We are hereby able to approximate the solution of a given nonlinear problem by a sequence of solutions of linear problems. This program will be carried out in detail for the nonlinear variational problem from section 2.2 (Example 4).

Let E and F be normed spaces and E_n and F_n finite-dimensional subspaces of E and F, respectively, with $\dim E_n = \dim F_n$. We now let A be a continuous, not necessarily linear operator mapping a subset

$D \subset E$ into F. We seek approximations to the solution $u \in D$ of

$$Au = 0. \tag{48}$$

Let P_n be a bounded, linear projection operator mapping F onto F_n; we then define a *projection method* by seeking a solution $u_n \in D_n$, $D_n \equiv D \cap E_n$, of the equation

$$P_n A u_n = 0. \tag{49}$$

The mapping $P_n A\colon D \to F_n$ is also continuous, since P_n is bounded. If $\{\phi_j\}_1^m$ and $\{\psi_j\}_1^m$ are bases in E_n and F_n, respectively, then (49) is clearly equivalent to

$$\langle Au_n, \Psi_j \rangle = 0, \quad j = 1,\dots,m, \tag{50}$$

where $\{\Psi_j\}$ is the (bi-)orthogonal basis in F_n^* defined as in Section 2.4. With $u_n = \Sigma_{k-1}^m c_k\phi_k$, equation (50) or (49) will in addition be equivalent to

$$f_{jn}(c_1,\dots,c_m) = 0, \quad j = 1,\dots,m, \tag{51}$$

where $f_{jn}(c_1,\dots,c_m)$ are continuous functions in m variables, defined by

$$f_{jn}(\xi_1,\dots,\xi_m) = \langle Av_n, \Psi_j \rangle, \quad v_n = \sum_{k=1}^m \xi_k\phi_k \in D_n, \quad j = 1,\dots,m.$$

The continuity of f_{jn} follows from that of $P_n A$.

If (as in Example 4 of Section 2.2) $E = F$ and $A = L-T$, where L is linear and bounded and T is a nonlinear and continuous mapping from E into itself, then (51) becomes

$$\sum_{k=1}^m \langle L\phi_k, \Psi_j \rangle c_k = \tau_{jn}(c_1,\dots,c_m), \quad j = 1,\dots,m, \tag{52}$$

where τ_{jn} is given by

$$\tau_{jn}(\xi_1,\dots,\xi_m) = \langle Tv_n, \Psi_j \rangle, \quad v_n = \sum_{k=1}^m \xi_k\phi_k \in E_n, \quad j = 1,\dots,m.$$

Properties of the Jacobian matrix $(\partial f_{j,n}/\partial c_k)$ are essential for solving the nonlinear system of equations in (51). We shall show in the following lemma how this Jacobian matrix can be represented by the projection of the Fréchet-derivative of A. The definition of the Fréchet-derivative from Section 2.1 carries over verbatim for mappings between normed linear spaces E and F. Furthermore, the mapping $A\colon D \subset E \to F$

is called *continuously Fréchet-differentiable* at $u \in D$ in case A is differentiable at each point of an open neighborhood U of u and the Frechet-derivative A' is continuous in the operator norm, i.e.,

$$\forall \varepsilon > 0, \ \exists \delta > 0 \ \ni: \forall v \in U: \ ||u-v|| \leq \delta$$
$$\text{and } \forall \phi \in E \Rightarrow ||(A'(v) - A'(u))\phi|| \leq \varepsilon||\phi||.$$

Lemma 2.17: Let $c = (c_1,\ldots,c_m) \in \mathbb{R}^m$, $u_n = \Sigma_{k=1}^m c_k\phi_k$ and A be continuously Fréchet-differentiable at u_n. Then $(f_{1n},\ldots,f_{mn})$ is continuously differentiable at $(c_1,\ldots,c_m)$ and the Jacobian matrix can be represented as

$$\frac{\partial f_{jn}}{\partial c_k}(c_1,\ldots,c_m) = \langle A'(u_n)\phi_k,\Psi_j\rangle, \quad 1 \leq j,k \leq m. \tag{53}$$

Proof: (i) With $c = (c_1,\ldots,c_m)$, we let $u_n = \Sigma_{k-1}^m c_k\phi_k$. Then, according to the chain rule, the partial derivatives of

$$f_{jn}(c_1,\ldots,c_n) = \langle Au_n,\Psi_j\rangle = \sum_{k=1}^{n} c_k\langle A\phi_k,\Psi_j\rangle$$

are given by (53). To prove the continuity of the partial derivatives, let $u_n = \Sigma_{k=1}^m c_k\phi_k$ and $v = u_n + h\phi_i$ with arbitrary i in $1 \leq i \leq m$. For every $\varepsilon > 0$, let $\delta > 0$ be the number occurring in the definition itself of the continuous Fréchet-differentiability of A at u_n. With $h \leq \delta/||\phi_i||$, we then have $||v-u_n|| = h||\phi_i|| \leq \delta$ and

$$\left|\frac{\partial f_{jn}}{\partial c_k}(c_1,\ldots,c_i + h,c_{i+1},\ldots,c_m) - \frac{\partial f_{jn}}{\partial c_k}(c_1,\ldots,c_i,\ldots,c_m)\right|$$
$$= |\langle(A'(v) - A'(u_n))\phi_k,\Psi_j\rangle| \leq ||\Psi_j|| \ ||(A'(v) - A'(u_n))\phi_k||$$
$$\leq \varepsilon||\Psi_j|| \ ||\phi_k||, \quad 1 \leq j,k \leq m,$$

thereby proving continuity of $\partial f_{jn}/\partial c_k$ with respect to the i-th argument. □

For results regarding the solvability of (51), we refer the reader to Ortega and Rheinboldt (1970). We shall later establish corresponding results in the course of our investigations on convergence questions.

To solve (51), iteratively, we can utilize Newton's method. Indeed, with the representation of the Jacobian in (53), this method can be expressed as

$$f_{jn}(c_1^{(t)},c_2^{(t)},\ldots,c_m^{(t)}) + \sum_{k=1}^{m} \langle A'(u_n^{(t)})\phi_k,\Psi_j\rangle(c_k^{(t+1)}-c_k^{(t)}) = 0 \qquad (54a)$$

or, equivalently, as

$$\langle A'(u_n^{(t)})(u_n^{(t+1)}-u_n^{(t)}),\Psi_j\rangle = -\langle Au_n^{(t)},\Psi_j\rangle, \quad j = 1,\ldots,m, \quad t = 0,1,\ldots, \qquad (54b)$$

or as

$$\langle A'(u_n^{(t)})u_n^{(t+1)},\Psi\rangle = \langle A'(u_n^{(t)})u_n^{(t)} - Au_n^{(t)},\Psi\rangle, \quad \Psi \in F_n^*, \quad t = 0,1,2,\ldots . \qquad (54c)$$

If we consider a sequence of increasing subspaces $E_{n-1} \subset E_n$, $n = 1,2,\ldots,$ and carry out only one iteration of Newton's method (54) for every n with $u_n^{(0)} = u_{n-1}$ as the starting vector, then we get

$$\langle A'(u_{n-1})u_n,\Psi\rangle = \langle A'(u_{n-1})u_{n-1}-Au_{n-1},\Psi\rangle, \quad \Psi \in F_n^*, \quad n = 1,2,\ldots . \qquad (55)$$

These equations, of course, make sense in case $u_{n-1} \notin E_n$ but with A differentiable at u_{n-1}. Equations (55) describe the *projective Newton's method*. We will obtain precisely the same equations if we first carry out a step of Newton's method for the given problem (48) and then approximate the resulting linear equation by a projection method in possibly different subspaces for every n. We shall not give results here on the convergence properties of Newton's method or of the projective Newton's method, but instead refer the reader to Ortega and Rheinboldt (1970), Witsch (1978), and the works cited therein.

Whenever $A = L-T$, the equations, analogous to (54b) and (55) for solving (52) iteratively, are of the respective form

$$\langle Lu_n^{(t+1)} - T'(u_n^{(t)})u_n^{(t+1)},\Psi_j\rangle = \langle Tu_n^{(t)} - T'(u_n^{(t)})u_n^{(t)},\Psi_j\rangle, \quad j = 1,\ldots,m, \quad t = 0,1,\ldots \qquad (56)$$

and (analogous to (55))

$$\langle Lu_n - T'(u_{n-1})u_n,\Psi\rangle = \langle Tu_{n-1}-T'(u_{n-1})u_{n-1},\Psi\rangle, \quad \Psi \in F_n^*, \quad n = 0,1,2,\ldots . \qquad (57)$$

We conclude our discussion by giving a concrete application of the idea just presented to Example 4 in Section 2.2. The mapping A is given by $A = \hat{A}-T$, where

$$(Tu,v)_1 = (f(\cdot,u),v)_0, \quad v \in V \quad (= H_0^1(0,1)),$$

and where $\hat{A}$ denotes the mapping of V into itself defined by $a(.,.)$: $V \times V \to \mathbb{R}$. The finite-dimensional spaces $E_n = F_n = [\phi_1,\ldots,\phi_n]$ have the "roof functions" as a basis. For our problem, we seek, as in the Ritz-Galerkin method, an approximate solution of (24) in the form $u_n = \Sigma_{k=1}^n c_k\phi_k$, with $c_k = u_n(x_k)$, $k = 1,\ldots,n$, determined by the nonlinear system,

$$\sum_{k=1}^{n} a(\phi_k,\phi_j)c_k = \int_0^1 f(x, \sum_{k=1}^{n} c_k\phi_k(x))\phi_j(x)dx, \quad j = 1,\ldots,n. \tag{58}$$

With the orthogonal (with respect to $(.,.)_1$) projections $P_n: V \to E_n$, solving (58) is clearly equivalent to solving $P_n\hat{A}u_n = P_nTu_n$ and is also equivalent to minimizing the functional $J(\cdot)$ (defined in (25)) over E_n. Theorem 2.12, applied to $A_n = P_n\hat{A}|E_n$ moreover shows that a solution of (58) is always unique whenever the assumptions of this theorem are satisfied.

In order to correctly formulate the projective Newton method, we first claim that, under the assumptions of Example 4 in Section 2.2, the derivative is given by $A'(u)v = \hat{A}v-T'(u)v$, $u,v \in V$, with $T'(u)$ defined by

$$(T'(u)v,\phi)_1 = (\Phi,\phi)_0, \quad \forall\phi \in V, \text{ and } \Phi(x) = f_y(x,u(x))v(x). \tag{59}$$

Proof of Claim (59): We must check the conditions of Fréchet-differentiability with respect to $||.||_1$, the underlying norm. Because of the Riesz Representation Theorem, $T'(u)$ is a well-defined mapping of V into itself for every $u \in V$. By definition of T and $T'(u)$, we have the relation

$$(T(u+v) - Tu - T'(v)v,\phi)_1 =$$
$$= \int_0^1 [f(x,(u+v)(x))-f(x,u(x))-f_y(x,u(x))v(x)]\phi(x)dx, \quad u,v,\phi \in V.$$

Since f is assumed to be differentiable with respect to its second argument with continuous partial derivative f_y, we have, for arbitrary $u \in V$ and $\varepsilon > 0$, a $\delta > 0$ such that for $||v||_{0,\infty} \leq \delta$ (with $||\cdot||_{0,\infty}$ the *supremum norm*)

$$\max_{x\in[0,1]} |f(x,(u+v)(x)) - f(x,u(x)) - f_y(x,u(x))v(x)| \leq \varepsilon||v||_{0,\infty}.$$

Next, let $\varepsilon > 0$ be arbitrary, with δ the number defined above, and let $v \in V$ with $||v||_1 \leq \delta$. In the proof of Lemma 2.1, we have shown, that for the supremum norm the inequality $||v||_{0,\infty} \le ||v||_1 \le \delta$ holds. We can thus conclude that

$$|(T(u+v)-Tu-T'(u)v,\phi)_1|$$

$$\le \max_{0\le x\le 1} |f(x,(u+v)(x)) - f(x,u(x)) - f_y(x,u(x))v(x)| \; ||\phi||_0$$

$$\le \varepsilon||v||_{0,\infty}||\phi||_0, \quad v,\phi \in V.$$

These inequalities show that (59) indeed gives the Fréchet-derivative of T. □

Because of the hypothesized continuity of $f_y(.,.)$, T is, moreover, continuously differentiable at every $u \in V$. The proof of this fact proceeds in a manner similar to that for showing (59). Newton's method for solving (58) thus has the following form (cf. (54a) or (56)),

$$\sum_{k=1}^{n} c_k^{(t+1)}\{a(\phi_k,\phi_j) - \int_0^1 f_y(x,u_n^{(t)}(x))\phi_k(x)\phi_j(x)dx\}$$
$$= \int_0^1 [f(x,u_n^{(t)}(x)) - \sum_{k=1}^{n} c_k^{(t)} f_y(x,u_n^{(t)}(x))\phi_k(x)]\phi_j(x)dx, \tag{60}$$
$$j = 1,\dots,n, \quad t = 0,1,\dots .$$

This is again a tridiagonal system of equations with the numbers $a(\phi_k,\phi_j)$ having been calculated in Section 2.3 for the roof functions. The following theorem shows that the Jacobian matrix in (60) is positive definite under the assumptions cited earlier, and thus that (60) is uniquely solvable.

Theorem 2.18. Under the assumptions of Example 4 in Section 2.2 (resp. (23)), the Jacobian matrix $B = (b_{jk})$ associated with the system of equations (58),

$$b_{jk} = a(\phi_k,\phi_j) - \int_0^1 f_y(x,u_n(x))\phi_k(x)\phi_j(x)dx, \quad 1 \le j,k \le n,$$

is positive definite for every $u_n = \Sigma_{i=1}^n c_i\phi_i$.

Proof: For given $c = (c_1,\dots,c_n) \neq 0$ in $\mathbb{R}^n$, let $v = \Sigma_{k=1}^n c_k\phi_k$. Then

$$(Bc,c) = \sum_{j,k=1}^{n} [a(\phi_k,\phi_j) - \int_0^1 f_y(x,u_n(x))\phi_k(x)\phi_j(x)dx]c_k c_j$$
$$= a(v,v) - \int_0^1 f_y(x,u_n(x))v^2(x)dx.$$

We obtain the following estimate for (Bc,c) via (23),

$$(Bc,c) \geq (\Lambda-\lambda)||v||_0^2,$$

where $(.,.)$ denotes the Euclidean scalar product. The linear independence of ϕ_j, $j = 1,\ldots,n$, implies that $v \neq 0$; and, since $\Lambda > \lambda$, $(Bc,c) > 0$. □

We can, of course, solve either (51), (52) or (58) for this example by Jacobi or Gauss-Seidel methods. Since $a(.,.)$ is positive definite in (58), we have positive diagonal elements $a(\phi_j,\phi_j) > 0$, and the Gauss-Seidel method becomes

$$c_j^{(t+1)} = \frac{1}{a(\phi_j,\phi_j)}\Big\{\tau_{jn}(c_1^{(t+1)},\ldots,c_{j-1}^{(t+1)}, c_j^{(t)},\ldots,c_n^{(t)}) - \sum_{k=1}^{j-1} a(\phi_j,\phi_k)c_k^{(t+1)} - \sum_{k=j+1}^{n} a(\phi_j,\phi_k)c_k^{(t)}\Big\}, \qquad (61)$$

$$j = 1,\ldots,n, \quad t = 0,1,\ldots,$$

where

$$\tau_{jn}(\xi_1,\ldots,\xi_n) = \int_0^1 f\Big(x, \sum_{k=1}^{n} \xi_k\phi_k(x)\Big)\phi_j(x)dx, \quad j = 1,\ldots,n.$$

REFERENCES

Agmon (1965), Aubin (1972,1979), Babuska & Aziz (1972), Böhmer (1974), de Boor (1978), Ciarlet (1978), Ciarlet, Schultz & Varga (1967)*, Collatz (1966), Dieudonné (1969), Fairweather (1978), Gallagher (1975), Gottlieb & Orszag (1977), Kantorovich & Akilov (1964), Krasnoselskii, Vainikko et al. (1972), Lions & Magenes (1972), Luenberger (1969), Marchuk (1975), Meis & Marcowitz (1981), Michlin (1969), Mikhlin & Smolitskiy (1967), Ortega & Rheinboldt (1970), Rektorys (1980), Stoer & Bulirsch (1978), Strang & Fix (1973), Varga (1962,1971), Witsch (1978)*.

*Article

Chapter 3
Approximation Methods for Integral Equations of the Second Kind

In this chapter, we introduce methods for solving numerically linear and nonlinear integral equations of the second kind. These methods can be subdivided into two classes: the first class consists of those methods whose approximate equations are also expressible as integral equations with the regions of integration, measures, and kernels perturbed from the corresponding quantities in the original equation. In particular, this class includes quadrature methods, e.g., Nyström methods and product integration methods. The second class constitutes projection methods which have already been introduced in Chapter 2. Examples of such methods are collocation methods and Galerkin methods. For the specific methods considered in this chapter, we derive the associated algebraic systems of equations and show how each method can be viewed as belonging to either of the two classes described above. We carry out this analysis for both linear and nonlinear problems. As preparation, we give some general results on the solvability of linear equations of the second kind and apply these to integral equations.

3.1. LINEAR INTEGRAL EQUATIONS OF THE SECOND KIND

The criteria for solving the integral equations in this chapter depend essentially on the function space setting where the analysis is cast. An important criterion for solvability is the compactness of the underlying integral operator. For example, when the underlying Banach space is the space of bounded and continuous functions, this property is shown by using the Arzela-Ascoli Theorem in conjunction with suitable assumptions on the kernel. In larger spaces, e.g., in L^p-spaces, weaker conditions on the kernel suffice to guarantee the compactness of the integral operator. For such criteria on the kernels, the reader will be referred to the literature.

In this section, we consider a *linear integral operator* K defined by

$$(Ku)(x) \equiv \int_G k(x,y)u(y)dy, \quad x \in G, \tag{1}$$

where G is a closed subset of a bounded closed subset M in $\mathbb{R}^d$ ($d \in \mathbb{N}$), and k and u are real-valued functions defined on $G \times G$ and G, respectively. The function k is called the *kernel* of the integral operator. For brevity, we define $k_x(y) \equiv k(x,y)$. We assume that u and k_x, for every $x \in G$, are Lebesgue integrable, so that (1) is well defined.

We are interested in the solutions $u \in C(G)$ - and approximations thereof - of the following linear *integral equation of the second kind* (also called a *Fredholm integral equation of the second kind)*,

$$u(x) - \int_G k(x,y)u(y)dy = w(x), \quad x \in G, \tag{2}$$

for given $w \in C(G)$. Here, $C(G)$ denotes the space of all bounded, continuous real-valued functions on G equipped with the supremum norm which we, in this chapter, simply denote by $||\cdot||$.

We note that a sufficient condition that K maps $C(G)$ into itself is that the kernel be continuous in its arguments. The assumptions of Theorem 3.4 below ensure that $Ku \in C(G)$ whenever $u \in C(G)$ for more general kernels.

One of the *first examples* of (2) is the integral equation form of the (Sturm-Liouville) boundary-value problem (cf. Example 1, Section 1.1)

$$-u'' + qu = f \quad \text{in} \quad [0,1], \quad u(0) = u(1) = 0.$$

This can be equivalently expressed as an integral equation via

$$u(x) + \int_0^1 G(x,y)q(y)u(y)dy = \int_0^1 G(x,y)f(y)dy, \quad 0 \leq x \leq 1,$$

where $G(x,y)$ is the *Green's Function* given by

$$G(x,y) = \begin{cases} x(1-y), & 0 \leq x \leq y \leq 1, \\ y(1-x), & 0 \leq y \leq x \leq 1. \end{cases}$$

Other nontrivial examples of integral equations occur by expressing so-called exterior boundary-value problems in $n \geq 2$ dimensions as integral equations with the integrals over the boundary of the region in question.

A concrete example of this from physics is the Robin problem of potential theory, where the distribution of charge is sought on the surface of a conductor. A large class of integral equations is represented by the linear transport equations, e.g., the linear Boltzmann equation, which is first expressed as an integro-differential equation and, in certain special cases, can then be written as an integral equation of the second kind.

We now cite an important criterion for solving equations of the second kind. A (not necessarily linear) operator K, mapping a normed space E into itself, is *compact* if the image of every bounded set is relatively compact. If this mapping is also continuous, then it is called *completely continuous*. We have the following result (cf. Dieudonné (1969), Ch. 11)

Theorem 3.1. Let E be a normed space and the linear operator $K: E \to E$ be completely continuous. Then $I-K$ is bijective with continuous inverse if, and only if, $I-K$ is injective. □

The injectivity of $I-K$ occurs if, and only if, $\lambda = 1$ is not an eigenvalue of the linear mapping K. The use of the associated adjoint operator K^* enables us to formulate a very general criterion (the "Fredholm Alternative") on the solvability of equations of the second kind; we shall not present this result, but instead refer the reader to the literature for details.

Another important criterion for solving linear equations of the second kind, with linear operators that are not necessarily compact, is provided by the following theorem whose proof relies on the Banach Fixed Point Theorem.

Theorem 3.2. Let E be a Banach space and T a bounded linear operator of E into itself satisfying $||T|| < 1$. Then $(I-T)^{-1}$ exists and is continuous with norm estimated by

$$||(I-T)^{-1}|| \leq (1 - ||T||)^{-1}. \quad \square$$

We proceed to give conditions which guarantee that the integral operator K defined in (2) is completely continuous. An essential tool in proving the complete continuity of K as a mapping from $C(G)$ into itself is the important Arzela-Ascoli Theorem (see, for example, Dieudonné (1969), 7.5.7, Kantorovich & Akilov (1964), Theorem I.3). For this theorem, G can be assumed to be a closed subset of a compact metric space M whose metric is denoted by $|\cdot,\cdot|$.

Lemma 3.3 (Theorem of Arzela-Ascoli): In order that a subset H of real- or complex-valued continuous functions be relatively compact in $C(G)$, the following two conditions are necessary and sufficient:

(3a) The set H is uniformly bounded, i.e.,

$$\exists C > 0, \ni: |u(x)| \le C \text{ for every } u \in H, x \in G.$$

(3b) The set H is equicontinuous, i.e.,

$$\forall \varepsilon > 0, \exists \delta > 0, \ni: \forall u \in H, \forall x,x' \in G: |x,x'| \le \delta \Rightarrow |u(x)-u(x')| \le \varepsilon. \quad \square$$

Condition (3b) is also called "uniform equicontinuity of H" since the δ is independent of $x \in G$. Because of the compactness of the set G, this uniform condition is equivalent to the often stated pointwise one where the δ depends on x.

The Arzela-Ascoli Theorem now enables us to prove the following criterion guaranteeing complete continuity of K.

Theorem 3.4. The integral operator K defined in (1) is a completely continuous mapping of $C(G)$ into itself, provided that

$$\sup_{x\in G} \int_G |k(x,y)|dy < \infty, \tag{4a}$$

$$\lim_{x'\to x} \int_G |k(x',y)-k(x,y)|dy = 0, \quad x \in G. \tag{4b}$$

In particular, these hypotheses are satisfied for continuous kernels.

Proof: We obviously have the following estimate,

$$|Ku(x)| \le ||u|| \int_G |k(x,y)|dy, \quad x \in G, \ u \in C(G).$$

With this estimate and (4a), we get the uniform boundedness of the image set $H = KB$ for every bounded set $B = \{u \in C(G): ||u|| \le c\}$. From condition (4b), we have that for arbitrary $x \in G$, and $\varepsilon > 0$, there is a $\delta > 0$ such that

$$\int_G |k(x',y) - k(x,y)|dy \le \varepsilon,$$

for all $x' \in G$ with $|x-x'| \le \delta$. Thus

$$|Ku(x') - Ku(x)| \le \varepsilon||u|| \le \varepsilon c$$

for all x' in $|x-x'| \le \delta$ and all $u \in B$. The preceding inequality implies that KB is pointwise equicontinuous, and hence uniformly equicontinuous because of the compactness of G. We now can appeal to the Arzela-Ascoli Theorem to conclude the relative compactness of H, and hence the compactness of $K: C(G) \to C(G)$. Every compact linear operator is also continuous, since the image of the unit ball is relatively compact and hence bounded. Thus the mapping K is proved to be completely continuous. Conditions (4a) and (4b) are clearly satisfied for a continuous kernel. □

In addition to continuous kernels, those kernels existing in product form, i.e., $k(x,y) = h(x,y)r(x,y)$, present an important class of examples. Here, h may be singular or weakly singular and r is continuous for $x \neq y$. Examples include *kernels of potential type*

$$k(x,y) = |x-y|^{-m} r(x,y).$$

Whenever $G = [a,b]$, then K is a completely continuous mapping of $C[a,b]$ into itself provided $m < 1$. Another example is a kernel in product form with $h(x,y) = \log(|x-y|)$.

Later, we shall consider integral operators and Fredholm integral equations in spaces larger than $C(G)$ - for example in $L^\infty(G)$ or $L^2(G)$. Sufficient conditions on the kernels guaranteeing complete continuity in these spaces can be found in, e.g., Kantorovich & Akilov (1964), Riesz & Nagy (1965). The complete continuity in $L^p(G)$, $1 \leq p < \infty$, is a consequence of (4a,b); in particular, continuous kernels will always generate completely continuous mappings on these spaces. A rather general treatment of the compactness properties of K as a mapping of $L^p(G)$ into $C(G)$ is given in a recent work by Graham & Sloan (1982); their treatise essentially relies on results attributed to Radon (1919) (cf. also Riesz & Nagy (1965), Sec. 90).

To conclude this section, we like to mention a certain generalization of the problems considered. We could allow an arbitrary compact metric space M, a closed subset G of M, and a (regular Borel) measure μ on G - which is identifiable with a bounded linear functional on $C(G)$ (cf. Rudin (1966), Dunford & Schwartz (1966)). Following Bourbaki (1965), Chapt. III, we call any bounded linear functional on $C(G)$ a *measure* on G. The value of the functional defined by μ at $u \in C(G)$ is the *integral with respect to the measure* μ expressed notationally by

$$\mu(u) = \int_G u d\mu, \quad \text{or by} \quad \mu(u) = \int_G u(x) d\mu(x).$$

The integral operator can then be defined via

$$(Ku)(x) = \mu(k_x u) = \int_G k(x,y)u(y)d\mu(y), \quad x \in G, \ u \in C(G).$$

A measure on G is called *nonnegative* (respectively, *positive*) if $\mu(u) \geq 0$ for $u \geq 0$ (respectively, $\mu(u) > 0$ for $0 \not\equiv u \geq 0$), $u \in C(G)$. For later purposes, we finally remark, that each measure μ on G possesses a *natural extension* $\hat{\mu}$ to a measure on M by defining

$$\hat{\mu}(v) \equiv \mu(v|G) = \int_G v d\mu, \quad v \in C(M). \tag{5}$$

The setting just described is basic for the work of Stummel (1974b) which, however, cannot be applied to weakly singular kernels. We therefore turn to the aforementioned problem setting - in particular, we require the conditions of Theorem 3.4 - but we need the notion of a measure and associated extensions for the approximation methods considered in the following section.

3.2. QUADRATURE METHODS

A *quadrature formula* for approximating integrals is a sum of the form

$$\mu_n(u) = \sum_{x \in G_n} u(x)\alpha_n(x). \tag{6}$$

Here, the G_n are finite subsets of M, the so-called *quadrature nodes*, and the real numbers $\alpha_n(x)$ are the *weights* of the points from G_n. Obviously, μ_n is a measure on G_n and (6) can also be expressed as

$$\mu_n(u) = \int_{G_n} u d\mu_n = \int_{G_n} u(x) d\mu_n(x), \quad u \in C(G_n). \tag{7}$$

The norm of μ_n is given by

$$||\mu_n|| = \sum_{x \in G_n} |\alpha_n(x)|. \tag{8}$$

The functions constituting $C(G_n)$ - i.e., the functions defined on finitely many discrete points - are denoted often as *grid* or *mesh functions*. It is trivial to see that grid functions are always continuous. The quadrature formula μ_n represents a nonnegative (respectively, positive) measure in case all the weights are nonnegative (respectively, positive).

In order to approximate the solution of (2), we seek a function $u_n \in C(G_n)$ which solves the algebraic system of equations

$$u_n(x) - \sum_{y \in G_n} k_n(x,y)u_n(y)\alpha_n(y) = w_n(x), \quad x \in G_n, \tag{9}$$

where k_n (respectively, w_n) represent approximations to k (respectively, w). In case $G_n \subset G$, we can define k_n and w_n as the restrictions of k and w, respectively, and (9) is then known as the *Nyström method* or the *Fredholm-Nyström method*. If $G_n \not\subset G$, then we can obtain the appropriate k_n, w_n by the following argument: Each continuous $v \in C(G)$ can be extended to a function $\hat{v} \in C(M)$ by the Tietze Extension Theorem (a corollary of Urysohn's Lemma - see, for example, Dieudonné (1969), Chapter IV.5, Kuratowski (1966), §14.IV) so that $\hat{v}|G = v$ and $\sup\{|\hat{v}(x)|: x \in M\} = \sup\{|v(x)|: x \in G\}$. (For this procedure, it suffices to require that G be a closed subset of a metric space M which does not necessarily have to be compact.) In a corresponding manner, we can extend a continuous kernel $k(.,.)$ to a continuous function $\hat{k} \in C(M \times M)$. We then choose, for our approximates k_n and w_n,

$$k_n = \hat{k}|G_n \times G_n, \quad w_n = \hat{w}|G_n.$$

We also designate this choice of k_n and w_n as a *Nyström method*. In both cases, we can express an approximation to the integral operator defined in (1) by

$$(K_n u_n)(x) = \int_{G_n} k_n(x,y)u_n(y)d\mu_n(y), \quad u_n \in C(G_n). \tag{10}$$

Approximating (2) by using the quadrature formulas considered in (6) leads to a finite-dimensional operator $K_n: C(G_n) \to C(G_n)$ which is clearly completely continuous. The solvability criterion of Theorem 3.1 will imply the well-known fact that a finite system of linear equations is uniquely solvable for each right-hand side whenever the associated homogeneous system possesses only the trivial solution.

In the course of our investigations in this section, we shall now introduce the so-called "product integration methods", which can be interpreted as an extension of the Nyström method to integral operators with special kernels which may be discontinuous. We assume that the underlying kernels can be represented in product form,

$$k(x,y) = h(x,y)r(x,y), \quad x,y \in G,$$

where now $G = [a,b]$, $b > a$, is a closed, bounded interval in $\mathbb{R}$, r is continuous on $G \times G$, and h is possibly singular or weakly singular. Let all functions considered be real-valued. Typical examples for $h(.,.)$ have already been given in Section 3.1. As before, the integral operator in (1) is to be understood as a mapping of $C(G)$ into itself.

A general description of the product integration method is as follows: Suppose that we have a finite set $G_n = \{y_1,\ldots,y_n\}$ of points from G along with continuous functions $\phi_1,\ldots,\phi_n$, for example, polynomials or piecewise polynomial functions. We assume that the matrix $(\phi_j(y_i))_{i,j}$ is nonsingular. With the ϕ_j, $1 \leq j \leq n$, we can then uniquely determine $\eta_j \in [\phi_1,\ldots,\phi_n]$ such that $\eta_j(y_i) = \delta_{ij}$, $1 \leq i,j \leq n$. We define *generalized weights* for functions defined on G by the formula

$$\omega_j(v) \equiv \int_a^b v(y)\eta_j(y)\,dy, \quad j = 1,\ldots,n. \tag{11}$$

The function v need not be necessarily continuous but regular enough so that the integral in (11) exists. Using the given functions ϕ_j, $1 \leq j \leq n$, we can easily determine the weights by the following system of equations,

$$\sum_{i=1}^{n} \omega_i(v)\phi_j(y_i) = \int_a^b v(y)\phi_j(y)\,dy, \quad j = 1,\ldots,n. \tag{12}$$

The *product integration method* now consists of approximating the integral operator

$$(Ku)(x) = \int_a^b h(x,y)r(x,y)u(y)\,dy$$

by

$$(K^{(n)}u)(x) \equiv \sum_{i=1}^{n} \omega_{n,i}r(x,y_i)u(y_i), \quad x \in G, \quad u \in C(G),$$

where $\omega_{n,i} = \omega_i(h_x)$, $i = 1,2,\ldots,n$, with $h_x(y) = h(x,y)$, $x,y \in G$. We assume here that the weights $\omega_{n,i}$ are given exactly.

The function $u^{(n)} \in C(G)$ which solves the following equation,

$$u^{(n)}(x) - \sum_{i=1}^{n} \omega_{n,i}r(x,y_i)u^{(n)}(y_i) = w(x), \quad x \in G, \tag{13}$$

is taken as an approximation to the solution of the integral equation (2). To determine $u^{(n)}$, we must evidently compute $u^{(n)}$ only at the quadrature nodes y_j, $j = 1,\ldots,n$, i.e., we must solve the following linear system,

$$u^{(n)}(y_j) - \sum_{i=1}^{n} \omega_{n,i}r(y_j,y_i)u^{(n)}(y_i) = w(y_j), \quad j = 1,2,\ldots,n, \tag{14}$$

and then set $u^{(n)}$ to be equal to

$$u^{(n)}(x) = w(x) + \sum_{i=1}^{n} \omega_{n,i}r(x,y_i)u^{(n)}(y_i), \quad a \leq x \leq b. \tag{15}$$

The latter equation is called the *natural interpolation formula* for the values of $u^{(n)}$ between the quadrature points.

We see immediately that (14) reduces to the Nystrom method in the case $h \equiv 1$, where the weights then are given by $\alpha_n(y_i) = \omega_i(1) = \int_a^b \eta_i(y)dy$. In the general case also, we can further explain the connection between the product integration method and quadrature methods of the form (9). Indeed, if we require that

$$\alpha_n(y_i) \equiv \int_a^b \eta_i(y)dy > 0, \quad i = 1,2,\dots,n, \tag{16}$$

and define

$$\begin{aligned} h_n(x,y_i) &\equiv \int_a^b h(x,y)\eta_i(y)dy/\alpha_n(y_i), \\ k_n(x,y_i) &\equiv h_n(x,y_i)r(x,y_i), \quad 1 \le i \le n, \end{aligned} \tag{17}$$

and

$$(K_n u_n)(x) \equiv \sum_{y \in G_n} \alpha_n(y)k_n(x,y)u_n(y), \quad x \in G_n, \quad u_n \in C(G_n), \tag{18}$$

then equation (14) determining $u_n(x) = u^{(n)}(x)$, $x \in G_n$, has exactly the form of (9) (with $w_n = w|G_n$). The operator K_n: $C(G_n) \to C(G_n)$, defined in (18), can also be expressed in the form (10) of an integral operator, where the associated positive measure μ_n on G_n is given by

$$\mu_n(u) = \sum_{x \in G_n} \alpha_n(x)u(x), \quad u \in C(G),$$

with the weights $\alpha_n(x)$ defined by (16). The weights $\alpha_n(y_i)$ can also be calculated as in (12) (with $v = 1$) by

$$\sum_{i=1}^n \alpha_n(y_i)\phi_j(y_i) = \int_a^b \phi_j(y)dy, \quad j = 1,2,\dots,n. \tag{19}$$

If the constant functions lie in the span of the ϕ_j, $j = 1,\dots,n$, then the weights must necessarily satisfy

$$\sum_{i=1}^n \alpha_n(y_i) = \sum_{i=1}^n \int_a^b \eta_i(y)dy = (b-a), \tag{20}$$

and therefore $||\mu_n|| = (b-a)$ (cf. (8)). This follows directly from (19) if we insert $1 = \Sigma_j \beta_j \phi_j$.

A simple first *example* of quadrature methods which fits into our framework here is the *composite trapezoidal rule*, whose quadrature nodes are not necessarily equidistant, and where the ϕ_j are chosen as basis functions for the space of functions which are piecewise linear on each $[y_{j-1},y_j]$, $j = 2,\ldots,n$, and continuous on all of $[a,b]$. For the weights, we have

$$\alpha_n(y_1) = \tfrac{1}{2}h_2, \quad \alpha_n(y_j) = \tfrac{1}{2}(h_j+h_{j+1}), \quad j = 2,\ldots,n-1, \quad \alpha_n(y_n) = \tfrac{1}{2}h_n,$$

where $h_j = y_j - y_{j-1}$. We obtain the *n-point Gauss rule* by choosing the y_j, $j = 1,\ldots,n$, to be the zeroes of the Legendre polynomial $P_n((2x-a-b)/(b-a))$ and $[\phi_1,\ldots,\phi_n]$ to span the space of all polynomials of degree less than or equal to $n-1$. The functions η_j, $1 \le j \le n$, are then the Lagrange interpolation polynomials at the points y_j, $1 \le j \le n$. We refer the reader to the work of Sloan (1980b,1981) for numerous examples and for results of more recent investigations on the choice of the quadrature points for product integration methods.

There are also other situations where an integral operator is approximated by an expression of the form of (10). This can occur, for example, if the kernel itself is defined by an integral and is approximated by a quadrature formula. Such an example is the "discrete-ordinates approximation" for a particularly simple form of the Boltzmann transport equation (see, for example, Anselone (1971), 3.8, Nelson & Victory (1979)).

3.3. PROJECTION METHODS

The projection methods introduced in Chapter 2 can be used in an analogous manner to the approximate solution of integral equations of the kind considered above. We begin our discussion by providing a general description of projection methods for linear equations of the second kind. Then we shall show that collocation methods and Galerkin methods for approximating solutions of Fredholm integral equations can be viewed as special cases.

Let K be a bounded linear operator mapping a normed space E into itself, and let E_o, E_n be subspaces of E, $n \in \mathbb{N}$. Further, let $P_n: E \to E_n$, $n \in \mathbb{N}$, denote associated bounded, linear projection operators. The restriction $K_o \equiv K|E_o$ is to be a mapping of E_o into itself. The given problem, then, is to determine a solution $u_o \in E_o$ of

$$u_o - K_o u_o = w_o, \tag{21}$$

where $w_o \in E_o$. Theorems 3.1 and 3.2 give results concerning solvability of (21). The projection method furnishes, as approximations to u_o, solutions $u_n \in E_n$, $n \in \mathbb{N}$, of

$$u_n - P_n K u_n = w_n, \tag{22}$$

where the $w_n \in E_n$, $n \in \mathbb{N}$, approximate w_o. In the simplest case, $w_n = P_n w_o$. If E_n is finite-dimensional, (22) is uniquely solvable for every $w_n \in E_n$ in case $A_n \equiv P_n(I-K)|E_n$ is injective. As in Section 2.4, we can write (22) as an algebraic system of equations when $\dim E_n < \infty$ which is tacitly assumed in this section. If $\{\phi_1,\ldots,\phi_m\}$ is a basis in E_n, then for every $v \in E$, there are uniquely determined numbers $\alpha_j(v)$, $j = 1,\ldots,m$, so that

$$P_n v = \sum_{i=1}^{m} \alpha_i(v)\phi_i. \tag{23}$$

The solution u_n of (22) can be expressed in the form $u_n = \Sigma_i \alpha_i(u_n)\phi_i$. Hence (22) is equivalent to the following system of equations for determining $\alpha_{ni} \equiv \alpha_i(u_n)$, $i = 1,\ldots,m$,

$$\alpha_{ni} - \sum_{j=1}^{m} \alpha_{nj}\alpha_i(\psi_j) = \alpha_i(w_n), \quad i = 1,\ldots,m, \tag{24}$$

where $\psi_j \equiv K\phi_j$, $j = 1,\ldots,m$. Indeed, we have $Ku_n = \Sigma_{j=1}^m \alpha_{nj}\psi_j$, and hence

$$P_n K u_n = \sum_{j=1}^{m} \alpha_{nj} P_n \psi_j = \sum_{i,j=1}^{m} \alpha_{nj}\alpha_i(\psi_j)\phi_i.$$

The linear independence of the ϕ_i, $1 \leq i \leq m$, insures, then, the equivalence of (22) and (24).

We consider once more the Fredholm integral equation (2), where G is a closed and bounded subset of $\mathbb{R}^d$, $d \in \mathbb{N}$. For the moment, we consider only continuous solutions of the integral equation and their approximations. We stipulate that the boundary of G have Lebesgue measure zero so that the Riemann integral will exist for functions continuous on G. Further, suppose the kernel satisfies condition (4a,b) relative to Riemann integration. Thus the integral operator defined in (1) constitutes a bounded, linear mapping of $E = L^\infty(G)$ into itself (see Kantorovich & Akilov (1964), X.2), with its restriction $K_o = K|E_o$, $E_o = C(G)$, a completely continuous operator from E_o into itself (cf. Theorem 3.3). In (24), we then set

$$\psi_i(x) = \int_G k(x,y)\phi_i(y)dy, \quad i = 1,\ldots,m,$$

where $E_n = [\phi_1,\ldots,\phi_m]$ denotes a subspace of E.

An important class of projection methods are *collocation methods*. We describe *collocation with piecewise continuous functions* in a manner which is adapted from a paper of Atkinson & Graham & Sloan (1983). Suppose that we are given a partition

$$a = s_0 < s_1 < \ldots < s_J = b$$

of pairwise distinct points and linear independent piecewise continuous functions $\phi_1,\ldots,\phi_n$ (with at most a finite number of discontinuities) such that the restrictions of ϕ_i to $I_j = (s_{j-1},s_j)$ are bounded and continuous for $i = 1,\ldots,n$ and $j = 1,\ldots,J$. There are further given *collocation points* x_i, $i = 1,\ldots,n$, each of which is assumed to lie in I_j for some j. As a consequence each x_i is a point of continuity of each basis function ϕ_j. By the Hahn-Banach Theorem, there exist *extended point evaluation functionals* d_{x_i} on $L^\infty(G)$ with the properties

$$||d_{x_i}|| = 1, \quad d_{x_i}(\phi_j) = \phi_j(x_i), \qquad i,j = 1,\ldots,n.$$

Note that the extended point evaluation functionals are not uniquely defined.

As a *collocation solution* we seek a function u_n as a linear combination of the ϕ_j, $j = 1,\ldots,n$, $u_n = \Sigma_{j=1}^n \alpha_{nj}\phi_j$, so that the system of equations

$$\sum_{j=1}^{n} \alpha_{nj}\{\phi_j(x_i) - \int_a^b k(x_i,y)\phi_j(y)dy\} = d_{x_i}(w), \; i = 1,\ldots,n, \tag{25}$$

is satisfied. In order to see that this method falls into the general framework of projection methods, we define a linear projection $P_n: E \to E_n = [\phi_1,\ldots,\phi_n]$ by

$$P_n v = \sum_{i=1}^{n} d_{x_i}(v)\eta_i, \qquad v \in E, \tag{26}$$

where the η_i, $i = 1,\ldots,n$, are a basis in E_n for which $\eta_i(x_j) = \delta_{ij}$, $1 \le i,j \le n$. Under the hypothesis $\det((\phi_j(x_i)) \neq 0$, the $\eta_i = \Sigma_\nu \beta_\nu^{(i)}\phi_\nu$, $i = 1,\ldots,n$, can be determined uniquely by the constraints

$$\sum_{\nu=1}^{n} \beta_\nu^{(i)}\phi_\nu(x_j) = \delta_{ij}, \qquad 1 \le i,j \le n. \tag{27}$$

Thus P_n is a well-defined linear operator from E_n into $L^\infty(G)$ having the property of a projection operator, $P_n^2 = P_n$, and with norm

$$||P_n|| = \operatorname*{ess\,sup}_{x\in G} \sum_{i=1}^{n} |\eta_i(x)|.$$

The system of equations (25) for determining the α_{nj}, $1 \le j \le n$, is equivalent to

$$u_{ni} - \sum_{j=1}^{n} u_{nj} \int_a^b k(x_i,y)\eta_j(y)\,dy = d_{x_i}(w), \quad i = 1,\dots,n. \tag{28}$$

for determining the $u_{ni} = u_n(x_i)$, $1 \leq i \leq n$, which is obvious by the representation $u_n = \Sigma_j u_{nj}\eta_j$. The system of equations in (28) is obviously equivalent to the projection method (22) with $w_n = P_n w$ and P_n defined as above.

A variant of the above solution procedure, analogous to the natural interpolation formula (15), occurs by setting

$$u_n'(x) \equiv w(x) + Ku_n(x) = w(x) + \sum_{i=1}^{n} \alpha_{ni} \int_a^b k(x,y)\phi_i(y)\,dy. \tag{29}$$

This approximation is called the *iterated collocation solution,* and the associated method is the *iterated collocation method.* The u_n and u_n' obviously agree at the collocation points but differ in the manner by which they are interpolated between these points. In general, u_n' will not naturally lie in E_n.

The iterated collocation method can be viewed as a special case of the product integration method, where $r(x,y) = 1$, $h(x,y) = k(x,y)$. We have assumed for the product integration method that the matrix $(\phi_j(x_i))_{i,j}$ is nonsingular. The iterated collocation solution then directly arises as the solution of the product integration method extrapolated to all x by the natural interpolation formula. This is also the same function extended to all x by natural interpolation from the values $u_n(x_j)$, $1 \leq j \leq n$, obtained by the Nyström method.

We now give *two examples* of collocation with piecewise polynomial functions. To define *collocation with piecewise constant functions*, let $a = s_o < s_1 < \dots < s_n = b$ be a (not necessarily equidistant) partition of $[a,b]$. We choose as collocation points exactly one point x_j from each I_j, where

$$I_j = (s_{j-1},s_j), \quad j = 1,\dots,n.$$

The associated functions η_j, $1 \leq j \leq n$, are taken to be the characteristic functions of the I_j, i.e.,

$$\eta_j(x) = 1, \quad x \in I_j, \quad \eta_j(x) = 0 \text{ otherwise}, \quad j = 1,\dots,n.$$

These functions clearly satisfy the condition $\eta_i(x_j) = \delta_{ij}$, $1 \leq i,j \leq n$, and are linearly independent. The associated *piecewise constant interpolation projection* is defined by

$$P_n u \equiv \sum_{j=1}^{n} u(x_j)\eta_j, \qquad u \in L^\infty(a,b),$$

and represents a step function which has the value $u(x_j)$ on I_j. System (28) has now the form

$$u_{ni} - \sum_{j=1}^{n} a_{ij}u_{nj} = d_{x_i}(w), \qquad i = 1,\dots,n,$$

with

$$a_{ij} = \int_{I_j} k(x_i,y)\,dy, \qquad 1 \le i,j \le n.$$

For *collocation with continuous, piecewise linear functions,* let the interval $[a,b]$ be partitioned as $a = x_o < x_1 < \dots < x_m = b$ where the nodes x_i are also the collocation points. We take as a basis for the space of all continuous, piecewise linear functions the $(m+1)$ "roof functions" (see also Section 2.3) defined by

$$\eta_o(x) = \begin{cases} (x_1-x)/h_1, & x \in I_1, \\ 0, & x \notin I_1, \end{cases}$$

$$\eta_m(x) = \begin{cases} (x-x_{m-1})/h_m, & x \in I_m, \\ 0, & x \notin I_m, \end{cases}$$

$$\eta_j(x) = \begin{cases} 0, & a \le x \le x_{j-1}, \\ (x-x_{j-1})/h_j, & x \in I_j, \\ (x_{j+1}-x)/h_{j+1}, & x \in I_{j+1}, \\ 0, & x_{j+1} \le x \le b, \end{cases}$$
$$j = 1,\dots,m-1,$$

where $I_j \equiv [x_{j-1},x_j]$, $h_j \equiv x_j-x_{j-1}$, $j = 1,2,\dots,m$. The *piecewise linear interpolatory projection* operator is defined by

$$P_n u \equiv \sum_{j=0}^{m} u(x_j)\eta_j.$$

The system (28) now has the form (for, say, a continuous w)

$$u_{ni} - \sum_{j=0}^{m} b_{ij}u_{nj} = w(x_j), \qquad j = 0,\dots,m,$$

where

$$b_{ij} = \int_{x_{j-1}}^{x_{j+1}} k(x_i,y)\eta_i(y)\,dy, \qquad i = 0,\dots,m, \quad j = 1,\dots,m-1,$$

$$b_{io} = \int_{x_o}^{x_1} k(x_i,y)\eta_o(y)\,dy, \qquad b_{im} = \int_{x_{m-1}}^{x_m} k(x_i,y)\eta_m(y)\,dy, \qquad i = 0,\dots,m.$$

Other classes of collocation methods can be obtained by basis functions defined by a single expression over the whole interval - for example, polynomials or trigonometric functions. For a detailed survey of these collocation methods, we refer the reader to the review article by Sloan (1980a) and the literature cited therein.

Still another class of projection methods is represented by *Galerkin methods* which we have already introduced in Section 2.4. In order to illustrate Galerkin methods for integral equations of the second kind, we cast our analysis in the Hilbert space $E = L^2(G)$, where G is a measurable subset of $\mathbb{R}^d$. E is endowed with the scalar product

$$(u,v)_o = \int_G u(y)\overline{v(y)}dy.$$

The integral operator K in (1) is a completely continuous mapping of E into itself, in case the kernel k is of Hilbert-Schmidt type, i.e.,

$$\int_G\int_G |k(x,y)|^2 dxdy < \infty.$$

We let $\{\phi_1,\ldots,\phi_m\}$ be a basis in E_n, a finite-dimensional subspace of E. The associated orthogonal projection operator $P_n: E \to E_n$ is defined by

$$u_n \equiv P_n u \in E_n: \quad (u_n,v)_o = (u,v)_o \quad \forall v \in E_n.$$

Galerkin methods for approximating the solution of (2) consist of determining the coefficients α_{ni}, $i = 1,\ldots,m$, in the representation $u_n = \Sigma_i \alpha_{ni}\phi_i$ of the approximate solution. These coefficients are solutions to the following system of equations,

$$\sum_{j=1}^{m} \{(\phi_j,\phi_i)_o - (K\phi_j,\phi_i)_o\}\alpha_{nj} = (w,\phi_i)_o, \quad i = 1,\ldots,m. \tag{30}$$

From the definition of the projection P_n, we see that (30) represents a projection method of the form (22).

Galerkin methods will, in general, require more expense than collocation methods, because the double integrals

$$(K\phi_j,\phi_i)_o = \int_G\int_G k(x,y)\phi_j(y)\overline{\phi_i(x)}dydx$$

on the left-hand side of (30) must be evaluated. Whenever the basis $\{\phi_j\}_j$ is orthonormal, (30) simplifies since $(\phi_j,\phi_i)_o = \delta_{ij}$, $1 \leq i,j \leq m$. The coefficients α_{nj} in (30) are then equal to $(u_n,\phi_j)_o$.

In Section 2.4 we have introduced several other modifications of Galerkin methods. Accordingly, we could more appropriately denote the method (30) as the Bubnov-Galerkin method. A variant is represented by the Petrov-Galerkin method, where the functions ϕ_i in (30) are replaced by elements ψ_i of a second basis $\{\psi_1,\ldots,\psi_m\}$ in E_n. We can select $\psi_i = K\phi_i$, which corresponds to the method of least squares. For the latter choice of ψ_i, we must stipulate that the integral operator K be injective.

3.4. APPROXIMATIONS OF NONLINEAR INTEGRAL EQUATIONS

There are no great conceptual difficulties in formally applying quadrature methods and projection methods to nonlinear integral equations. We shall devote ourselves to this task in this section and shall derive the associated systems of nonlinear equations. In addition, we shall provide conditions which will be shown to guarantee the differentiability of both the integral operator and the approximate operators.

Again, let $G \subset M$ be a bounded, closed subset of $\mathbb{R}^d$. With a kernel $k(.,.,.)\colon G \times G \times \mathbb{R} \to \mathbb{R}$, we define the associated *nonlinear integral operator* by

$$(Ku)(x) \equiv \int_G k(x,y,u(y))dy, \quad x \in G, \quad u \in C(G), \tag{31}$$

where the integral is defined in Lebesgue sense. The definition of K makes sense, and K is a mapping of $C(G)$ into itself provided that $k(x,.,u(.))$ is Lebesgue integrable for every fixed $x \in G$ and every $u \in C(G)$, and that

$$\lim_{x'\to x} \int_G (k(x',y,u(y)) - k(x,y,u(y)))dy = 0, \quad x \in G, \quad u \in C(G).$$

These conditions are satisfied, for example, if the kernel k is continuous in all arguments. For a given $w \in C(G)$, we now consider the following *nonlinear integral equation of the second kind* in $C(G)$,

$$u(x) - \int_G k(x,y,u(y))dy = w(x), \quad x \in G, \tag{32}$$

also termed a *Urysohn equation* or a *nonlinear Fredholm integral equation*. With kernels of the special form $k(x,y,z) = m(x,y)n(y,z)$, (32) is sometimes called an equation of *Hammerstein type*.

An example of a nonlinear integral equation arises from the following boundary-value problem

$$u''(x) = f(x,u(x)), \quad x \in (0,1), \quad u(0) = u(1) = 0,$$

expressed in equivalent integral equation form,

$$u(x) + \int_0^1 G(x,y)f(y,u(y))dy = 0, \qquad 0 \leq x \leq 1,$$

where $G(.,.)$ is the Green's Function given in Section 3.1.

Let us first consider approximating equations for (32) in a setting which is analogous to that of Section 3.2. To this end, let $k_n(.,.,.)$: $G_n \times G_n \times \mathbb{R} \to \mathbb{R}$ be a continuous function with G_n a bounded closed subset of $M \subset \mathbb{R}^d$ and with μ_n a measure on G_n. Then the operator K_n, given by

$$K_n u_n(x) \equiv \int_{G_n} k_n(x,y,u_n(y))d\mu_n(y), \qquad x \in G_n, \quad u_n \in C(G_n), \tag{33}$$

is an integral operator on $C(G_n)$. As an approximation of the solution of (32) we seek a solution $u_n \in C(G_n)$ of

$$u_n(x) - \int_{G_n} k_n(x,y,u_n(y))d\mu_n(y) = w_n(x), \qquad x \in G_n, \tag{34}$$

where $w_n \in C(G_n)$ represents an approximation for w.

As in Section 3.2, *quadrature methods* can be also treated in the setting outlined here. Indeed, for a continuous, nonlinear kernel k, the Nyström method can be described as follows:

Nyström Method:
$$\mu_n(u) = \sum_{x \in G_n} \alpha_n(x)u(x), \qquad k_n(x,y,z) = \hat{k}(x,y,z), \; x,y \in G_n, \quad z \in \mathbb{R}, \tag{35}$$

where $\hat{k}: M \times M \times \mathbb{R} \to \mathbb{R}$ is a continuous extension of a continuous kernel k. If k, furthermore, has product form,

$$k(x,y,z) = h(x,y)r(x,y,z), \quad x,y \in G = [a,b], \quad z \in \mathbb{R},$$

where r is continuous and h is possibly singular or weakly singular, then we can describe the product integration method in the following way:

Product Integration Method:
$$\mu_n(u) = \sum_{x \in G_n} \alpha_n(x)u(x), \qquad k_n(x,y,z) = h_n(x,y)r(x,y,z), \; x,y \in G_n, \; z \in \mathbb{R}, \tag{36}$$

where α_n and h_n are defined as in (16) and (17), respectively. Here, as before, $k_n(.,.,.)$ is a continuous function with respect to all three arguments - this is trivially valid with respect to x and y, since G_n is a discrete set of points. As we have noted for linear problems, there are also other examples of nonlinear problems where an integral equation of the type (32) is approximated by an equation of the form (34).

As in the earlier chapters, we now proceed to derive the Fréchet-derivatives of the approximating nonlinear operator K_n given by (33). The space $C(G_n)$ is also equipped with the maximum norm which, for simplicity, is again denoted by $||\cdot||$. The continuous differentiability at some $v_n \in C(G_n)$ is then guaranteed by the following condition on the kernel:

(37) For some number $\rho > 0$, the kernel $k_n(.,.,.)$ is differentiable with respect to the third argument in the region

$$U_n \equiv \{(x,y,z) \in G_n \times G_n \times \mathbb{R}: |z - v_n(y)| \leq \rho\}$$

and the partial derivative $\partial k_n/\partial z$ is continuous at every point in U_n with respect to all arguments.

At this point, we would like to mention that U_n is a compact subset of $\mathbb{R}^{d+2}$, and hence the continuity of the partial derivative holds uniformly in U_n. We are therefore able to prove the following result.

Lemma 3.5. Whenever condition (37) is satisfied for some $v_n \in C(G_n)$, then the integral operator given by (33) is continuously differentiable at v_n with the Fréchet-derivative given by

$$(K_n'(v_n)h_n)(x) = \int_{G_n} \frac{\partial k_n}{\partial z}(x,y,v_n(y))h_n(y)d\mu_n(y), \quad x \in G_n, \quad h_n \in C(G_n). \tag{38}$$

Proof: By definition, the measure μ_n is a bounded linear functional on $C(G_n)$, i.e., for some $\gamma_n > 0$ we have $|\mu_n(\phi)| \leq \gamma_n||\phi||$, $\phi \in C(G_n)$. From (37), we have the following relation for the kernel k_n,

$$k_n(x,y,z') = k_n(x,y,z) + \frac{\partial k_n}{\partial z}(x,y,z)(z'-z) + r_n(x,y,z;z'-z),$$

$$(x,y,z),(x,y,z') \in U_n,$$

where, according to (37), the remainder term r_n satisfies the condition:

$$\forall\varepsilon > 0, \ \exists\delta > 0, \ \ni: \forall(x,y,z), \ (x,y,z') \in U_n: \ |z-z'| \leq \delta$$

$$\Rightarrow |r_n(x,y,z;z'-z)| \leq \frac{\varepsilon}{\gamma_n} |z' - z|.$$

This uniform differentiability property is a consequence of the Mean Value Theorem yielding

$$k_n(x,y,z') - k_n(x,y,z) = (z-z')\left(\frac{\partial k_n}{\partial z}(x,y,z+\upsilon(z'-z)) - \frac{\partial k_n}{\partial z}(x,y,z)\right),$$

for all (x,y,z), $(x,y,z') \in U_n$ (with some $\upsilon \in [0,1]$), and the hypothesized continuity of $\partial k_n/\partial z$ which is uniform on the compact set U_n (cf. also Vainberg (1964), p. 45). For arbitrary $\varepsilon > 0$, $z_n \in C(G_n)$ with $||z_n - v_n|| < \rho$ and $h_n \in C(G_n)$ with $||h_n|| \leq \min(\rho - ||z_n - v_n||, \delta)$, we have the relation

$$k_n(x,y,z_n(y)+h_n(y)) = k_n(x,y,z_n(y)) + \frac{\partial k_n}{\partial z}(x,y,z_n(y))h_n(y)$$
$$+ r_n(x,y,z_n(y);\, h_n(y)), \quad x,y \in G_n.$$

Here z_n, h_n are continuous functions on G_n; the kernel $k_n(.,.,.)$ is continuous in $G_n \times G_n \times \mathbb{R}$, and $\partial k_n/\partial z(.,.,.)$ is continuous in U_n. The remainder term r_n is then a continuous function with respect to $x,y \in G_n$. For fixed $x \in G_n$, we can therefore apply the measure μ_n as a functional on this continuous function in $y \in G_n$ to get (see (33) and (38) with z_n in place of u_n and v_n, respectively)

$$K_n(z_n+h_n) = K_n z_n + K_n'(z_n)h_n + R_n(z_n;h_n)$$

where

$$R_n(z_n;h_n)(x) \equiv \int_{G_n} r_n(x,y,z_n(y);h_n(y))d\mu_n(y), \quad x \in G_n.$$

Because $||h_n|| \leq \min(\rho - ||z_n - v_n||, \delta)$, we have

$$|r_n(x,y,z_n(y);h_n(y))| \leq \frac{\varepsilon}{\gamma_n}|h_n(y)| \leq \frac{\varepsilon}{\gamma_n}\,||h_n||, \quad x,y \in G_n,$$

and hence

$$|R_n(z_n;h_n)(x)| \leq ||\mu_n||\,\frac{\varepsilon}{\gamma_n}\,||h_n|| \leq \varepsilon||h_n||, \quad x \in G_n.$$

We have thus shown that K_n is differentiable at every function z_n in a neighborhood of v_n, with the derivative $K_n'(z_n)$ having the representation (38) (with z_n substituted for v_n). The continuity of $\partial k_n/\partial z$ implies that $K_n'(z_n)$ is a bounded linear mapping of $C(G_n)$ into itself. The continuity of the derivative follows easily from the hypothesized continuity of $\partial k_n/\partial z$ in U_n. □

For the linear integral operator defined in (38), $k_n'(x,y) \equiv \partial k_n/\partial z(x,y,v_n(y))$ is a continuous kernel, and thus $K_n'(v_n)$ is a completely continuous mapping of $C(G_n)$ into itself according to Theorem 3.4. The continuity required in (37) is trivially present with respect to the first and second arguments, in case G_n is a finite set of points.

(39) We now give some results guaranteeing the differentiability of the given nonlinear integral operator itself. The following condition plays an important role:

> For some number $\rho > 0$, the kernel $k(.,.,.)$ is to be differentiable with respect to the third argument in the region
>
> $$U \equiv \{(x,y,z) \in G \times G \times \mathbb{R}: |z - v(y)| \leq \rho\},$$
>
> with the resulting partial derivative $\partial k/\partial z(x,y,.)$ continuous in $\{z \in \mathbb{R}: |z - v(y)| \leq \rho\}$ uniformly for all $x,y \in G$.

For our next result, we let $B_\rho^o(v)$ denote the open ball with radius ρ and center v (with respect to the maximum norm). Note that, in contrast to Lemma 3.5, we do not assume the continuity of $\partial k/\partial z$ with respect to all arguments.

<u>Lemma 3.6</u>. Suppose (39) is valid for some $v \in C(G)$, and, moveover, suppose that

$$\sup_{x\in G} \int_G \left|\frac{\partial k}{\partial z}(x,y,\phi(y))\right| dy < \infty, \tag{40a}$$

$$\lim_{x'\to x} \int_G \left|\frac{\partial k}{\partial z}(x',y,\phi(y)) - \frac{\partial k}{\partial z}(x,y,\phi(y))\right| dy = 0, \quad x \in G, \tag{40b}$$

for all $\phi \in B_\rho^o(v)$. Then K is continuously differentiable at v with derivative

$$(K'(\phi)h)(x) = \int_G \frac{\partial k}{\partial z}(x,y,\phi(y))h(y)dy, \; x \in G, \; h \in C(G), \; \phi \in B_\rho^o(v), \tag{41}$$

and, in addition, $K'(\phi)$ is a completely continuous mapping of $C(G)$ into itself. In particular, (40a,b) are satisfied whenever $\partial k/\partial z$ is continuous with respect to all three arguments in U.

<u>Proof</u>: For an arbitrary $\varepsilon > 0$, there exists a $\delta > 0$ depending only on ε such that for all $\phi \in B_\rho^o(v)$ and $h \in C(G)$ with $||h|| \leq \min(\rho - ||\phi-v||,\delta)$, we can estimate the remainder term associated with the kernel by

$$|r(x,y,\phi(y);h(y))| \leq \varepsilon|h(y)|, \quad x,y \in G.$$

(Note that the differentiability in (39) holds uniformly in U since U is compact and $\partial k/\partial z$ is assumed to be continuous in z uniformly with respect to x,y; the same argument is used in the proof of Lemma 3.5 in connection with condition (37).) By (40a), the following expression is well-defined,

$$R(\phi;h)(x) \equiv \int_G r(x,y,\phi(y);h(y))dy, \quad x \in G,$$

and can be estimated in terms of the Lebesgue measure of G in the following way,

$$|R(\phi;h)(x)| \leq \varepsilon||h||\text{meas}(G), \quad x \in G,$$

whenever ϕ, h satisfy the above restrictions. This shows the differentiability of K at all $\phi \in B_\rho^o(v)$ along with representation (41) of the derivative. The continuity of the derivative follows from the hypothesized continuity property of $\partial k/\partial z$. Finally, Theorem 3.4, in conjunction with (40a,b), shows the complete continuity of the integral operators given in (41) as mappings of $C(G)$ into itself. □

It should be noted here that, under the assumptions of Lemma 3.6, the integral operator K is itself continuous at all $\phi \in B_\rho^o(v)$, because of the results to be presented in Chapter 6.

If the underlying kernel is continuous, then the following condition is clearly sufficient for the differentiability condition (39) to hold at $v = \hat{v}|G$, where $\hat{v} \in C(M)$.

(42) For some number $\rho > 0$, the continuous extension $\hat{k}(.,.,.)$ of the continuous kernel $k(.,.,.)$ is differentiable with respect to the third argument in

$$\hat{U} \equiv \{(x,y,z) \in M \times M \times \mathbb{R}: |z - \hat{v}(y)| \leq \rho\},$$

with the partial derivative $\partial\hat{k}/\partial z(x,y,.)$ continuous in $\{z \in \mathbb{R}: |z-\hat{v}(y)| \leq \rho\}$ uniformly for all $x,y \in M$.

This is also sufficient for the Nyström method (35) to satisfy the differentiability condition (37). For the product integration method (36), (37) is satisfied whenever (39) is valid for the factor $r(.,.,.)$. The continuity of $\partial k_n/\partial z$ with respect to the first two arguments is again trivially apparent for both methods, since G_n consists of only finitely many points.

After having presented general quadrature methods specified by (34), we now give a general description of *projection methods* for approximating the nonlinear integral equation (32). Let E be, as before, a normed space of functions on G - e.g., $E = L^\infty(G)$ or $E = L^2(G)$ - and let $E_o = C(G)$ be a subspace of E. The integral operator (31) is to map E into itself, and, moreover, its restriction $K_o = K|E_o$ is to map E_o into itself. Further, let $E_n \subset E$ be a finite-dimensional subspace of E with a basis $\{\phi_1,\ldots,\phi_m\}$ and an associated bounded, linear projection $P_n: E \to E_n$. As in Section 3.3, we let $\alpha_i(\phi)$, $1 \le i \le m$, denote the coefficients of the unique representation $P_n\phi = \Sigma_i\, \alpha_i(\phi)\phi_i$, $\phi \in E$. Then the projection method approximates the solution of (32) by the solution $u_n \in E_n$ of

$$u_n - P_nKu_n = w_n, \tag{43}$$

where $w_n \in E_n$ approximates $w \in E_o$. With the notation,

$$\kappa_i(\xi_1,\xi_2,\ldots,\xi_m) \equiv \alpha_i(Kv_n), \quad i = 1,\ldots,m,$$

where $v_n = \Sigma_j\, \xi_j\phi_j \in E_n$, we see that (43) is equivalent to computing the coefficients of $u_n = \Sigma_j\, \alpha_{nj}\phi_j$ from the following nonlinear system of equations,

$$\alpha_{ni} - \kappa_i(\alpha_{n1},\ldots,\alpha_{nm}) = \alpha_i(w_n), \quad i = 1,\ldots,m. \tag{44}$$

The approximate operator $K_n \equiv P_nK|E_n$ can then be expressed as

$$K_nv_n = \sum_{i=1}^{m} \kappa_i(\xi_1,\ldots,\xi_m)\phi_i, \quad v_n = \sum_{j=1}^{m} \xi_j\phi_j \in E_n. \tag{45}$$

In particular, for *collocation methods,* we set $G = [a,b]$, $E = L^\infty(G)$, or $E = E_o = C(G)$ (the latter for collocation with continuous functions), and then we have (with the extended point evaluation functionals d_{x_i})

$$\begin{aligned} &\alpha_i(\phi) = d_{x_i}(\phi), \quad \alpha_{ni} = u_n(x_i), \\ &\kappa_i(\alpha_{n1},\ldots,\alpha_{nm}) = \int_a^b k(x_i,y,\sum_{j=1}^{m} \alpha_{nj}\phi_j(y))dy, \quad i = 1,\ldots,m, \end{aligned} \tag{46}$$

where, for simplicity, $\phi_j = \eta_j$, i.e., $\phi_j(x_i) = \delta_{ij}$, $1 \le i,j \le m$.

For *Galerkin methods,* we set $G = [a,b]$, $E = L^2(G)$, and

$$\begin{aligned} &\alpha_i(\phi) = (\phi,\phi_i)_o, \quad \alpha_{ni} = (u_n,\phi_i)_o, \\ &\kappa_i(\alpha_{n1},\ldots,\alpha_{nm}) = (Ku_n,\phi_i)_o = \int_a^b\int_a^b k(x,y,\sum_{j=1}^{m} \alpha_{nj}\phi_j(y))\phi_i(x)dydx, \end{aligned} \tag{47}$$

where $\{\phi_i\}$ is assumed to be an orthonormal basis for E_n.

With projection methods, the form of the derivatives of the approximating operator $K_n = P_n K|E_n$ (cf. (45)) is clear provided that the derivative of K itself is known. That is, if K is differentiable at $v_n \in E_n$ and P_n is a bounded linear operator, then $K_n'(v_n)$ is expressed in terms of $K'(v_n)$ by

$$K_n'(v_n) = P_n K'(v_n)|E_n. \tag{48}$$

For the Jacobian matrix of the functions κ_i defined by the formula following (43), we get analogous to Lemma 2.17

$$\frac{\partial \kappa_i}{\partial \xi_j}(\xi_1,\ldots,\xi_m) = \alpha_i(K'(v_n)\phi_j), \quad 1 \le i,j \le m. \tag{49}$$

The Fréchet-derivative (48) is then expressible as

$$K_n'(v_n)h_n = \sum_{i,j=1}^{m} \frac{\partial \kappa_i}{\partial \xi_j}(\xi_1,\ldots,\xi_m)\zeta_j\phi_i,$$

where $v_n = \Sigma_k\ \xi_k\phi_k$, $h_n = \Sigma_k\ \zeta_k\phi_k$. We shall not delve into the question of when the given integral operator (31) is differentiable in spaces larger than $C(G)$. Under appropriate continuity and differentiability assumptions, we can, of course, express the Jacobian matrices for the special methods (46) and (47) by:

Collocation Method: $$\frac{\partial \kappa_i}{\partial \xi_j}(\xi_1,\ldots,\xi_m) = \int_a^b \frac{\partial k}{\partial z}(x_i,y,\Sigma_k\xi_k\phi_k(y))\phi_j(y)\,dy;$$

Galerkin Method: $$\frac{\partial \kappa_i}{\partial \xi_j}(\xi_1,\ldots,\xi_m) = \int_a^b\int_a^b \frac{\partial k}{\partial z}(x,y,\Sigma_k\xi_k(y))\phi_j(y)\phi_i(x)\,dydx.$$

REFERENCES

Anselone (1971), Atkinson (1976), Baker (1977), Bourbaki (1965), Dieudonné (1969), Dunford & Schwartz (1966), Graham & Sloan (1982)*, Ikebe (1972)*, Kantorovich & Akilov (1964), Krasnoselskii (1964), Krasnoselskii, Vainikko et al. (1972), Kuratowski (1966), Mikhlin & Smolitskiy (1967), Nelson & Victory (1979)*, Radon (1919)*, Riesz & Nagy (1965), Rudin (1966), Sloan (1980a,1980b,1981)*, Sloan, Noussair & Burn (1979)*, Smirnow (1964), Smithies (1958), Stummel (1973b), Stummel (1974b,1975)*.

---

*Article(s)

Chapter 4
Approximation Methods for Initial Value Problems in Partial Differential Equations

We begin this rather extensive chapter by presenting several numerical methods for solving the heat equation and the wave equation (Section 4.1 to 4.3), which are typical examples of parabolic and hyperbolic problems, respectively. The methods we discuss comprise not only finite-difference methods but also Galerkin methods; our methods are either explicit or implicit and include so-called multilevel (more precisely, three-level) methods. In Section 4.4, we present finite-difference and Galerkin methods for approximating various classes of nonlinear initial value problems and discuss the solvability of the associated systems of nonlinear equations. Finally, we show in Section 4.5 how the problems considered in the previous sections - along with their approximating equations - can be viewed as operator equations in appropriate function spaces.

A purpose of this chapter is to assemble many of the concrete methods which we shall repeatedly refer to in Part IV of the text where our convergence analysis is carried out by means of the theory developed in Part II. A more practical purpose is to derive the system of equations associated with each method which ultimately must be solved numerically. We give results on their solvability and discuss practical numerical techniques for solving them.

The text assumes that the reader has a knowledge of the classification of partial differential equations and, moreover, has a general knowledge of the rudimentary existence and uniqueness results for several of the prototype examples (e.g., the heat equation, the wave equation, etc.) For general background material on these matters, we refer to the books by Meis & Marcowitz (1981) and by John (1980).

The methods we discuss are only certain representatives of various classes of methods. The numerous possibilities for constructing still other methods are made clear only at a few places in the text. We shall often refer to the literature for a discussion of variants of the methods presented here and of still other classes of methods. At this juncture, we would like to point out that the classical characteristic method for hyperbolic systems of first order will not be treated in the text, but instead we refer the reader to the discussion in Törnig (1979), 17.2.

4.1. DIFFERENCE METHODS FOR THE HEAT EQUATION

A typical example of a parabolic partial differential equation is the inhomogeneous heat equation in one spatial variable,

$$u_t(x,t) = au_{xx}(x,t) + s(x,t), \quad x \in G, \quad t > 0, \tag{1a}$$

where the conductivity coefficient $a > 0$ is considered constant and where $s(x,t)$ represents a source term. Our spatial domain G is either $G = \mathbb{R}$, or $G = [0,\infty)$ or $G = [0,1]$. In all three cases, we prescribe the initial temperature distribution

$$u(x,0) = u_0(x), \quad x \in G, \tag{1b}$$

but in the second or third choices for G, we prescribe respectively boundary conditions at $x = 0$ or at $x = 0$ and $x = 1$. The first choice for G leads to a pure IVP (IVP = initial value problem), whereas the other choices for G lead to initial-boundary-value problems (abbreviated IBVP).

In this section, we shall concentrate on the third choice for G and prescribe boundary conditions of the following type,

$$\begin{aligned} \alpha_0 u(0,t) - \alpha_1 u_x(0,t) &= \gamma_0(t) \\ \beta_0 u(1,t) + \beta_1 u_x(1,t) &= \gamma_1(t), \quad t > 0. \end{aligned} \tag{1c}$$

As with (time-independent) boundary-value problems, we distinguish among the following cases (cf. Section 1.1):

Dirichlet Boundary Conditions: $\alpha_0 = \beta_0 = 1, \ \alpha_1 = \beta_1 = 0,$

Neumann Boundary Conditions: $\alpha_0 = \beta_0 = 0, \ \alpha_1 = \beta_1 = 1,$

Boundary Conditions of the Third Type: $|\alpha_0| + |\alpha_1| > 0,$
$|\beta_0| + |\beta_1| > 0.$

The functions u_0, s, γ_0, and γ_1 are assumed continuous in all their arguments. A (classical) solution of (1) is to be at least once continuously differentiable in t for each x and twice continuously differentiable in x for each t and, moreover, to satisfy (1a-c). Implicit in our requirements on a solution of (1) are the conditions that $\gamma_0(0) = u_0(0)$ and $\gamma_1(0) = u_0(1)$ whenever, say, Dirichlet boundary conditions are imposed. We shall only consider the solvability of (1a-c) on a finite time interval $[0,T]$.

For computing numerical approximations, we subdivide the spatial and time intervals into subintervals of equal length and denote the grid points by

$$x_j \equiv jh, \quad j = 0,\ldots,J,$$

$$t_k \equiv k\tau, \quad k = 0,\ldots,N,$$

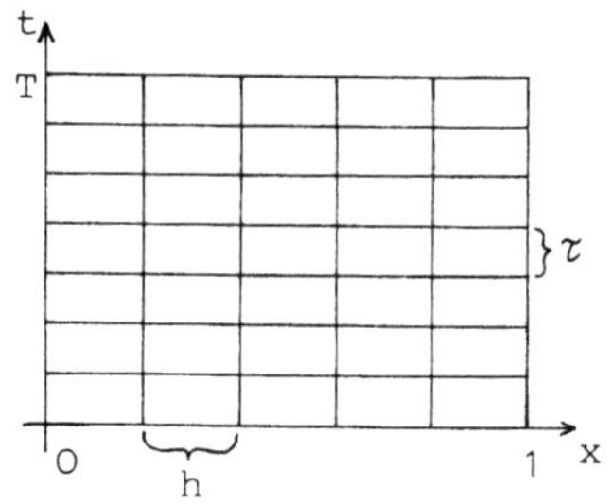

where $h = 1/J$ and $\tau = T/N$ denote the *mesh widths* in the x- and t-directions, respectively. τ is also called the *step width* or *step size*. The grid points are labeled as $p_{j,k} = (x_j,t_k)$ for convenience.

The derivatives in (1a) will now be approximated by difference quotients. If we choose the forward difference quotient of first order in the t-direction, and the central difference quotient of second order in the x-direction, then we get as an approximation of (1a),

$$\frac{1}{\tau}(v(x,t+\tau) - v(x,t)) = \frac{a}{h^2}(v(x+h,t) - 2v(x,t) + v(x-h,t)) + s(x,t). \tag{2a}$$

If we let $r = \tau/h^2$ denote the *mesh ratio*, then (2a) is equivalent to

$$v(x,t+\tau) = (1-2ar)v(x,t) + ar(v(x+h,t) + v(x-h,t)) + \tau s(x,t). \tag{2b}$$

The grid points occurring in (2a,b) are depicted graphically in the adjacent figure. We see that the approximation to the solution of the heat equation (1a), with initial condition (1b) and Dirichlet boundary conditions, can be computed explicitly from the following equations,

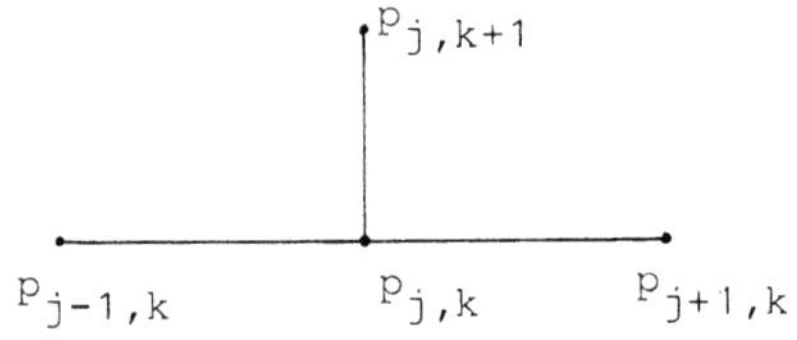

$$v_j^0 = u_{0,j}, \quad j = 0,1,\ldots,J; \quad v_0^k = \gamma_0^k; \quad v_J^k = \gamma_1^k, \quad k = 1,\ldots,N,$$

$$v_j^{k+1} = (1-2ar)v_j^k + ar(v_{j+1}^k+v_{j-1}^k) + \tau s_j^k, \; j = 1,\ldots,J-1, \quad k = 0,\ldots,N-1. \tag{3}$$

Here, $v_j^k = v(x_j,t_k)$, $u_{0,j} = u_0(x_j)$, $\gamma_\ell^k = \gamma_\ell(t_k)$, $\ell = 0,1$, etc. For the case of boundary conditions of the third kind, we approximate the derivatives with respect to x by central difference quotients of first order, and obtain, as an approximation to the boundary conditions,

$$\alpha_0 v_0^k - \alpha_1 \frac{v_1^k - v_{-1}^k}{2h} = \gamma_0^k, \quad \beta_0 v_J^k + \beta_1 \frac{v_{J+1}^k - v_{J-1}^k}{2h} = \gamma_1^k. \tag{4a}$$

The grid points $p_{-1,k}$ and $p_{J+1,k}$ lie outside the intervals considered; the associated approximations v_{-1}^k and v_{J+1}^k can, however, be eliminated by means of the equations in (4a),

$$\begin{aligned} \alpha_1 v_{-1}^k &= \alpha_1 v_1^k - 2h(\alpha_0 v_0^k - \gamma_0^k), \\ \beta_1 v_{J+1}^k &= \beta_1 v_{J-1}^k - 2h(\beta_0 v_J^k - \gamma_1^k), \quad k = 1,\ldots,N. \end{aligned} \tag{4b}$$

In order to compute the approximations v_0^{k+1} and v_J^{k+1}, we insert (4b) into (2b) and obtain the values v_j^{k+1}, $j = 1,\ldots,J-1$, by using (3). Altogether, we get an approximation to the solution of the IBVP (1a,b,c) via

$$\begin{aligned} v_j^0 &= u_{0,j}, \quad j = 0,\ldots,J, \\ \alpha_1 v_0^{k+1} &= [\alpha_1 - 2ar(\alpha_1+\alpha_0 h)]v_0^k + 2ar(\alpha_1 v_1^k + h\gamma_0^k) + \tau\alpha_1 s_0^k, \\ v_j^{k+1} &= (1-2ar)v_j^k + ar(v_{j+1}^k+v_{j-1}^k) + \tau s_j^k, \quad j = 1,\ldots,J-1, \\ \beta_1 v_J^{k+1} &= [\beta_1 - 2ar(\beta_1+h\beta_0)]v_J^k + 2ar(\beta_1 v_{J-1}^k + h\gamma_1^k) + \tau\beta_1 s_J^k, \\ & \qquad k = 0,1,\ldots,N-1. \end{aligned} \tag{5}$$

By this procedure, we can simultaneously treat the cases of Dirichlet-, Neumann-, and general boundary conditions of the third kind. In all cases, we are able to explicitly compute the approximations on the following time level by those on the previous one. For this reason, the method defined by (2) is an example of an *explicit finite-difference method.*

If we approximate the partial derivative with respect to t by the backward difference quotient, then we get an *implicit difference method*

$$\frac{1}{\tau}(v(x,t) - v(x,t-\tau)) = \frac{a}{h^2}(v(x+h,t) - 2v(x,t) + v(x-h,t)) + s(x,t). \tag{6}$$

The grid points appearing in (6) are arranged as in the adjacent diagram. The approximations for the solution of the heat equation cannot be explicitly calculated, but are solutions of the following system of equations (in case of Dirichlet boundary conditions):

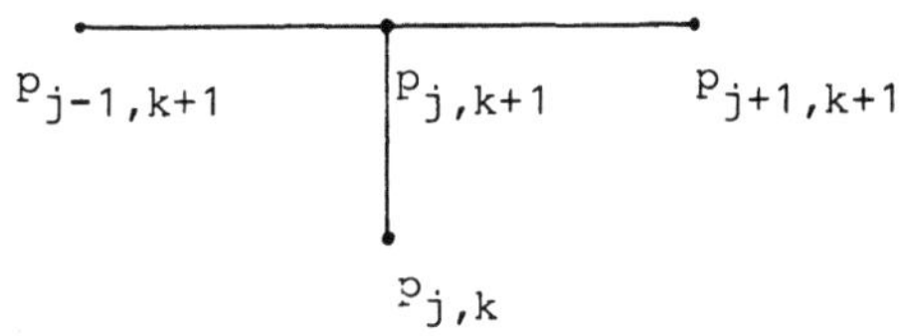

$$v_j^0 = u_{0,j},\ j = 0,\dots,J,\ v_0^k = \gamma_0^k,\ v_J^k = \gamma_1^k,\ k = 1,\dots,N,$$

$$-arv_{j+1}^{k+1} + (1+2ar)v_j^{k+1} - arv_{j-1}^{k+1} = v_j^k + \tau s_j^{k+1}, \tag{7}$$

$$j = 1,\dots,J-1,\ k = 0,1,\dots,N-1.$$

A combination of methods (2) and (6) provides the following scheme,

$$\begin{aligned}(1+2ar\Theta)v(x,t+\tau) &- ar\Theta(v(x+h,t+\tau) + v(x-h,t+\tau)) \\ &= (1-2ar(1-\Theta))v(x,t) + ar(1-\Theta)(v(x+h,t) + v(x-h,t)) \\ &\quad + \tau(\Theta s(x,t+\tau) + (1-\Theta)s(x,t)),\end{aligned} \tag{8}$$

where Θ is a parameter with values in $[0,1]$. For $\Theta = 0$ and $\Theta = 1$ we have the explicit method (2) and the implicit method (6), respectively. When $\Theta = 1/2$, we get the well-known *Crank-Nicolson method*. For an arbitrary $\Theta \in (0,1)$, method (8) involves values at six grid points as drawn in the adjacent figure. With Dirichlet boundary conditions, we must solve the following system,

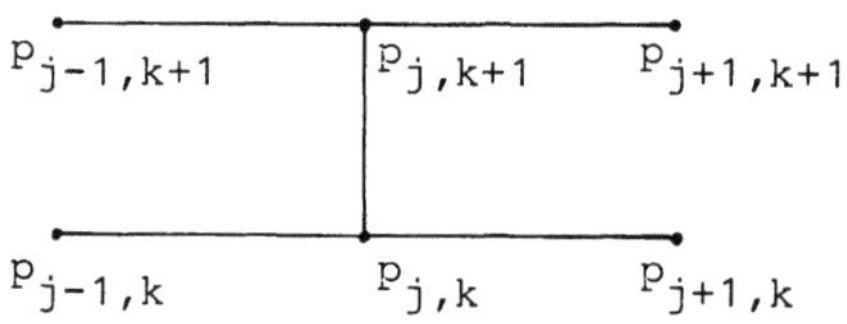

$$\begin{aligned}&v_j^0 = u_{0,j},\ j = 0,1,\dots,J,\quad v_0^k = \gamma_0^k,\quad v_J^k = \gamma_1^k,\ k = 1,\dots,N, \\ &(1+2ar\Theta)v_j^{k+1} - ar\Theta(v_{j+1}^{k+1} + v_{j-1}^{k+1}) \\ &\quad = (1-2ar(1-\Theta))v_j^k + ar(1-\Theta)(v_{j+1}^k + v_{j-1}^k) \\ &\qquad + \tau(\Theta s_j^{k+1} + (1-\Theta)s_j^k),\ j = 1,\dots,J-1,\ k = 0,\dots,N-1.\end{aligned} \tag{9}$$

The analogous system of equations for general boundary conditions of the third kind arises by eliminating the values v_{-1}^ν, v_{J+1}^ν, $\nu = k,k+1$, as in (4b), and computing v_0^{k+1} and v_J^{k+1} from (8). We get

$$
\begin{aligned}
& v_j^0 = u_{0,j}, \quad j = 0,1,\dots,J, \\
& [\alpha_1 + 2ar\Theta(\alpha_1 + \alpha_0 h)]v_0^{k+1} - 2ar\Theta\alpha_1 v_1^{k+1} \\
& \quad = [\alpha_1 - 2ar(1-\Theta)(\alpha_1 + \alpha_0 h)]v_0^k + 2ar(1-\Theta)\alpha_1 v_1^k \\
& \quad + 2arh(\Theta\gamma_0^{k+1} + (1-\Theta)\gamma_0^k) + \alpha_1\tau(\Theta s_0^{k+1} + (1-\Theta)s_0^k), \\
& (1 + 2ar\Theta)v_j^{k+1} - ar\Theta(v_{j+1}^{k+1} + v_{j-1}^{k+1}) \\
& \quad = (1 - 2ar(1-\Theta))v_j^k + ar(1-\Theta)(v_{j+1}^k + v_{j-1}^k) \\
& \quad + \tau(\Theta s_j^{k+1} + (1-\Theta)s_j^k), \quad j = 1,\dots,J-1, \\
& [\beta_1 + 2ar\Theta(\beta_1 + \beta_0 h)]v_J^{k+1} - 2ar\Theta\beta_1 v_{J-1}^{k+1} \\
& \quad = [\beta_1 - 2ar(1-\Theta)(\beta_1 + \beta_0 h)]v_J^k + 2ar(1-\Theta)\beta_1 v_{J-1}^k \\
& \quad + 2arh(\Theta\gamma_1^{k+1} + (1-\Theta)\gamma_1^k) + \beta_1\tau(\Theta s_J^{k+1} + (1-\Theta)s_J^k), \quad k = 0,\dots,N-1.
\end{aligned} \tag{10}
$$

The equations for the case of Dirichlet boundary conditions result when we insert $v_0^0 = u_{0,0} = \gamma_0^0$ and $v_J^0 = u_{0,J} = \gamma_1^0$ (cf. (9)). Solvability of (9) is a consequence of the weak row sum criterion satisfied by the underlying matrix (for the definition, see Section 1.2). For the general case of mixed boundary conditions, we can apply the same criterion in (10); for $j = 0$ this criterion is satisfied if $\alpha_1 > 0$, $\alpha_0 \geq 0$, and correspondingly, for $j = J$, if $\beta_1 > 0$, $\beta_0 \geq 0$.

Those methods requiring only points from two time levels are known as *two-level schemes*. It is quite possible to incorporate in our numerical schemes points and approximations from other, and earlier, time levels. A classic example of a three-level scheme is the *Du Fort-Frankel method*,

$$
\begin{aligned}
\frac{1}{2\tau}(v(x,t+\tau) - v(x,t-\tau)) = \frac{a}{h^2}(&v(x+h,t) - v(x,t+\tau) - v(x,t-\tau) \\
&+ v(x-h,t)) + s(x,t),
\end{aligned} \tag{11a}
$$

or equivalently,

$$
\begin{aligned}
&(1+2ar)v(x,t+\tau) = (1-2ar)v(x,t-\tau) \\
&+ 2ar(v(x+h,t) + v(x-h,t)) + 2\tau s(x,t).
\end{aligned} \tag{11b}
$$

The mesh points occurring in (11b) are illustrated in the figure on the next page. This method is also an explicit one, since there is only one point on the last time level.

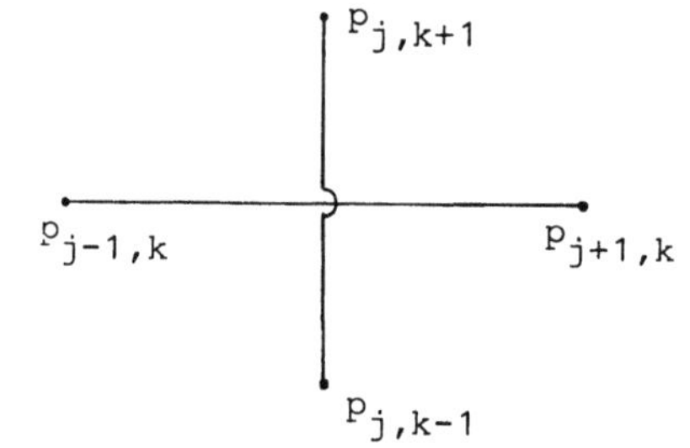

We can always express multilevel schemes as two-level systems which can then be treated in a unified manner. If we set $w(x,t) = v(x,t-\tau)$, then the Du Fort-Frankel method is equivalent to the following system (of two equations) which then represents a two-level scheme,

$$(1+2ar)v(x,t+\tau) = 2ar(v(x+h,t) - v(x-h,t)) + (1-2ar)w(x,t) + 2\tau s(x,t),$$
$$w(x,t+\tau) = v(x,t). \tag{11c}$$

This method is clearly suitable for calculating approximations starting at $t = t_2$. We can, for example, use the explicit method defined by (2) to calculate values v_j^1, $j = 0,\dots,J$, or use (2) to calculate values at $\tau/2$ and then use (11) to compute approximate values for $k = 1,2,\dots$. If the first option is selected, then approximate values for the solution to the heat equation, with initial conditions and Dirichlet boundary conditions, can be calculated as follows,

$$\begin{aligned} v_j^0 &= u_{0,j},\ j = 0,1,\dots,J;\quad v_0^k = \gamma_0^k,\quad v_J^k = \gamma_1^k,\ k = 1,2,\dots,N, \\ v_j^1 &= (1-2ar)v_j^0 + ar(v_{j+1}^0+v_{j-1}^0) + \tau s_j^0, \\ v_j^{k+1} &= (1+2ar)^{-1}\{2ar(v_{j+1}^k+v_{j-1}^k) + (1-2ar)v_j^{k-1} + 2\tau s_j^k\}, \\ &\qquad j = 1,\dots,J-1,\ k = 1,\dots,N-1. \end{aligned} \tag{12}$$

The formulas for the second option are similarly derived. The derivation of the corresponding equations for the case of boundary conditions of the third kind is left to the reader.

Up to now, we have only considered *finite-difference methods* with *constant coefficients* - i.e., the coefficients occurring in the equations associated with either explicit or implicit methods depend on neither position nor time. This is due to the conductivity coefficient occurring in the heat equation itself which is assumed to be constant. If this coefficient is positive and dependent on x and t, then we obtain in a manner analogous to (2) an explicit *finite-difference method* with *variable coefficients*,

$$v(x,t+\tau) = (1-2ra(x,t))v(x,t) + a(x,t)(v(x+h,t) + v(x-h,t)) + \tau s(x,t). \tag{13}$$

In addition to the methods just introduced for the heat equation, there are numerous other methods (also appropriate for equations with variable coefficients) which can be found in the literature cited at the end of this chapter.

In the final portions of this section, we consider the *two-dimensional heat equation*

$$\frac{\partial u}{\partial t}(x,t) = a\left(\frac{\partial^2 u}{\partial x_1^2} + \frac{\partial^2 u}{\partial x_2^2}\right)(x,t) + s(x,t), \quad x \in G \subset \mathbb{R}^2, \quad t \in [0,T], \tag{14a}$$

with an initial condition,

$$u(x,0) = u_0(x), \quad x \in G, \tag{14b}$$

and Dirichlet boundary conditions,

$$u(x,t) = g(x,t), \quad x \in \partial G, \quad t \in [0,T]. \tag{14c}$$

For the sake of simplicity, we shall restrict our analysis to the unit square $G = (0,1) \times (0,1)$ and consider no other types of boundary conditions in the subsequent discussion.

For deriving finite-difference methods, we let $h_1 = 1/J$ and $h_2 = 1/M$ be the equidistant mesh widths in the x_1- and x_2-directions, respectively; we let

$$\bar{G}_h \equiv \{x = (jh_1, mh_2),\ j = 0,1,\ldots,J,\ \ m = 0,1,\ldots,M\},$$

$G_h = \bar{G}_h \cap G$, and $\partial G_h = \bar{G}_h \cap \partial G$ denote the sets of mesh points in $\bar{G}, G$, and ∂G, respectively. As before, we subdivide the time interval into equidistant intervals with lengths $\tau = T/N$. A Padé approximation to the exponential function now leads to the following finite-difference approximation (for a derivation, see Varga (1962), Ch. 8, Marchuk (1975), Ch. 4),

$$(I - \tfrac{1}{2} a_1 D_{1,h}^2)(I - \tfrac{1}{2} a_2 D_{2,h}^2)v(x,t_{k+1}) = (I + \tfrac{1}{2} a_1 D_{1,h}^2)(I + \tfrac{1}{2} a_2 D_{2,h}^2)v(x,t_k) + \tau s(x,t_{k+1/2}), \quad x \in G_h, \tag{15}$$

where $D^2_{\nu,h}$ denotes the central difference quotient of second order in the x_ν-direction, $\nu = 1,2$, and

$$a_\nu = ar_\nu, \quad r_\nu = \tau/h_\nu^2, \quad \nu = 1,2, \quad t_{k+1/2} = (t_k+t_{k+1})/2.$$

For every k, (15) is a system of $(J-1) \times (M-1)$ equations and unknowns. Distinct orderings of the grid points in G_h lead to different types of equations for $v(x,t_{k+1})$ (cf. also Sec. 1.5). We interpret the values of $v^k = v(\cdot,t_k)$ at the mesh points as a $(J-1) \times (M-1)$ dimensional vector and order its components as follows:

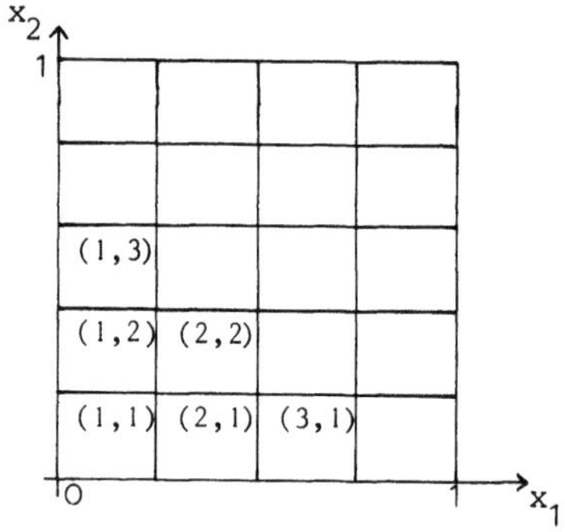

$$v^k = (v^k_{1,1},v^k_{2,1},\ldots,v^k_{J-1,1}, v^k_{1,2},\ldots,v^k_{J-1,2}, v^k_{1,3},\ldots,v^k_{J-1,M-1})^T.$$

Then (15) can be expressed as a system of equations having block structure,

$$\begin{pmatrix} A_{11} & A_{12} & \cdots & & 0 \\ A_{21} & A_{22} & A_{23} & & \\ \vdots & & \ddots & & \\ & & & \ddots & \\ 0 & \cdots & & A_{M-1,M-2} & A_{M-1,M-1} \end{pmatrix} v^{k+1} = z^k.$$

The matrices A_{im} are $(J-1) \times (J-1)$ tridiagonal matrices of the form

$$A_{ii} = \begin{pmatrix} b & -c & & 0 \\ -c & b & -c & \\ & \ddots & & \\ & & \ddots & \\ 0 & & -c & b \end{pmatrix}, \quad A_{i,i+1}=A_{i,i-1} = \begin{pmatrix} -d & e & & 0 \\ e & -d & e & \\ & \ddots & & \\ & & \ddots & \\ 0 & & e & -d \end{pmatrix}$$

where $b = (1+a_1)(1+a_2)$, $c = \frac{1}{2} a_1(1+a_2)$, $d = \frac{1}{2} a_2(1+a_1)$ and $e = \frac{1}{4} a_1 a_2$. The right-hand side z^k comes from the right-hand side of (15) and the boundary values $g_{j,m}^k$, $g_{j,m}^{k+1}$. The entire matrix is sparse and banded. The system of equations is uniquely solvable whenever, e.g., $a_1 a_2 \le 1/2$ and its solution can then be computed rather efficiently by Gaussian elimination. We can also use iterative methods such as Block-Jacobi, Block-Gauss-Seidel, or Block-Relaxation methods.

We can also employ ADI methods (ADI = Alternating Direction Implicit) for numerically solving (15) which entails a splitting into horizontal and vertical directions. One of the best known splittings of (15) is the *Peaceman-Rachford method*,

$$
\begin{aligned}
(I - \tfrac{1}{2} a_1 D_{1,h}^2) v^*(x,t_{k+1}) &= (I + \tfrac{1}{2} a_2 D_{2,h}^2) v(x,t_k) + \tfrac{\tau}{2} s(x,t_{k+1/2}), \\
(I - \tfrac{1}{2} a_2 D_{2,h}^2) v(x,t_{k+1}) &= (I + \tfrac{1}{2} a_1 D_{1,h}^2) v^*(x,t_{k+1}) + \tfrac{\tau}{2} s(x,t_{k+1/2}),
\end{aligned}
\qquad (16)
$$

$$x \in G_h.$$

When written out, these equations are

$$
\begin{aligned}
&(1+a_1) v_{j,m}^{*,k+1} - \tfrac{1}{2} a_1 (v_{j-1,m}^{*,k+1} - v_{j+1,m}^{*,k+1}) \\
&\quad = (1-a_2) v_{j,m}^k + \tfrac{1}{2} a_2 (v_{j,m-1}^k + v_{j,m+1}^k) + \tfrac{\tau}{2} s_{j,m}^{k+1/2}, \\
&(1+a_2) v_{j,m}^{k+1} - \tfrac{1}{2} a_2 (v_{j,m-1}^{k+1} + v_{j,m+1}^{k+1}) \\
&\quad = (1-a_1) v_{j,m}^{*,k+1} + \tfrac{1}{2} a_1 (v_{j-1,m}^{*,k+1} + v_{j+1,m}^{*,k+1}) + \tfrac{\tau}{2} s_{j,m}^{k+1/2},
\end{aligned}
$$

$$j = 1,2,\dots,J-1, \quad m = 1,2,\dots,M-1,$$

where, for brevity, $v_{j,m}^{*,k+1} = v^*(jh_1, mh_2, t_{k+1})$, $s_{j,m}^{k+1/2} = s(jh_1, mh_2, t_{k+1/2})$. The following figures show which of the values at the points on the t-levels for k and $(k+1)^*$ are needed to compute values at the $(k+1)^*$-th and $(k+1)$-th levels, respectively:

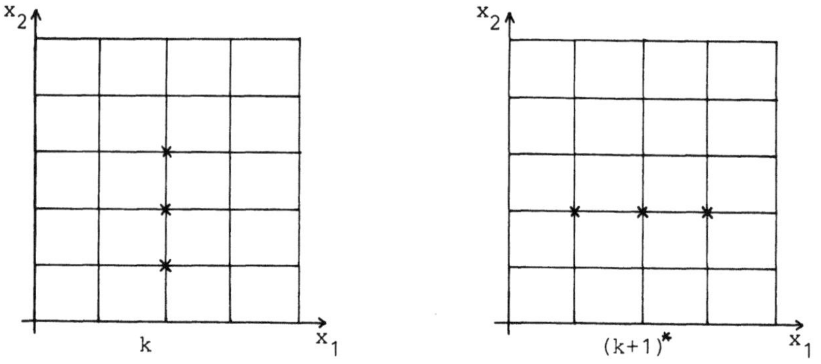

The solvability of the system of the equations in (16) is clearly guaranteed without any restriction on a_1 or a_2, which is a first advantage over the system defined by (15). For general ADI methods, the values of the auxiliary function $v^{*,k+1}$ are not necessarily approximate values for the solution of the given IVP at an intermediate value of t. Nevertheless, we shall see later that, for this particular method, $v^{*,k+1}$ is an approximation to $u(\cdot,t_{k+1/2})$. In order to calculate $v^{*,k+1}$ from the first equation in (16), we need the boundary values $v^{*,k+1}_{0,m}$, $v^{*,k+1}_{J,m}$, $m = 1,\dots,M-1$, which we must obtain from the given boundary condition (14c). This can be done by subtracting the two equations in (16) which then yields

$$v^{*,k+1}_{j,m} = \frac{1}{2}(I - \frac{1}{2} a_2 D^2_{2,h}) g^{k+1}_{j,m} + \frac{1}{2}(I + \frac{1}{2} a_2 D^2_{2,h}) g^k_{j,m}, \qquad m = 1,\dots,M-1, \quad j = 0,J. \tag{17}$$

We are now in a position to compute the values of $v^{*,k+1}$ row-wise (more precisely, parallel to the x_1-axis) and the values of v^{k+1} columnwise (more precisely, parallel to the x_2-axis), where at each stage we can employ either Gaussian elimination or an iteration method.

Other decompositions or splittings of (15) yield other ADI methods; other Padé approximations yield other systems of equations than that in (15), which can again be split so as to define other ADI methods. We refer to the books of Forsythe and Wasow (1960), Yanenko (1971), Marchuk (1975), Mitchell (1969), Mitchell and Griffiths (1980).

4.2. GALERKIN METHODS FOR THE HEAT EQUATION

We have already demonstrated in Chapter 2 how we can utilize appropriate variational formulations of boundary-value problems to derive finite element methods. This can be done analogously in the case of initial-boundary-value problems. Indeed, in this setting, a variational form is also approximated in finite-dimensional subspaces, and the resulting methods are again labeled as Galerkin methods. Among these methods we distinguish between semidiscrete (or continuous-time) and discrete-time methods. The latter methods arise by additionally approximating the partial derivative in t in a suitable manner, whereas the former methods leave the time variable undiscretized, thereby leading to a system of ordinary differential equations (in t). There are many possibilities of deriving suitable variational formulations, and each of these can be approximated in numerous concrete subspaces. In this section, we derive the standard Galerkin method for the one- and two-dimensional heat equation, and indicate possibilities for constructing Galerkin methods based on still other variational formulations.

We restrict our analysis to the case of Dirichlet boundary conditions, and consider, at the outset, the one-dimensional heat equation (1a,b). The solution clearly satisfies (after integration by parts) the equations

$$(u_t,\phi)_0 + (au',\phi')_0 = (s,\phi)_0, \quad t \in [0,T], \quad \phi \in V, \tag{18a}$$

$$(u,\phi)_0\big|_{t=0} = (u_0,\phi)_0, \quad \phi \in V, \tag{18b}$$

where $(.,.)_0$ is the usual L^2-inner product and $V = H_0^1(0,1)$ is the completion of $\{v \in C^2[0,1]: v(0) = v(1) = 0\}$ with respect to $||\cdot||_1$ (cf. Sec. 2.2). The conductivity coefficient a is taken to be a continuous function of x and satisfies $0 < \alpha_0 \le a(x)$, $x \in [0,1]$. As in the case of elliptic equations, equation (18a) is called a *variational formulation* of the heat equation (1a); the prime denotes the weak derivative with respect to x.

The initial value problem (18a,b) is to be interpreted in the following manner: For given $u_0 \in H_0^1(0,1)$, $s(t) \in L^2(0,1)$, $t \in [0,T]$, a solution $u: [0,T] \to H_0^1(0,1)$ is sought which is continuous in t and differentiable as a mapping from $[0,T]$ into $L^2(0,1)$ endowed with the usual (norm) topology. We can interpret (18a,b) more generally in the distributional sense, for which we refer the reader to Lions and Magenes (1972), Ch. 4. Results on existence and uniqueness of solutions in the more classical setting underlying the treatment here can be found, e.g., in Aubin (1979), Ch. 4.

For any Banach space E (with norm $|\cdot|_E$) we denote by $C^1([0,T],E)$ the space of all functions v: $[0,T] \to E$ which are *differentiable* in $[0,T]$, i.e., for all $t \in [0,T]$, there exists a $v'(t) \in E$ such that

$$\lim_{\substack{\tau \to 0 \\ t+\tau \in [0,T]}} \frac{1}{|\tau|} |v(t+\tau) - v(t) - \tau v'(t)|_E = 0.$$

The function v is called *continuously differentiable* if it is differentiable and if the derivative $v'(\cdot)$ is continuous on $[0,T]$. We label $C([0,T],E)$ the space of all continuous functions v: $[0,T] \to E$. With these notations, we thus seek as a solution of (18a,b) a function $u \in C([0,T],H_0^1(0,1)) \cap C^1([0,T],L^2(0,1))$.

As with Galerkin methods in approximating solutions to boundary-value problems, we seek approximations in finite-dimensional subspaces E_n of $H_0^1(0,1)$. A *semidiscrete* (or *continuous-time*) *Galerkin method* consists of determining a continuously differentiable functions v: $[0,T] \to E_n$ which satisfies

$$(v_t,\phi)_0 + (av',\phi')_0 = (s,\phi)_0, \quad \phi \in E_n, \quad t \in [0,T]. \tag{19a}$$

For approximating the initial data, an obvious choice for $v(0) \in E_n$ is

$$(v(0),\phi)_0 = (u_0,\phi)_0, \quad \phi \in E_n. \tag{19b}$$

With a basis $\{\phi_1,\phi_2,\ldots,\phi_m\}$ for E_n, we express v as

$$v(x,t) = \sum_{i=1}^{m} c_i(t)\phi_i(x), \quad 0 \le x \le 1, \quad 0 \le t \le T,$$

where the coefficients c_i are continuously differentiable functions of t. Substituting this representation of v into (19a,b), we see that $c(t) = (c_1(t),\ldots,c_m(t))$ is the solution of the following system of ordinary differential equations,

$$\sum_{i=1}^{m} c_i'(t)(\phi_i,\phi_j)_0 + \sum_{i=1}^{m} c_i(t)(a\phi_i',\phi_j')_0 = (s(t),\phi_j)_0, \quad 0 \le t \le T, \quad j = 1,\ldots,m, \tag{20a}$$

with initial values determined by

$$\sum_{i=1}^{m} c_i(0)(\phi_i,\phi_j)_0 = (u_0,\phi_j)_0, \quad j = 1,\ldots,m. \tag{20b}$$

For E_n, we take, for example, spaces of piecewise polynomial functions associated with a mesh $0 = x_0 < x_1 < \ldots < x_J = 1$ in the spatial interval. Since we have only discretized the x-variable and left t undiscretized, the procedure described by (20a,b) is also called *method of lines*.

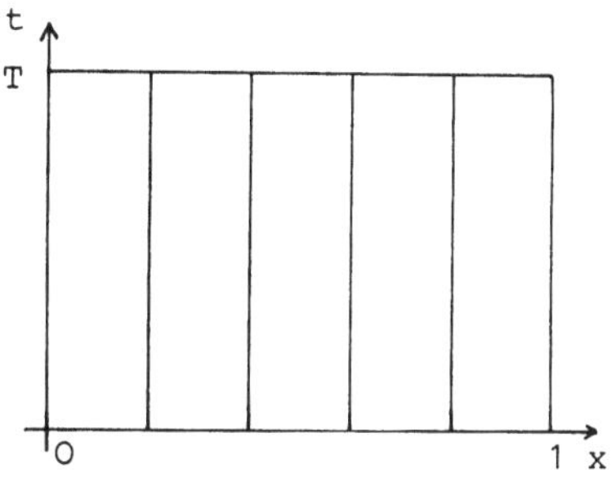

If we discretize also in the t-direction by a uniform mesh with width $\tau = T/N$, $t_k \equiv k\tau$, $k = 1, \ldots, N$, we must use an appropriate approximation for the derivative with respect to t. The *Crank-Nicolson-Galerkin method* selects the arithmetic mean of the forward and backward difference quotients as in the Crank-Nicolson method (cf. (8)). Thus, we seek functions $v^k \in E_n$, $k = 0,\ldots,N$, which satisfy

$$\left(\frac{v^{k+1}-v^k}{\tau},\phi\right)_0 + a\left(\tfrac{1}{2}(v^k+v^{k+1}),\phi\right) = \left(\tfrac{1}{2}(s^{k+1}+s^k),\phi\right)_0,$$

$$\phi \in E_n, \quad k = 0,\ldots,N-1, \tag{21}$$

$$(v^0,\phi)_0 = (u_0,\phi)_0, \quad \phi \in E_n.$$

(For brevity, we write $a(\psi,\phi) \equiv (a\psi',\phi')_0$, $\phi,\psi \in V$, and set $s^\nu \equiv s(\cdot,t_\nu)$, $\nu = k,k+1$.) By means of the basis functions ϕ_i, $i = 1,2,\ldots,m$, each v can be represented as

$$v^k(x) = \sum_{i=1}^m c_i^{(k)}\phi_i(x), \quad 0 \le x \le 1.$$

The system of equations (21) can thus be expressed as

$$\frac{1}{\tau}\sum_{i=1}^m (c_i^{(k+1)} - c_i^{(k)})(\phi_i,\phi_j)_0 + \sum_{i=1}^m c_i^{(k+1/2)}a(\phi_i,\phi_j)$$

$$= (s^{k+1/2},\phi_j)_0, \quad j = 1,\ldots,m, \quad k = 0,1,\ldots,N-1,$$

$$\sum_{i=1}^m c_i^{(0)}(\phi_i,\phi_j)_0 = (u_0,\phi_j)_0, \quad j = 1,\ldots,m,$$

where

$$c_i^{(k+1/2)} \equiv (c_i^{(k)} + c_i^{(k+1)})/2, \quad s^{k+1/2} \equiv (s^k+s^{k+1})/2.$$

We consider the system of equations (21) for the special case where E_n consists of continuous, piecewise linear functions with zero boundary values determined by a uniform mesh $x_j = jh$, $j = 0,\dots,J$ $(h = 1/J)$. The coefficients of the resulting system of equations have already been computed in Section 2.3 as

$$(\phi_i,\phi_j)_0 = \begin{cases} h/6\ , & i = j-1,\ j+1, \\ 2h/3, & i = j, \\ 0\ , & \text{otherwise}, \end{cases}$$

$$(\phi_i',\phi_j')_0 = \begin{cases} -1/h, & i = j-1,\ j+1, \\ 2/h\ , & i = j, \\ 0\ , & \text{otherwise}. \end{cases}$$

For a constant $a > 0$, (21) yields, after multiplication by τ/h, the following system of equations for computing $v_j^{k+1} = c_j^{(k+1)} = v^{k+1}(x_j)$:

$$\begin{aligned} &(\tfrac{2}{3} + ar)v_j^{k+1} + (\tfrac{1}{6} - \tfrac{1}{2}\,ar)(v_{j+1}^{k+1} + v_{j-1}^{k+1}) \\ &\qquad = (\tfrac{2}{3} - ar)v_j^k + (\tfrac{1}{6} + \tfrac{1}{2}\,ar)(v_{j+1}^k + v_{j-1}^k) + \tfrac{\tau}{h}\,(s^{k+1/2},\phi_j)_0, \\ &\qquad\qquad j = 1,\dots,J-1,\ k = 0,1,\dots,N-1, \\ &\tfrac{1}{6}(v_{j-1}^0 + 4v_j^0 + v_{j+1}^0) = \tfrac{1}{h}(u_0,\phi_j)_0,\quad j = 1,\dots,J-1. \end{aligned} \tag{22}$$

The inner products on the right-hand sides can be computed as in the Ritz method by a quadrature method, e.g., by Simpson's rule or by Gaussian quadrature formulas. The strong row sum criterion is again satisfied for the system (22), and this system can be numerically solved by using Gaussian elimination for tridiagonal matrices.

The solvability of the Crank-Nicolson-Galerkin Method (21) can be shown in general for arbitrary basis functions $\phi_1,\dots,\phi_m$ of finite-dimensional subspaces $E_n \in V$. With the bilinear form

$$\hat{a}(\psi,\phi) = (\psi,\phi)_0 + \tfrac{1}{2}\,\tau a(\psi,\phi),\quad \psi,\phi \in V,$$

and the linear functionals

$$f_k(\phi) = (v^k,\phi)_0 - \tfrac{1}{2}\,\tau a(v^k,\phi) + (s^{k+1/2},\phi)_0,\quad \phi \in V,\quad k = 0,\dots,N-1,$$

we see that the Crank-Nicolson-Galerkin method is equivalent to

$$\hat{a}(v^{k+1},\phi) = f_k(\phi),\quad \phi \in E_n,\quad k = 0,\dots,N-1,$$

or, with $v^k = \sum_{i=1}^m c_i^{(k)}\phi_i$, to

$$\sum_{i=1}^{m} c_i^{(k+1)} \hat{a}(\phi_i,\phi_j) = f_k(\phi_j), \quad j = 1,\dots,m, \quad k = 0,\dots,N-1.$$

The symmetric, bounded bilinear form $\hat{a}(.,.)$ is also elliptic on V with respect to $||\cdot||_1$ since

$$|\hat{a}(\phi,\phi)| = |\phi|_0^2 + \frac{\tau}{2} a(\phi,\phi) \geq \min(1, \frac{\tau}{2}\alpha_0)||\phi||_1^2, \quad \phi \in V.$$

Moreover, f_k is a bounded functional on V with respect to $||\cdot||_1$ so that a unique solution $v^{k+1} \in E_n$ is assured by the Lax-Milgram Lemma for every $k = 0,\dots,N-1$. From the ellipticity of $\hat{a}(.,.)$, the positive definiteness of the matrix $(\hat{a}(\phi_i,\phi_j))$ follows as in Theorem 2.18, which thus again ensures unique solvability.

At this point, we would like to stress that the variational formulation (18a) is by no means the only possible fundamental equation from which a Galerkin procedure can be derived. We now introduce still another variational formulation which will lead to other methods and which was originally suggested by Thomee and Wahlbin (1975). The essential point to notice here is that, with homogeneous Dirichlet boundary conditions, $u_t(0,t) = u_t(1,t) = 0$ necessarily for all t. By integrating by parts, we see that the solution of (1a) with homogeneous Dirichlet boundary conditions also satisfies the equations

$$(u_t',\phi')_0 + (au'',\phi'')_0 + (s,\phi'')_0 = 0, \quad t \in (0,T), \tag{23}$$

for all $\phi \in E \equiv H^2(0,1) \cap H_0^1(0,1)$. ($H^2(0,1)$ is the Sobolev space of all functions possessing generalized derivatives up to second order in $L^2(0,1)$). Equation (23) is now our *second possible variational formulation* of the heat equation, and can also serve as a basis for constructing approximating schemes. With finite-dimensional subspaces $E_n \subset E$, we obtain a *semidiscrete Galerkin method* by

$$(v_t',\phi')_0 + (av'',\phi'')_0 = -(s,\phi'')_0, \quad \phi \in E_n. \tag{24a}$$

A suitable approximation of the initial data is provided by

$$(v''(0),\phi'')_0 = (u_0'',\phi'')_0, \quad \phi \in E_n. \tag{24b}$$

In analogy to our treatment of the Crank-Nicolson scheme, we obtain a *discrete-time Galerkin method* by

$$\frac{1}{\tau}(v_x^{k+1}-v_x^k,\phi')_0 + (av_{xx}^{k+1/2},\phi'')_0 = -(s^{k+1/2},\phi'')_0, \quad \phi \in E_n, \quad k = 0,\dots,N-1, \tag{25a}$$

$$(v^0_{xx},\phi'')_0 = (u''_0,\phi'')_0, \quad \phi \in E_n. \tag{25b}$$

We also write v^k_x and v^k_{xx} to denote $(v^k)'$ and $(v^k)''$, respectively. The space of continuous, piecewise linear functions determined by a mesh in $[0,1]$ is not a subspace of $H^2(0,1)$ and, for this reason, is not appropriate for this particular Galerkin method. A possible subspace is that of the cubic splines, from which an associated banded system of equations arises with bandwidth 7. We shall not discuss the approximating properties of splines and of subspaces of general piecewise polynomials, but refer the reader, for example, to Aubin (1972), Böhmer (1974), Fairweather (1978), Chapter 2, Meis & Marcowitz (1981), II.15. With regard to constructing Galerkin methods for other types of boundary conditions, the reader should consult Babuska & Aziz (1972), Ch. 11, and Fairweather (1978), Sec. 4.6, and the references cited therein.

From the construction of Galerkin methods for the (2-dimensional) Poisson equation in Section 2.4, we can now easily derive Galerkin methods for the two-dimensional heat equation (14a,b) with homogeneous Dirichlet boundary conditions (i.e., $g = 0$ in (14c)). Corresponding to (18a), our variational formulation is

$$(u_t,\phi)_0 + (a\nabla u,\nabla\phi)_0 = (s,\phi)_0, \quad \phi \in H^1_0(G), \tag{26}$$

where, for brevity, we write $(a\nabla u,\nabla\phi)_0 = \int_G a\nabla u\cdot\nabla v \, dx$. With subspaces $E_n \subset H^1_0(G)$, we get a *semidiscrete (continuous-time) Galerkin method* by

$$(v_t,\phi)_0 + (a\nabla v,\nabla\phi)_0 = (s,\phi)_0, \quad \phi \in E_n, \quad t \in [0,T], \tag{27}$$

and the (discrete-time) *Crank-Nicolson-Galerkin method* by

$$\left(\frac{v^{k+1}-v^k}{\tau},\phi\right)_0 + (a\nabla v^{k+1/2},\nabla\phi)_0 = (s^{k+1/2},\phi)_0, \quad \phi \in E_n, \quad k = 0,\dots,N-1. \tag{28}$$

For a rectangular region G and a decomposition into rectangles with continuous, piecewise bilindar functions determined by these rectangles, the associated system of equations can be immediately deduced from those for the Poisson equation in Section 2.4.

The other variational form (23) for the heat equation is immediately extendible to the two-dimensional setting (with Δu in lieu of u''), and, correspondingly, we get associated semidiscrete and discrete-time Galerkin methods. In addition, the treatment in this section can be directly generalized to constructing Galerkin methods for heat equations in arbitrary (spatial) dimensions.

4.3. NUMERICAL METHODS FOR THE WAVE EQUATION

The wave equation is a typical example of a hyperbolic partial differential equation. For solving this equation numerically, we introduce three finite-difference methods - Friedrichs' method, the Courant-Isaacson-Rees method, and the Lax-Wendroff method - along with several Galerkin methods. The finite-difference methods rely on representing the wave equation as a system of two first order (hyperbolic) equations. We obtain Galerkin methods by approximating appropriate variational formulations of the given differential equation in finite-dimensional subspaces, i.e., in spaces of, say, continuous, piecewise polynomial functions. As with parabolic equations, there are semidiscrete (continuous-time) and discrete-time Galerkin methods. The first of these methods leads to an initial value problem for a system of ordinary differential equations of second order. The latter methods, i.e., the fully discrete-time Galerkin methods, require the solution of a system of equations at each (discrete) time level, where in the right-hand side will appear the approximate values at points from two preceding time levels.

The *initial value problem* (or the *Cauchy problem*) for the *wave equation* is (with constant $c > 0$)

$$u_{tt} - c^2 u_{xx} = 0, \quad -\infty < x < \infty, \quad t > 0, \tag{29a}$$

$$\begin{cases} u(x,0) = f_0(x), \\ u_t(x,0) = g_0(x), \end{cases} \quad -\infty < x < \infty. \tag{29b}$$

It is well-known that the (classical) solution of (29) is expressible by d'Alembert's formula

$$u(x,t) = \frac{1}{2}\,(f_0(x+ct) + f_0(x-ct)) + \frac{1}{2c}\int_{x-ct}^{x+ct} g_0(s)\,ds,$$

provided that g_0 is continuously differentiable and f_0 is twice continuously differentiable. Problem (29a,b) is equivalent to

$$\begin{cases} \dfrac{\partial u_1}{\partial t} - c\,\dfrac{\partial u_2}{\partial x} = 0, \\[2ex] \dfrac{\partial u_2}{\partial t} - c\,\dfrac{\partial u_1}{\partial x} = 0, \end{cases} \quad -\infty < x < \infty, \quad t > 0, \tag{30a}$$

$$\begin{cases} u_1(x,0) = f_0(x), \\ u_2(x,0) = \frac{1}{c}\int_0^x g_0(s)ds + C_0, \end{cases} \qquad -\infty < x < \infty, \tag{30b}$$

where C_0 is an arbitrary constant and $u_1 = u$ is the above solution to the wave equation. With the vector $\underline{u} = (u_1,u_2)^T$ and the matrix $A = c\begin{pmatrix} 0 & 1 \\ 1 & 0 \end{pmatrix}$, (30a) can be expressed as

$$\frac{\partial}{\partial t}\underline{u} = A\frac{\partial}{\partial x}\underline{u}, \quad -\infty < x < \infty, \quad t > 0. \tag{31}$$

By adding and subtracting the two equations in (30a), we obtain the following equivalent system for $\underline{w} = (w_1,w_2)^T$, $w_1 = u_1 + u_2$, $w_2 = u_1 - u_2$,

$$\frac{\partial}{\partial t}\underline{w} = D\frac{\partial}{\partial x}\underline{w}, \quad -\infty < x < \infty, \quad t > 0, \tag{32}$$

where $D = c\begin{pmatrix} 1 & 0 \\ 0 & -1 \end{pmatrix}$. The system (32) is called the *characteristic form of the wave equation* and $u_1 \pm u_2$ are the components of its solution. In going from (31) to (32), we have done nothing other than to diagonalize A by a matrix consisting of eigenvectors of A. Here, D is given by $D = PAP^{-1}$, with

$$P = \begin{pmatrix} 1 & 1 \\ 1 & -1 \end{pmatrix}, \qquad P^{-1} = \frac{1}{2}\begin{pmatrix} 1 & 1 \\ 1 & -1 \end{pmatrix}.$$

For deriving finite-difference approximations for (31) or (32), we must first investigate approximate solutions to their scalar counterparts. We do not go into this matter here, but instead cite three well-known methods for systems of the form (31). We subdivide the x- and t-axis into equidistant subintervals of widths h and τ, respectively, and define $x_j \equiv jh$, $j = 0,\pm1,\pm2,\ldots$, and $t_k \equiv k\tau$, $k = 0,1,\ldots$. *Friedrichs' method* has the form ($\underline{v} = (v_1,v_2)^T$):

$$\begin{aligned} &\frac{1}{\tau}[\underline{v}(x,t+\tau) - \frac{1}{2}(\underline{v}(x+h,t) + \underline{v}(x-h,t))] \\ &\qquad = \frac{1}{2h}A(\underline{v}(x+h,t) - \underline{v}(x-h,t)), \quad (x,t) = (x_j,t_k). \end{aligned} \tag{33a}$$

With the *mesh ratio* $\lambda = \tau/h$, (33a) can be expressed as

$$\underline{v}(x,t+\tau) = \frac{1}{2}(I-\lambda A)\underline{v}(x-h,t) + \frac{1}{2}(I+\lambda A)\underline{v}(x+h,t), \quad (x,t) = (x_j,t_k). \tag{33b}$$

Expressing the above system in terms of the components (v_1,v_2), we see that Friedrichs' method for the wave equation has the form,

$$\begin{aligned} v_1(x,t+\tau) &= \tfrac{1}{2}(v_1(x+h,t) + v_1(x-h,t)) + \tfrac{1}{2}\, c\lambda(v_2(x+h,t) \\ &\qquad - v_2(x-h,t)), \\ v_2(x,t+\tau) &= \tfrac{1}{2}(v_2(x+h,t) + v_2(x-h,t)) + \tfrac{1}{2}\, c\lambda(v_1(x+h,t) \\ &\qquad - v_1(x-h,t)). \end{aligned} \tag{33c}$$

The *Courant-Isaacson-Rees method* proceeds from the characteristic form (32) and is given by

$$\begin{aligned} &\frac{1}{\tau}(\underline{z}(x,t+\tau) - \underline{z}(x,t)) \\ &\quad = \frac{1}{h}(D^+(\underline{z}(x+h,t) - \underline{z}(x,t)) + D^-(\underline{z}(x,t) - \underline{z}(x-h,t)), \\ &\qquad (x,t) = (x_j,t_k), \end{aligned} \tag{34a}$$

where $\underline{z} = (z_1,z_2)^T$ and $D = \text{diag}(\alpha_j)$, $D^+ = \text{diag}(\max(\alpha_i,0))$, $D^- = \text{diag}(\min(\alpha_i,0))$ and $\alpha_1 = c$, $\alpha_2 = -c$ for (32). Multiplying by τ yields

$$\underline{z}(x,t+\tau) = (I - \lambda(D^+-D^-))\underline{z}(x,t) + \lambda(D^+\underline{z}(x+h,t) - D^-\underline{z}(x-h,t)). \tag{34b}$$

We note that $D^+ - D^- = \text{diag}(|\alpha_i|)$. The vectors $\underline{u}$ and $\underline{w}$ in (31) and (32) satisfy the relation $\underline{w} = P\underline{u}$. For $\underline{v} = P^{-1}\underline{z}$, we get from (34b) - after multiplying by P^{-1} - the equation

$$\underline{v}(x,t+\tau) = (I - \lambda(A^+-A^-))\underline{v}(x,t) + \lambda(A^+\underline{v}(x+h,t) - A^-\underline{v}(x-h,t)), \tag{34c}$$

where $A^{\pm} = P^{-1}D^{\pm}P$. For the wave equation, we compute

$$A^+ = \frac{c}{2}\begin{pmatrix}1 & 1\\ 1 & 1\end{pmatrix}, \quad A^- = \frac{-c}{2}\begin{pmatrix}1 & -1\\ -1 & 1\end{pmatrix}, \quad A^+ - A^- = cI$$

so that (34c) yields the following equations for the components of $\underline{v} = (v_1,v_2)^T$:

$$\begin{aligned} v_1(x,t+\tau) &= (1-c\lambda)v_1(x,t) + \tfrac{1}{2}\, c\lambda(v_1(x-h,t) + v_1(x+h,t)) \\ &\qquad + \tfrac{1}{2}\, c\lambda(v_2(x+h,t) - v_2(x-h,t)), \\ v_2(x,t+\tau) &= (1-c\lambda)v_2(x,t) + \tfrac{1}{2}\, c\lambda(v_2(x-h,t) + v_2(x-h,t)) \\ &\qquad + \tfrac{1}{2}\, c\lambda(v_1(x+h,t) - v_1(x-h,t)). \end{aligned} \tag{34d}$$

We observe immediately that both methods are also well defined for systems of the form (31) possessing more than two equations with variable coefficient matrices.

As a third finite-difference method we present the *Lax-Wendroff method*, which is especially suitable for constant coefficients as we shall see later. It has the following form,

$$\begin{aligned} \underline{v}(x,t+\tau) &= \underline{v}(x,t) + \frac{1}{2}\lambda A(\underline{v}(x+h,t) - \underline{v}(x-h,t)) \\ &\quad + \frac{1}{2}\lambda^2 A^2(\underline{v}(x+h,t) - 2\underline{v}(x,t) + \underline{v}(x-h,t)). \end{aligned} \tag{35a}$$

When written out for the wave equation, (35a) becomes

$$\begin{aligned} v_1(x,t+\tau) &= v_1(x,t) + \frac{1}{2}c^2\lambda^2(v_1(x+h,t) - 2v_1(x,t) + v_1(x-h,t)) \\ &\quad + \frac{1}{2}c\lambda(v_2(x+h,t) - v_2(x-h,t)), \\ v_2(x,t+\tau) &= v_2(x,t) + \frac{1}{2}c^2\lambda^2(v_2(x+h,t) - 2v_2(x,t) + v_2(x-h,t)) \\ &\quad + \frac{1}{2}c\lambda(v_1(x+h,t) - v_1(x-h,t)). \end{aligned} \tag{35b}$$

At this point, we would like to mention the books by Gladwell and Wait (1979) and by Mitchell and Griffiths (1980) where numerous finite-difference methods for hyperbolic partial differential equations are presented and analyzed. We also mention the treatment in Törnig (1979), 17.2, for the well-known "characteristic method" for the wave equation.

We now turn to deriving *Galerkin methods* for the *generalized wave equation*

$$\frac{\partial^2 u}{\partial t^2} - \frac{\partial}{\partial x}\left(a(x)\frac{\partial u}{\partial x}\right) = s(x,t), \quad 0 \le x \le 1, \quad t \in [0,T], \tag{36a}$$

with initial conditions,

$$u(x,0) = f_0(x), \quad \frac{\partial u}{\partial t}(x,0) = g_0(x), \quad 0 \le x \le 1, \tag{36b}$$

and boundary conditions,

$$u(0,t) = u(1,t) = 0, \quad 0 \le t \le T. \tag{36c}$$

The functions a, f_0, g_0, and s are taken to be at least continuous in their arguments, with $0 < \alpha_0 \le a(x)$, $x \in [0,1]$. After integration by parts, we see that every solution of (36a) and (36b) also satisfies the following *variational formulation* of the generalized wave equation,

$$(u_{tt},\phi)_0 + (au',\phi')_0 = (s,\phi)_0, \quad \phi \in H_0^1(0,1), \quad t \in [0,T]. \tag{37a}$$

For the initial conditions, we have correspondingly

$$(u,\phi)_0|_{t=0} = (f_0,\phi)_0, \quad (u_t,\phi)_0|_{t=0} = (g_0,\phi)_0, \quad \phi \in H_0^1(0,1). \tag{37b}$$

We refer the reader to Lions & Magenes (1972), Sec. 3.8, for a rigorous interpretation of the variational formulation in the distributional sense and for results on the existence of solutions to (37a,b) under weaker assumptions on the problem data than imposed here.

For deriving our approximation schemes, we choose the same notations as in our treatment of the Galerkin methods in Section 4.2. For brevity, we let $a(\psi,\phi)$ again denote $(a\psi',\phi')_0$, $\psi,\phi \in H^1(0,1)$. For the *standard semidiscrete Galerkin method,* we seek a twice continuously differentiable function (in t) v: $[0,T] \to E_n$, which satisfies

$$(v_{tt},\phi)_0 + a(v,\phi) = (s,\phi)_0, \quad \phi \in E_n, \quad t \in [0,T], \tag{38a}$$

$$(v(0),\phi)_0 = (f_0,\phi)_0, \quad (v_t(0),\phi)_0 = (g_0,\phi)_0, \quad \phi \in E_n, \tag{38b}$$

where again E_n is a finite-dimensional subspace of $H_0^1(0,1)$. The equations (38a,b) constitute an initial value problem for a system of linear ordinary differential equations of second order. Indeed, with $\phi_1,\ldots,\phi_m$ in E_n as a set of basis functions, the solution of (38) can be represented by

$$v(x,t) = \sum_{i=1}^{m} c_i(t)\phi_i(x), \quad 0 \le x \le 1, \quad 0 \le t \le T,$$

with twice continuously differentiable coefficient functions $c_i(t)$, $i = 1,\ldots,m$, satisfying

$$\sum_{i=1}^{m} \{c_i''(t)(\phi_i,\phi_j)_0 + c_i(t)a(\phi_i,\phi_j)\} = (s(t),\phi_j)_0, \quad 0 \le t \le T,$$

$$\sum_{i=1}^{m} c_i(0)(\phi_i,\phi_j)_0 = (f_0,\phi_j)_0,$$

$$\sum_{i=1}^{m} c_i'(0)(\phi_i,\phi_j)_0 = (g_0,\phi_j)_0, \quad j = 1,\ldots,m.$$

With the matrices $A = (a(\phi_i,\phi_j))$, $B = ((\phi_i,\phi_j)_0)$, we can equivalently express the initial value problem as follows,

$$Bc''(t) + Ac(t) = S(t), \quad t \in [0,T],$$

$$Bc(0) = F_0, \quad Bc'(0) = G_0,$$

where now

$$S(t) \equiv (s_1(t),\ldots,s_m(t)), \quad s_j(t) \equiv (s(t),\phi_j)_0, \quad 1 \le j \le m,$$

$$F_0 \equiv ((f_0,\phi_1)_0,\ldots,(f_0,\phi_m)_0), \quad G_0 \equiv ((g_0,\phi_1)_0,\ldots,(g_0,\phi_m)_0).$$

The matrices A and B are positive definite - and are thereby nonsingular - and hence the initial value problem (38a,b) is uniquely solvable.

A *discrete-time Galerkin method* suggested by Dupont (1973) determines approximations $v^k \in E_n$ for $u(.,t_k)$, $t_k = k\tau$, $k = 0,\ldots,N$, $(\tau = T/N)$ from the equations

$$\begin{aligned}(D_\tau^2 v^k,\phi)_0 &+ \frac{1}{2}\, a(v^{k+1/2} + v^{k-1/2},\phi) \\ &= \frac{1}{2}(s^{k+1/2} + s^{k-1/2},\phi)_0, \quad \phi \in E_n, \quad k = 1,2,\ldots,N-1,\end{aligned} \tag{39a}$$

where $D_\tau^2 v^k = (v^{k+1} - 2v^k + v^{k-1})/\tau^2$ (in analogy to the central difference quotient of second order). Suitable approximations for v^0,v^1 can be found in the above mentioned original work. We shall also present other approximations to v^0,v^1 later in Chapter 13. If we define

$$v^{k,\Theta} = \Theta v^{k+1} + (1-2\Theta)v^k + \Theta v^{k-1}, \qquad 0 \le \Theta \le 1,$$

then $v^{k,1/4} = (v^{k+1/2} + v^{k-1/2})/2$; correspondingly, $s^{k,\Theta}$ is defined using $s^\nu = s(.,t_\nu)$, $\nu = k,k\pm1$. With basis functions $\phi_1,\ldots,\phi_m$ of E_n, (39a) is equivalent to determining the coefficients in the representation of $v^{k+1} = \sum_i c_i^{(k+1)}\phi_i$ via

$$\begin{aligned}\frac{1}{\tau^2}\sum_{i=1}^{m} (c_i^{(k+1)}-2c_i^{(k)}+c_i^{(k-1)})(\phi_i,\phi_j)_0 &+ c_i^{(k),1/4} a(\phi_i,\phi_j) \\ &= (s^{k,1/4},\phi_j)_0, \quad j = 1,\ldots,m, \quad k = 1,\ldots,N-1,\end{aligned} \tag{39b}$$

where $c_i^{(k),\Theta} \equiv \Theta(c_i^{(k+1)} + c_i^{(k-1)}) + (1-2\Theta)c_i^{(k)}$, $i = 1,\ldots,m$. For the wave equation (i.e., $a(x) = c^2 > 0$, $s = 0$), with E_n consisting of continuous piecewise linear functions determined by an equally spaced mesh, $x_j = jh$, $j = 0,\ldots,J$, and having the roof functions as the set of basis functions, we get for $v_j^k = c_j^{(k)} = v(x_j,t_k)$ (after multiplying by τ^2/h),

$$\begin{aligned}\frac{1}{6}(v_{j-1}^{k+1} + 4v_j^{k+1} + v_{j+1}^{k+1}) &- \frac{1}{3}(v_{j-1}^k + 4v_j^k + v_{j+1}^k) \\ + \frac{1}{6}(v_{j-1}^{k-1} + 4v_j^{k-1} + v_{j+1}^{k-1}) &- \lambda^2 c^2(v_{j-1}^{k,1/4} - 2v_j^{k,1/4} + v_{j+1}^{k,1/4}) = 0,\end{aligned} \tag{40}$$

$$j = 1,\ldots,J-1, \quad k = 1,\ldots,N-1.$$

We put all the terms of the k-th and (k-1)-th level on the right-hand side to obtain

$$\begin{aligned}(\tfrac{2}{3} + \tfrac{1}{2}\lambda^2c^2)v_j^{k+1} &+ (\tfrac{1}{6} - \tfrac{1}{4}\lambda^2c^2)(v_{j-1}^{k+1} + v_{j+1}^{k+1}) \\ &= (\tfrac{4}{3} - \lambda^2c^2)v_j^k + (\tfrac{1}{3} + \tfrac{1}{2}\lambda^2c^2)(v_{j-1}^k + v_{j+1}^k) \\ &- (\tfrac{2}{3} + \tfrac{1}{2}\lambda^2c^2)v_j^{k-1} - (\tfrac{1}{6} - \tfrac{1}{4}\lambda^2c^2)(v_{j-1}^{k-1} + v_{j+1}^{k-1}), \\ &\qquad j = 1,\ldots,J-1, \quad k = 1,\ldots,N-1.\end{aligned} \tag{40b}$$

For later purposes, it is advantageous to express the above three-level scheme (since values from three time levels occur) as a two-level method. This can be brought about as with the Du Fort-Frankel method by setting $w_j^k = v_j^{k-1}$. The resulting system of two equations can be immediately deduced from (40b). Of course, we could have performed such a splitting already for (39a). Another splitting, however, is more advantageous for latter study. If we set $u^k = v^{k+1/2}$ and $w^k = (v^{k+1} - v^k)/\tau$, (39a) is equivalent to

$$\begin{aligned}&\frac{1}{\tau}(w^k - w^{k-1},\phi)_0 + \frac{1}{2}a(u^k + u^{k-1},\phi) = (s^{k,1/4},\phi)_0, \quad \phi \in E_n, \\ &\frac{1}{\tau}(u^k - u^{k-1}) = w^{k-1/2}, \quad k = 1,\ldots,N-1.\end{aligned} \tag{39c}$$

In a similar manner, Baker (1976) proposed the following (2-level) *discrete-time Galerkin method* which is clearly different from the previous one (cf. also Fairweather (1978), Sec. 5.6):

$$\begin{aligned}&\frac{1}{\tau}(w^k - w^{k-1},\phi)_0 + a(v^{k-1/2},\phi) = (s^{k-1/2},\phi)_0, \quad \phi \in E_n, \\ &\frac{1}{\tau}(v^k - v^{k-1}) = w^{k-1/2}, \quad k = 1,\ldots,N.\end{aligned} \tag{41}$$

Here, it is rather clear how we select the initial value approximations - namely by

$$(v^0,\phi)_0 = (f_0,\phi)_0, \quad (w^0,\phi)_0 = (g_0,\phi)_0, \quad \phi \in E_n.$$

We now proceed to discuss how periodic solutions of hyperbolic equations can be approximated by means of Galerkin methods. To this end, we consider the *scalar hyperbolic initial value problem of first order*,

$$\begin{aligned}&u_t(x,t) = c(x)u_x(x,t) + s(x,t), \quad x \in \mathbb{R}, \quad t \in [0,T], \\ &u(x,0) = u_0(x), \quad x \in \mathbb{R}, \quad u(0,t) = u(1,t), \quad t \in [0,T].\end{aligned} \tag{42}$$

Here, we seek a solution u periodic in x (with period 1), and we assume that the functions c, u_0, s occurring in (42) are periodic (in x) also with period 1. The completion of the space $C^1_\#$ of all continuously differentiable functions with period 1 with respect to the norm $||\cdot||_1$ (defined in Section 2.1, Example 1) is denoted as $H^1_\#$; such a function space is clearly isomorphic to

$$H = \{\phi \in H^1(0,1):\ \phi(0) = \phi(1)\}.$$

A *variational formulation* associated with (42) is

$$(u_t - cu_x,\phi)_0 = (s,\phi)_0,\quad \phi \in H. \tag{43}$$

In the *standard semidiscrete Galerkin method*, we seek a continuously differentiable function $v\colon [0,T] \to E_n$, where E_n is a finite-dimensional subspace of H, by solving

$$(v_t - cv_x,\phi)_0 = (s,\phi)_0,\quad \phi \in E_n. \tag{44}$$

If we employ a discretization in the time variable - as with the Crank-Nicolson method - then we get the following *discrete-time Galerkin method* for determining approximations v^k of $u(.,t_k)$, where $t_k = k\tau$, $k = 1,\ldots,N$, $\tau = T/N$:

$$\left(\frac{v^{k+1}-v^k}{\tau} - cv_x^{k+1/2},\phi\right)_0 = (s^{k+1/2},\phi)_0,\quad \phi \in E_n,\ k = 0,1,\ldots,N-1. \tag{45a}$$

As before, we write v^ν_x for $(v^\nu)'$, $\nu = k,k+1$. An obvious choice for approximating the initial function is

$$(v^0,\phi)_0 = (u_0,\phi)_0,\quad \phi \in E_n. \tag{45b}$$

With $\phi_1,\ldots,\phi_m$ as a basis for E_n, we obtain, using (45a), a system of m equations (by setting $\phi = \phi_j$, $j = 1,\ldots,m$) for the m unknowns $c_1^{(k+1)},\ldots,c_m^{(k+1)}$ in $v^{k+1} = \Sigma_{i=1}^m c_i^{(k+1)}\phi_i$. The solvability of such a system is assured for τ sufficiently small and for $c \in C^1_\#$. Indeed, we have the relation

$$(c\psi',\psi)_0 = -\frac{1}{2}(c'\psi,\psi)_0,\quad \psi \in H,$$

so that the solution $\psi_n \in E_n$ of the homogeneous system

$$(\psi_n - \frac{\tau}{2}\, c\psi_n',\phi)_0 = 0,\quad \phi \in E_n,$$

satisfies

$$((1 + \frac{\tau}{4} c')\psi_n,\psi_n)_0 = 0.$$

The conclusion that $\psi_n = 0$ necessarily follows, in case, say, $\tau|c'|_{0,\infty} \leq 2$.

We close this section by presenting in detail the system of equations (45a) when E_n is the space of continuous, piecewise linear functions determined by a uniform mesh. Let $x_j = jh$, $j = 0,\ldots,J$, with $h = 1/J$. As a set of basis functions, we again take the roof functions $\phi_1,\ldots,\phi_{J-1}$ defined in Section 2.3 where in the setting of periodic functions we additionally need

$$\phi_J(x) = \begin{cases} (x-x_{J-1})/h, & x \in I_J, \\ (x_1-x)/h, & x \in I_1, \\ 0, & \text{otherwise}. \end{cases}$$

We illustrate our set of basis functions in the following figure:

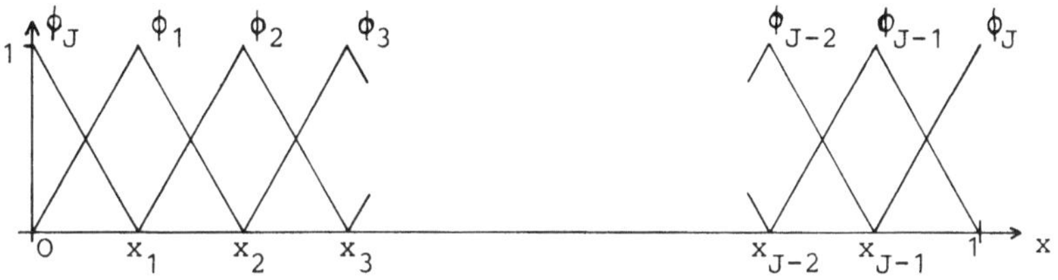

With E_n as the span of $\{\phi_1,\ldots,\phi_J\}$, we then seek solutions $v^{k+1} = \sum_{i=1}^{J} v^{k+1}(x_i)\phi_i$ in E_n to the following system of equations (where we have set $v_i^\nu = v^\nu(x_i)$, $i = 1,\ldots,J$, $\nu = k,k+1/2,k+1$):

$$\frac{1}{\tau}\sum_{i=1}^{J} \{(v_i^{k+1}-v_i^k)(\phi_i,\phi_j)_0 - v_i^{k+1/2}(c\phi_i',\phi_j)_0\} = (f^{k+1/2},\phi_j)_0, \qquad j = 1,\ldots,J. \tag{46a}$$

For this particular basis, we have essentially calculated $(\phi_i,\phi_j)_0$ in Section 2.3 (cf. also Section 4.2); we additionally need

$$(\phi_J,\phi_1)_0 = (\phi_1,\phi_J)_0 = \frac{h}{6}, \quad (\phi_J,\phi_J)_0 = \frac{2}{3}h,$$

and, for $c_{j,i} \equiv (c\phi_i',\phi_j)_0$, we have

$$c_{J,1} = \frac{1}{h}\int_{I_1} c\phi_J dx, \quad c_{1,J} = -\frac{1}{h}\int_{I_1} c\phi_1 dx, \quad c_{J,J} = -(c_{J,J-1} + c_{J,1}),$$

and

$$c_{j,j-1} = \frac{-1}{h}\int_{I_j} c\phi_j dx, \quad x_{j,j+1} = \frac{1}{h}\int_{I_{j+1}} c\phi_j dx, \quad c_{j,j} = -(c_{j,j-1}+c_{j,j+1}),$$

$$j = (1),2,\ldots,J-1,(J).$$

The formulas in the last row are to also hold for $c_{1,1}$, $c_{1,2}$, $c_{J,J-1}$ whereas the remaining quantities (namely, $c_{J,1}$, $c_{1,J}$, $c_{J,J}$) are given in the row above. With this notation, we can express the system of equations (46a) as (after multiplying by $\lambda = \tau/h$)

$$\left(\frac{2}{3} - \frac{\lambda}{2} c_{1,1}\right)v_1^{k+1} + \left(\frac{1}{6} - \frac{\lambda}{2} c_{1,2}\right)v_2^{k+1} + \left(\frac{1}{6} - \frac{\lambda}{2} c_{1,J}\right)v_J^{k+1}$$

$$= \left(\frac{2}{3} + \frac{\lambda}{2}c_{1,1}\right)v_1^k + \left(\frac{1}{6} + \frac{\lambda}{2}c_{1,2}\right)v_2^k + \left(\frac{1}{6} + \frac{\lambda}{2}c_{1,J}\right)v_J^k + \lambda(s^{k+1/2},\phi_1)_0,$$

$$\left(\frac{1}{6} - \frac{\lambda}{2}c_{j,j-1}\right)v_{j-1}^{k+1} + \left(\frac{2}{3} - \frac{\lambda}{2}c_{j,j}\right)v_j^{k+1} + \left(\frac{1}{6} - \frac{\lambda}{2}c_{j,j+1}\right)v_{j+1}^{k+1}$$

$$= \left(\frac{1}{6} + \frac{\lambda}{2}c_{j,j-1}\right)v_{j-1}^k + \left(\frac{2}{3} + \frac{\lambda}{2}c_{j,j}\right)v_j^k + \left(\frac{1}{6} + \frac{\lambda}{2}c_{j,j+1}\right)v_{j+1}^k \tag{46b}$$

$$+ \lambda(s^{k+1/2},\phi_j)_0, \quad j = 2,\ldots,J-1,$$

$$\left(\frac{1}{6} - \frac{\lambda}{2}c_{J,1}\right)v_1^{k+1} + \left(\frac{1}{6} - \frac{\lambda}{2}c_{J,J-1}\right)v_{J-1}^{k+1} + \left(\frac{2}{3} - \frac{\lambda}{2}c_{J,J}\right)v_J^{k+1}$$

$$= \left(\frac{1}{6} + \frac{\lambda}{2}c_{J,1}\right)v_1^k + \left(\frac{1}{6} + \frac{\lambda}{2}c_{J,J-1}\right)v_{J-1}^k + \left(\frac{2}{3} + \frac{\lambda}{2}c_{J,J}\right)v_J^k + \lambda(s^{k+1/2},\phi_J)_0.$$

For the case of constant c, we have

$$c_{J,1} = -c_{1,J} = \frac{c}{2}, \quad c_{j,j+1} = -c_{j,j-1} = \frac{c}{2}, \quad c_{j,j} = 0,$$

$$j = (1),2,\ldots,J-1,(J).$$

In order to numerically solve this system of equations, we first note that we can convert the associated matrix to a banded matrix of band width 5 by means of a permutation transformation determined by (1,3,5,7, ...,8,6,4,2). The following figure makes this clear. Using Gaussian elimination for banded matrices, we can solve the permuted system rather efficiently.

It is clear how we can extend the above construction of Galerkin approximations for the scalar equation (42) to systems of first order equations, and thereby to the wave equation.

$$
\begin{pmatrix}
x & x & 0 & & & 0 & x \\
x & x & x & & & 0 & 0 \\
0 & x & x & x & & & 0 \\
 & & & \ddots & & & \\
0 & & & x & x & x & 0 \\
0 & 0 & & & x & x & x \\
x & 0 & & & 0 & x & x
\end{pmatrix}
\xrightarrow{\text{Perm.}}
\begin{pmatrix}
x & x & x & & & & & & 0 \\
x & x & 0 & x & & & & & \\
x & 0 & x & 0 & x & 0 & & & \\
 & & & \ddots & & & & & \\
 & & & & x & 0 & x & 0 & x \\
 & & & & & x & 0 & x & x \\
0 & & & & & 0 & x & x & x
\end{pmatrix}
$$

4.4. THE NUMERICAL SOLUTION OF NONLINEAR INITIAL VALUE PROBLEMS

Our earlier approach in treating nonlinear boundary-value problems in Sections 1.4 and 2.5, combined with the techniques used so far in this chapter, clearly motivate corresponding approximations for nonlinear initial value problems.

First, we give the appropriate finite-difference approximations for the following *semilinear parabolic initial-boundary-value problem*:

$$u_t = a(x,t)u_{xx} + F(x,t,u,u_x), \quad 0 \leq x \leq 1, \quad 0 \leq t \leq T, \tag{47a}$$

$$\begin{aligned} &u(x,0) = u_0(x), \quad x \in [0,1], \quad u(0,t) = \gamma_0(t), \\ &u(0,t) = \gamma_0(t), \quad u(1,t) = \gamma_1(t), \quad t \in [0,T]. \end{aligned} \tag{47b}$$

We assume that u_0, γ_0, γ_1, a, and F are at least continuous in all their variables, with a positive; further assumptions will be later specified in the text. We approximate the derivatives in x by central difference quotients, and the time derivative u_t by either the forward or backward difference quotient. As with the methods (8) and (9) for treating the heat equation, we again treat both possibilities here by introducing a parameter Θ in $0 \leq \Theta \leq 1$, and thereby obtain the following nonlinear system for approximating the solution of (47a,b) on equidistant mesh points $x_j = jh$, $t_k = k\tau$:

$$\begin{aligned} &v_j^{k+1} - \Theta\, ra(x_j,t_{k+1})(v_{j+1}^{k+1} - 2v_j^{k+1} + v_{j-1}^{k+1}) \\ &\quad - \Theta\, \tau F(x_j,t_{k+1},v_j^{k+1},(v_{j+1}^{k+1} - v_{j-1}^{k+1})/(2h)) \\ &= v_j^k + (1-\Theta)ra(x_j,t_k)(v_{j+1}^k - 2v_j^k + v_{j-1}^k) \\ &\quad + (1-\Theta)\tau F(x_j,t_k,v_j^k,(v_{j+1}^k - v_{j-1}^k)/(2h)), \; j = 1,\dots,J-1, \; k = 0,\dots,N-1, \end{aligned} \tag{48a}$$

$$v_j^0 = u_0(x_j),\ j = 0,\dots,J,\quad v_0^k = \gamma_0(t_k),\quad v_J^k = \gamma_1(t_k),\qquad k = 1,\dots,N. \tag{48b}$$

We use the same notations as in Section 4.1; in particular, $r = \tau/h^2$. When $\Theta = 0$, we have an *explicit method* which corresponds to approximating u_t by a forward difference quotient. For each $\Theta > 0$, we get an *implicit finite-difference method*, and $\Theta = 1$ is associated with the backward difference quotient (for the time derivative). Analogously, we can label the case when $\Theta = 1/2$ as the *Crank-Nicolson method*. This method is, however, different from the nonlinear Crank-Nicolson method discussed in Mitchell and Griffiths (1980), Section 2.22, where the differential equation (47a) is approximated by,

$$\frac{1}{\tau}(v_j^{k+1} - v_j^k) = a(x_j, t_{k+1/2})\frac{1}{h^2}(v_{j+1}^{k+1/2} - 2v_j^{k+1/2} + v_{j-1}^{k+1/2})$$
$$+ F(x_j, t_{k+1/2}, v_j^{k+1/2}, (v_{j+1}^{k+1/2} - v_{j-1}^{k+1/2})/(2h)), \tag{49}$$
$$j = 1,\dots,J-1,\quad k = 0,\dots,N-1.$$

The right-hand side of (47a) is approximated by using the arithmetic mean values $t_{k+1/2} = (t_k + t_{k-1})/2$, $v_j^{k+1/2} = (v_j^k + v_j^{k+1})/2$, and the correspondingly averaged central difference quotients of first and second order. Both of these Crank-Nicolson methods are viable numerical schemes, as our later investigations will show (cf. also Marchuk (1975), Ch. 4).

In the following, we examine iterative methods for solving (48a,b); analogous methods can be applied to (49). If we set $v^k = (v_0^k,\dots,v_J^k)$, $k = 0,\dots,N$, then we have (in case $\Theta > 0$) a nonlinear system of equations to be solved for v^{k+1}, $k = 0,1,\dots,N-1$,

$$g_{j,k}(v^{k+1}) = 0,\quad j = 0,\dots,J. \tag{50}$$

The functions $g_{j,k}(\xi_0,\dots,\xi_J)$ depend, for each $j = 1,\dots,J-1$, only on ξ_μ, $\mu = j-1,j,j+1$, and also on the previously calculated approximate values v_μ^k, $\mu = j,j\pm1$; for $j = 0$ and $j = J$, we have

$$g_{0,k}(\xi_0,\dots,\xi_J) = \xi_0 - \gamma_0(t_{k+1}),\quad g_{J,k}(\xi_0,\dots,\xi_J) = \xi_J - \gamma_1(t_{k+1}).$$

The solvability of (50), and the convergence of an iterative method to its numerical solution, depend on properties of the associated Jacobian matrix. For representing this matrix, we let F_y and F_z denote the partial derivatives of $F(x,t,y,z)$ with respect to the third and fourth arguments, respectively, and set $\xi = (\xi_0,\ldots,\xi_J) \in \mathbb{R}^{J+1}$, $a_j^{k+1} = a(x_j,t_{k+1})$. The dependence of F_z and F_y on $(x_j,t_{k+1},\xi_j,(\xi_{j+1}-\xi_{j-1})/(2h))$ is not explicitly indicated. The associated Jacobian matrix then has the form:

$$\frac{\partial g_{j,k}}{\partial \xi_j}(\xi) = (1 + 2\Theta r a_j^{k+1}) - \Theta\tau F_y, \quad j = 1,\ldots,J-1,$$

$$\frac{\partial g_{j,k}}{\partial \xi_j}(\xi) = 1, \quad j = 0,J,$$

$$\frac{\partial g_{j,k}}{\partial \xi_{j-1}}(\xi) = -\Theta r a_j^{k+1} + \frac{1}{2}\, rhF_z, \quad j = 1,\ldots,J-1,$$

$$\frac{\partial g_{j,k}}{\partial \xi_{j+1}}(\xi) = -\Theta r a_j^{k+1} - \frac{1}{2}\, rhF_z, \quad j = 1,\ldots,J-1,$$

$$\frac{\partial g_{j,k}}{\partial \xi_i}(\xi) = 0, \quad \text{otherwise}.$$

Under appropriate assumptions on the Jacobian matrix, we can utilize Newton's method to compute v^{k+1} iteratively (see, for example, the assumptions in the following theorem). One particular simplification is the following (one-step) SOR-Newton method (cf. Ortega & Rheinboldt (1970), 7.4, 13.5), which incorporates the ideas of the SOR-methods with those of the Newton method. In case the diagonal elements of the Jacobian matrix are nonzero, the (s+1)-iteration

$$v_{(s+1)}^{k+1} = (v_{0,s+1}^{k+1},\ldots,v_{J,s+1}^{k+1})$$

for v^{k+1} is given by

$$v_{j,s+1}^{k+1} = v_{j,s}^{k+1} - \omega g_{j,k}(v_{(s)}^{k+1,j})/\frac{\partial g_{j,k}}{\partial \xi_j}(v_{(s)}^{k+1,j}), \qquad j = 0,\ldots,J, \quad s = 0,1,2,\ldots \tag{51}$$

where $\omega > 0$ is a (real) relaxation parameter, and

$$v_{(s)}^{k+1,j} = (v_{0,s+1}^{k+1},\ldots,v_{j-1,s+1}^{k+1},\, v_{j,s}^{k+1},\ldots,v_{J,s}^{k+1})^T, \quad j = 0,\ldots,J.$$

A useful starting value is given by

$$v_{(0)}^{k+1} = (\gamma_0(t_{k+1}),v_1^k,\ldots,v_{J-1}^k,\gamma_1(t_{k+1})).$$

Thus, we do not have to solve in each iteration step a (tridiagonal) system of equations as in Newton's method, but obtain each single component explicitly. If (50) is actually a linear system of equations, i.e.,

$$g_{j,k}(\xi) = \sum_{\nu=0}^{J} a_{j\nu}\xi_\nu - b_j, \quad j = 0,\dots,J,$$

with a matrix $A = (a_{j\nu})$ and vector $b = (b_0,\dots,b_J)$, then (51) is the usual SOR-method which is the Gauss-Seidel method when $\omega = 1$.

For every $k = 0,\dots,N-1$, the iterative method described by (51) converges whenever $\omega \in (0,1]$, and the Jacobian matrix is an M-matrix (cf. Ortega and Rheinboldt (1970), Chapter 7 and 13, Törnig (1979), 8.1, 16.4). The latter condition is guaranteed whenever

$$0 < \alpha_0 \le a(x,t), \quad F_y \text{ bounded from above}, \quad |F_z| \text{ bounded},$$

and h,τ are chosen so small so that $h|F_z| \le 2\alpha_0\Theta$ and $\tau\Theta F_y < 1$, respectively. Then

$$\frac{\partial g_{j,k}}{\partial \xi_{j\pm1}} \le 0, \quad \frac{\partial g_{j,k}}{\partial \xi_j} > 0, \quad \text{and} \quad \left(\frac{\partial g_{j,k}}{\partial \xi_i}\right) \text{ is strictly diagonally dominant,}$$

and the associated matrix is therefore an M-matrix (cf. Ortega and Rheinboldt (1970), 2.4).

We next consider another type of nonlinearity, namely the *quasilinear parabolic partial differential equation* given by

$$u_t = (a(x,u)u_x)_x + F(x,t,u) \quad \text{in} \quad [0,1] \times [0,T], \tag{52}$$

with initial and boundary conditions given as in (47b). The functions a and F are assumed continuous in all their arguments, with $0 < \alpha_0 \le a(x,y) \le \alpha_1 < \infty$. By approximating $(au_x)_x$ analogous to 1.(14) and taking arithmetic averages of the forward and backward difference quotients in the t-direction (as with the Crank-Nicolson method), we obtain the following *finite-difference approximation* for (52),

$$\begin{aligned}\frac{1}{\tau}(v_j^{k+1} - v_j^k) = \frac{1}{2h^2}\{&a_{j-1/2}^{k+1}v_{j-1}^{k+1} - (a_{j-1/2}^{k+1} + a_{j+1/2}^{k+1})v_j^{k+1} + a_{j+1/2}^{k+1}v_{j+1}^{k+1} \\ &+ a_{j-1/2}^{k}v_{j-1}^{k} - (a_{j-1/2}^{k} + a_{j+1/2}^{k})v_j^{k} + a_{j+1/2}^{k}v_{j+1}^{k}\} \\ &+ \tfrac{1}{2}[F(x_j,t_{k+1},v_j^{k+1}) + F(x_j,t_k,v_j^k)], \quad j = 1,\dots,J-1,\end{aligned} \tag{53}$$

where $a_{j\pm1/2}^\nu = a(x_{j\pm1/2},(v_j^\nu + v_{j\pm1}^\nu)/2)$, $\nu = k,k+1$. This can be equivalently expressed (after multiplication by τ) as

$$[1 + \frac{r}{2}(a_{j-1/2}^{k+1} + a_{j+1/2}^{k+1})]v_j^{k+1} - \frac{r}{2}(a_{j-1/2}^{k+1}v_{j-1}^{k+1} + a_{j+1/2}^{k+1}v_{j+1}^{k+1})$$

$$- \frac{\tau}{2} F(x_j, t_{k+1}, v_j^{k+1})$$

$$= [1 - \frac{r}{2}(a_{j-1/2}^{k} + a_{j+1/2}^{k})]v_j^{k} + \frac{r}{2}(a_{j-1/2}^{k}v_{j-1}^{k} + a_{j+1/2}^{k}v_{j+1}^{k})$$

$$+ \frac{\tau}{2} F(x_j, t_k, v_j^k), \quad j = 1,\ldots,J-1.$$

We leave to the reader the tasks of computing the associated Jacobian matrix, of constructing the appropriate Newton or SOR-Newton methods, and of determining under what assumptions the solvability of (53) will be assured (e.g., by ascertaining appropriate conditions guaranteeing that the Jacobian matrix is an M-matrix).

We now turn to approximating the nonlinear IVP (52) by a Galerkin method, which can be derived by employing the ideas used in constructing the Crank-Nicolson-Galerkin method for the heat equation. Without loss of generality, we consider homogeneous Dirichlet boundary conditions. For E_n a finite-dimensional subspace of $H_0^1(0,1)$ (for example, continuous, piecewise linear functions determined by a mesh in the spatial interval $[0,1]$), we first obtain a *semidiscrete Galerkin method*, by which we seek a continuously differentiable function $v: [0,T] \to E_n$ as a solution of

$$(v_t, \phi)_0 + (\alpha(v)v', \phi')_0 = (\Phi(t,v), \phi)_0, \quad \phi \in E_n, \quad t \in [0,T]. \tag{54a}$$

For brevity, we have set $\alpha(\psi) \equiv a(.,\psi(\cdot))$ and $\Phi(t,\psi) \equiv F(.,t,\psi(\cdot))$ for $\psi \in H_0^1(0,1)$. The initial approximation $v(0) \in E_n$ may again be suitably chosen as in (19b). With $\phi_1,\ldots,\phi_m$ a set of basis functions for E_n, we set $v(x,t) = \sum_{i=1}^{m} c_i(t)\phi_i(x)$, with the result that (54a) becomes an initial value problem for determining $c(t) = (c_1(t),\ldots,c_m(t))$ expressed by the following nonlinear system of ordinary differential equations:

$$\begin{aligned} &\sum_{i=1}^{m} c_i'(t)(\phi_i,\phi_j)_0 + \sum_{i=1}^{m} c_i(t)(\alpha(\sum_{s=1}^{m} c_s(t)\phi_s)\phi_i',\phi_j')_0 \\ &\quad = (\Phi(t, \sum_{s=1}^{m} c_s(t)\phi_s), \phi_j)_0, \quad j = 1,\ldots,m, \quad t \in [0,T]. \end{aligned} \tag{54b}$$

Under additional assumptions concerning the differentiability of $a(.,.)$ and $F(.,.,.)$, we can utilize the existence and uniqueness results from the theory of ordinary differential equations to deduce the unique solvability of (54b) (cf. e.g,, Fairweather (1978), 4.2, Hartman (1964), Walter (1976)).

By discretizing the time variable t by $t_k = k\tau$, $\tau = T/N$, an approach similar to (21) leads to the following *nonlinear Crank-Nicolson-Galerkin method:*

$$\left(\frac{v^{k+1}-v^k}{\tau},\phi\right)_0 + \frac{1}{2}(\alpha(v^{k+1})v_x^{k+1} + \alpha(v^k)v_x^k,\phi')_0$$
$$= \frac{1}{2}(\Phi(t_{k+1},v^{k+1}) + \Phi(t_k,v^k),\phi)_0, \quad \phi \in E_n, \quad k = 0,1,\ldots,N-1. \tag{55a}$$

By representing v^ν, $\nu = k,k+1$, as $v^\nu = \Sigma_i c_i^\nu \phi_i$, we see that (55a) is equivalent to the following nonlinear system of equations for determining the vector of coefficients $c^{(k+1)} = (c_1^{(k+1)},\ldots,c_m^{(k+1)})$:

$$\sum_{i=1}^m c_i^{(k+1)}(\phi_i,\phi_j)_0 + \frac{\tau}{2}\sum_{i=1}^m c_i^{(k+1)}(\alpha(\sum_{s=1}^m c_s^{(k+1)}\phi_s)\phi_i',\phi_j')_0$$
$$= \sum_{i=1}^m c_i^{(k)}(\phi_i,\phi_j)_0 - \frac{\tau}{2}\sum_{i=1}^m c_i^{(k)}(\alpha(\sum_{s=1}^m c_s^{(k)}\phi_s)\phi_i',\phi_j')_0$$
$$+ \frac{\tau}{2}(\Phi(t_{k+1},\sum_{s=1}^m c_s^{(k+1)}\phi_s) + \Phi(t_k,\sum_{s=1}^m c_s^{(k)}\phi_s),\phi_j)_0, \tag{55b}$$
$$j = 1,\ldots,J-1, \quad k = 0,\ldots,N-1.$$

If we define the matrices $B = ((\phi_i,\phi_j)_0)$, $A(\xi) = ((\alpha(\Sigma_s\xi_s\phi_s)\phi_i',\phi_j')_0)$ and the vector-valued function $\Psi(t,\xi) = (\Psi_1(t,\xi),\ldots,\Psi_m(t,\xi))$, where

$$\Psi_j(t,\xi) = (\Phi(t,\sum_{s=1}^m \xi_s\phi_s),\phi_j)_0, \quad j = 1,\ldots,m,$$

then we observe that (55b) is also equivalent to

$$Bc^{(k+1)} + \frac{\tau}{2}A(c^{(k+1)})c^{(k+1)} - \frac{\tau}{2}\Psi(t_{k+1},c^{(k+1)})$$
$$= Bc^{(k)} - \frac{\tau}{2}A(c^{(k)})c^{(k)} + \frac{\tau}{2}\Psi(t_k,c^{(k)}), \quad k = 0,1,\ldots,N-1. \tag{55c}$$

The matrix B is symmetric and positive definite (cf. Theorem 2.18 with $A = I$). Because of the assumption that $a(x,y) \geq \alpha_0 > 0$, $0 \leq x \leq 1$, $y \in \mathbb{R}$, $A(\xi)$ is also symmetric and positive definite for every vector $\xi \in \mathbb{R}^m$. Indeed, the Poincaré-Friedrichs inequality implies that $|\phi'|_0$ is a norm on $H_0^1(0,1)$, and that, for arbitrary $0 \neq \zeta = (\zeta_1,\ldots,\zeta_m)$ with $\phi = \sum_i \zeta_i\phi_i$ and the Euclidean scalar product $(.,.)$,

$$(A(\xi)\zeta,\zeta) = \sum_{i,j=1}^{m} (\alpha(\sum_{s=1}^{m} \xi_s\phi_s)\phi_i',\phi_j')_0\zeta_i\zeta_j$$

$$= (\alpha(\sum_{s=1}^{m} \xi_s\phi_s)\phi',\phi')_0 \geq \alpha_0|\phi'|_0^2 > 0, \quad \xi \in \mathbb{R}^m.$$

Using an approach similar to that used to construct the second Crank-Nicolson method (49) for approximating the nonlinear parabolic differential equation (47a), we can obtain yet another scheme (cf. Fairweather (1978), Sec. 4.4), namely

$$\begin{aligned}(\frac{v^{k+1}-v^k}{\tau},\phi)_0 &+ (\alpha(v^{k+1/2})v_x^{k+1/2},\phi')_0 \\ &= (\Phi(t_{k+1/2},v^{k+1/2}),\phi)_0, \quad \phi \in E_n, \quad k = 0,\ldots,N-1,\end{aligned} \tag{56a}$$

or, equivalently,

$$\begin{aligned}B\,\frac{c^{(k+1)}-c^{(k)}}{\tau} + A(c^{(k+1/2)})c^{(k+1/2)} &= \Psi(t_{k+1/2},c^{(k+1/2)}), \\ k &= 0,1,\ldots,N-1,\end{aligned} \tag{56b}$$

where the mean values $t_{k+1/2}, c^{(k+1/2)}$ are defined in the usual manner.

In order to avoid calculating the solution of the nonlinear system of equations in (55) (or also in (56)), say, by Jacobi or Gauss-Seidel iterative methods, we indicate still other possibilities for numerically solving such systems - the so-called *predictor-corrector methods*. (In particular, such schemes are widely used for numerically solving initial value problems in ordinary differential equations in conjunction with multistep methods.) Proceeding from (55c), we can develop appropriate modifications of such schemes to apply to the problems at hand which lead to the following *Crank-Nicolson Predictor-Corrector method* for calculating an approximation $c^{(k+1)}$ to $u(.,t_{k+1})$:

$$\begin{aligned}B\tilde{c}^{(k+1)} + \tau A(c^{(k)})\frac{\tilde{c}^{(k+1)}+c^{(k)}}{2} &= Bc^{(k)} + \tau\Psi(t_k,c^{(k)}), \\ Bc^{(k+1)} + \tau A(\tilde{c}^{(k+1)})\frac{c^{(k+1)}+c^{(k)}}{2} &= Bc^{(k)} + \frac{\tau}{2}(\Psi(t_{k+1},\tilde{c}^{(k+1)}) + \Psi(t_k,c^{(k)})), \\ k &= 0,\ldots,N-1.\end{aligned} \tag{57}$$

Thus, in the first (predictor) step we compute an auxiliary veetor $\tilde{c}^{(k+1)}$ and then, in the second step, insert this into the nonlinear part of (55c) in place of $c^{(k+1)}$. A modification of the usual predictor-corrector methods for initial value problems consists of using $(\tilde{c}^{(k+1)}+c^{(k)})/2$

in place of $c^{(k)}$ in the predictor step and $(c^{(k+1)} + c^{(k)})/2$ in place of $\tilde{c}^{(k+1)}$ in the corrector step. Because of the positive definiteness of B and of $A(\xi)$, both (linear) systems in (57) are uniquely solvable. We refer the reader to Fairweather (1978), (4.43), for an analogous predictor-corrector method based on the equations of (56a) or (56b).

As the last class of examples studied in this section, we consider systems of *quasilinear hyperbolic initial value problems* having the form,

$$u_t(x,t) - D(x,t,u(x,t))u_x(x,t) = s(x,t,u(x,t)), \quad x \in \mathbb{R},\ t \in [0,T],$$
$$u(x,0) = u_0(x),\ x \in \mathbb{R}. \tag{58}$$

Concerning results on existence of solution and unique solvability of such problems, we refer to the treatises by John (1980) and by Törnig (1979), Sec. 17.3, and to the references cited therein. Without loss of generality, we restrict our analysis to the case where $D = \mathrm{diag}(d_i)$ is a diagonal $\iota \times \iota$-matrix, since we can, by definition, diagonalize a hyperbolic system of first order and convert it to its normal (or characteristic) form. The functions $u(x,t)$ and $s(x,t,y)$ are to be vector-valued, each with ι components, and $d_i = d_i(x,t,y)$, $i = 1,\dots,\iota$.

The finite-difference methods, given in Section 4.3 for the wave equation, can be defined in a similar manner for the nonlinear problems we consider here. With equidistant mesh points $x_j = jh$, $j = 0,\pm1,\pm2,\dots$, and $t_k = k\tau$, $k = 0,\dots,N$, with mesh widths $h > 0$ and $\tau = T/N$ in the x- and t-directions, respectively, we obtain the following methods:

Friedrichs method:

$$v_j^{k+1} = \tfrac{1}{2}(1 - \lambda D(x_j,t_k,v_j^k))v_{j-1}^k$$
$$+ \tfrac{1}{2}(1 + \lambda D(x,t_k,v_j^k))v_{j+1}^k + \tau s(x_j,t_k,v_j^k), \tag{59}$$

Courant-Isaacson-Rees method:

$$v_j^{k+1} = -\lambda D^-(x_j,t_k,v_j^k)v_{j-1}^k + \lambda D^+(x_j,t_k,v_j^k)v_{j+1}^k$$
$$+ [1 - \lambda(D^+-D^-)(x_j,t_k,v_j^k)]v_j^k + \tau s(x_j,t_k,v_j^k), \tag{60}$$

Lax-Wendroff method:

$$v_j^{k+1} = v_j^k + \frac{\lambda}{2} D(x_j,t_k,v_j^k)\{(v_{j+1}^k - v_{j-1}^k)$$
$$+ \lambda D(x_j,t_k,v_j^k)(v_{j+1}^k-2v_j^k + v_{j-1}^k)\} + \tau s(x_j,t_k,v_j^k). \tag{61}$$

Here, as before, $\lambda = \tau/h$ denotes the ratio of mesh widths, and $D^{\pm} = \mathrm{diag}(d_i^{\pm})_{i=1,\ldots,l}$, with

$$d_i^{+}(x,t,y) = \max(0,d_i(x,t,y)), \quad d_i^{-}(x,t,y) = \min(0,d_i(x,t,y)).$$

The above methods can be expressed in the form

$$v_j^{k+1} = \sum_{\mu=-1}^{1} B_\mu(x_j,t_k,v_j^k)v_{j+\mu}^k + \tau s(x_j,t_k,v_j^k), \qquad j = 0,\pm 1,\ldots, \quad k = 0,1,\ldots,N-1, \tag{62}$$

where the matrices $B_\mu(x,t,y)$, $\mu = 0,\pm 1$ can be discerned by comparing (62) with each of the expressions for the individual methods.

A simple example of a problem having the form (58) is the scalar, semilinear hyperbolic initial value problem,

$$u(x,0) = u_0(x), \quad u_t - u_x = 2u^2, \quad x \in \mathbb{R}, \quad t \in [0,T].$$

For a periodic initial function u_0 such that $|u_0|_{0,\infty} < 1/(2T)$, we can express the solution explicitly as

$$u(x,t) = u_0(x)/(1 - 2tu_0(x)).$$

4.5. PURE INITIAL VALUE PROBLEMS AND A GENERAL REPRESENTATION OF APPROXIMATION METHODS

The examples we have considered of parabolic and hyperbolic initial value problems are either pure initial value problems or initial-boundary-value problems. We begin this section by briefly indicating how quasilinear initial-boundary-value problems in one spatial dimension with inhomogeneous Dirichlet boundary conditions can be viewed as pure initial value problems in an appropriate vector space. With these observations, we shall see that the study of pure initial value problems, and of the appropriate approximation methods, will encompass a large class of initial-boundary-value problems. The approximation methods introduced in the previous sections of this chapter can be commonly expressed in operator notation, in which the approximations to the boundary conditions of the given initial-boundary-value problems are also included. On the basis of such a representation, we shall carry out the convergence analysis of these methods in Part IV of this book.

Let us now consider the general, one-(spatial) dimensional, quasilinear differential equation of second order,

$$au_{xx} + 2bu_{xt} + cu_{tt} = s \quad \text{in} \quad (0,1) \times (0,T],$$

where appropriate initial conditions and inhomogeneous Dirichlet boundary conditions,

$$u(0,t) = \gamma_0(t), \quad u(1,t) = \gamma_1(t), \quad 0 < t \leq T,$$

are given. The functions $a,b,c,$ and s may depend on x,t,u,u_x. The spatial interval $(0,1)$ is not restrictive for the finite interval case, since any other finite interval can be transformed to the unit interval by a continuous, one-to-one mapping. Without loss of generality, we shall always assume homogeneous Dirichlet boundary conditions, since, otherwise, we can solve a similar problem for

$$v(x,t) = u(x,t) - (1-x)\gamma_0(t) - x\gamma_1(t)$$

(cf. also Section 1.1).

In many cases, initial-boundary-value problems can be rewritten as pure initial value problems in appropriate spaces. For example, in case of homogeneous Dirichlet boundary conditions (on a finite spatial interval) one may choose the space of all continuous, periodic (with period 2), odd functions because they necessarily vanish at $x = 0$ and $x = 1$. Similar considerations apply to problems with mixed boundary conditions for which we refer to Meis & Marcowitz (1981), Sec. I.8.

We notice that, in the variational formulations of our examples of initial-boundary-value problems, homogeneous Dirichlet boundary conditions are incorporated in the underlying spaces, so that pure initial value problems occur. This is also the case with other types of boundary conditions - in particular, Neumann boundary conditions.

The initial value problems we have investigated in the previous section - now viewed as pure initial value problems - can be formulated in the following general manner. Let F be a Banach space with norm $|\cdot|_F$ with E a subspace of F equipped with a norm $|\cdot|_E$ which is continuously embedded in F, i.e.,

$$|g|_F \leq c_E|g|_E, \quad g \in E.$$

Further, suppose for every $t \in [0,T]$, $(T > 0)$, $A(t)$ is a (not necessarily linear) mapping from E into F. Also, let $u_0 \in E$, $w \in C([0,T],F)$ denote the problem data. The domain of definition of $A(t)$ in the space E is assumed to be independent of t, and is denoted by $D(A)$.

We seek a function $u \in C([0,T],E) \cap C^1([0,T],F)$ satisfying the following differential-operational equation

$$u(0) = u_0, \quad \frac{du}{dt}(t) - A(t)u(t) = w(t), \qquad t \in [0,T]. \tag{63}$$

In our examples, $A(t)$, $t \in [0,T]$, are differential operators in the spatial variable x, and E and F are spaces of functions of x satisfying certain regularity conditions. For example, for the semilinear parabolic initial value problem (47a,b), $A(t)$ is given by

$$(A(t)g)(x) = a(x,t)g''(x) + F(x,t,g(x),g'(x)), \quad x \in \mathbb{R}, \quad t \in [0,T].$$

For the quasilinear parabolic initial value problem (52), which we have approximated by Galerkin methods, we discern $A(t)$ from the underlying variational formulation. We can therefore view equation (52) as acting in $L^2(0,1)$, with $A(t)$ given by the right-hand side in (52), where the spatial derivatives are interpreted in the weak sense. Concerning the domain of definition, we shall make no precise statements yet. For nonlinear problems, we shall always assume their solvability and ensure local uniqueness by imposing additional requirements (if necessary).

In the variational formulation (18a) of the heat equation, we let $E = H_0^1(0,1)$ and $F = L^2(0,1)$; the operator A is independent of t and is defined by the elliptic bilinear form $a(\psi,\phi) = (a\psi',\phi')_0$, with domain $D(A) = E$ and range in F. The boundary conditions in this case are incorporated in the domain space E. The wave equation, expressed as an equivalent system of two equations, is another particular example of the class of problems of the form (63). We can also convert the variational form (37a) into a first order system in the vector $\underline{u} = (u,v)$, by setting $v = u_t$.

For nonlinear problems of the form (63), we can clearly incorporate $w(t)$ into the mapping $A(t)\colon D(A) \subset E \to F$. We have then a *semihomogeneous IVP*,

$$u(0) = u_0, \quad \frac{du}{dt}(t) = A(t)u(t), \quad t \in [0,T]. \tag{64}$$

We say that a semihomogeneous initial value problem is *continuously solvable* on $D \subset E$, if, for every $t \in [0,T]$, there exists a continuous mapping $S(t)$ of D in E and, for every $u_0 \in D$, the unique solution of (64) is given by

$$u(t) = S(t)u_0, \qquad t \in [0,T].$$

We can express the semihomogeneous IVP as

$$Tu = (u_0, 0)$$

where T is the mapping $T: D(T) \to E \times C([0,T],F)$ defined by

$$Tv = (v(0), \frac{dv}{dt} - A(.)v),$$

with

$$D(T) = \{v \in C([0,T],E) \cap C^1([0,T],F): v(t) \in D(A), t \in [0,T]\}.$$

Continuous solvability on D thus means that a nonempty subset Z of $D(T)$ exists such that the restriction $\tilde{T}$ of T to Z, having image $R(\tilde{T}) = D \times \{0\}$, is bijective and continuously invertible. (From the context, it will be always clear whether the operator T or the equally denoted final time $T > 0$ is meant.)

For linear mappings $A(t)$, we have the concept (due to Hadamard) of a *properly posed initial value problem on* D of the form (64) which means that D is (at least) dense in E and that (64) is continuously solvable on D (cf. Richtmyer & Morton (1967), Sec. 4.3, Meis & Marcowitz (1981), Ch. 1). If D is a dense subspace of the Banach space E, then the continuous linear mappings $S(t): D \to E$ can be extended uniquely to mappings $\hat{S}(t): E \to E$, with $||\hat{S}(t)|| = ||S(t)||$, $t \in [0,T]$ (see, e.g., Kantorovich & Akilov (1964), Chap. IV). If $u_0 \in E$, then $\hat{u}(t) = \hat{S}(t)u_0$, $t \in [0,T]$, is called the *generalized solution* of (64).

If, in addition, A is independent of t and is a closed linear operator on a dense subspace $D(A)$ of $E = F$, satisfying

$$\exists M \geq 0, \quad \forall\lambda > 0, \quad \forall m \geq 1, \quad \exists (A-\lambda I)^{-m} \in B(E,E): \quad ||(A-\lambda I)^{-m}|| \leq M/\lambda^m, \tag{65}$$

then $-A$ is the infinitesimal generator of an equibounded semigroup $\{S(t)\}$, given by

$$S(t) = \exp(tA) = \lim_{m\to\infty} (I - \frac{t}{m} A)^{-m}, \quad t \geq 0.$$

We can still weaken condition (65) by requiring it to apply only for $\lambda > \beta$ with bound $M(\lambda-\beta)^{-m}$; in such a case, we obtain a quasibounded semigroup. If A itself is a bounded operator of E into itself, then the associated semigroup is given by the usual exponential operator $S(t) = \sum_m t^m A^m/m!$. By utilizing the semigroup (associated with a linear mapping A independent of t), we can express the solution of the inhomogeneous IVP (63) as

$$u(t) = S(t)u_0 + \int_0^t S(t-s)w(s)ds, \quad t \in [0,T],$$

in case $u_0 \in D(A)$ and w satisfies certain restrictions (cf. Richtmyer & Morton (1967), Sec. 3.7, Aubin (1979), Ch. 14, Sec. 3). Using the extensions of $S(t)$, $t \in [0,T]$, we can define

$$\hat{u}(t) = \hat{S}(t)u_0 + \int_0^t \hat{S}(t-s)w(s)ds, \quad t \in [0,T],$$

with $u_0 \in E$, $w \in C([0,T],E)$, as the *generalized solution* of the inhomogeneous linear IVP (for a detailed treatment of semigroups, see Hille & Phillips (1957), Kato (1966), Chapter 9, Aubin (1979), Chapter 14).

For nonlinear, inhomogeneous IVP, we are fundamentally concerned with questions of local uniqueness of the solution - whose existence we have assumed - and with its continuous dependence on the data u_0 and w in a sense to be defined later.

For a general representation of the approximation methods introduced in the previous sections, let the time interval be equally subdivided into N equal subintervals with $\tau = T/N$ denoting the mesh width. At the outset, we assume that we have a null sequence τ_n, $n \in I$ (= $\mathbb{N}$) of positive mesh widths, and we describe associated meshes in $[0,T]$ by

$$[0,T]_n = \{t \in [0,T]: t = t_k \equiv k\tau_n, \quad k = 0,\dots,N\},$$

$$[0,T]_n' = [0,T]_n - \{0\}.$$

The number N of discrete time values depends then on n, but we shall omit this dependence in the notation we employ. We have that $N\tau_n = T$ and $N \to \infty$ (as $n \to \infty$). The discrete-time approximation methods we have considered up to now can be expressed in the following general way. Let $E_n, F_n, n \in I$, be Banach spaces, and let $C_n^{(\ell)}(t)\colon D_n^{(\ell)} \subset E_n \to F_n$, $n \in I$, $\ell = 0,1$, be mappings whose domains of definition are assumed (for the sake of simplicity) to be independent of t. In the approximate problems, we seek, for given $u_{0,n} \in E_n$, and $w_n(t) \in F_n$, $t \in [0,T]_n'$, $n \in I$, solutions $u_n(t) \in E_n$, $t \in [0,T]_n$, of

$$u_n(0) = u_{n,0}, \quad C_n^{(0)}(t)u_n(t) = C_n^{(1)}(t)u_n(t-\tau_n) + \tau_n w_n(t), \qquad t \in [0,T]_n', \quad n \in I. \tag{66}$$

If $C_n^{(0)}(t) = I$ (= identity), (66) is called an *explicit method*.

For the linear examples, the finite-difference methods have the following form,

$$\sum_{|\mu|\le N_0} B_{n,\mu}^{(0)}(\cdot,t)T_n^{\mu}u_n(t) = \sum_{|\mu|\le N_1} B_{n,\mu}^{(1)}(\cdot,t)T_n^{\mu}u_n(t-\tau_n) + \tau_n w_n(t), \quad t \in [0,T]_n'. \tag{67}$$

Here $B_{n,\mu}^{(\ell)}(x,t)$ are $\iota \times \iota$-matrices with entries depending on $x \in \mathbb{R}^d$ and $t \in [0,T]$, and T_n^{μ} denotes the *shift operator*

$$(T_n^{\mu}f)(x) = f(\xi_1+\mu_1 h_n^{(1)},\ldots,\xi_d + \mu_d h_n^{(d)}), \quad x \in (\xi_1,\ldots,\xi_d) \in G_n, \quad f \in C(G_n), \tag{68}$$

with $C(G_n)$, the space of all bounded vector-valued mesh functions defined on a uniform mesh $G_n \subset \mathbb{R}^d$ (with equidistant mesh widths $h_n^{(1)},\ldots,h_n^{(d)}$ in the corresponding variables). Also $\mu = (\mu_1,\ldots,\mu_d)$ is a multi-index with modulus $|\mu| = |\mu_1| + \ldots + |\mu_d|$. In this class of examples, we thus have

$$(C_n^{(\ell)}(t)g_n)(x) = \sum_{|\mu|\le N_\ell} B_{n,\mu}^{(\ell)}(x,t)g_n(x+\mu h_n), \quad x \in G_n, \quad g_n \in C(G_n), \quad n \in I, \quad \ell = 0,1. \tag{69}$$

If the inverses $C_n^{(0)}(t)^{-1}$ exist, the equations in (66) are solvable and we can also express this method in explicit form as

$$u_n(0) = u_{n,0}, \quad u_n(t) = C_n(t)u_n(t-\tau_n) + \tau_n C_n^{(0)}(t)^{-1}w_n(t), \quad t \in [0,T]_n', \quad n \in I, \tag{70}$$

where $C_n(t) \equiv C_n^{(0)}(t)^{-1}C_n^{(1)}(t)$.

For our concrete *examples*, we have in detail:

The *One-Dimensional Heat Equation:*

Explicit Method (2): $\iota = 1$, $N_0 = 0$, $N_1 = 1$,

$B_{n,0}^{(0)} = 1$, $B_{n,-1}^{(1)} = B_{n,1}^{(1)} = ar$, $B_{n,0}^{(1)} = 1-2ar$, $w_n(x,t) = s(x,t-\tau_n)$;

Method (8) (Crank-Nicolson Method, in case $\Theta = 1/2$):

$\iota = 1$, $N_0 = N_1 = 1$,

$$B_{n,-1}^{(0)} = B_{n,1}^{(0)} = -ar\Theta, \quad B_{n,0}^{(0)} = 1 + 2ar\Theta,$$

$$B_{n,-1}^{(1)} = B_{n,1}^{(1)} = ar(1-\Theta), \quad B_{n,0}^{(1)} = 1 - 2ar(1-\Theta),$$

$$w_n(x,t) = \Theta s(x,t) + (1-\Theta)s(x,t-\tau_n);$$

Du Fort-Frankel Method (11): $\iota = 2$, $N_0 = 0$, $N_1 = 1$,

$$B_{n,0}^{(0)} = \frac{1}{2}\begin{pmatrix}1+2ar & 0\\ 0 & 1\end{pmatrix},$$

$$B_{n,-1}^{(1)} = B_{n,1}^{(1)} = \begin{pmatrix}ar & 0\\ 0 & 0\end{pmatrix}, \quad B_{n,0}^{(1)} = \frac{1}{2}\begin{pmatrix}0 & 1-2ar\\ 1 & 0\end{pmatrix},$$

$$w_n(x,t) = (s(x,t-\tau_n),0), \quad t = t_k, \quad k = 2,3,\ldots;$$

Crank-Nicolson-Galerkin Method (22) (with continuous, piecewise linear basis functions): $\iota = 1$, $N_0 = N_1 = 1$,

$$B_{n,-1}^{(0)} = B_{n,1}^{(0)} = \frac{1}{6} - \frac{1}{2}\,ar, \quad B_{n,0}^{(0)} = \frac{2}{3} + ar,$$

$$B_{n,-1}^{(1)} = B_{n,1}^{(1)} = \frac{1}{6} + \frac{1}{2}\,ar, \quad B_{n,0}^{(1)} = \frac{2}{3} - ar,$$

$$w_n(x_j,t_{k+1}) = \frac{1}{h_n}\,(s^{k+1/2},\phi_j)_0.$$

In all these methods, the coefficients $B_{n,\mu}^{(1)}$ do not depend on x,t - nor even on n; for the Du Fort-Frankel method, we must, in any case, choose another approximation for $t = t_1$ by using, say, an appropriate explicit method. If, however, the conductivity coefficient depends on x and t, then we get, for example, the following method:

Explicit Method (13) (with variable coefficients):

$$\iota = 1, \quad N_0 = 0, \quad N_1 = 1, \quad B_{n,0}^{(0)}(x,t) = 1,$$

$$B_{n,-1}^{(1)}(x,t) = B_{n,1}^{(1)}(x,t) = a(x,t), \quad B_{n,0}^{(1)}(x,t) = 1 - 2ra(x,t).$$

We also obtain methods with coefficients dependent on x for the case where boundary conditions containing derivatives are approximated. For the case of Neumann boundary conditions, we now give the coefficients of the Crank-Nicolson method (for general $0 \le \Theta \le 1$).

Method (10) (for the case of Neumann boundary conditions, i.e., $\alpha_0 = \beta_0 = 0$, $\alpha_1 = \beta_1 = 1$):

$B^{(\ell)}_{n,\mu}(x_j,t)$, $w_n(x_j,t)$ as with method (8) for $j = 1,\dots,J-1$,

$B^{(0)}_{n,0}(x_0,t) = B^{(0)}_{n,0}(x_J,t) = 1 + 2ar\Theta$,

$B^{(0)}_{n,1}(x_0,t) = B^{(0)}_{n,-1}(x_J,t) = -2ar\Theta$,

$B^{(1)}_{n,0}(x_0,t) = B^{(1)}_{n,0}(x_J,t) = 1 - 2ar(1-\Theta)$,

$B^{(1)}_{n,1}(x_0,t) = B^{(1)}_{n,-1}(x_J,t) = 2ar(1-\Theta)$,

$B^{(\ell)}_{n,\mu}(x_0,t) = 0$ otherwise, $B^{(\ell)}_{n,\mu}(x_J,t) = 0$ otherwise,

$w_n(x_0,t_{k+1}) = \frac{2a}{h}(\Theta\gamma_0^{k+1} + (1-\Theta)\gamma_0^k) + (\Theta s_0^{k+1} + (1-\Theta)s_0^k)$,

$w_n(x_J,t_{k+1}) = \frac{2a}{h}(\Theta\gamma_1^{k+1} + (1-\Theta)\gamma_1^k) + (\Theta s_J^{k+1} + (1-\Theta)s_J^k)$.

The *Two-Dimensional Heat Equation:*

Method (15): $\iota = 1$, $N_0 = N_1 = 1$,

$B^{(0)}_{n,0,0} = (1+a_1)(1+a_2)$, $B^{(0)}_{n,-1,0} = B^{(0)}_{n,1,0} = -\frac{1}{2}a_1(1+a_2)$,

$B^{(0)}_{n,0,-1} = B^{(0)}_{n,0,1} = -\frac{1}{2}a_2(1+a_1)$,

$B^{(0)}_{n,-1,-1} = B^{(0)}_{n,1,-1} = B^{(0)}_{n,-1,1} = B^{(0)}_{n,1,1} = \frac{1}{4}a_1a_2$,

$B^{(0)}_{n,\mu_1,\mu_2} = 0$ otherwise,

$B^{(1)}_{n,0,0} = (1-a_1)(1-a_2)$, $B^{(1)}_{n,-1,0} = B^{(1)}_{n,1,0} = \frac{1}{2}a_1(1-a_2)$

$B^{(1)}_{n,0,-1} = B^{(1)}_{n,0,1} = \frac{1}{2}a_2(1-a_1)$,

$B^{(1)}_{n,-1,-1} = B^{(1)}_{n,1,-1} = B^{(1)}_{n,-1,1} = B^{(1)}_{n,1,1} = \frac{1}{4}a_1a_2$,

$B^{(1)}_{n,\mu_1,\mu_2} = 0$ otherwise.

ADI Method (16) of *Peaceman-Rachford:* This method is a so-called "product method" where the associated finite-difference operators are given by

$$C_n(t) = (C_n^{(0),2})^{-1}C_n^{(1),2}(C_n^{(0),1})^{-1}C_n^{(1),1}, \quad t \in [0,T]_n', \quad n \in I.$$

The $C_n^{(\ell),\nu}$, $\nu = 1,2$, $\ell = 0,1$, are defined by

$$C_n^{(0),1} = \sum_{|\mu|\le 1} B^{(0),1}_{n,\mu}T_n^{\mu,0}, \quad C_n^{(1),1} = \sum_{|\mu|\le 1} B^{(1),1}_{n,\mu}T_n^{0,\mu},$$

where

$$B_{n,0}^{(0),1} = 1 + a_1, \quad B_{n,-1}^{(0),1} = B_{n,1}^{(0),1} = -\frac{1}{2} a_1,$$
$$B_{n,0}^{(1),1} = 1 - a_2, \quad B_{n,-1}^{(1),1} = B_{n,1}^{(1),1} = \frac{1}{2} a_2,$$

and

$$C_n^{(0),2} = \sum_{|\mu|\le 1} B_{n,\mu}^{(0),2} T_n^{0,\mu}, \quad C_n^{(1),2} = \sum_{|\mu|\le 1} B_{n,\mu}^{(1),2} T_n^{\mu,0},$$

with

$$B_{n,0}^{(0),2} = 1 + a_2, \quad B_{n,-1}^{(0),2} = B_{n,1}^{(0),2} = -\frac{1}{2} a_2,$$
$$B_{n,0}^{(1),2} = 1 - a_1, \quad B_{n,-1}^{(1),2} = B_{n,1}^{(1),2} = \frac{1}{2} a_1.$$

The *Wave Equation* (29a) (or (31)): (always $w_n = 0$)

Friedrichs' Method (33): $\iota = 2$, $N_0 = 0$, $N_1 = 1$,

$$B_{n,0}^{(0)} = I, \quad B_{n,0}^{(1)} = 0, \quad B_{n,\pm 1}^{(1)} = \frac{1}{2}(I \pm \lambda A);$$

Courant-Isaacson-Rees Method (34): $\iota = 2$, $N_0 = 0$, $N_1 = 1$,

$$B_{n,0}^{(0)} = I, \quad B_{n,0}^{(1)} = (1 - c\lambda) I, \quad B_{n,-1}^{(1)} = -\lambda A^-, \quad B_{n,1}^{(1)} = \lambda A^+;$$

Lax-Wendroff Method (36): $\iota = 2$, $N_0 = 0$, $N_1 = 1$,

$$B_{n,0}^{(0)} = I, \quad B_{n,0}^{(1)} = I - \lambda^2 A^2, \quad B_{n,-1}^{(1)} = -\frac{1}{2}\lambda A(I - \lambda A),$$
$$B_{n,1}^{(1)} = \frac{1}{2}\lambda A(I + \lambda A);$$

Discrete-Time Galerkin Method (40) (as a 2-level method): $\iota = 2$, $N_0 = N_1 = 1$,

$$B_{n,-1}^{(0)} = B_{n,1}^{(0)} = \begin{pmatrix} \frac{1}{6} - \frac{1}{4}\lambda^2 c^2 & 0 \\ 0 & 0 \end{pmatrix}, \quad B_{n,0}^{(0)} = \begin{pmatrix} \frac{2}{3} + \frac{1}{2}\lambda^2 c^2 & 0 \\ 0 & 1 \end{pmatrix},$$

$$B_{n,-1}^{(1)} = B_{n,1}^{(1)} = \begin{pmatrix} \frac{1}{3} + \frac{1}{2}\lambda^2 c^2 & -(\frac{1}{6} - \frac{1}{4}\lambda^2 c^2) \\ 0 & 0 \end{pmatrix},$$

$$B_{n,0}^{(1)} = \begin{pmatrix} \frac{4}{3} - \lambda^2 c^2 & -(\frac{2}{3} + \frac{1}{2}\lambda^2 c^2) \\ 1 & 0 \end{pmatrix};$$

Discrete-Time Galerkin Method (46) (for the scalar hyperbolic equation (42) with constant c): $\iota = 1$, $N_0 = N_1 = J$,

$$B_{n,-1}^{(0)} = \frac{1}{6} + \frac{1}{4}\lambda c,\quad B_{n,0}^{(0)} = \frac{2}{3},\quad B_{n,1}^{(0)} = \frac{1}{6} - \frac{1}{4}\lambda c,\quad B_{n,j}^{(0)} = 0 \text{ otherwise.}$$

$$B_{n,-1}^{(1)} = \frac{1}{6} - \frac{1}{4}\lambda c,\quad B_{n,0}^{(1)} = \frac{2}{3},\quad B_{n,1}^{(1)} = \frac{1}{6} + \frac{1}{4}\lambda c,\quad B_{n,j}^{(1)} = 0 \text{ otherwise,}$$

$$w_n(x_j,t_k) = \frac{1}{h_n}(s^{k+1/2},\phi_j)_0.$$

(These definitions are to be understood for all mesh points $x_j = jh_n$ where $j \equiv \nu(\text{mod } J)$.)

For the nonlinear finite-difference methods introduced in the previous section, we can define the associated finite-difference operators in an analogous way. For example, we have:

Nonlinear Crank-Nicolson Method (48) (for approximating the semilinear parabolic IVP (47)):

$$\begin{aligned}(C_n^{(0)}(t_{k+1})g_n)(x_j) &= (1 + 2\Theta ra(x_j,t_{k+1}))g_n(x_j)\\ &\quad - \Theta ra(x_j,t_{k+1})(g_n(x_{j-1}) + g_n(x_{j+1}))\\ &\quad - \Theta\tau_n F(x_j,t_{k+1},g_n(x_j),(g_n(x_{j+1}) - g_n(x_{j-1}))/(2h_n)),\\ (C_n^{(1)}(t_{k+1})g_n)(x_j) &= (1 - 2(1-\Theta)ra(x_j,t_k))g_n(x_j)\\ &\quad + (1-\Theta)ra(x_j,t_k)(g_n(x_{j-1}) + g_n(x_{j+1}))\\ &\quad + (1-\Theta)\tau_n F(x_j,t_k,g_n(x_j),(g_n(x_{j+1}) - g_n(x_{j-1}))/(2h_n)),\end{aligned}$$

$$w_n = 0.$$

It is clear how we should define nonlinear finite-difference operators associated with method (53), with the nonlinear Friedrichs method, with the Courant-Isaacson-Rees method, and with the Lax-Wendroff method.

We now come to the Galerkin methods which, in case of continuous, piecewise linear trial and test functions, can be viewed as finite-difference methods. For the linear examples, the approximating operators $C_n^{(\ell)}(t)$, defined via the associated variational formulation, are mappings of each E_n into itself where the E_n are finite-dimensional subspaces of an underlying Hilbert space E. The functions $w_n(t) \in E_n$ are projections of the given inhomogeneous right-hand side. With the L^2- and H^1-scalar products - $(.,.)_0$ and $(.,.)_1$, respectively - we get the following description of the Galerkin methods for the examples we consider:

Crank-Nicolson-Galerkin Method (21) (for the heat equation):
$E = H_0^1(0,1)$,

$$(C_n^{(0)}(t)g_n,\phi_n)_1 = (g_n,\phi_n)_0 + \frac{1}{2}\tau_n a(g_n,\phi_n),$$

$$(C_n^{(1)}(t)g_n,\phi_n)_1 = (g_n,\phi_n)_0 - \frac{1}{2}\tau_n a(g_n,\phi_n), \quad g_n,\phi_n \in E_n,$$

$$(w_n(t_{k+1}),\phi_n)_1 = (s^{k+1/2},\phi_n)_0, \quad \phi_n \in E_n.$$

Discrete-Time Galerkin Method (45) (for the scalar hyperbolic IVP (42)): $E = H$ (isomorph of $H_\#^1$)

$$(C_n^{(0)}(t)g_n,\phi_n)_0 = (g_n - \frac{a}{2}\tau_n g_n',\phi_n)_0,$$

$$(C_n^{(1)}(t)g_n,\phi_n)_0 = (g_n + \frac{a}{2}\tau_n g_n',\phi_n)_0, \quad g_n,\phi_n \in E_n,$$

$$(w_n(t_{k+1}),\phi_n)_0 = (s^{k+1/2},\phi_n)_0, \quad \phi_n \in E_n.$$

For the nonlinear Galerkin methods, the approximating operators are to be defined analogously. We shall display these operators when we analyze the methods in Chapter 11 (see, in particular, Section 11.1).

Finally, we transform the general approximating equation (66) further by defining $X_n \equiv C([0,T]_n,E_n)$, $Y_n \equiv E_n \times C([0,T]_n',F_n)$ and

$$T_n v_n \equiv (v_n(0),T_n^1 v_n),$$
$$(T_n^1 v_n)(t) \equiv \frac{1}{\tau_n}\{C_n^{(0)}(t)v_n(t)-C_n^{(1)}(t)v_n(t-\tau_n)\}, \quad t \in [0,T]_n', \ n \in I. \tag{71}$$

The domain of definition $D(T_n)$ of T_n includes all $v_n \in X_n$ for which $v_n(t) \in D_n^{(0)} \cap D_n^{(1)}$, $t \in [0,T]_n$. With this notation, (66) is equivalent to

$$u_n(0) = u_{n,0}, \quad (T_n^1 u_n)(t) = w_n(t), \text{ or to } T_n u_n = (u_{n,0},w_n). \tag{72}$$

REFERENCES

Ames (1977), Ansorge (1978), Ansorge & Haas (1970), Aubin (1972,1979), Babuska & Aziz (1972)*, Babuska, Prager & Vitasek (1966), Baker (1976)*, Böhmer (1974), Collatz (1966), Dupont (1973)*, Fairweather (1978), Forsythe & Wasow (1960), Gladwell & Wait (1979), Gottlieb & Orszag (1977), Hartman (1964), Hille & Phillips (1957), van der Houwen (1968,1977),

*Article

Isaacson & Keller (1966), John (1967,1982), Kantorovich & Akilov (1964), Kato (1966), Lions & Magenes (1972), Marchuk (1975), Meis & Marcowitz (1981), Mikhlin & Smolitskiy (1967), Mitchell (1969), Mitchell & Griffiths (1980), Mitchell & Wait (1977), Ortega & Rheinboldt (1970), Parter (1980)*, Richtmyer & Morton (1967), Thomée & Wahlbin (1975)*, Törnig (1979), Trotter (1958)*, Varga (1962), Walter (1976), Yanenko (1971).

*Article

Part II
Convergence Theory

The following abstract setting underpins the special problems we have considered up to now. To begin with, suppose a given problem can be formulated as an operator equation

$$Au = w,$$

with A, a mapping between suitable spaces E and F. Then, suppose we can construct approximate solutions, which again are expressed as solutions of operator equations

$$A_n u_n = w_n, \quad n \in \mathbb{N},$$

where A_n, $n \in \mathbb{N}$, represents a sequence of mappings between spaces E_n and F_n.

The concrete settings arising in the problems considered in Part I of this book can be depicted graphically by the following typical figure,

$$\begin{array}{ccc} E & \xrightarrow{\;A\;} & F \\ {\scriptstyle R_n^E}\downarrow & & \downarrow{\scriptstyle R_n^F} \\ E_n & \xrightarrow{\;A_n\;} & F_n \end{array}$$

E_n and F_n are either subspaces of E and F, respectively, e.g., in case of projection methods where R_n^E and R_n^F can be chosen as projection operators (cf. Chapter 2 and Section 3.3). Or, E_n and F_n are spaces of grid functions where, obviously, no subspace situation is present, and

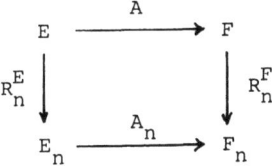

R_n^E and R_n^F are essentially pointwise restrictions (cf. Chapter 1 and Section 3.2). Moreover, a more complicated situation arises in case of initial value problems where we have termed the spaces X, X_n, Y, Y_n and the corresponding R_n^X, R_n^Y will be specified in Chapter 11.

The solvability of the given problem and that of the approximate problems have already been discussed for the most part. Now, three fundamental questions arise:

(i) How do the spaces E_n and F_n, $n \in \mathbb{N}$, approximate the spaces E and F, respectively?

(ii) In what sense does the sequence A_n, $n \in \mathbb{N}$, represent an approximation to A?

And, most importantly,

(iii) In what sense are the elements of the sequence u_n, $n \in \mathbb{N}$, approximations to the solution u of the given problem?

To address these questions, we first define a concept of convergence, which is meaningful for all the problems and approximation methods that we have considered up to now. This aim can be achieved by the concept of "discrete convergence", which is introduced in Chapter 5 and illustrated by means of a series of examples. In Chapter 6, we develop the salient theorems on the discrete convergence of mappings and solutions to linear and nonlinear operator equations. In Chapter 7, we prove sufficient and in some cases - necessary convergence criteria formulated in terms of compactness properties associated with the sequence of operators A_n, $n \in \mathbb{N}$.

In order to make the study of the material presented in this part of the book easier, we want to point out that only Section 5.1 from Chapter 5 is needed for the stability and convergence theory developed in Chapters 6 and 7. The other sections in Chapter 5 treat examples of discrete approximations - and verify the corresponding properties - which, however, present the underlying framework for the problem areas analyzed in Parts III and IV of this book. For a first understanding of the theory of discrete approximations and discrete convergence, we therefore suggest to study Section 5.1 and the rather general, but trivial examples in Section 5.2 and skip Sections 5.3 and 5.4 for the time being.

Chapter 5
The Concepts of Discrete Convergence and Discrete Approximations

The concepts of discrete convergence and discrete approximation make the idea of approximating elements or spaces by sequences of elements or spaces more precise in a general setting. These concepts are introduced in Section 5.1, and explained by means of introductory examples. In Section 5.2, we show how discrete approximations are constructed by restriction and imbedding operators. On the other hand, we demonstrate that restriction operators can always be defined from a given discrete approximation. In Section 5.3, we illustrate these concepts by studying bounded continuous functions defined on a domain G and associated discrete approximations by sequences of functions defined on perturbations of G. The properties of a discrete approximation are then characterized by conditions on the perturbed domains of definition. Since the supremum norm is the underlying norm, we can also term this convergence "discretely uniform convergence", but we wish to point out that this convergence is nothing other than discrete convergence in the sense of Section 5.1 with respect to a particular norm. In Section 5.4, we explore the concept of discrete convergence further by considering the example of discrete approximations of L^p-spaces over perturbed regions of integration in $\mathbb{R}^d$, $d \in \mathbb{N}$. The concept of discrete approximation can be characterized in this setting by conditions on the regions of integration, or, equivalently, by the weak convergence of the Lebesgue measures associated with each (perturbed) region of integration. We explore the latter condition in some detail by providing results on the weak convergence of arbitrary measures supported on subsets of metric spaces. We then consider as an application further discrete approximations of L^p-spaces and of spaces of continuous functions, where the discrete convergence is explained with the help of convergent quadrature formulas. The weak convergence of measures is closely connected with the concept of discrete convergence of mappings studied in Section 6.3, and is needed in Chapter 10 for the convergence analysis of approximation methods for integral equations.

The concept of discrete convergence was developed by Stummel (1970) for linear mappings defined on normed linear spaces, and generalized later by him (1973a) to include mappings defined on arbitrary sets. By using this concept, Stummel was able to obtain a generalization of the numerous approximation schemes presented in the open literature (cf. the literature cited at the end of this chapter), and thereby was able to develop a unifying perspective for considering approximation methods for operator equations. The material in this chapter is principally taken from Stummel's work, but includes also results from the work of R. D. Grigorieff and the author. For the sake of brevity, we shall omit the proofs at a few places in the text and refer the reader to the original literature instead.

5.1. DEFINITIONS, BASIC PROPERTIES, AND FIRST EXAMPLES

The following, general mathematical setting is basic for introducing the concepts of discrete convergence and discrete approximation. Let $I = \mathbb{N}$ be the set of positive integers and let subsequences of integers from I be denoted by I', I'' etc. Suppose, moreover, that we are given normed linear spaces E, respectively E_n, $n \in I$, whose norms are expressed as $||\cdot||$, respectively $||\cdot||_n$, $n \in I$. A mapping lim is called a *discrete convergence,* if it maps a nonempty subset of the set of all sequences of elements $u_n \in E_n$, $n \in I$, into E and possesses the additional

Property 5.1. For every pair of sequences (u_n), (v_n), the following condition is true:

$$\lim(u_n) = u \text{ and } ||u_n - v_n||_n \to 0 \Longleftrightarrow \lim(u_n) = \lim(v_n). \tag{1}$$

We write, moreover, $\lim u_n = u$ or $u_n \to u(n \in I)$, and say that the element u is the (*discrete*) *limit* of the sequence (u_n). As a mapping, lim is well-defined, a fact which directly implies the uniqueness of the limit, i.e.,

$$u_n \to u, \quad u_n \to v \quad (n \in I) \Rightarrow u = v.$$

One direction ($\Rightarrow$) in (1) furnishes a test for $\lim v_n = u$. The reverse implication means that sequences with equal limits necessarily satisfy the condition that $||u_n - v_n||_n \to 0$ $(n \in I)$.

The set of all sequences is designated by $\pi_n E_n$. The set of images of lim, R(lim), consists of all elements from E which are limits of sequences $(u_n) \in \pi_n E_n$. If the mapping lim is surjective, i.e., R(lim) = E, then the triple $E, \pi_n E_n$, lim is called a *discrete approximation* and is written as $A(E, \pi_n E_n, \lim)$.

We provide now several simple examples of discrete approximations.

Example 1. Let E be a normed linear space and let $E_n = E$, $n \in I$. As discrete convergence, we take the usual convergence in E, i.e.,

$$u_n \to u \ (n \in I) \iff ||u_n - u|| \to 0 \quad (n \in I). \tag{2}$$

Then the above properties are obviously satisfied, and the discrete convergence is surjective. With the concept of discrete convergence, we have thus found an extension of the concept of usual convergence in normed spaces. □

Example 2. Suppose we are given subspaces $E_0, E_n \subset E$, $n \in I$, and suppose, moreover, the discrete convergence is defined precisely as in (2). Then we again obtain a well-defined mapping which satisfies (1). The surjectivity of lim as a mapping onto E_0, and the existence of a discrete approximation $A(E_0, \pi_n E_n, \lim)$ will follow if, and only if,

$$|u, E_n| \equiv \inf\{||u - u_n||: u_n \in E_n\} \to 0 \quad (n \in I), \quad u \in E_0. \tag{3}$$

If (3) is true, then, by definition of the infimum, we have for every $u \in E_0$ a sequence $u_n \in E_n$, $n \in I$, satisfying (2). If, conversely, the surjectivity of the limit mapping is satisfied, then we easily conclude from $u_n \to u$ $(n \in I)$, that

$$|u, E_n| \le ||u - u_n|| \to 0 \ (n \in I), \quad u \in E_0.$$

Condition (3) is satisfied, for example, if the subspaces E_n, $n \in I$, form an increasing sequence of subspaces and their union Φ is dense relative to E_0, i.e.,

$$E_1 \subset E_2 \subset \ldots \subset \Phi \equiv \bigcup_{n \in I} E_n \quad \text{and} \quad E_0 \subset \overline{\Phi}.$$

In other words, there is to each $u \in E_0$ and each $\varepsilon > 0$, a $\phi \in \Phi$ such that $||u - \phi|| \le \varepsilon$, and accordingly a number $\nu \in I$ so that $\phi \in E_n$ for all $n \ge \nu$. Thus it follows that

$$|u, E_n| \le ||u - \phi|| + |\phi, E_n| \le \varepsilon, \quad n \ge \nu. \qquad □$$

Example 3. The case of projection methods can be classified as a special case of Example 2. Condition (3) is obviously satisfied for a sequence of subspaces $E_n \subset E$, $n = 0,1,2,\ldots$ and associated projection operators $P_n: E \to E_n$, $n = 0,1,2,\ldots$, in the case that $(P_n)_{n\in I}$ converges strongly to P_0, i.e.

$$||P_n u - P_0 u|| \to 0 \ (n \in I), \quad u \in E. \tag{4}$$

Whenever $E_n \subset E_0$, $n \in I$, and the sequences of projection operators $P_n: E_0 \to E_n$, $n \in I$, is uniformly bounded - i.e. there exists a $\beta \geq 0$ so that $||P_n u|| \leq \beta||u||$, $n \in I$, $u \in E_0$ - then (3) is even equivalent to

$$||P_n v - v|| \to 0 \ (n \in I), \quad v \in E_0. \tag{5}$$

The necessity of (5) is apparent since, as in Example 2, we have the existence of a sequence $u_n \in E_n$, $n \in I$, for each $v \in E_0$ such that $||u_n - v|| \to 0$ $(n \in I)$. Together with $P_n u_n = u_n$, $n \in I$, we deduce that

$$||P_n v - v|| \leq ||P_n(v-u_n)|| + ||u_n - v|| \leq (1+\beta)||u_n - v|| \to 0 \ (n \in I).$$

Finally, we wish to point out the important principle of uniform boundedness, which implies that a strongly convergent sequence of bounded linear mappings on a Banach space E is necessarily uniformly bounded. □

The cases of discrete convergences and discrete approximations that we have considered in the above examples have still other interesting features. For one thing, a linear combination of sequences converges to the corresponding linear combination of the limits. In this case, we call the discrete convergence *asymptotically linear*, i.e., it possesses

Property 5.2. From $u_n \to u$, $v_n \to v$ $(n \in I)$, $\alpha,\beta \in \mathbb{K}$ $(= \mathbb{R}$ or $\mathbb{C})$ follows

$$\alpha u_n + \beta v_n \to \alpha u + \beta v \quad (n \in I). \tag{6}$$

The case of an asymptotically linear discrete convergence yields the following characterization of discrete convergence of null sequences:

$$u_n \to 0 \Longleftrightarrow ||u_n||_n \to 0 \ (n \in I). \tag{7}$$

To see this, we deduce from (6) that the null elements $0_n \in E_n$, $n \in I$, converge discretely to $0 \in E$. The equivalence of the two statements in (7) follows from Property 5.1 by an appropriate assignment of values to u_n and v_n.

Another important feature of discrete convergences, which is satisfied in the above examples, is the following *convergence property* of the norms:

Property 5.3. If $u_n \to u$ $(n \in I)$, then $||u_n||_n \to ||u||$ $(n \in I)$.

When the discrete convergence is asymptotically linear and the norms satisfy Property 5.3, the following criterion is useful for deducing the discrete convergence of a sequence of elements.

Lemma 5.4. Under the Properties 5.2 and 5.3, the discrete convergence of u_n to u occurs if, and only if, (i) $u \in R(\lim)$ and (ii) there is, for each $\varepsilon > 0$, an element $v \in E$ and a sequence $v_n \to v$ $(n \in I)$ with the properties

$$||u - v|| \leq \varepsilon, \quad \limsup_{n \in I} ||u_n - v_n||_n \leq \varepsilon. \tag{8}$$

Proof: If, indeed, $u_n \to u$ $(n \in I)$, then u is obviously in $R(\lim)$, and the condition of the lemma trivially follows with $v = u$, $v_n = u_n$, $n \in I$. Conversely, let $u_n' \in E_n$, $n \in I$, be a sequence which discretely converges to u, i.e., $u_n' \to u$ $(n \in I)$. Such a sequence exists by virtue of the hypothesis that $u \in R(\lim)$. For every $\varepsilon > 0$, Properties 5.2 and 5.3 applied to the given sequence $v_n \to v$ $(n \in I)$ allows us to conclude that

$$\limsup_{n \in I} ||u_n - u_n'||_n \leq \limsup_{n \in I} (||u_n - v_n||_n + ||v_n - u_n'||_n) \leq \varepsilon + ||u - v|| \leq 2\varepsilon.$$

Therefore the convergence $||u_n - u_n'||_n \to 0$ $(n \in I)$ holds, which is tantamount to the assertion that $u_n \to u$ $(n \in I)$, because of Property 5.1. The latter holds per definition for a discrete convergence. □

5.2. RESTRICTION AND IMBEDDING OPERATORS

Frequently, discrete approximations are defined by sequences of restriction operators, and we now proceed to define this important class of operators. Toward this end, let E be a normed linear space and E_n, $n \in I$, other normed spaces with (not necessarily linear) mappings $R_n: E \to E_n$, $n \in I$. Then the sequence (R_n) is called a *sequence of restriction operators* in case the following relation holds:

$$||R_n u - R_n v||_n \to 0 \ (n \in I) \Rightarrow u = v, \quad u, v \in E. \tag{9}$$

A discrete convergence is then obtained by defining

$$u_n \to u \ (n \in I) \Leftrightarrow ||u_n - R_n u||_n \to 0 \ (n \in I), \tag{10}$$

and hence $A(E, \pi_n E_n, \lim)$ is a discrete approximation. By virtue of condition (9), the limits are unique, and Property 5.1 will follow from the triangle inequality. Surjectivity is trivial for (10). The additional

Property 5.2 of asymptotic linearity is here equivalent to the condition

$$||\alpha R_n u + \beta R_n v - R_n(\alpha u + \beta v)||_n \to 0 \ (n \in I). \tag{11}$$

For the special case when R_n, $n \in I$, is a sequence of linear mappings, this condition is always satisfied, and moreover (9) is equivalent to

$$||R_n u||_n \to 0 \ (n \in I) \Rightarrow u = 0, \quad u \in E. \tag{12}$$

For the case of restriction operators, Property 5.3 of the discrete convergence of the norms is equivalent to

$$||R_n u||_n \to ||u|| \ (n \in I), \ u \in E. \tag{13}$$

From time to time in our analysis, we shall require a stability property for the sequence (R_n). A sequence of restriction operators $R_n: E \to E_n$, $n \in I$, is called *stable*, if, for every sequence $u^{(n)}$, $n = 0,1,2,\ldots$, from E, we have

$$||u^{(n)} - u^{(0)}|| \to 0 \Rightarrow ||R_n u^{(n)} - R_n u^{(0)}||_n \to 0 \quad (n \to \infty). \tag{14}$$

For linear restriction operators, property (14) holds if, and only if, R_n is bounded for almost all n and $\limsup_{n \in I} ||R_n|| < \infty$ (cf. Lemma 6.9 in Section 6.2).

The examples from Section 5.1 also provide us with examples of restriction operators giving discrete approximations $A(E_0, \pi_n E_n, \lim)$ when E_0, E_n, $n \in I$, are subspaces of E.

<u>Example 1</u> (cf. Example 2 in Section 5.1): For the case where $E_0, E_n \subset E$, $n \in I$, we define by means of the norm convergence in E (see (2)) a discrete convergence, which, in addition, is surjective whenever $|u, E_n| \to 0$ $(n \in I)$, $u \in E_0$ (see (3)). In this setting, we are always able to define a sequence of restriction operators $r_n: E_0 \to E_n$, $n \in I$. To see this, let $(\varepsilon_n)_{n \in I}$ be a null sequence of positive numbers; for each $u \in E_0$, and additionally, for each $n \in I$, choose elements $v_n \in E_n$ so that $||u - v_n|| \le |u, E_n| + \varepsilon_n$, $n \in I$. Then a sequence of restriction operators r_n is defined by $r_n u = v_n$, $n \in I$, since relation (9) is a consequence of

$$||u - r_n u|| \le |u, E_n| + \varepsilon_n \to 0 \ (n \in I). \tag{15}$$

The convergence of sequences $u_n \in E_n$, $n \in I$, to an element $u \in E_0$ with respect to the norm in E is clearly equivalent to the convergence of the sequence of positive numbers $||u_n - r_n u||$ to zero. The restriction operators thereby defined have the properties (11) and (13), since the usual norm convergence will trivially possess the feature of asymptotic linearity and satisfy the Convergence Property 5.3. □

Example 2 (cf. Example 3 in 5.1): By means of a strongly convergent sequence of projection operators (cf. (4)), a special sequence of restriction operators is defined by $R_n u = P_n u$, $u \in E_0$, $n \in I$. Property (9) follows from the strong convergence, and properties (11) and (13) are trivially satisfied. Bounded linear projection operators form a stable sequence of restriction operators (in the sense of (14)), if, and only if, they are uniformly bounded. □

These examples have already indicated that a sequence of restriction operators (R_n) can always be defined whenever a discrete approximation $A(E,\pi_n E_n,\lim)$ exists. To show that this is true in general, we only need to choose for every $u \in E$ an arbitrary, but fixed, sequence $u_n' \to u$ $(n \in I)$, and define R_n by

$$R_n u = u_n', \quad n \in I, \ u \in E.$$

From Property 5.1 we know that the discrete convergence of a sequence is equivalent to

$$u_n \to u \ (n \in I) \Longleftrightarrow ||u_n - R_n u||_n \to 0 \ (n \in I).$$

Moreover the conditions (11) and (13) are equivalent to Property 5.2 and 5.3, respectively. The latter equivalence indicates that in order to deduce the discrete convergence of the norms we need only show the convergence of $||u_n'||_n$ to $||u||$ for a single sequence u_n', $n \in I$, converging discretely to u, because then convergence of the norms occurs automatically for all sequences converging discretely to u.

We wish to point out that we just have defined a sequence of restriction operators for a given discrete approximation, whereas before we have obtained discrete approximations with the aid of restriction operators. We now turn to a special problem occurring in the construction of discrete approximations. For certain examples, the interesting question arises of whether a discrete approximation, constructed by using restriction operators defined on a dense subspace, can be defined on the whole space via suitable extensions of the restriction operators. The reader should think of the example of the dense subspace $C_0^\infty(a,b)$ in $L^2(a,b)$, for which (pointwise) restrictions to mesh points in (a,b) can be defined in a natural manner (cf. the following Section 5.3). To answer this question, we first establish that a discrete convergence with convergent norms, defined by means of restriction operators, is uniquely determined by specifying sequences converging to elements belonging to a dense subspace D in E. Our uniqueness theorem is formulated as follows:

Theorem 5.5. Let $R_n: E \to E_n$, $r_n: E \to E_n$, $n \in I$, be two sequences of restriction operators which satisfy (11) and (13), and let $A(E,\pi_n E_n,\lim^R)$, respectively $A(E,\pi_n E_n,\lim^r)$, be the associated discrete approximations. If there exists a dense subset $D \subset E$ so that for all $\phi \in D$, $\phi_n \in E_n$, $n \in I$,

$$\|\phi_n - R_n\phi\|_n \to 0 \Longleftrightarrow \|\phi_n - r_n\phi\|_n \to 0 \ (n \in I), \tag{16}$$

then $\lim^r u_n = u$ if, and only if, $\lim^R u_n = u$ for all $u \in E$, $u_n \in E_n$, $n \in I$.

Proof: We show that, for arbitrary $u \in E$, $u_n \in E_n$, $n \in I$, $\lim^r u_n = u$ implies $\lim^R u_n = u$. For every $\varepsilon > 0$, there exists a $v \in D$ so that $\|u - v\| \le \varepsilon$. The convergence of the norm along with property (11) allows us to conclude that

$$\lim_{n\in I} \|u_n - r_n v\|_n = \|u - v\| \le \varepsilon.$$

By hypothesis (with $\phi_n = r_n v$), we have $\|r_n v - R_n v\|_n \to 0$ $(n \in I)$, so that $\lim^R r_n v = v$ is also true. Now, Lemma 5.4 applied to $\lim^R$ (with $v_n = r_n v$) yields the assertion $\lim^R u_n = u$. □

Condition (16) indicates that the sets of the sequences, which converge discretely to elements in D relative to $\lim^r$ and $\lim^R$, agree with each other. We could have also considered in the last theorem restriction operators $r_n: D \to E_n$, $n \in I$, defined only on dense subsets D of E. Then we would be able to conclude analogously that $\lim^R$ is determined uniquely by $\lim^r$, in case (16) is valid. According to Lemma 5.4, discretely convergent sequences (relative to $\lim^R$) can be also characterized by sequences convergent relative to $\lim^r$ as follows (cf. Stummel (1970), 4.1.(6)):

$$\lim^R u_n = u \Longleftrightarrow \forall\varepsilon > 0,\ \exists\phi \in D \ \ni: \|u - \phi\| \le \varepsilon, \quad \limsup_{n\in I} \|u_n - r_n\phi\|_n \le \varepsilon. \tag{17}$$

We are now in a position to extend a discrete approximation $A(D,\pi_n E_n,\lim^D)$ of a dense subspace $D \subset E$ to $A(E,\pi_n E_n,\lim^E)$ by defining $\lim^E$ as in (17) (with $\lim^E = \lim^R$). As mentioned above, the extension is unique. We formulate the existence theorem for uniquely determined extensions in the following way.

Theorem 5.6. Let $A(D,\pi_n E_n,\lim^D)$ be a discrete approximation of a dense subspace D in E, which is defined by a sequence of restriction operators $r_n: D \to E_n$, $n \in I$, and satisfies properties (11) and (13). Then there exists a uniquely determined discrete approximation $A(E,\pi_n E_n,\lim^E)$ of the normed space E with the Properties 5.2 and 5.3 which is characterized by (17) (with $\lim^E = \lim^R$) and $\lim^E u_n = \phi$ if, and only if, $\lim^D u_n = \phi$ for all $\phi \in D$, $u_n \in E_n$, $n \in I$. □

We shall not provide the somewhat long, but elementary proof from Stummel (1970), 4.1(9) but refer the reader to the original literature. At this point, we should remark that the hypothesized presence of restriction operators represents no restriction in the last two theorems. As indicated in the paragraphs following Example 2, these operators are indeed always present with discrete approximations.

Finally, we can also construct discrete approximations by using so-called imbedding operators. We label a sequence of mappings $J_n: E_n \to E$, $n \in I$, between normed spaces E,E_n, $n \in I$, a *sequence of imbedding operators* in case the following relation

$$||u_n - v_n||_n \to 0 \iff ||J_n u_n - J_n v_n|| \to 0 \ (n \in I) \tag{18}$$

is satisfied for each pair of sequences $(u_n),(v_n) \in \pi_n E_n$. Then a discrete convergence can be defined by

$$u_n \to u \ (n \in I) \iff ||J_n u_n - u|| \to 0 \ (n \in I). \tag{19}$$

Condition (19) ensures primarily the uniqueness of discrete limits; in conjunction with (18), (19) yields the equivalent statements of Property 5.1 of the definition of a discrete convergence. In order to insure that a discrete convergence so defined has a nonempty domain of definition, we always tacitly assume, in connection with imbeddings, that at least one sequence (u_n) will exist for which $(J_n u_n)$ converges.

Condition (18) is satisfied, for example, if numbers $\alpha_1 > 0$, $\alpha_2 > 0$, $\nu \geq 0$ exist with the property

$$\alpha_1 ||u_n - v_n||_n \leq ||J_n u_n - J_n v_n|| \leq \alpha_2 ||u_n - v_n||_n, \qquad u_n,v_n \in E_n, \quad n \geq \nu. \tag{20}$$

The discrete convergence lim, defined by (19), is surjective and

$A(E,\pi_n E_n,\lim)$ is a discrete approximation if, in addition, mappings $R_n: E \to E_n$, $n \in I$, exist so that

$$||J_n R_n u - u|| \to 0 \ (n \in I), \quad u \in E. \tag{21}$$

If we are given injective, linear operators J_n, $n \in I$, then we can always define a discrete convergence according to (19) by introducing the following norms in E_n,

$$||u_n||_n \equiv ||J_n u_n||, \quad u_n \in E_n, \quad n \in I.$$

In this case, the discrete convergence is trivially asymptotically linear and the norms satisfy the Convergence Property 5.3.

Whenever $J_n: E_n \to E$, $n \in I$, are linear operators, we can even show that a discrete convergence with the Property 5.2 can be defined by (19) if, and only if, (J_n) represents a sequence of imbedding operators (i.e., if (18) is true) which additionally is equivalent to the two-sided inequality in (20) (with $v_n = 0_n$). These results are consequences of our further studies in Section 6.2 concerning the stability and inverse stability of linear mappings.

5.3. DISCRETE UNIFORM CONVERGENCE OF CONTINUOUS FUNCTIONS

As a further, nontrivial example, we consider the spaces $E = C(G)$, $E_n = C(G_n)$, $n \in I$, of bounded, continuous functions over sets G and G_n, $n \in I$, respectively, belonging to a metric space M (with metric denoted by $|.,.|$). We do not necessarily assume that G_n, $n = 1,2,\ldots$, are subsets of G. The spaces $C(G)$ and $C(G_n)$, $n \in I$, are endowed with the supremum norm, which we denote by $||\cdot||_{0,\infty}$ without any distinction. Using the notation of Sections 5.1 and 5.2, we have here

$$||\cdot|| = ||\cdot||_{0,\infty} \text{ in } E = C(G), \text{ and } ||\cdot||_n = ||\cdot||_{0,\infty} \text{ in } E_n = C(G_n), \quad n \in I.$$

We recall that, if G is a closed subset of M, each function $u \in C(G)$ can be extended to be a continuous function on all of M according to the Tietze-Urysohn Extension Theorem. We thus attempt to define a discrete convergence as follows:

$$u_n \to u \ (n \in I) \iff \exists \hat{u} \in C(M): \hat{u}|G = u \text{ and } \sup_{x \in G_n} |u_n(x) - \hat{u}(x)| \to 0 \quad (n \in I). \tag{22}$$

We first pose the question of whether the mapping defined in this manner is indeed a well-defined one, i.e., of whether the discrete limits are unique. Then the question comes to mind of whether the discrete convergence is surjective and finally of whether the norms converge. All these properties can be characterized by requirements on the subsets G, G_n, $n \in I$. To formulate these conditions, we need still further the concepts of the *closed limit superior* and the *closed limit inferior* of sets G_n, $n \in I$, defined by

$$\operatorname{Lim\,sup} G_n = \{x \in M: \liminf_{n\in I} |x,G_n| = 0\}, \tag{23a}$$

$$\operatorname{Lim\,inf} G_n = \{x \in M: \limsup_{n\in I} |x,G_n| = 0\}. \tag{23b}$$

The following inclusion is obvious:

$$\operatorname{Lim\,inf} G_n \subset \operatorname{Lim\,sup} G_n.$$

These sets are always closed. In the case where the closed limit inferior equals the closed limit superior, then (G_n) is said to be *convergent* to the set

$$G = \operatorname{Lim\,sup} G_n = \operatorname{Lim\,inf} G_n$$

and we hereby express $G = \operatorname{Lim} G_n$ for the *limit*. The elements of the set $\operatorname{Lim\,sup} G_n$, respectively $\operatorname{Lim\,inf} G_n$, are characterized by being limits of subsequences, respectively of entire sequences of elements from G_n,

$$x \in \operatorname{Lim\,sup} G_n \iff \exists (x_n) \in \pi_n G_n,\ \exists I' \subset I: |x,x_n| \to 0\ (n \in I') \tag{24a}$$

$$x \in \operatorname{Lim\,inf} G_n \iff \exists (x_n) \in \pi_n G_n: |x,x_n| \to 0\ (n \in I). \tag{24b}$$

The following lemma provides other equivalent characterizations of the concepts of closed limit superior and closed limit inferior of a sequence of sets G_n, $n \in I$. For the sake of brevity, we omit the proof, but refer the reader to Stummel (1973b, 1975).

Lemma 5.7. The following properties are equivalent for subsets G, G_n, $n \in I$, of a metric space M:

$$G \subset \operatorname{Lim\,sup} G_n \iff \forall v \in C(M): ||v|G|| \le \limsup_{n\in I} ||v|G_n||_n, \tag{25}$$

$$G \subset \operatorname{Lim\,inf} G_n \iff \forall v \in C(M): ||v|G|| \le \liminf_{n\in I} ||v|G_n||_n. \tag{26}$$

If M is compact, then

$$\operatorname{Lim\,sup} G_n \subset \overline{G} \Leftrightarrow \forall v \in C(M):\ \limsup_{n\in I} \|v|G_n\|_n \le \|v|G\|. \tag{27}$$

If M is compact, and G is, in addition, closed, then

$$\operatorname{Lim} G_n = G \Leftrightarrow \forall v \in C(M):\ \lim_{n\in I} \|v|G_n\|_n = \|v|G\|. \quad \square \tag{28}$$

Whenever we refer to the conditions (25),...,(28) in the following, we mean the relations between the regions which are expressed at the left of the equivalence signs. The property of a discrete convergence for the above definition (22) can be now characterized in an equivalent manner by relations between G and G_n, $n \in I$. First, however, we proceed to show under which conditions a well-defined mapping is defined by (22).

Lemma 5.8. Condition (25) is necessary and sufficient for defining a mapping lim from $\pi_n C(G_n)$ into $C(G)$ by (22). In the case G is closed, this mapping will then be surjective.

Proof: (i) Suppose (25) is valid. Then let $(u_n) \in \pi_n C(G_n)$, $u,v \in C(G)$ be such that extensions $\hat{u},\hat{v} \in C(M)$ exist for which

$$\hat{u}|G = u,\quad \hat{v}|G = v,\quad \|u_n - \hat{u}|G_n\|_n \to 0,\quad \|u_n - \hat{v}|G_n\|_n \to 0\ (n \in I).$$

For $w = \hat{u} - \hat{v} \in C(M)$, then

$$\|w|G_n\|_n \le \|u_n - \hat{u}|G_n\|_n + \|u_n - \hat{v}|G_n\|_n \to 0\ (n \in I).$$

Using the equivalence in (25) from Lemma 5.7, we conclude that $\|w|G\| = 0$, i.e., $u = v$.

(ii) Conversely, we assume that (25) is not true. Then there exists a point $z \in G$, a number $\rho > 0$ and an integer $\nu \in I$, such that $|z,G_n| \ge \rho$ for all $n \ge \nu$ (cf. Definition (23a)). The Tietze-Urysohn Extension Theorem furnishes a function $w \in C(M)$ with the properties

$$w(z) = 1,\quad w(x) = 0 \ \text{ for } \ |x - z| \ge \rho,\quad |w(x)| \le 1 \ \text{ in } \ M.$$

Thus $w|G_n = 0$ for all $n \ge \nu$, and

$$\|w|G_n - 0|G_n\|_n \to 0 \quad (n \in I).$$

There are therefore two functions, namely w and 0, of the type that - with $u_n = w|G_n$, $n \in I$ - the convergence criterion (at the right of the equivalence sign) in (22) is satisfied. By hypothesis, the equality $w|G = 0|G$ must be true, thereby contradicting $w(z) = 1$ and $z \in G$.

(iii) The surjectivity with closed G follows from the Tietze-Urysohn Extension Theorem because, with the extension $\hat{u} \in C(M)$ of $u \in C(G)$ and with $u_n = \hat{u}|G_n$, $n \in I$, we conclude that $u_n \to u$ $(n \in I)$ in the sense of (22). □

The following theorem provides a criterion for Property 5.1 to be satisfied. Then a discrete convergence will be defined by (22).

Theorem 5.9. The relation

$$\limsup_{n\in I} ||v|G_n||_n = ||v|G||, \quad v \in C(M), \tag{29}$$

is necessary and sufficient for a discrete convergence to be defined by (22). Under condition (29), the discrete convergence (22) is independent of the extension $\hat{u}$.

Proof: (i) Suppose (29) is true. Lemma 5.8, in connection with Lemma 5.7, insures us that a well-defined mapping is defined by (22). In order to prove Property 5.1, we let $(u_n), (v_n) \in \pi_n C(G_n)$ with $u_n \to u$ $(n \in I)$ and $||u_n - v_n||_n \to 0$ $(n \in I)$. Thus there exists a function $\hat{u} \in C(M)$ with $\hat{u}|G = u$ and $||u_n - \hat{u}|G_n||_n \to 0$ $(n \in I)$. For (v_n) we have,

$$||v_n - \hat{u}|G_n||_n \leq ||u_n - v_n||_n + ||u_n - \hat{u}|G_n||_n \to 0 \ (n \in I),$$

so that (v_n) also converges discretely to u, $u = \lim u_n = \lim v_n$. Conversely, let $\lim u_n = \lim v_n$, i.e., suppose there exist functions $\hat{u}, \hat{v} \in C(M)$ with $\hat{u}|G = u$, $\hat{v}|G = v$ and

$$||u_n - \hat{u}|G_n||_n \to 0, \quad ||v_n - \hat{v}|G_n||_n \to 0 \ (n \in I).$$

For $w = \hat{u} - \hat{v}$, we have $w|G = 0$ and, according to (29), $||w|G_n||_n \to 0$ $(n \in I)$. This shows

$$||u_n - v_n||_n \leq ||u_n - \hat{u}|G_n||_n + ||w|G_n||_n + ||\hat{v}|G_n - v_n||_n \to 0 \ (n \in I).$$

(ii) Suppose the discrete convergence lim is well defined by (22) and satisfies the equivalence relation (1) from Property 5.1. We already know from Lemma 5.8 along with Lemma 5.7 that "$\geq$" is valid in (29). In order to prove the reverse inequality, let $v \in C(M)$ and let $\varepsilon > 0$ be arbitrary. We define

$$U_\varepsilon = \{x \in M: |v(x)| < ||v|G|| + \varepsilon\}.$$

Then U_ε is an open neighborhood of $\overline{G}$. The Tietze-Urysohn Theorem furnishes a function $w \in C(M)$ with

$$w|\overline{G} = 0, \quad w|(M - U_\varepsilon) = 1, \quad 0 \leq w(x) \leq 1, \quad x \in M.$$

We easily observe that $w|G_n \to w|G$ $(n \in I)$ and that $0|G_n \to 0|G$ $(n \in I)$, and hence we can conclude

$$||w|G_n||_n \to 0 \ (n \in I)$$

by virtue of condition (1) and the fact that $w|G = 0$. Therefore there exists an index $\nu \in I$, so that

$$|w(x)| \leq ||w|G_n||_n < 1 \quad \text{for all} \quad x \in G_n, \ n \geq \nu.$$

By definition of w, $G_n \subset U_\varepsilon$ for all $n \geq \nu$; and, by construction of U_ε,

$$\limsup_{n \in I} ||v|G_n||_n \leq ||v|G|| + \varepsilon$$

follows. The reverse inequality is shown.

(iii) Let (u_n) be an arbitrary sequence discretely convergent to u, i.e., suppose there exists a $\hat{u} \in C(M)$ with $\hat{u}|G = u$ and $||u_n - \hat{u}|G_n||_n \to 0$ $(n \in I)$. Let $\hat{v}$ be any other continuation of u. With w defined by $w = \hat{u} - \hat{v}$, we have $w|G = 0$, and, by virtue of (29), we can conclude

$$\big|\, ||u_n - \hat{u}|G_n||_n - ||u_n - \hat{v}|G_n||_n \big| \leq ||w|G_n||_n \to 0 \ (n \in I).$$

Therefore we see that the discrete convergence is independent of the extension so selected. □

With the equivalence (27) from Lemma 5.7, we have immediately the following corollary to Theorem 5.9.

Theorem 5.10. Let M be a compact metric space and let $G \subset M$ be closed. Then the condition $\text{Lim sup } G_n = G$ is necessary and sufficient for a discrete convergence lim to be defined by (22) and for $A(C(G), \pi_n C(G_n), \lim)$ to represent a discrete approximation. □

By retaining the same assumptions, we can reformulate this theorem in such a manner that condition $\text{Lim sup } G_n = G$ is equivalent to generating a sequence of restriction operators via

$$r_n u = \hat{u}|G_n, \quad n \in I,$$

for every $u \in C(G)$ and every arbitrary extension $\hat{u} \in C(M)$. The discrete convergence (22) can then be expressed as

$$u_n \to u \iff ||u_n - r_n u||_n \to 0 \ (n \in I).$$

We easily see that, without further assumptions, Property 5.2 of asymptotic linearity is present with the discrete convergence (22). Finally, we wish to point out that the last equivalence (28) from Lemma 5.7 specifies, with $\operatorname{Lim} G_n = G$, an equivalent condition for the Convergence Property 5.3 of the norms.

Since (22) represents an extension of the concept of uniform convergence of continuous functions, the convergence described in (22) is also labeled as "discretely uniform convergence". It is, however, nothing other than the definition of discrete convergence given in the previous section with respect to special spaces and norms.

As an *example,* we consider now a compact interval $M = [a_0,b_0] \subset \mathbb{R}$ and a closed subinterval $G = [a,b]$ in M. The sets G_n are to be sets consisting of a finite number of points in M,

$$G_n = \{x_1^{(n)},x_2^{(n)},\ldots,x_N^{(n)}\} \subset M,\ n \in I, \text{ with } x_1^{(n)} \le x_2^{(n)} \le \ldots \le x_N^{(n)}.$$

The following theorem establishes the relationship to the closed limit inferior and closed limit superior. We shall designate $a_n = x_1^{(n)}$, $b_n = x_N^{(n)}$.

Theorem 5.11. Under the assumption

$$h_n \equiv \max_{1\le j\le N-1} |x_{j+1}^{(n)} - x_j^{(n)}| \to 0 \quad (n \to \infty) \tag{30}$$

we have the following equivalences:

$$[a,b] \subset \operatorname{Lim\,sup} G_n \iff \liminf_{n\in I} a_n \le a, \quad \limsup_{n\in I} b_n \ge b. \tag{31}$$

$$\operatorname{Lim\,sup} G_n \subset [a,b] \iff \liminf_{n\in I} a_n \ge a, \quad \limsup_{n\in I} b_n \le b. \tag{32}$$

$$\operatorname{Lim} G_n = [a,b] \iff \lim_{n\in I} a_n = a, \quad \lim_{n\in I} b_n = b. \tag{33}$$

Proof: (i) In case $[a,b] \subset \operatorname{Lim\,sup} G_n$, then according to definition (23a), there are subsequences $I',I'' \subset I$ with $x_n \in G_n$, $n \in I'$, and $y_n \in G_n$, $n \in I''$, such that $x_n \to a$ $(n \in I')$ and $y_n \to b$ $(n \in I'')$. For the sequences $\hat{x}_n = x_n$, $n \in I'$, $\hat{x}_n = a_n$, $n \in I-I'$, respectively $\hat{y}_n = y_n$, $n \in I''$, $y_n = b_n$, $n \in I-I''$, then we see that

$$\liminf_{n\in I} a_n \le \liminf_{n\in I} \hat{x}_n \le \lim_{n\in I'} x_n = a,$$

and

$$\limsup_{n\in I} b_n \geq \limsup_{n\in I} \hat{y}_n \geq \lim_{n\in I''} y_n = b,$$

respectively. For the converse, we determine, using (30), that there is a subsequence $x_n \in G_n$, $n \in I'$, such that $x_n \to x$ $(n \in I')$ for every x in the interval $[\liminf a_n, \limsup b_n]$. Hence, $x \in \operatorname{Lim\,sup} G_n$. Each element x from $\operatorname{Lim\,sup} G_n$ trivially satisfies the inequality $\liminf a_n \leq x \leq \limsup b_n$, so that we ultimately can conclude

$$[\liminf_{n\in I} a_n, \limsup_{n\in I} b_n] = \operatorname{Lim\,sup} G_n.$$

If, in addition, the condition on the right-side of (31) is valid, then $[a,b] \subset \operatorname{Lim\,sup} G_n$.

(ii) In the case $\operatorname{Lim\,sup} G_n \subset [a,b]$, it is not difficult to see that each sequence of elements $x_n \in G_n$, $n \in I$, satisfies the inequalities $\liminf x_n \geq a$ and $\limsup x_n \leq b$. In particular, this shows the condition on the right in (32). Conversely, we can immediately conclude $[a,b] \supset \operatorname{Lim\,sup} G_n$ by utilizing the above representation of $\operatorname{Lim\,sup} G_n$.

(iii) For the case $\operatorname{Lim} G_n = [a,b]$, we see initially that $\liminf a_n = a$ and $\limsup b_n = b$ (from (31) and (32)). Since also $a \in \operatorname{Lim\,inf} G_n$, there is a sequence $x_n \in G_n$, $n \in I$, with $x_n \to a$ $(n \in I)$. Now let us assume $x_0 \equiv \limsup a_n > a$. Then there is a subsequence $I' \subset I$ with $a_n \to x_0$ $(n \in I')$. Because $a_n \leq x_n$, $n \in I'$, then we must have $x_0 \leq a$ - which establishes a contradiction. Thus $\liminf a_n = \limsup a_n = \lim a_n = a$. We conclude that a corresponding result holds for (b_n). As in part (i), we see, because of (30), that there exists for each x in $\lim a_n \leq x \leq \lim b_n$, a sequence $x_n \in G_n$, $n \in I$, with $x_n \to x$ $(n \in I)$. This enables us to conclude that

$$[\lim a_n, \lim b_n] \subset \operatorname{Lim\,inf} G_n,$$

and $\operatorname{Lim} G_n = [a,b]$ follows by additionally appealing to (31) and (32) in case the right-hand side of (33) holds. □

In the concluding paragraphs of this section, we investigate the question of whether the spaces $X = C_E[0,T]$, $X_n = C_{E_n}[0,T]_n$, $n \in I$, form a discrete approximation if a discrete approximation $A(E,\pi_n E_n, \lim^E)$ exists by using restrictions $R_n^E\colon E \to E_n$, $n \in I$. We let the meshes be defined by

$$[0,T]_n \equiv \{t \in [0,T]\colon t = t_k \equiv k\tau_n,\ k = 0,\dots,N_n\},\quad \tau_n = T/N_n,\ n \in I.$$

This is precisely the situation for the approximation methods for initial value problems discussed in Section 4.4 of Chapter 4. The notation $|\cdot|_n$ will indicate the norms in E_n, and $||\cdot||_n = ||\cdot||_{\infty,n}$ will denote the maximun norms in X_n,

$$||v_n||_{\infty,n} \equiv \max_{t\in[0,T]_n} |v_n(t)|_n, \quad v_n \in X_n, \quad n \in I.$$

By prescribing

$$(R_n^X u)(t) = R_n^E u(t), \quad t \in [0,T]_n, \quad u \in X, \quad n \in I, \tag{34}$$

we can define mappings $R_n^X\colon X \to X_n$, $n \in I$, which now must be checked to see whether they satisfy the condition (9) for restriction mappings. For the sake of simplicity, we assume that the R_n^E, $n \in I$, are linear. Then the R_n^X, $n \in I$, will also be linear mappings.

Theorem 5.12. Let the sequence of mesh widths τ_n, $n \in I$, converge to zero. A sequence of restriction operators will be defined by (34) if, and only if, the following relation holds:

$$\max_{t\in[0,T]_n} |R_n^E u(t)|_n \to 0 \Rightarrow \max_{t\in[0,T]_n} |u(t)| \to 0 \ (n \in I), \ u \in X. \tag{35}$$

Condition (35) is satisfied, in case R_n^E, $n \in I$, represents a stable sequence of restriction operators.

Proof: (i) Suppose (35) is true. What needs to be shown is relation (12). From $||R_n^X u_n||_{\infty,n} \to 0$ $(n \in I)$,

$$\max_{t\in[0,T]_n} |u(t)| \to 0 \ (n \in I)$$

follows by appealing to (35). Together with $\tau_n \to 0$, a little thought convinces us that $u = 0$. Conversely, if (12) is true, then (35) is trivially satisfied.

(ii) Let $||R_n^X u||_{\infty,n} \to 0$ $(n \in I)$. Also let us assume

$$\max_{t\in[0,T]_n} |u(t)| \not\to 0 \ (n \to \infty);$$

then there exist a subsequence $I' \subset I$, points $t_n \in [0,T]_n$, $n \in I'$, and a number $\varepsilon_0 > 0$, with $|u(t_n)| > \varepsilon_0$, $n \in I'$. Because of the boundedness of $[0,T]$, there is a point $t_0 \in [0,T]$ and a subsequence $I'' \subset I'$ with $t_n \to t_0$ $(n \in I'')$. The continuity of u implies that $u(t_n) \to u(t_0)$ $(n \in I'')$, and therefore that $|u(t_0)| \geq \varepsilon_0$. The stability of the restrictions R_n^E, $n \in I$, yields the result that $|R_n^E u(t_0)|_n \to 0$ $(n \in I'')$, since

$$|R_n^E u(t_0)|_n \leq |R_n^E u(t_0) - R_n^E u(t_n)|_n + |R_n^E u(t_n)|_n$$
$$\leq \beta |u(t_0) - u(t_n)| + ||R_n^X u||_{\infty,n} \to 0 \quad (n \in I'').$$

By (12), $u(t_0)$ must be zero, thereby contradicting $|u(t_0)| \geq \varepsilon_0 > 0$. □

5.4. DISCRETE APPROXIMATIONS OF L^p-SPACES AND WEAK CONVERGENCE OF MEASURES

We begin by investigating the problem of when L^p-spaces, defined on perturbed regions of integration, form discrete approximations. The goal of this work is precisely the same as in the previous section, namely to characterize discrete approximations by conditions on the domains (cf. Theorem 5.13 and Lemma 5.14). We further show that the conditions so obtained can be equivalently expressed in terms of the weak convergence of Lebesgue measures.

Subsequently, we study the weak convergence of arbitrary measures on regions in $\mathbb{R}^d$. This convergence can be characterized by the concepts of stability and consistency. These investigations serve on the one hand as a preparation for the convergence analysis of approximate solutions of integral equations in Chapter 10. On the other hand, we obtain examples of discrete approximations for spaces of continuous functions and for L^p-spaces, where the discrete convergence is explained with the use of convergent quadrature formulas.

For a measurable set $G \subset \mathbb{R}^d$, $d \in \mathbb{N}$, let $L^p(G)$ denote for p in $[1,\infty)$ the space of all functions defined on G with values in $\mathbb{K} = \mathbb{R}$ or $\mathbb{C}$ for which $|u(\cdot)|^p$ is Lebesgue integrable on G. The norm in $L^p(G)$ is the usual norm, given by

$$||u||_{o,p} = \left(\int_G |u(x)|^p dx\right)^{1/p}$$

We write L^p for $L^p(\mathbb{R}^d)$. For every measurable subset $G \subset \mathbb{R}^d$, we let $L_0^p(G)$ be the subspace of functions in L^p such that $u = 0$ almost everywhere in $\mathbb{R}^d - G$. The restriction of $w \in L^p$ on G is again denoted as $w|G$. Also, we write $||w||_{o,p,G}$ for $||w|G||_{o,p}$. For every $u \in L^p(G)$, we let $u^o \in L^p$ denote the extension of u defined by $u^o = u$ in G, $u^o = 0$ in $\mathbb{R}^d - G$. The convergence in L^p is taken to be the usual norm convergence, defined by

$$u_n \to u \ (n \in I) \Longleftrightarrow ||u_n - u||_{o,p} \to 0 \ (n \in I) \tag{36}$$

for $u, u_n \in L^p$.

We now let G, G_n, $n \in I$, be measurable subsets of $\mathbb{R}^d$. Then we can describe the existence of a discrete approximation $A(L_0^p(G), \pi_n L_0^p(G_n), \lim)$, with the norm convergence (36) as the discrete convergence, by the approximability condition (3) from Example 2 in Section 5.1, which can moreover be characterized by a requirement on the sets G, G_n, $n \in I$. In the following discussion, we shall denote the Lebesgue measure of measurable sets in $\mathbb{R}^d$ by "meas".

Theorem 5.13. With the norm convergence (36) as the discrete convergence lim, $A(L_0^p(G), \pi_n L_0^p(G_n), \lim)$ is a discrete approximation if, and only if,

$$|u, L_0^p(G_n)| \to 0 \ (n \in I), \quad u \in L_0^p(G), \tag{37}$$

which is moreover equivalent to the requirement that

$$\text{meas}(M \cap (G - G_n)) \to 0 \ (n \in I) \tag{38}$$

for every bounded closed d-dimensional interval $M \subset \mathbb{R}^d$.

Proof: The equivalence of (37) and the presence of a discrete approximation is clear from Example 2, Section 5.1. Now suppose (38) is true. For each $u \in L_0^p(G)$, $u_n \equiv (u|G_n)^o \in L^p(G_n)$, $n \in I$, and we have

$$||u_n - u||_{o,p} = ||u||_{o,p,G-G_n} = \left(\int_{G-G_n} |u(x)|^p dx\right)^{1/p}$$

The hypothesis (38) yields $||u_n - u||_{o,p} \to 0 \ (n \in I)$, which in particular proves (37). Conversely, suppose (37) is true, and let M be an arbitrary bounded closed d-dimensional interval of $\mathbb{R}^d$. We define

$$w(x) = 1, \quad x \in M \cap G, \quad w(x) = 0, \quad x \in \mathbb{R}^d - (M \cap G).$$

Then $w \in L_0^p(G)$, and from (37) we know there exists a sequence $w_n \in L_0^p(G_n)$, $n \in I$, with $||w - w_n||_{o,p} \to 0 \ (n \in I)$. Obviously, $w - w_n = 1$ on $M \cap (G - G_n)$ so that

$$\text{meas}(M \cap (G - G_n)) = \int_{M \cap (G - G_n)} 1 \, dx \le ||w - w_n||_{o,p}^p \to 0 \ (n \in I). \qquad \square$$

This result characterizes the existence of a discrete approximation, with norm convergence in L^p as discrete convergence, by conditions on the regions of integration. In Example 3 from Section 5.1 we have seen that a sufficient condition for the discrete approximation of subspaces with projection operators has been given by the pointwise (strong) convergence of the projections. Here, bounded linear projections $P_n: E_n \to E$, $n = 0,1,2,\ldots$, with $G_0 = G$, $E = L_0^p(G)$, $E_n = L_0^p(G_n)$, $n = 0,1,2,\ldots$, can be defined in the following way,

$$P_n u = u|G_n \text{ in } G_n, \quad P_n u = 0 \text{ in } \mathbb{R}^d - G_n, \quad u \in L^p, \quad n = 0,1,2,\ldots . \tag{39}$$

The sequence (P_n) is uniformly bounded, since $||P_n|| \leq 1$, $n \in I$. The pointwise or strong convergence (4), $||P_n w - P_0 w||_{o,p} \to 0$ $(n \in I)$, is clearly equivalent to the relation

$$\lim_{n \in I} (w|G_n)^o = (w|G)^o, \quad w \in L^p. \tag{40}$$

The latter can again be characterized by a condition on the sets G, G_n, $n \in I$. To describe such a condition, we recall the definition of *symmetric difference*, i.e., for the sets G and G_n,

$$G \Delta G_n = (G-G_n) \cup (G_n-G), \quad n \in I.$$

Lemma 5.14. The condition (40) is satisfied if, and only if,

$$\text{meas}(M \cap (G \Delta G_n)) \to 0 \ (n \in I) \tag{41}$$

for all bounded closed d-dimensional intervals $M \subset \mathbb{R}^d$.

Proof: From the formula,

$$||(w|G_n)^o - (w|G)^o||_{o,p} = ||w||_{o,p,G\Delta G_n}, \quad n \in I, \quad w \in L^p,$$

We can easily conclude that relation (41) proves (40). The converse will likewise result from the preceding formula if we define, for an arbitrary bounded closed d-dimensional interval M, $w(x) = 1$ for $x \in M \cap (G \Delta G_n)$ and $w(x) = 0$ elsewhere. □

In Example 3 from Section 5.1, we have seen that (40) is also equivalent to (37), in case $E_n \subset E_0$, $n \in I$. Such an inclusion means here that $G_n \subset G$, and hence $G \Delta G_n = G - G_n$, $n \in I$. With Lemma 5.14 and Theorem 5.13, we have once again shown the equivalence of (37) and (40) for the special case of the projection operators defined by (39).

We now restrict our scope somewhat and consider the Lebesgue integral only for continuous functions. We write

$$\nu(u) \equiv \int_G u(x)dx, \quad u \in C(G), \quad \nu_n(u_n) \equiv \int_{G_n} u_n(x)dx, \tag{42}$$
$$u_n \in C(G_n), \quad n \in I,$$

where now G, G_n, $n \in I$, represent closed, nonempty subsets of a compact set $M \subset \mathbb{R}^d$. By means of (42), we have bounded linear functionals on $C(G)$, respectively $C(G_n)$, $n \in I$, i.e., there exist numbers α, α_n, $n \in I$, such that

$$|\nu(u)| \le \alpha||u||_{0,\infty}, \quad u \in C(G), \quad |\nu_n(u_n)| \le \alpha_n||u_n||_{0,\infty}, \tag{43}$$
$$u_n \in C(G_n), \quad n \in I.$$

As before, the notation $||\cdot||_{0,\infty}$ in the above expressions specifies the maximum norm in $C(G)$ and $C(G_n)$, $n \in I$. Let us call the functions ν and ν_n, $n \in I$, defined in (42) the *Lebesgue measures on* G and G_n, $n \in I$, respectively (for measures, cf. the concluding paragraphs of Section 3.1). The inequalities in (43) are obviously true with $\alpha = \text{meas}(G)$, $\alpha_n = \text{meas}(G_n)$, and we have

$$|\nu(1)| = \text{meas}(G), \quad |\nu_n(1)| = \text{meas}(G_n), \quad n \in I.$$

The Lebesgue measures are nonnegative, i.e., $\nu(u) \ge 0$ for $u \ge 0$, $u \in C(G)$; as are the measures ν_n, $n \in I$. Moreover, ν is positive, i.e., $\nu(u) > 0$ for $0 \not\equiv u \ge 0$, $u \in C(G)$, in case G possesses a nonempty interior. We shall assume that this is indeed the case in the following. Finally, we recall that for the natural extensions $\hat{\nu}$ and $\hat{\nu}_n$ ($n \in I$), given by

$$\hat{\nu}(w) \equiv \nu(w|G), \quad \hat{\nu}_n \equiv \nu_n(w|G_n), \quad n \in I, \quad w \in C(M),$$

we have $||\hat{\nu}|| = ||\nu||$, $||\hat{\nu}_n|| = ||\nu_n||$, $n \in I$, and that the weak convergence $\hat{\nu}_n \rightharpoonup \hat{\nu}$ ($n \in I$) is defined by $|\hat{\nu}_n(w) - \hat{\nu}(w)| \to 0$ ($n \in I$), $w \in C(M)$.

After these preliminary remarks, we are now in the position to characterize the important condition (41) by means of the weak convergence of the Lebesgue measures $\hat{\nu}_n$ to $\hat{\nu}$. We should point out in advance that the sets occurring in (41) satisfy the relations

$$M \cap (G \,\Delta\, G_n) = G \,\Delta\, G_n = (G \cap (M-G_n)) \cup ((M-G) \cap G_n)$$

in the present situation where $G, G_n \subset M$.

<u>Theorem 5.15</u>. With the above assumptions on M, G, G_n, $n \in I$, the following equivalence holds:

$$\hat{\nu}_n \rightharpoonup \hat{\nu} \iff \text{meas}(G \,\Delta\, G_n) \to 0 \ (n \in I). \tag{44}$$

Proof: (⇐) Because of

$$G_n = [G_n \cap (M-G)] \cup (G_n \cap G), \quad G = [G \cap (M-G_n)] \cup (G \cap G_n),$$

we have the representation

$$\begin{aligned}\hat{\nu}_n(w) - \hat{\nu}(w) &= \int_{G_n} w\,dx - \int_G w\,dx \\ &= \int_{G_n \cap (M-G)} w\,dx - \int_{G \cap (M-G_n)} w\,dx = \int_{G \Delta G_n} w\chi_n\,dx, \quad n \in I,\end{aligned}$$

where $\chi_n(x) = 1$, $x \in G_n \cap (M-G)$, $\chi_n(x) = -1$, $x \in G \cap (M-G_n)$. Because of (43), it therefore follows from $\mathrm{meas}(G \Delta G_n) \to 0$ $(n \in I)$, that

$$|\hat{\nu}_n(w) - \hat{\nu}(w)| \le \int_{G \Delta G_n} |w|\,dx \le \mathrm{meas}(G \Delta G_n) \sup_{x \in M} |w(x)| \to 0 \ (n \in I),$$

$$w \in C(M).$$

(⇒) First we show by a contradiction argument that $\mathrm{meas}(G \cap (M-G_n)) \to 0$ $(n \in I)$. To this end, let us assume that a subsequence $I' \subset I$ and a number $m > 0$ exist such that $\mathrm{meas}(G \cap (M-G_n)) \ge m$, $n \in I'$ Let O be an open set encompassing G, $O \supset G$, with $\mathrm{meas}(O - G) \le m/2$ (see, for example, Smirnow (1971), 35). According to the Tietze-Urysohn extension theorem, there is a function $u \in C(\mathbb{R}^d)$ with the properties $u|G = 1$, $u|(\mathbb{R}^d - O) = 0$, $0 \le u(x) \le 1$, $x \in \mathbb{R}^d$ Therefore the following relations hold,

$$\begin{aligned}|\nu(u|G) - \nu_n(u|G_n)| &= \left|\int_G u\,dx - \int_{G_n} u\,dx\right| \\ &= \left|\int_{G \cap (M-G_n)} u\,dx - \int_{G_n \cap (M-G)} u\,dx\right| \\ &\ge \int_{G \cap (M-G_n)} u\,dx - \int_{O-G} u\,dx \ge \mathrm{meas}(G \cap (M-G_n)) - \mathrm{meas}(O-G).\end{aligned}$$

The penultimate inequality utilizes the facts that

$$G_n \cap (M-G) = \{[G_n \cap (M-G)] \cap O\} \cup \{(\mathbb{R}^d - O) \cap [G_n \cap (M-G)]\},$$

$$u|(\mathbb{R}^d - O) = 0 \quad \text{and} \quad [G_n \cap (M-G)] \cap O \subset (M-G) \cap O \subset O-G.$$

Then, we obtain

$$|\nu(u|G) - \nu_n(u|G_n)| \ge m - m/2 = m/2 > 0, \quad n \in I',$$

in contradiction to $\hat{\nu}_n \rightharpoonup \hat{\nu}$ $(n \in I)$.

It remains to show that $\text{meas}(G_n \cap (M-G)) \to 0$ $(n \in I)$. With $w = 1$ in the representation

$$\hat{\nu}_n(w) - \hat{\nu}(w) = \int_{G_n \cap (M-G)} w\,dx - \int_{G \cap (M-G_n)} w\,dx$$

we see that

$$|\hat{\nu}_n(w) - \hat{\nu}(w)| \geq \text{meas}(G_n \cap (M-G)) - \text{meas}(G \cap (M-G_n)).$$

Since $\text{meas}(G \cap (M-G_n)) \to 0$ $(n \in I)$ has already been shown, the desired assertion follows from the weak convergence of $\hat{\nu}_n$ to $\hat{\nu}$ $(n \in I)$. □

We now consider a more general situation of measures on subsets of a compact metric space and characterize their weak convergence. As in Section 3.1, we call any bounded linear functional μ on $C(M)$, (M a compact metric space), a *measure on* M. Thus, for any measure μ, there exists a number $\alpha \geq 0$, such that $|\mu(u)| \leq \alpha||u||_{0,\infty}$, $u \in C(M)$. The definitions of nonnegative and positive measures, and of the natural extension of measures are given in Section 3.1. We denote the value of μ at u as the *integral* of u with respect to the measure μ and write $\mu(u) = \int_M u\,d\mu = \int_M u(x)d\mu(x)$.

Now, let G, G_n, $n \in I$, be closed subsets of a compact metric space M and μ and μ_n, $n \in I$, be measures on G and G_n, $n \in I$, respectively. We say that μ, μ_n, $n \in I$, is *consistent* if, to each $u \in C(G)$, there is an extension $\hat{u} \in C(M)$ such that

$$\lim_{n \in I} \mu_n(\hat{u}|G_n) = \mu(u). \tag{45}$$

The sequence μ_n, $n \in I$, is called *stable*, if the norms of the μ_n, $n \in I$, are uniformly bounded, i.e.,

$$|\mu_n(u_n)| \leq \beta||u_n||_{0,\infty}, \quad u_n \in C(G_n), \quad n \in I. \tag{46}$$

The next fundamental result characterizes the weak convergence of the natural extensions $\hat{\mu}_n \rightharpoonup \hat{\mu}$ $(n \in I)$. In order to formulate this result properly, we utilize the definition of closed limits of subsets of metric spaces presented in Section 5.3 (cf. (23), (24)). We choose here a formulation the assumptions of which include the stipulation that $\text{Lim sup}\, G_n = G$. From Theorem 5.10, we know that this requirement is equivalent to the existence of a discrete approximation $A(C(G), \pi_n(G_n), \lim)$ associated with the discrete uniform convergence lim. For another formulation of the following result, we refer the reader to Stummel (1973b), §9.

Theorem 5.16. The weak convergence

$$\hat{\mu}_n \rightharpoonup \hat{\mu} \ (n \in I) \tag{47a}$$

implies that

$$\mu, \mu_n, \ n \in I, \text{ is consistent and } (\mu_n) \text{ is stable.} \tag{47b}$$

Conversely, (47a) is also implied by (47b) in case $\operatorname{Lim\,sup} G_n = G$. Whenever μ is a positive measure, either (47a) together with $\operatorname{Lim\,sup} G_n \subset G$ or (47b) together with $\operatorname{Lim\,sup} G_n = G$ yields that $\operatorname{Lim} G_n = G$.

Proof: (i) Because $\hat{\mu}_n \rightharpoonup \hat{\mu}$, the sequence $|\mu_n(w)|$, $n \in I$, is bounded for every $w \in C(M)$. Now, the principle of uniform boundedness implies that $||\hat{\mu}_n|| \le \beta$, $n \in I$. Because $||\mu_n|| = ||\hat{\mu}_n||$, we see that (μ_n) is stable. The consistency will result if we take the Tietze-Urysohn extension $\hat{u}$ of $u \in C(G)$ and let $w = \hat{u}$ in the definition of weak convergence.

(ii) For an arbitrary $w \in C(M)$, let $u = w|G$, $u_n = w|G_n$, $n \in I$. Because of the consistency, there exists an extension $\hat{u} \in C(M)$ of u such that $\mu_n(u|G_n) \to \mu(u)$ $(n \in I)$. Since we have assumed that $\operatorname{Lim\,sup} G_n = G$, a discrete convergence is defined by (22); and, according to Lemma 5.8, $w|G_n = u_n \to u = w|G$ and $\hat{u}|G_n \to u$ $(n \in I)$ hold. From Property 5.1 of a discrete convergence, we then have

$$||u_n - \hat{u}|G_n||_{0,\infty} \to 0 \ (n \in I).$$

Using the stability of (μ_n), we see that

$$|\hat{\mu}_n(w) - \hat{\mu}(w)| \le |\mu_n(u_n - \hat{u}|G_n)| + |\mu_n(\hat{u}|G_n) - \mu(u)|$$

$$\le ||\mu_n||\ ||u_n - \hat{u}|G_n||_{0,\infty} + |\mu_n(\hat{u}|G_n) - \mu(u)| \to 0 \ (n \in I).$$

(iii) Now we assume that μ is positive and that $\operatorname{Lim\,sup} G_n \subset G$, and using (47a) we proceed to show indirectly that $G \subset \operatorname{Lim\,inf} G_n$. Let us further assume that there is a point $z \in G$ with $z \notin \operatorname{Lim\,inf} G_n$, or equivalently $|z, G_n| \ge \varepsilon_0 > 0$, $n \in I'$, for some subsequence $I' \subset I$ (cf. (23b)). From the Tietze-Urysohn extension theorem, there is a function $w \in C(M)$ defined by

$$w(x) = \begin{cases} 1 & \text{for } x = z, \\ 0 & \text{for } x \text{ such that } |x,z| \ge \varepsilon_0, \\ 0 \le w(x) \le 1 & \text{for } x \text{ otherwise in } M. \end{cases}$$

Therefore,

$$||w|G_n||_{0,\infty} = 0, \quad n \in I', \text{ and } \hat{\mu}_n(w) = \hat{\mu}_n(w|G_n) = 0, \quad n \in I'.$$

Because of the weak convergence $\hat{\mu}_n \rightharpoonup \hat{\mu}$ $(n \in I)$, we conclude $\mu(w|G) = \hat{\mu}(w) = 0$ in contradiction to the fact that $\mu(w|G) > 0$ (since $0 \neq w|G \geq 0$ and μ is positive). The remaining assertion follows from the equivalence of (47a) and (47b) under the hypothesis $\operatorname{Lim\,sup} G_n = G$. □

From the theorem of Banach-Steinhaus, we see that the weak convergence of a sequence of functionals can be characterized by the stability of the sequence itself and by its weak convergence on a dense subset. The condition (47a) - and the equivalent condition (47b) under the assumption $\operatorname{Lim\,sup} G_n = G$ - are therefore equivalent to the following statement:

(47c) (μ_n) is stable and there exists a dense subset $D \subset C(M)$ with the property that $\mu_n(\phi|G_n) \to \mu(\phi|G)$ $(n \in I)$ for all $\phi \in D$.

In Theorem 5.15, we have already studied a *first example* of weak convergent sequences of measures when Lebesgue measures are the underlying measures. As a *second example* we consider the approximation of the Riemann integral by quadrature formulas. For a compact subset M of $\mathbb{R}^d$ and a nonempty closed subset $G \subset M$ with boundary having Lebesgue measure zero, the Riemann integral exists for every $u \in C(G)$, and is denoted by

$$\mu(u) \equiv \int_G u(x)\,dx.$$

This integral defines a positive measure on G in case the interior of G is nonempty (see, for example, Smirnow (1964), Sec. 90-96). It is well known that the Riemann integral is the limit of upper and lower approximating (Riemann) sums. Let now

$$\mu_n(u_n) \equiv \sum_{x \in G_n} u_n(x)\alpha_n(x), \qquad u_n \in C(G_n), \quad n \in I,$$

be a sequence of quadrature formulas, where G_n denote finite sets of nodes in M and $\alpha_n(x) \in \mathbb{R}$, $x \in G_n$, $n \in I$, are the associated weights. The norms of μ_n, $n \in I$, are then given by

$$||\mu_n|| = \sum_{x \in G_n} |\alpha_n(x)|, \quad n \in I.$$

The weak convergence $\hat{\mu}_n \rightharpoonup \hat{\mu}$ $(n \in I)$ is in this context equivalent to

$$\lim_{n\in I} \sum_{x\in G_n} w(x)\alpha_n(x) = \int_G w(x)dx, \quad w \in C(M). \tag{48}$$

From the Weierstrass approximation theorem, we know that the space $P(x_1,\ldots,x_d)$ of polynomials in $x = (x_1,\ldots,x_d) \in M$ is dense in $C(M)$. The remark following Theorem 5.16 (cf. condition (47c)) immediately provides the following characterization of the convergence (48) of quadrature formulas.

Theorem 5.17. The convergence (48) of quadrature formulas is equivalent to the convergence

$$\lim_{n\in I} \sum_{x\in G_n} p(x)\alpha_n(x) = \int_G p(x)dx \quad \text{for all} \quad p \in P(x_1,\ldots,x_d)$$

and the existence of a number $\beta \geq 0$ such that

$$\sum_{x\in G_n} |\alpha_n(x)| \leq \beta, \quad n \in I. \tag{49}$$

If, moreover, $\operatorname{Lim\,sup} G_n \subset G$, and the Riemann integral represents a positive measure on G, then (48) is equivalent to (49) together with the consistency of μ, μ_n, $n \in I$. □

For $E = C(G)$, $E_n = C(G_n)$, $n \in I$, $1 \leq p < \infty$, we define

$$||u||_{o,p} \equiv \left(\int_G |u(x)|^p dx\right)^{1/p}, \quad u \in E, \tag{50a}$$

$$||u_n||_{o,p} \equiv \left(\sum_{x\in G_n} |u_n(x)|^p \alpha_n(x)\right)^{1/p}, \quad u_n \in E_n, \quad n \in I. \tag{50b}$$

Since the Riemann integral represents a positive measure under the assumptions specified above, $||\cdot||_{o,p}$ defines a norm on $C(G)$. For nonnegative weights $\alpha_n(x)$, $||\cdot||_{o,p}$ in (50b) defines a seminorm on $C(G_n)$, and even a norm on $C(G_n)$ in case the weights are positive. We need for the present, however, no assumptions on the weights. The following theorem defines and characterizes a discrete convergence different from the discrete uniform convergence that we have considered in Section 5.3.

Theorem 5.18. For $u \in C(G)$, $u_n \in C(G_n)$, $n \in I$, a discrete convergence, satisfying Properties 5.2 and 5.3, will be defined by stipulating

$$\begin{aligned} u_n \to n \ (n \in I) \iff \exists \hat{u} \in C(M):\ \hat{u}|G = u \ \text{and} \\ \sum_{x\in G_n} |u_n(x) - \hat{u}(x)|^p \alpha_n(x) \to 0 \quad (n \in I), \end{aligned} \tag{51}$$

if and only if, the quadrature formulas converge (i.e., if (48) is true or $\hat{\mu}_n \rightharpoonup \hat{\mu}$ $(n \in I)$ holds).

Proof: (i) Suppose $\hat{\mu}_n \rightharpoonup \hat{\mu}$ $(n \in I)$. We first show that the discrete convergence (51) is indeed a well-defined mapping. Let $u_n \in C(G_n)$, $n \in I$, and $\hat{u},\hat{v} \in C(M)$ be extensions of $u,v \in C(G)$ with $||u_n - \hat{u}|G_n||_{o,p} \to 0$ and $||u_n - \hat{v}|G||_{o,p} \to 0$ $(n \in I)$. For $\hat{w} = \hat{u} - \hat{v}$, we have, since $\hat{\mu}_n \rightharpoonup \hat{\mu}$ $(n \in I)$, and $|\hat{w}(\cdot)|^p \in C(M)$, that

$$||\hat{w}|G||_{o,p}^p = \int_G |\hat{w}(x)|^p dx = \lim_{n\in I} \sum_{x\in G_n} |\hat{w}(x)|^p \alpha_n(x) = \lim_{n\in I} ||\hat{w}|G_n||_{o,p}^p.$$

We have then that

$$||\hat{w}|G||_{o,p} = \lim_{n\in I} ||\hat{w}|G_n||_{o,p} \le \lim_{n\in I} (||\hat{u}|G_n - u_n||_{o,p} + ||u_n - \hat{v}|G_n||_{o,p}) = 0,$$

from which we conclude, by the definiteness property of the norm $||\cdot||_{o,p}$ on $C(G)$, that $\hat{w}|G = 0$, or that $u = v$. To prove condition (1), we easily see that $||v_n - \hat{u}|G_n||_{o,p} \to 0$ $(n \in I)$, provided that both $||u_n - \hat{u}|G_n||_{o,p}$, $n \in I$, and $||v_n - u_n||_{o,p}$, $n \in I$, converge to zero. Conversely, it follows that

$$\hat{\mu}_n(|\hat{u} - \hat{v}|^p) = ||(\hat{u} - \hat{v})|G_n||_{o,p} \to 0 \ (n \in I)$$

- and hence that

$$||u_n - v_n||_{o,p} \le ||v_n-\hat{v}|G_n||_{o,p} + ||u_n-\hat{u}|G_n||_{o,p} + ||(\hat{u}-\hat{v})|G_n||_{o,p} \to 0 \text{ -}$$

provided that $\hat{u}|G = \hat{v}|G$ and both $||u_n - \hat{u}|G_n||_{o,p}$, $n \in I$, and $||v_n - \hat{v}|G_n||_{o,p}$, $n \in I$, approach zero. The asymptotic linearity of the discrete convergence defined by (51) is obvious. A final task is to show the convergence of the norms (cf. Property 5.3). To this end, let $u_n \to u$ $(n \in I)$ in the sense of (51). Since $\hat{\mu}_n \rightharpoonup \hat{\mu}$ $(n \in I)$, we can easily see that

$$||\hat{u}|G_n||_{o,p} \to ||\hat{u}|G||_{o,p} = ||u||_{o,p} \quad (n \in I)$$

for the extension $\hat{u}$ of u. Together with the definition (51), we also get $||u_n||_{o,p} \to ||u||_{o,p}$ $(n \in I)$, since

$$|\ ||u_n||_{o,p} - ||\hat{u}|G_n||_{o,p}| \le ||u_n - \hat{u}|G_n||_{o,p} \to 0 \ (n \in I).$$

(ii) In order to prove the necessity of $\hat{\mu}_n \rightharpoonup \mu$ $(n \in I)$, let $w \in C(M)$ be arbitrary. Because of the Convergence Property 5.3 of the norms, we have that

$$\sum_{x \in G_n} |w(x)|^p \alpha_n(x) \to \int_G |w(x)|^p dx \quad (n \in I)$$

since trivially $w|G_n \to w|G$ $(n \in I)$. For any nonnegative function $w_+ \in C(M)$, we know $w_+^{1/p} \in C(M)$ and therefore that

$$\sum_{x \in G_n} w_+(x)\alpha_n(x) \to \int_G w_+(x)dx \ (n \in I).$$

Finally, we note that, with $\mathbb{K} = \mathbb{R}$, every function $w \in C(M)$ can be written $w = w_+ - w_-$ with

$$w_+ \equiv \frac{1}{2}(w + |w|), \quad w_- \equiv \frac{1}{2}(|w| - w).$$

Also $w_+ \geq 0$, $w_- \geq 0$, and both w_+ and w_- are in $C(M)$. Hence,

$$\sum_{x \in G_n} w(x)\alpha_n(x) = \sum_{x \in G_n} w_+(x)\alpha_n(x) - \sum_{x \in G_n} w_-(x)\alpha_n(x)$$
$$\to \int_G w_+(x)dx - \int_G w_-(x)dx = \int_G w(x)dx \quad (n \in I).$$

For the case $\mathbb{K} = \mathbb{C}$, we can show the assertion by decomposing w into real and imaginary parts and applying the above result to each part. □

The surjectivity of the discrete convergence defined by (51) - and thereby the existence of a discrete approximation $A(C(G),\pi_n C(G_n),\lim)$ - is guaranteed by hypothesis (48) because of the closedness of G and because of the use of the Tietze-Urysohn extension theorem. Moreover, the discrete convergence (51) is then independent of the special extension, i.e., assuming (48), we have the characterization

$$u_n \to u \ (n \in I) \iff \forall \hat{u} \in C(M): \ \hat{u}|G = u,$$
$$\sum_{x \in G_n} |u_n(x) - \hat{u}(x)|^p \alpha_n(x) \to 0 \ (n \in I).$$

Indeed, for any two extensions $\hat{u}$, $\hat{v}$ of u, we have from (48) that

$$\sum_{x \in G_n} |\hat{u}(x) - \hat{v}(x)|^p \alpha_n(x) \to \int_G |(\hat{u} - \hat{v})(x)|^p dx = 0 \ (n \in I).$$

If $||\cdot||_{o,p}$ consitute norms on $C(G_n)$, $n \in I$, e.g., by means of positive weights, then the preceding example can be viewed in the context of Section 5.2. Namely, we can define mappings $R_n: C(G) \to C(G_n)$, $n \in I$, by

$$(R_n u)(x) \equiv \hat{u}(x), \quad x \in G_n, \quad u \in C(G), \quad n \in I,$$

with $\hat{u} \in C(M)$ an arbitrary but fixed extension of $u \in C(G)$. Theorem 5.18 directly shows that (R_n) is a sequence of restriction operators, which satisfy in addition condition (13), i.e., the Convergence Property 5.3 of the norms. Since $C(G)$ lies dense in $L^p(G)$, we can obtain immediately from Theorem 5.6 a discrete approximation $A(L^p(G), \pi_n C(G_n), \lim)$, where the discrete convergence of $u_n \in C(G_n)$, $n \in I$, to $u \in L^p(G)$ is specified by (cf. (17))

$$u_n \to u \ (n \in I) \Longleftrightarrow \forall\, \varepsilon > 0, \ \exists \hat{\phi} \in C(M): \int_G |u(x) - \hat{\phi}(x)|^p dx \le \varepsilon^p$$
$$\text{and} \quad \limsup_{n \in I} \sum_{x \in G_n} |u_n(x) - \hat{\phi}(x)|^p \alpha_n(x) \le \varepsilon^p.$$

REFERENCES

Anselone (1965)*, Anselone (1971), Anselone & Ansorge (1981)*, Aubin (1967a)*, Aubin (1972), Browder (1967)*, Chartres & Stepleman (1972)*, Grigorieff (1969,1973a)*, Grigorieff (1973b), Petryshyn (1968a)*, Reinhardt (1975a)*, Smirnow (1971), Stummel (1970,1973a,1974a,1974b,1975)*, Stummel (1973b), Stummel & Reinhardt (1973)*, Vainikko (1976).

*Article(s)

Chapter 6
Discrete Convergence of Mappings and Solutions of Equations

In this chapter, we establish the fundamental convergence results for solutions of sequences of nonlinear equations with differentiable operators. To prepare the reader for the analysis in this chapter, we examine in Section 6.1 the relationship between the continuity of a mapping on the one hand and the differentiability and boundedness of its derivatives on the other. The most important result in 6.1 is a quantitative formulation of the Inverse Function Theorem (see Theorem 6.7). If we apply this result to sequences of differentiable mappings, then we obtain equivalent characterizations of the concepts of stability and inverse stability (see Section 6.2, Theorem 6.12). In Section 6.3, we introduce the concepts of consistency and of discrete convergence of sequences of mappings. It turns out that discrete convergence is equivalent to stability together with consistency (cf. Theorem 6.13). By virtue of the characterizations of stability to be discussed in Section 6.2, we are able to obtain equivalent conditions for the discrete convergence of differentiable mappings, along with error estimates (cf. Theorem 6.14). The concluding Section 6.4 establishes and characterizes the discrete convergence of solutions by using the concept of inverse discrete convergence. The most important result of this section is Theorem 6.21 which gives equivalent conditions for biconvergence. With an appropriate choice of underlying norms, we are able to state another important result (cf. Theorem 6.23) which allows us to infer from the inverse stability of equicontinuously equidifferentiable mappings, the local solvability, and the convergence of the approximate solutions.

Most of the material of this section is adapted from Stummel (1976b). Many of the basic results, which occur without the assumption of differentiability on the operators, are due to Stummel & Reinhardt (1973).

6.1. CONTINUITY AND DIFFERENTIABILITY OF MAPPINGS AND THEIR INVERSES

To prepare the reader for the results on (countably infinite) sequences of mappings in Section 6.2 below, we state and prove theorems in this section which establish the relation between the continuity of a mapping and its (Fréchet-) derivative, and develop analogous results for the inverse. It should be pointed out that the definition of a Fréchet-derivative at a point does not specifically require its continuity (as a linear mapping).

In the following, let E and F be normed spaces, and u^0 denote an arbitrary but fixed vector in E. We label the norms in E and F by the common symbol $||\cdot||$. It is well-known that a linear mapping $L: D(L) \subset E \to F$ (with *domain of definition* $D(L)$) is continuous if and only if it is bounded. A linear, injective mapping L is called *continuously invertible* if the inverse $L^{-1}: R(L) \to E$ satisfies

$$||L^{-1}|| \equiv \sup_{0 \neq w \in R(L)} \frac{||L^{-1}w||}{||w||} < \infty,$$

where $R(L)$ is the *image* or *range* of L in F. A linear, injective mapping is called *bicontinuous*, in case it is both continuous and continuously invertible. We let $N(L) \equiv \{u \in D(L): Lu = 0\}$ designate the *null space* or *kernel* of L. Also we write $L: E \to F$ whenever $D(L) = E$ and $R(L) \subset F$. The set of bounded, linear mappings from E to F is denoted as $B(E,F)$.

In the following, we let A be a (not necessarily linear) mapping with domain of definition $D(A) \subset E$ and image $R(A)$ F. The notation $B_\rho^0(u)$ $(B_\rho(u))$ means the open (closed) ball of radius ρ with center $u \in E$.

For the sake of better comprehension, we repeat at this particular juncture the definition of a differentiable and continuously differentiable mapping. A mapping A is said to be (*Fréchet-*) *differentiable at the point* u^0 if there exists a linear mapping $A'(u^0): E \to F$ with the following property:

$$\forall \varepsilon > 0,\ \exists\, \delta > 0\ \ni: \forall h \in E: ||h|| \le \delta$$
$$\Rightarrow u^0 + h \in D(A) \quad \text{and} \quad ||R^A(u^0;h)|| \le \varepsilon ||h||,$$

where $R^A(u^0;h)$ is the *remainder term* given by

$$R^A(u^0;h) \equiv A(u^0 + h) - AU^0 - A'(u^0)h.$$

$A'(u^0)$ is called the *(Frechet-) derivative* of A at u^0 which is unique. The mapping A is said to be *continuously differentiable at the point* u^0 if A is differentiable at every point of a neighborhood of u^0 and if the following condition holds:

$$\forall \varepsilon > 0,\ \ \exists \eta > 0\ \ni: \forall u \in B_\eta(u^0),\ \ \forall h \in E$$
$$\Rightarrow ||(A'(u^0) - A'(u))h|| \le \varepsilon ||h||.$$

The point u^0 is necessarily an interior point of the domain of definition of the mapping A. We should point out here that boundedness of the derivative is not stipulated in the definition.

The following lemma shows that the derivative of a differentiable mapping is continuous if, and only if, the mapping itself is continuous at the corresponding point of differentiability. In (1c), the appropriate assertions for continuously differentiable mappings are stated in parentheses. Also, $D(A')$ denotes the domain of definition of A' (i.e., the set of points where A is differentiable).

Theorem 6.1. Let A be differentiable (respectively, continuously differentiable) at a point u^0. Then the following statements are equivalent:

(1a) A is continuous at u^0.

(1b) $A'(u^0)$ is continuous.

(1c) $||A'(u^0)|| < \infty$, and for every $\alpha > ||A'(u^0)||$, there exists a $\rho > 0$ such that $B_\rho(u^0) \subset D(A)$ (respectively, $B_\rho(u^0) \subset D(A')$) and

$$||Au - Au^0|| \le \alpha ||u - u^0||, \qquad u \in B_\rho(u^0).$$

(respectively, $||Au - Av|| \le \alpha ||u - v||, \qquad u,v \in B_\rho(u^0)$.)

Proof: (1a) $\Rightarrow$ (1b). The continuity and differentiability of A at u^0 furnishes, for every arbitrary $\varepsilon > 0$, a $\delta > 0$ so that $B_\delta(u^0) \subset D(A)$ and

$$||Au - Au^0|| \leq \varepsilon, \quad ||R^A(u^0;h)|| \leq \varepsilon\delta$$

for all $u = u^0 + h$, $||h|| \leq \delta$. The boundedness of $A'(u^0)$ easily follows, since

$$||A'(u^0)h'|| \leq \frac{1}{\delta}(||Au - Au^0|| + ||R^A(u^0;h')||) \leq \frac{1}{\delta}(\varepsilon + \varepsilon\delta)$$

for arbitrary $h' \in E$ with $||h'|| \leq 1$ and $u = u^0 + \delta h'$.

(1b) $\Rightarrow$ (1c). The differentiability of A at u^0 and the boundedness of $A'(u^0)$ furnishes, for any $\varepsilon > 0$, a $\delta > 0$, such that $B_\delta(u^0) \subset D(A)$ and

$$\begin{aligned} ||Au - Au^0|| &\leq ||A'(u^0)(u-u^0)|| + ||R^A(u^0;u-u^0)|| \\ &\leq (||A'(u^0)|| + \varepsilon)||u - u^0||, \end{aligned}$$

for all $u \in B_\delta(u^0)$. Choosing $\varepsilon = \alpha - ||A'(u^0)||$ and $\rho = \delta$, we obtain (1c).

(1c) $\Rightarrow$ (1a). Trivial.

Whenever A is continuously differentiable at u^0, we must show that the strengthened conditions in (1c) follow from (1b). Indeed, for every $\varepsilon > 0$, there is a $\delta > 0$ such that

$$||A'(u) - A'(u^0)|| \leq \varepsilon, \text{ for all } u \in B_\delta(u^0) \subset D(A').$$

Since $A'(u^0)$ is bounded, there is associated to every $\alpha > ||A'(u^0)||$ and $\varepsilon = \alpha - ||A'(u^0)||$ a $\delta = \rho > 0$ such that

$$||A'(u)|| \leq ||A'(u^0)|| + \varepsilon = \alpha, \quad u \in B_\rho(u^0).$$

Finally, using the well-known Mean Value Theorem, we see that the assertion is proved by the following estimates,

$$\begin{aligned} ||Au - Av|| &\leq ||u - v|| \sup_{0 \leq t \leq 1} ||A'(u + t(v-u))|| \\ &\leq \alpha||u - v||, \quad u,v \in B_\rho(u^0). \end{aligned}$$ □

For the inverse, we have an analogous result.

Theorem 6.2. Let A be differentiable (respectively, continuously differentiable) at u^0 with $A'(u^0)$ injective. Then the following statements are equivalent:

(2a) $A'(u^0)$ is continuously invertible.

(2b) $||A'(u^0)^{-1}|| < \infty$, and to every β in $0 < \beta < ||A'(u^0)^{-1}||^{-1}$, there exists a $\rho > 0$ such that $B_\rho(u^0) \subset D(A)$ (respectively, $B_\rho(u^0) \subset D(A'))$ and

$$\beta||u - u^0|| \leq ||Au - Au^0||, \quad u \in B_\rho(u^0)$$

(respectively,

$$\beta||u - v|| \leq ||Au - Av||, \quad u,v \in B_\rho(u^0)).$$

<u>Proof</u>: (2a) ➡ (2b). According to the hypothesis and (2a), there is, for every $\varepsilon > 0$, a $\delta > 0$ such that

$$\beta'||u - u^0|| \leq ||A'(u^0)(u - u^0)|| \leq ||Au - Au^0|| + ||R^A(u^0;u - u^0)||$$

$$\leq ||Au - Au^0|| + \varepsilon||u - u^0||, \quad u \in B_\delta(u^0) \subset D(A),$$

where $\beta' \equiv ||A'(u^0)^{-1}||^{-1}$. For arbitrary $\beta \in (0,\beta')$, (2b) will result if we set $\varepsilon = \beta' - \beta$ and $\rho = \delta$.

(2b) ➡ (2a). Using the inequality in (2b), we have that

$$\beta \leq ||A'(u^0)h|| + \frac{1}{s}||R^A(u^0;sh)||$$

whenever $u = u^0 + sh$, $s \in (0,\rho]$, and $||h|| = 1$. The differentiability of A at u^0 easily implies that $||R^A(u^0;sh)||/s \to 0$ $(s \to 0)$ and hence $\beta \leq ||A'(u^0)h||$ for all $h \in E$ with $||h|| = 1$. The continuous invertibility of $A'(u^0)$ is thereby shown.

Whenever A is continuously differentiable at u^0, we must show in analogy to Theorem 6.1 that the strengthened criterion (2b) results from (2a). First, to $\beta' \equiv ||A'(u^0)^{-1}||^{-1}$ (> 0) and to every $\beta \in (0,\beta')$ there is an $\eta > 0$ such that A is differentiable in a neighborhood encompassing $B_\eta(u^0)$ and that

$$||A'(v) - A'(u^0)|| \leq \beta' - \beta$$

for all $v \in B_\eta(u^0)$. Secondly, as a consequence of the Mean Value Theorem, the inequalities

$$||Au-Av-A'(u^0)(u-v)|| \leq ||u-v|| \sup_{0\leq t\leq 1} ||A'(v+t(u-v)) - A'(u^0)||$$

$$\leq (\beta' - \beta)||u - v||$$

hold for all $u,v \in B_\eta(u^0)$. Thus,

$$\beta'||u - v|| \le ||A'(u^0)(u-v)|| \le ||A'(u^0)(u-v) - (Au-Av)|| + ||Au-Av||$$
$$\le (\beta' - \beta)||u - v|| + ||Au - Av||$$

for all $u,v \in B_\eta(u^0)$ which proves (2b) (with $\rho = \eta$) for the case of continuous differentiability. □

A combination of Theorem 6.1 and 6.2 furnishes immediately a corresponding result (with two-sided inequalities) for bicontinuous $A'(u^0)$.

In order to derive results about the differentiability of inverse mappings, we need in addition the concept of local bijectivity. A mapping $A: D(A) \subset E \to F$ is called *locally injective* at $u^0 \in D(A)$, if a neighborhood $B_\rho^0(u^0) \subset D(A)$ exists such that the restriction $\tilde{A} = A|B_\rho^0(u^0)$: $B_\rho^0(u^0) \to AB_\rho^0(u^0)$ is injective. A is called *locally bijective* at u^0, in case A is locally injective at u^0 and there is a $\sigma > 0$ so that $B_\sigma^0(Au^0) \subset AB_\rho^0(u^0)$. The concept of an open mapping is closely related to the concept of local bijectivity, Indeed, a mapping is called *open* at $u^0 \in D(A)$, if the image of every neighborhood $B_\rho^0(u^0)$ of u^0 is a neighborhood of $w^0 = Au^0$; more precisely, there is to every $\rho > 0$, a $\alpha > 0$ such that $B_\sigma^0(w^0) \subset AB_\rho^0(u^0)$. The following lemma shows the relationship between open mappings and mappings with continuous inverses. Its proof is evident.

Lemma 6.3. Let A be locally bijective at u^0 and $\tilde{A}$ be the associated local restriction. Then the inverse $\tilde{A}^{-1}$ is continuous at $w^0 = Au^0$ if, and only if, A is open at u^0. □

In order to prove the following interesting result, we must assume that the spaces E and F are complete.

Theorem 6.4. Let A be continuous, differentiable and open at u^0, and suppose that E and F are Banach spaces. Then the bijectivity and bicontinuity of $A'(u^0)$ are consequences of its injectivity and continuous invertibility.

Proof: First, it follows from Theorem 6.1 that $A'(u^0)$ is continuous, since A is continuous and differentiable at u^0. Furthermore, we see from Theorem 6.2 that condition (2b) is satisfied because of the continuous invertibility of $A'(u^0)$. Now, let β be an arbitrary number in $0 < \beta < ||A'(u^0)^{-1}||^{-1}$, and let ρ be the number associated with β in (2b). From the definition of differentiability of A at u^0, we see that there exists, for arbitrary $\varepsilon > 0$, a $\delta \in (0,\rho]$ such that $B_\delta(u^0) \subset D(A)$ and, moreover, that

$$\|Av - Au^0 - A'(u^0)(v - u^0)\| \le \varepsilon\|v - u^0\|$$

$$\le \frac{\varepsilon}{\beta}\|Av - Au^0\|, \quad v \in B_\delta(u^0).$$

Let ε be chosen so small that $q \equiv \varepsilon/\beta < 1$. Since A is open at u^0, there exists a $\sigma > 0$ with $B_\sigma(w^0) \subset AB_\delta(u^0)$, where $w^0 = Au^0$. For an arbitrary vector k, with $\|k\| \le \sigma$, we then have $w^0 + k \in B_\sigma(w^0)$; and hence a $v \in B_\delta(u^0)$ such that $Av = w^0 + k$. Moreover, for $h = v - u^0$ we have (since $k = Av - Au^0$),

$$\|k - A'(u^0)h\| \le q\|k\| \quad \text{and} \quad \|h\| \le \frac{1}{\beta}\|k\|. \tag{3}$$

It can be easily seen that every multiple λh, λk, $\lambda \in \mathbb{R}$, of h, k will also satisfy this inequality. Hence for each $k \in F$, there is an $h \in E$ so that (3) is satisfied with a $q < 1$. The surjectivity of $A'(u^0)$ will finally result from a well-known theorem (see, for example, Kantorovich & Akilov (1964), XIV.1). □

The relation between the derivative of the inverse and the inverse of the derivative is described by the following lemma.

Lemma 6.5. Let A be differentiable, open, and locally injective at u^0, with the derivative $A'(u^0)$ bijective and continuously invertible. Then $\tilde{A}^{-1}$ is continuous and differentiable at $w^0 = Au^0$, and the derivative has the representation $(\tilde{A}^{-1})'(w^0) = A'(u^0)^{-1}$.

Proof: For every $\varepsilon > 0$, there exists a $\delta > 0$ such that $\|R^A(u^0;h)\| \le \varepsilon\|h\|$ for all $\|h\| \le \delta$. Without loss of generality, let $B_\delta(u^0) \subset U \equiv D(A)$. With $w^0 = Au^0$, we define

$$S(w^0;k) \equiv \tilde{A}^{-1}(w^0 + k) - \tilde{A}^{-1}w^0 - A'(u^0)^{-1}k$$

for all $w = Au \in W \equiv R(\tilde{A})$, $k = w - w^0$. We now show that $S(w^0;k)$ is the remainder term in differentiating $\tilde{A}^{-1}$ at w^0. Obviously, $S(w^0;k) = -A'(u^0)^{-1}R^A(u^0;h)$ with $h = \tilde{A}^{-1}(w - w^0) = u - u^0$, $u \in B_\delta(u^0)$. Since A is open at u^0, A is locally bijective at u^0; and, by Lemma 6.3, $\tilde{A}^{-1}$ is continuous at w^0. There is thus an $\eta > 0$ so that

$$\|u - u^0\| = \|\tilde{A}^{-1}w - \tilde{A}^{-1}w^0\| \le \delta \quad \text{for all} \quad \|k\| = \|w - w^0\| \le \eta.$$

Combined with the inequality in condition (2b) from Theorem 6.2, the above inequality yields the estimates

$$\|S(w^0;k)\| \le \frac{1}{\beta'}\|R^A(u^0;h)\| \le \frac{\varepsilon}{\beta'}\|u-u^0\| \le \frac{\varepsilon}{\beta\beta'}\|Au-Au^0\| = \frac{\varepsilon}{\beta\beta'}\|k\|$$

for all k, $||k|| \leq \eta$. Therefore we see that $S(w^0;k)$ satisfies the properties of a remainder term, and $(\tilde{A}^{-1})'(w^0) = A'(u^0)^{-1}$. □

Using the results from Lemma 6.3, we can replace the openness condition in the hypothesis of the preceding lemma. We show further that a continuously differentiable mapping with a bijective and bicontinuous derivative is always open.

Theorem 6.6. Let E, F be Banach spaces, and let A be differentiable in a neighborhood of u^0 with the derivative A' continuous at u^0 and $A'(u^0)$ bijective and bicontinuous. Then A is open in a neighborhood of u^0.

Proof: Let ε be arbitrarily selected in $0 < \varepsilon < \beta'$, where $\beta' = ||A'(u^0)^{-1}||^{-1}$, and $\delta > 0$ be the quantity from the continuity of A' at u^0 (i.e., A is differentiable in $B_\delta(u^0)$ with $||A'(u) - A'(u^0)|| \leq \varepsilon$ for all $u \in B_\delta(u^0)$). We define now $G(z;w)$ by

$$G(z;w) \equiv z - A'(u^0)^{-1}(Az-w), \quad z \in B_\delta(u^0), \quad w \in F.$$

Then the mapping $H_w = G(\cdot;w)$ is a contraction for every $w \in F$, since

$$||H_w'(z)|| = ||A'(u^0)^{-1}(A'(u^0) - A'(z))|| \leq \frac{\varepsilon}{\beta'} < 1, \quad z \in B_\delta(u^0).$$

Now, for each $u^1 \in B_\delta^0(u^0)$, $\delta_1 \equiv \delta - ||u^1 - u^0|| > 0$ and moreover $B_{\delta_1}(u^1) \subset B_\delta(u^0)$. Also, for every η in $0 < \eta \leq \delta_1$, with σ given by $\sigma \equiv (\beta'-\varepsilon)\eta$, and for every $w \in B_\sigma(w^1)$, with $w^1 = Au^1$, we see that the ball $B_\eta(u^1)$ is mapped into itself by H_w, as the following inequality shows,

$$\begin{aligned} ||G(u;w) - u^1|| &\leq ||G(u;w) - G(u^1;w)|| + ||G(u^1;w) - u^1|| \\ &\leq \frac{\varepsilon}{\beta'}||u - u^1|| + ||A'(u^0)^{-1}(Au^1 - w)|| \\ &\leq \frac{\varepsilon}{\beta'}||u - u^1|| + \frac{1}{\beta'}||w^1-w|| \leq \frac{\varepsilon}{\beta'}\eta + \frac{\sigma}{\beta'} = \eta, \\ &\qquad u \in B_\eta(u^1). \end{aligned}$$

We solve the equation $Av = w$ by appealing to the Banach Fixed Point Theorem to yield a fixed point $v \in B_\eta(u^1)$ of $v = G(v;w)$ for every $w \in B_\sigma(w^1)$. Hence $B_\sigma(w^1) \subset AB_\eta(u^1)$ for all $u^1 \in B_\delta^0(u^0)$ which shows conclusively that A is open at each vector in a neighborhood of u^0. □

Using the above results, we can now show the main theorem of this section. In essence, this theorem is the Inverse Function Theorem, which is a special case of the Implicit Function Theorem (see, for example, Dieudonné (1969), X.2, Ortega & Rheinboldt (1970), 5.2). In addition to the usual formulation of this theorem, we provide and show some two-sided inequalities.

Theorem 6.7. Let E, F be Banach spaces, and suppose A is continuously differentiable at u^0 and set $w^0 = Au^0$. Then the following statements are equivalent:

(4a) A is locally bijective at u^0; the associated restriction $\tilde{A}$ is bicontinuous at u^0; and $\tilde{A}^{-1}$ is continuously differentiable at w^0.

(4b) $A'(u^0)$ is bijective and bicontinuous.

(4c) $A'(u^0)$ is bijective, $||A'(u^0)|| < \infty$, $||A'(u^0)^{-1}|| < \infty$, and to arbitrary α, β for which $\alpha > ||A'(u^0)||$, $0 < \beta < ||A'(u^0)^{-1}||^{-1}$, there exists a positive number ρ such that

$$B_\rho(u^0) \subset D(A'), \quad A \text{ is open in } B^0_\rho(u^0), \text{ and}$$

$$\frac{1}{\alpha}||Au - Av|| \le ||u-v|| \le \frac{1}{\beta}||Au-Av||, \quad u,v \in B_\rho(u^0).$$

Proof: (4a) ➡ (4b). Application of Theorem 6.1 to $\tilde{A}^{-1}$ provides the existence of positive numbers σ, β with

$$||\tilde{A}^{-1}w - \tilde{A}^{-1}w^0|| \le \frac{1}{\beta}||w - w^0||, \quad w \in B_\sigma(w^0).$$

Since A is continuous at $u^0 = \tilde{A}^{-1}w^0$, Lemma 6.3 applied to $\tilde{A}^{-1}$ shows the openness of $\tilde{A}^{-1}$ at w^0. Therefore, there is a ball $B_\rho(u^0)$ with $B_\rho(u^0) \subset \tilde{A}^{-1}B_\sigma(w^0)$, and the bijectivity of $\tilde{A}$ yields in addition that

$$\beta||u - u^0|| \le ||Au - Au^0||, \quad u \in B_\rho(u^0) \subset \tilde{A}^{-1}B_\sigma(w^0).$$

From Theorem 6.2 we conclude that $\beta||h|| \le ||A'(u^0)h||$ for all $h \in E$. The injectivity and continuous invertibility of $A'(u^0)$ are thereby shown. Because $\tilde{A}^{-1}$ is continuous at w^0, we see that A is open at u^0 (cf. Lemma 6.3). Statement (4b) now follows by appealing to Theorem 6.4.

(4b) ➡ (4c). From (4b), we obtain the statements (1c) and (2b) when A is continuously differentiable at u^0. Additionally, it follows from Theorem 6.6 that A is open in a neighborhood $B^0_\rho(u^0)$ of u^0.

(4c) ⇒ (4a). The two-sided inequality in (4c) immediately shows that A is locally bijective and continuous in $B_\rho(u^0)$, with the restriction $\tilde{A} = A|B_\rho^0(u^0): B_\rho^0(u^0) \to AB_\rho^0(u^0)$ both bijective and bicontinuous. What must be shown now is that $\tilde{A}$ is continuously differentiable at w^0. According to Lemma 6.3, A is open at every point $v \in B_\rho^0(u^0)$, since $\tilde{A}^{-1}$ is continuous at Av. Moreover, there is to every $u^1 \in B_\rho^0(u^0)$ a ball $B_{\rho_1}(u^1) \subset B_\rho^0(u^0) \subset B_\rho(u^0)$, such that by (4c) we have

$$\frac{1}{\alpha}||Au - Au^1|| \le ||u - u^1|| \le \frac{1}{\beta}||Au - Au^1||, \quad u \in B_{\rho_1}(u^1).$$

As in the first part of this proof, we infer from the second inequality that $A'(u^1)$ is injective and continuously invertible. From Theorem 6.4, we see that $A'(u^1)$ is both bijective and bicontinuous. Applying Lemma 6.5 to u^1, we see that A^{-1} is differentiable at $w^1 = Au^1$ having derivative $(\tilde{A}^{-1})'(w^1) = A'(u^1)^{-1}$. Finally, we see from this representation and from the continuity of A' at u^0 that $(\tilde{A}^{-1})'$ is continuous at w^0. □

Using a result of John (1968) one can state that the maximal radius in (4c) is given by $\sigma = \beta\rho$.

6.2. STABILITY AND INVERSE STABILITY OF SEQUENCES OF MAPPINGS

The results of the previous section can be directly applied to sequences of mappings. For example, the continuity of a sequence of mappings, considered as a single mapping is tantamount to the equicontinuity of the sequence itself. This property turns out to be equivalent to the concept of stability, which can be characterized immediately by Theorem 6.1 in terms of the uniform boundedness of the derivatives. Corresponding results occur for the inverses and for the naturally associated concept of "inverse stability". The main theorem of this section is a reformulation of Theorem 6.7 for sequences of mappings.

Given a sequence E_n, $n \in I$ (= $\mathbb{N}$), of Banach spaces, we denote as before the linear subspace of bounded sequences $\underset{\sim}{u} = (u_n)$ in $\pi_n E_n$ as $\ell^\infty(E_n)$. We now let the norm of a sequence (u_n) of elements $u_n \in E_n$, $n \in I$, be defined by

$$||(u_n)|| = ||\underset{\sim}{u}|| \equiv \sup_{n \in I} ||u_n||_n,$$

where $||\cdot||_n$ is the norm in each E_n, $n \in I$. Equipped with such a norm,

$\ell^\infty(E_n)$ is then a Banach space. We designate the open ball in $\ell^\infty(E_n)$ by the notation,

$$B_\rho^0((u_n)) = B_\rho^0(\underset{\sim}{u}) = \{\underset{\sim}{v} = (v_n) \in \pi_n E_n: \ ||\underset{\sim}{u} - \underset{\sim}{v}|| < \rho\}.$$

A sequence (A_n) of mappings between Banach spaces E_n, F_n, $n \in I$, defines a mapping $A = (A_n)$ with domain of definition in $\pi_n E_n$ and range in $\pi_n F_n$. We do not distinguish between the norms in E_n and F_n and denote both by $||\cdot||_n$. Obviously, $\underset{\sim}{A} = (A_n)$ is injective (respectively, bijective) if, and only if, each A_n, $n \in I$, is injective (respectively, bijective). The domain of definition and the range of $\underset{\sim}{A} = (A_n)$ are described in terms of the domains and ranges of each A_n, $n \in I$, by $D(\underset{\sim}{A}) = \pi_n D(A_n)$, $R(\underset{\sim}{A}) = \pi_n R(A_n)$.

We consider now a sequence of linear mappings $L_n: E_n \to F_n$, $n \in I$. Then the continuity of $\underset{\sim}{L} = (L_n)$ is obviously equivalent to the uniform boundedness of the sequence L_n, $n \in I$, which in turn is equivalent to its equicontinuity, i.e.,

$$\underset{\sim}{L} \text{ continuous} \Longleftrightarrow \sup_{n\in I} ||L_n|| < \infty \Longleftrightarrow (L_n) \text{ equicontinuous.}$$

In case L_n, $n \in I$, is a sequence of injective, linear mappings, then we have correspondingly the equivalence between continuous invertibility of $\underset{\sim}{L}$ and equicontinuous invertibility of (L_n). For nonlinear mappings A_n, $n \in I$, the differentiability (respectively, continuous differentiability) of $\underset{\sim}{A}$ at $\underset{\sim}{u}^0 = (u_n^0)$ is equivalent to the equidifferentiability (respectively, equicontinuous equidifferentiability) of A_n, $n \in I$, at (u_n). We shall employ throughout the remaining text the terminology appropriate for sequences of mappings - for example, terms such as equicontinuous equidifferentiable, etc. Because of the importance of equidifferentiability (equicontinuous equidifferentiability) we give here once more a comprehensive definition of these concepts. (A_n) is called *equidifferentiable* at $(u_n^0) \in \ell^\infty(E_n)$, if there exists a sequence of linear mappings $A_n'(u_n^0)$: $E_n \to F_n$, $n \in I$, such that

$$\forall\varepsilon > 0, \ \exists\delta > 0 \ \ni: u_n^0 + h_n \in D(A_n), \ ||R_n^A(u_n^0;h_n)||_n \le \varepsilon||h_n||_n \quad (5)$$

for every $n \in I$ and h_n in the closed ball $||h_n||_n \le \delta$. We note that the mappings $A_n'(u_n^0)$ are the (Fréchet-) derivatives of A_n at u_n^0 and the R_n^A, $n \in I$, defined by

$$R_n^A(u_n^0;h_n) \equiv A_n(u_n^0 + h_n) - A_n u_n^0 - A_n'(u_n^0)h_n, \quad n \in I,$$

are the associated remainder terms. The sequence (A_n) is defined to be *equicontinuously equidifferentiable* at $(u_n^0) \in \ell^\infty(E_n)$ if (A_n) is equidifferentiable at each $(v_n) \in B_\rho^0((u_n^0))$, for some $\rho > 0$ (independent of n) and, moreover, if

$$\forall \varepsilon > 0, \quad \exists \eta > 0 \quad \ni: \; ||(A_n'(u_n^0) - A_n'(u_n))h_n||_n \leq \varepsilon ||h_n||_n, \tag{6}$$

for every $n \in I$, $h_n \in E_n$ and every $u_n \in E_n$ with $||u_n^0 - u_n||_n \leq \eta$.

Theorem 6.1 now directly provides two characterizations of equicontinuity for sequences of differentiable mappings. Using a result of Stummel & Reinhardt (1973), we show first the equivalence of equicontinuity to an apparently weaker condition which we denote as *stability*. Indeed a sequence (A_n) is called *stable* at (u_n) if the following relation is valid for every sequence $v_n \in D(A_n)$, $n \in I$,

$$||u_n - v_n||_n \to 0 \Rightarrow ||A_n u_n - A_n v_n||_n \to 0 \quad (n \in I). \tag{7}$$

(We emphasize that for brevity, "$(n \in I)$" will again be written instead of "$(n \to \infty)$". Likewise, the notation "$(n \in I')$" is used in lieu of "$(n \to \infty,\ n \in I')$" for subsequences $I' \subset I$.)

We now turn to proving the following theorem giving equivalent characterizations of stability.

Theorem 6.8. Let the sequence (A_n) be equicontinuously equidifferentiable (respectively, equidifferentiable) at (u_n^0). Then the following statements are equivalent:

(8a) (A_n) is stable at (u_n^0) and every A_n is continuous at u_n^0, $n \in I$.

(8b) (A_n) is equicontinuous at (u_n^0).

(8c) The sequence $||A_n'(u_n^0)||$, $n \in I$, is bounded.

(8d) $\alpha' \equiv \sup_n ||A_n'(u_n^0)|| < \infty$; and to every $\alpha > \alpha'$, there is a $\rho > 0$ such that $B_\rho(u_n^0) \subset D(A_n')$ (respectively, $\subset D(A_n)$) and

$$||A_n u_n - A_n v_n||_n \leq \alpha ||u_n - v_n||_n, \quad u_n, v_n \in B_\rho(u_n^0),$$
$$\text{(respectively, } v_n = u_n^0), \quad n \in I.$$

Proof: The equivalence of (8b), (8c), and (8d) follows from Theorem 6.1. The continuity of every single A_n at u_n^0, $n \in I$, along with the stability condition (7) follows from (8b). Conversely, assume (8a). We now show by a contradiction argument that (A_n) is equicontinuous at (u_n^0).

Indeed, let us assume that

$$\exists\varepsilon_0 > 0 \quad \ni: \forall\delta > 0 \quad \exists n \in I,\ v_n \in E_n:\ ||v_n - u_n^0||_n \le \delta$$
$$\text{and} \quad ||A_n v_n - A_n u_n^0||_n > \varepsilon_0.$$

With $\delta = 1/t$, $t = 1,2,\ldots$, we have the existence of indices $n_t \in I$ and of elements $v_{n_t}^{(t)} \in E_{n_t}$ such that

$$||v_n^{(t)} - u_n^0||_n \le 1/t \quad \text{and} \quad ||A_n v_n^{(t)} - A_n u_n^0||_n > \varepsilon_0, \quad n = n_t, \quad t = 1,2,\ldots .$$

If $\{n_t\colon t = 1,2,\ldots\}$ contains a subsequence $I' \subset I$ of pairwise different indices, then we define

$$\hat{v}_n = v_n^{(t)}, \quad n = n_t \in I', \quad \hat{v}_n = u_n^0, \quad n \in I-I'.$$

By construction, $||\hat{v}_n - u_n^0||_n \to 0\ (n \in I)$; and, because of the stability (7), the sequence $||A_n\hat{v}_n - A_n u_n^0||_n$, $n \in I$, must also converge to zero, thereby contradicting $||A_n\hat{v}_n - A_n u_n^0||_n > \varepsilon_0$, $n \in I'$. If $\{n_t\colon t = 1,2,\ldots\}$ consists of only finitely many different indices, then there is an $\nu_0 \in I$ and a subsequence $I' \subset I$ so that $\nu_0 = n_t$, $t \in I'$. We then have

$$||v_{\nu_0}^{(t)} - u_{\nu_0}^0||_{\nu_0} \to 0\ (t \in I') \quad \text{and} \quad ||A_{\nu_0} v_{\nu_0}^{(t)} - A_{\nu_0} u_{\nu_0}^0||_{\nu_0} > \varepsilon_0,\ t \in I',$$

thereby contradicting the continuity of A_{ν_0}. □

As is evident in the proof, we do not need the hypothesis of equicontinuous equidifferentiability for the equivalence of statements (8a) and (8b).

We now give a characterization of stability for *linear mappings*.

Lemma 6.9. A sequence of linear operators L_n, $n \in I$, is stable at (u_n^0) if, and only if, L_n is bounded for almost all $n \in I$ and

$$\limsup_{n\in I} ||L_n|| < \infty. \tag{9}$$

Proof: From (9), stability follows immediately. Let now (L_n) be stable at (u_n^0) and suppose (9) is not true, i.e., $||L_n|| \to \infty$ $(n \in I')$ for a subsequence $I' \subset I$. By definition of the norm of a linear operator, there exist elements $v_n \in E_n$, $n \in I'$, with $||L_n v_n||_n/||v_n||_n \to \infty\ (n \in I')$. If we define $\hat{v}_n \equiv v_n/||L_n v_n||_n$, $n \in I'$, then

$$||L_n\hat{v}_n||_n = 1, \quad n \in I', \quad \text{and} \quad ||\hat{v}_n||_n \to 0\ (n \in I').$$

If we now set $u_n = u_n^0 + \hat{v}_n$, $n \in I'$, $u_n = u_n^0$, $n \in I-I'$, then

$$||u_n - u_n^0||_n \to 0 \ (n \in I), \quad \limsup_{n \in I} ||L_n u_n - L_n u_n^0||_n = 1,$$

thus contradicting the stability at (u_n^0). □

Since the characterization of stability in Lemma 6.9 does not depend on the sequence (u_n^0), we have the following corollary to Lemma 6.9.

Lemma 6.10. A sequence (L_n) of linear operators is stable at each (u_n) if, and only if, it is stable at some (u_n^0). □

From Lemmas 6.9 and 6.10, we have immediately the following theorem giving equivalent conditions for the stability of a sequence of bounded, linear operators.

Theorem 6.11. For every $n \in I$, let L_n be linear and bounded. Then the following statements are equivalent:

(10a) (L_n) is stable at each sequence (u_n).
(10b) (L_n) is stable at some sequence (u_n^0).
(10c) $\sup_{n \in I} ||L_n|| < \infty$. □

These equivalent statements also follow from Theorem 6.8, if we take into consideration, that $L_n'(u_n) = L_n$, $u_n \in E_n$, whenever the L_n, $n \in I$, are linear. An obvious consequence of the latter results is that statement (8c) in Theorem 6.8 - and hence each of the statements (8a-d) - are furthermore equivalent to the stability of the sequence $(A_n'(u_n^0))$ along with the boundedness of each individual $A_n'(u_n^0)$, $n \in I$.

At this point, we provide several *examples* of stable sequences of mappings:

Example 1. Let E be a Banach space, F a normed space, with $E_n = E$, $F_n = F$, $n \in I$, and let $L_n: E_n \to F_n$, $n \in I$, be a sequence of bounded linear mappings. In our terminology, the *principle of uniform boundedness* means precisely that the stability of (L_n) is a consequence of the pointwise boundedness of the sequence (L_n), $n \in I$, i.e.,

$$\sup_{n \in I} ||L_n u|| < \infty, \quad u \in E. \quad □$$

Example 2. Let E and F be normed spaces with F_n, $n \in I$, subspaces of F and $P_n: F \to F_n$, $n \in I$, a stable sequence of linear, bounded projection operators (i.e., there is a $\gamma \geq 0$ such that $||P_n|| \leq \gamma$, $n \in I$). Further, let $A: D(A) \subset E \to F$ be a continuous mapping. Then the sequence

$A_n = P_nA$, $n \in I$, is stable at every sequence $u_n \in D(A)$, $n \in I$, which converges to some $u \in D(A)$. Indeed, it follows from $||u_n - u|| \to 0$ and $||u_n - v_n|| \to 0$ $(n \in I)$, that $||v_n - u|| \to 0$ $(n \in I)$; with $u \in D(A)$, the continuity of A yields

$$||A_nu_n - A_nv_n|| \leq \gamma||Au_n - Av_n|| \to 0 \ (n \in I). \qquad \square$$

Example 3. Let G_n, $n \in I$, be closed sets of a compact metric space M, with $C(G_n)$, $n \in I$, the corresponding Banach spaces of continuous functions on G_n. Let μ_n, $n \in I$, be measures on each respective G_n, $n \in I$, as defined in Section 5.4. With $E_n = C(G_n)$, $n \in I$, and $F_n = \mathbb{K}$ $(= \mathbb{R}$ or $\mathbb{C})$, we see that the sequence $(\mu_n)_{n\in I}$ generates a sequence of bounded linear mappings, $\mu_n: E_n \to F_n$, $n \in I$. The stability of μ_n, $n \in I$, (i.e., the existence of a $C \geq 0$ independent of n such that

$$\left|\int_{G_n} u_n d\mu_n\right| \leq C||u_n||_n, \quad u_n \in C(G_n), \quad n \in I,$$

with the maximum norm $||\cdot||_n$ taken in each $C(G_n)$) is equivalent to the uniform boundedness of the norms $||\mu_n||$, $n \in I$. From Theorem 6.11 we see that this uniform boundedness property is equivalent to stability as defined by (7). The natural extensions $\hat{\mu}_n$, $n \in I$, of μ_n to measures on M are stable, if $|\hat{\mu}_n(w)|$, $n \in I$, is bounded for every $w \in C(M)$. Such a conclusion follows again from the Uniform Boundedness Principle. $\square$

As before, we let (A_n) be a sequence of (not necessarily linear) mappings. We call (A_n) *inversely stable* at (u_n), if

$$||A_nu_n - A_nv_n||_n \to 0 \Rightarrow ||u_n - v_n||_n \to 0 \ (n \in I), \tag{11}$$

for every sequence $v_n \in D(A_n)$, $n \in I$. If (A_n) is locally bijective at (u_n), then the inverse stability at (u_n) is obviously equivalent to the stability of $(\tilde{A}_n^{-1})$ at $(w_n) = (A_nu_n)$. Because of this fact, we can immediately formulate a theorem analogous to Theorem 6.8 which gives equivalent conditions for the inverse stability whenever $(\tilde{A}_n^{-1})$ is assumed to be equicontinuously equidifferentiable. We leave the statement and proof of this result to the reader.

For linear mappings, we can show as in Lemma 6.9 that the inverse stability at an arbitrary sequence (u_n) is equivalent to

$$||L_nv_n||_n \leq \gamma_0||v_n||_n, \quad v_n \in E_n, \quad n \geq \nu, \tag{12}$$

for some constant $\gamma_0 > 0$ and an index $\nu \in I$. Consequently, we can obtain a result analogous to Lemma 6.10. For continuously invertible linear mappings L_n, $n \in I$, the inverse stability is equivalent to the validity of (12) for all n. For brevity, we shall say that a sequence of linear operators is *stable* (respectively, *inversely stable*) if it is stable (respectively, inversely stable) at each sequence.

A sequence (A_n) of (not necessarily linear) mappings is called *bistable* at (u_n), if (A_n) is stable and inversely stable at (u_n), i.e.,

$$||u_n - v_n||_n \to 0 \Longleftrightarrow ||A_n u_n - A_n v_n||_n \to 0 \ (n \in I)$$

for every sequence $v_n \in D(A_n)$, $n \in I$. Now Theorem 6.7 applied to (A_n) immediately yields the following result providing equivalent conditions for the bistability of a sequence of equicontinuously equidifferentiable mappings.

<u>Theorem 6.12</u>. Let (A_n) be equicontinuously equidifferentiable at (u_n^0) and let $(w_n^0) = (A_n u_n^0)$. Then the following statements are equivalent:

(13a) (A_n) is locally bijective at (u_n^0); the associated restrictions $\tilde{A}_n$ are bijective and bicontinuous at u_n^0 for every $n \in I$; the sequence (A_n) is bistable at (u_n^0); and the sequence of inverses $(\tilde{A}_n^{-1})$ is equicontinuously equidifferentiable at (w_n^0).

(13b) (A_n) is locally bijective at (u_n^0) with the sequence of restrictions $(\tilde{A}_n)$ equibicontinuous at (u_n^0), and $(\tilde{A}_n^{-1})$ is equicontinuously equidifferentiable at (w_n^0).

(13c) The derivatives $A_n'(u_n^0)$ are bijective for every $n \in I$; and both of the sequences $||A_n'(u_n^0)||$, $n \in I$, and $||A_n'(u_n^0)^{-1}||$, $n \in I$, are bounded.

(13d) $A_n'(u_n^0)$ is bijective for every $n \in I$;

$$\alpha' \equiv \sup_{n \in I} ||A_n'(u_n^0)|| < \infty, \quad \beta' \equiv \inf_{n \in I} ||A_n'(u_n^0)^{-1}||^{-1} > 0;$$

for every $\alpha > \alpha'$ and every $\beta \in (0,\beta')$, there is a number $\rho > 0$ such that $B_\rho((u_n^0)) \subset D((A_n'))$, (A_n) is open in $B_\rho^0((u_n^0))$, and

$$\frac{1}{\alpha}||A_n u_n - A_n v_n||_n \le ||u_n - v_n||_n \le \frac{1}{\beta}||A_n u_n - A_n v_n||_n,$$

$$u_n, v_n \in B_\rho(u_n^0), \quad n \in I. \quad \square$$

6.3. CONSISTENCY AND DISCRETE CONVERGENCE OF MAPPINGS

In this section, we introduce the concepts of consistency and discrete convergence of mappings. The fundamental Theorem 6.13 states that discrete convergence is equivalent to stability and consistency. In addition, we have from Theorem 6.8 several characterizations of stability at our disposal. The concepts introduced in this section are explained by preliminary examples. Finally, we show that the limit itself of a discretely convergent sequence of mappings is necessarily continuous.

Let $A(E,\pi_n E_n,\lim^E)$ and $A(F,\pi_n F_n,\lim^F)$ be discrete approximations of normed spaces E,E_n,F,F_n, $n \in I$, and let $A\colon D(A) \subset E \to F$, $A_n\colon D(A_n) \subset E_n \to F_n$, $n \in I$, be (not necessarily linear) mappings. The sequence (A_n) is called *discretely convergent to* A *at* $u \in D(A)$ (in symbols: $A_n \to A$ $(n \in I)$ at u), if for every sequence $u_n \in D(A_n)$, $n \in I$, the following relation holds:

$$u_n \to u \ (n \in I) \Rightarrow A_n u_n \to Au \ (n \in I).$$

We say (A_n) is *discretely convergent* to A (in symbols: $A_n \to A$ $(n \in I)$) if $A_n \to A$ $(n \in I)$ at every $u \in D(A)$. If we consider linear mappings, then we tacitly assume that Property 5.2 of asymptotic linearity holds for both discrete convergences $\lim^E$ and $\lim^F$. We shall state explicitly whether Property 5.3 of the norms is required.

The following simple *examples* show that the definition of discretely convergent sequences is a generalization of the usual concept of convergence.

Example 1. Let $E_n = E$, $F_n = F$, $n \in I$, and let $\lim^E$, $\lim^F$ denote the norm convergence in E and F, respectively. Suppose moreover that $A_n = A$, $n \in I$. Then the discrete convergence of (A_n) to A at u is tantamount to the continuity of the mapping A at u. □

Example 2. Let $E_n = E$, $F_n = F$, $n \in I$, be as in Example 1, with the mappings A_n possibly different from A. Then $A_n \to A$ $(n \in I)$ directly means the "continuous convergence" $A_n \underset{s}{\to} A$ $(n \in I)$ in the sense of Rinow (1961), §9. The continuity of A is, of course, necessary for the continuous convergence of $A_n \underset{s}{\to} A$ $(n \in I)$. □

Example 3. Let E be a Banach space, F a normed space with $E_n = E$, $F_n = F$. Let $A_n\colon E \to F$, $n \in I$, be a sequence of bounded, linear mappings, and let A be a mapping of E into F. Then $A_n \to A$ $(n \in I)$ if, and only if, A_n, $n \in I$, converges pointwise to A, i.e.,

$$A_n u \to Au \ (n \in I), \quad u \in E.$$

The pointwise convergence namely guarantees, by the Uniform Boundedness Principle, the existence of a constant $\alpha \geq 0$ such that $||A_n|| \leq \alpha$, $n \in I$. From this, we see that $A_n u_n \to Au$ $(n \in I)$ for every sequence $u_n \to u$ $(n \in I)$, since we have the estimates

$$||A_n u_n - Au|| \leq \alpha ||u_n - u|| + ||A_n u - Au||, \quad n \in I.$$

A is necessarily bounded and linear. Moreover, we know from the Banach-Steinhaus Theorem that A_n, $n \in I$, converges pointwise to A, if, and only if, the sequence A_n, $n \in I$, is uniformly bounded and

$$A_n \phi \to A\phi \quad (n \in I)$$

for all ϕ in some dense subspace $\Phi \subset E$. □

<u>Example 4</u>. Let E_n, $n \in I$, be subspaces of a normed space E with the property that

$$|u, E_n| \to 0 \quad (n \in I), \quad u \in E.$$

Then a discrete convergence is given by the convergence relative to the norm in E (cf. 5.(3)). For subspaces F_n of a normed space F, let P_n: $F \to F_n$, $n \in I$, be bounded linear projection operators with the property that

$$P_n w \to w \ (n \in I), \ w \in F, \text{ and } \ ||P_n|| \leq \gamma < \infty, \ n \in I.$$

Then a discrete convergence is likewise defined by the convergence in F (cf. 5.(4)). Finally, assume that A: $D(A) \subset E \to F$ is continuous. If we define $A_n = P_n A|E_n$, $n \in I$, then (A_n) converges discretely to A, since the discrete convergence of an arbitrary sequence u_n, $n \in I$, to u implies $Au_n \to Au$ $(n \in I)$ and hence

$$||A_n u_n - Au|| \leq \gamma ||Au_n - Au|| + ||P_n Au - Au|| \to 0 \ (n \in I). \quad □$$

<u>Example 5</u>. Let G, G_n, $n \in I$, be closed subsets of a compact metric space M, with $C(G)$, $C(G_n)$, $n \in I$, the associated Banach spaces of continuous functions and with μ, μ_n, $n \in I$, measures on G and G_n, $n \in I$, respectively. For the natural extensions $\hat{\mu}, \hat{\mu}_n$, $n \in I$ of μ, μ_n, $n \in I$, to M, $\hat{\mu}_n \to \hat{\mu}$ $(n \in I)$ in the sense of discrete convergence if, and only if, $\hat{\mu}_n \rightharpoonup \hat{\mu}$ $(n \in I)$ in the sense of weak convergence (cf. Section 5.4). This is a special case of Example 3 with $E = E_n = C(M)$ and $F = F_n = \mathbb{K}$.

The discrete convergence of the measures themselves is characterized in Example 6 at the end of this section. □

In essence, these examples anticipate the characterization of discrete convergence in Theorem 6.13 stated later. For the formulation of this theorem, we still need the concept of consistency. The sequence A, A_n, $n \in I$, is called *consistent* at $u \in D(A)$, if a sequence $u_n' \in D(A_n)$, $n \in I$, exists such that

$$u_n' \to u, \quad A_n u_n' \to Au \ (n \in I). \tag{14}$$

We label the sequence (u_n') a *consistency sequence* for u. The sequence A, A_n, $n \in I$, is termed *consistent* if it is consistent at every $u \in D(A)$. The preceding examples immediately provide *examples* of consistent mappings. In Example 3, the pointwise convergence $A_n u \to Au$ $(n \in I)$, $u \in E$, is exactly the consistency at u with the constant consistency sequence $u_n' = u$, $n \in I$. In Example 4, every sequence $u_n' \to u$ $(n \in I)$ is a consistency sequence, and the existence of such a sequence $u_n' \in E_n$, $n \in I$, follows from $|u,E_n| \to 0$ $(n \in I)$; then the sequence A, A_n, $n \in I$, is obviously consistent at u. For the measures μ, μ_n, $n \in I$ (cf. Example 5), we have already defined a concept of consistency in Section 5.4. Under the hypothesis $\operatorname{Lim\,sup} G_n = G$ we know that there exists a discrete approximation $A(C(G), \pi_n C(G_n), \lim)$ (cf. Section 5.3). The consistency of measures in the sense of 5.(45) then directly means the consistency of μ, μ_n, $n \in I$, at every $u \in C(G)$ with consistency sequence $\hat{u}|G_n$, $n \in I$.

Other, nontrivial exmaples of consistent operators have already been discussed for finite-difference approximations for boundary-value problems (cf. Chapter 1). There, in view of Section 5.3, the truncation errors were defined by $\tau_h = A_h r_h u - r_h Au$, where the boundary conditions are taken into account in A and its approximations A_h. The operators r_h are the restrictions to a mesh with equidistant mesh points. As we have seen in Section 5.3, a discrete convergence is given by

$$u_h \to u \Longleftrightarrow ||r_h u - u_h|| \to 0 \quad (h \to 0,\ h \in \Lambda)$$

with a null sequence $\Lambda = (h_1, h_2, \ldots)$ of step widths. The convergence of the truncation errors to zero, $\tau_h \to 0$ $(h \in \Lambda)$, then corresponds directly to the consistency of A, A_h, $h \in \Lambda$, at u, where the consistency sequence is specially given by $u_h' = r_h u$, $h \in \Lambda$.

In the remainder of this section, we shall always assume that $A(D(A), \pi_n D(A_n), \lim^E)$ forms a discrete approximation. This will be the case for consistent A, A_n, $n \in I$. It is then trivial to see that a discretely convergent sequence is necessarily consistent. The following theorem shows that the stability (7) is precisely the needed condition to characterize discrete convergence.

Theorem 6.13. The discrete convergence $A_n \to A$ $(n \in I)$ at $u \in D(A)$ is equivalent tc the consistency of A, A_n, $n \in I$, at u together with the stability of (A_n) at any consistency sequence.

Proof: If $A_n \to A$ $(n \in I)$ at $u \in D(A)$, then A, A_n, $n \in I$, is necessarily consistent at u. Let (u_n') be a consistency sequence and $v_n \in D(A_n)$, $n \in I$, such that $||u_n - v_n||_n \to 0$ $(n \in I)$. Then $v_n \to u$ $(n \in I)$, and, by assumption, $A_n v_n \to Au$ together with $||A_n v_n - A_n u_n'||_n \to 0$ $(n \in I)$. Therefore, (A_n) is stable at (u_n'). To prove the converse, let us assume $u_n \to u$ $(n \in I)$. Because of consistency, there exists a sequence (u_n') with $u_n' \to u$, $A_n u_n' \to Au$ $(n \in I)$. The stability at (u_n') then allows us to infer that $||A_n u_n' - A_n u_n||_n \to 0$ and hence that $A_n u_n \to Au$ $(n \in I)$. □

Using the characterization of stability given in Theorem 6.8, we are able to formulate the following theorem which provides equivalent conditions for the discrete convergence of differentiable mappings. For simplicity, we assume here and in the remainder of the chapter that E, E_n, F, F_n, $n \in I$, are Banach spaces, although we will not need their completeness at every point in the presentation.

Theorem 6.14. Let the assumptions of Theorem 6.8 be valid, and suppose that $u_n^0 \to u^0$ $(n \in I)$ for a $u^0 \in D(A)$. Then the following statements are equivalent:

(15a) $A_n \to A$ $(n \in I)$ at u^0 and A_n is continuous at u_n^0 for every $n \in I$.

(15b) A, A_n, $n \in I$, is consistent at u^0 with the consistency sequence (u_n^0), and one of the statements (8a) to (8d) is true.

If one of the statements (15a) or (15b) is met, then the following error estimate holds for every sequence $w_n^1 \to Au^0$ $(n \in I)$,

$$||A_n u_n - w_n^1||_n \leq \alpha ||u_n - u_n^0||_n + ||A_n u_n^0 - w_n^1||_n, \qquad u_n \in B_\rho(u_n^0), \quad n \in I. \tag{15c}$$

Proof: (15a) ⇒ (15b). According to Theorem 6.13, $A_n \to A$ $(n \in I)$ at u^0 implies the consistency of A, A_n, $n \in I$, at u^0 and the stability of (A_n) at (u_n^0). Theorem 6.8 can now be used to conclude the validity of each of the statements (8a) to (8d).

(15b) $\Rightarrow$ (15c). If now one of the statements (8a) to (8d) is valid, we know from Theorem 6.8 that all are valid. If we utilize the triangle in equality and (8d) with (v_n) defined as $(v_n) = (u_n^0)$, we get the estimate in (15c).

(15b) $\Rightarrow$ (15a). The inequality in (8d) implies the stability of (A_n) at the consistency sequence. Theorem 6.13 thus yields the discrete convergence $A_n \to A$ $(n \in I)$ at u^0. Finally, (8a) ensures the second statement of (15a). □

We now show that A is continuous if $A_n \to A$ $(n \in I)$. This result is to be expected, since it is true for the special cases considered in Examples 1 and 2. In order to prove the following theorem, we have to assume that Properties 5.2 and 5.3 of asymptotic linearity and of convergent norms are satisfied for the discrete approximations $A(E,\pi_n E_n,\lim^E)$ and $A(F,\pi_n F_n,\lim^F)$.

Theorem 6.15. If there is a sequence (A_n) with $A_n \to A$ $(n \in I)$, then A is continuous.

Proof: We prove the assertion by contradiction. Let us assume that there exist elements $u, u^{(j)} \in D(A)$, $j \in I$, and an $\varepsilon_0 > 0$ such that

$$||u - u^{(j)}|| \to 0 \ (j \to \infty) \quad \text{and} \quad ||Au^{(j)} - Au|| \geq \varepsilon_0, \quad j \in I.$$

Let (u_n), $(u_n^{(j)})$ be sequences such that $u_n \to u$, $u_n^{(j)} \to u^{(j)}$. Because $A_n \to A$ $(n \in I)$, there are indices $\nu_j > \nu_{j-1}$, $j = 2,3,\ldots$, so that

$$||u_n^{(j)} - u_n||_n \leq ||u^{(j)} - u|| + \frac{1}{j}, \quad ||A_n u_n^{(j)} - A_n u_n||_n \geq \varepsilon_0/2,$$
$$n \geq \nu_j, \quad j \in I.$$

If we define

$$v_n = u_n^{(j)}, \quad \nu_j \leq n < \nu_{j+1}, \ j \in I, \quad v_n = u_n^{(1)}, \quad 1 \leq n < \nu_1,$$

then

$$||v_n - u_n||_n \to 0 \ (n \in I) \quad \text{and} \quad ||A_n v_n - A_n u_n||_n \geq \varepsilon_0/2, \quad n \in I,$$

in contradiction to $A_n v_n \to Au$, $A_n u_n \to Au$ $(n \in I)$. □

At this point, we would like to point out further that, for a continuous mapping, consistency at each element in a dense subset $\Phi \subset D(A)$ is equivalent to consistency at every element in $D(A)$ itself. In order to see this, we again need to make use of Properties 5.2 and 5.3.

The preceding remark, used in conjunction with Theorem 6.11 and Theorem 6.13, gives the following theorem which specifies equivalent conditions for discrete convergence of bounded linear mappings. The proof is left to the reader.

Theorem 6.16. Let $L \in B(E,F)$, $L_n \in B(E_n,F_n)$, $n \in I$. Then the discrete convergence of L_n to L is equivalent to the uniform boundedness of the L_n, $n \in I$, and the consistency of L,L_n, $n \in I$. If the Convergence Property 5.3 for the norms holds, then consistency at any element in E follows from consistency at each element in a dense subset of E. □

In passing, we investigate the discrete convergence of measures.

Example 6. Let μ,μ_n, $n \in I$, be measures on $C(G)$, $C(G_n)$, $n \in I$, where G,G_n are closed subsets of a compact metric space M. Under the assumption that $\operatorname{Lim\,sup} G_n = G$, there exists a discrete approximation $A(C(G), \pi_n C(G_n),\lim)$ with the discrete (uniform) convergence 5.(22) (see Theorem 5.10). According to Theorem 6.16, the discrete convergence of the measures $\mu_n \to \mu$ $(n \in I)$ is equivalent to the uniform boundedness of (μ_n) together with the consistency of μ,μ_n, $n \in I$. The consistency of μ,μ_n, $n \in I$, can obviously be replaced by the special consistency condition 5.(45), since $\hat{u}|G_n \to u$ $(n \in I)$ for every extension $\hat{u} \in C(M)$ of $u \in C(G)$. Theorem 5.16 from Section 5.4, along with Theorem 6.16, shows that the weak convergence of the natural extension $\hat{\mu}_n \rightharpoonup \hat{\mu}$ $(n \in I)$ is sufficient for the discrete convergence $\mu_n \to \mu$ $(n \in I)$, and is, moreover, necessary whenever $\operatorname{Lim\,sup} G_n = G$. If μ is positive, then we additionally get $\operatorname{Lim} G_n = G$. This ensures the convergence of the maximum norms, so that consistency for a dense subset of $C(G)$ (e.g., polynomials) then implies consistency for all of $C(G)$. □

6.4. DISCRETE CONVERGENCE OF SOLUTIONS AND BICONVERGENCE

In this section, we examine the central question of whether the solutions of

$$A_n u_n = w_n, \quad n \in I,$$

converge (discretely) to the solution of

$$Au = w$$

in the sense of the definition given in Chapter 5, whenever the right-hand sides w_n, $n \in I$, converge to w. Theorem 6.17 will show that this convergence is present in case the sequence A,A_n, $n \in I$, is consistent and (A_n) is inversely stable. In case the inverses exist, the convergence relation

$$w_n \to w \Rightarrow u_n \to u \ (n \in I)$$

is nothing else than the discrete convergence $A_n^{-1} \to A^{-1}$ $(n \in I)$ at w. The characterizations of inverse stability in the preceding section furnish a series of equivalent statements on the convergence of solutions including error estimates. Using the characterizations of bistability, we prove a theorem giving equivalent conditions for biconvergence. By means of a suitable choice of norms, this particular theorem can be applied to the case where (A_n) is not necessarily stable. Also in this case, we obtain results on the solvability of the approximate equations as well as convergence of the approximate solutions.

As before, let $A(E,\pi_n E_n,\lim^E)$ and $A(F,\pi_n F_n,\lim^F)$ be discrete approximations of Banach spaces. We say that (A_n) is *inversely discretely convergent to* A *at* $u \in D(A)$, if

$$A_n u_n \to Au \ (n \in I) \Rightarrow u_n \to u \ (n \in I) \tag{16}$$

for every sequence $u_n \in D(A_n)$, $n \in I$. (A_n) is called *inversely discretely convergent to* A if (16) is valid for every $u \in D(A)$. For bijective mappings A, A_n, $n \in I$, (A_n) is inversely discretely convergent to A at u if, and only if, $A_n^{-1} \to A^{-1}$ $(n \in I)$ at Au. We now present a result analogous to Theorem 6.13 which is a generalization of the well-known Lax Equivalence Theorem for the present setting of discrete approximations.

<u>Theorem 6.17</u>. Let A, A_n, $n \in I$, be consistent at $u \in D(A)$ with consistency sequence (u_n'). Then (A_n) is inversely discretely convergent to A at u if, and only if, (A_n) is inversely stable at the consistency sequence.

<u>Proof</u>: Suppose (A_n) is inversely discretely convergent to A at u. For an arbitrary sequence $v_n \in D(A_n)$, $n \in I$, with $||A_n v_n - A_n u_n'||_n \to 0$ $(n \in I)$, we have $A_n v_n \to Au$, and both $v_n \to u$ and $||v_n - u_n'||_n \to 0$ $(n \in I)$ because of (16). Therefore (A_n) is inversely stable at (u_n'). Conversely, suppose the inverse stability condition is valid for the consistency sequence, and let $A_n u_n \to Au$ $(n \in I)$ for an arbitrary sequence $u_n \in D(A_n)$, $n \in I$. Therefore $||A_n u_n - A_n u_n'||_n \to 0$; and, on account of the inverse stability at (u_n'), we must have that $||u_n - u_n'||_n \to 0$ and hence $u_n \to u$ $(n \in I)$. □

We can deduce consistency from inverse discrete convergence, if an additional approximation property is satisfied.

Lemma 6.18. If there exists a sequence $w_n \in R(A_n)$, $n \in I$, such that $w_n \to Au$ $(n \in I)$, then A, A_n, $n \in I$, is consistent at $u \in D(A)$, provided that (A_n) is inversely discretely convergent to A at u.

Proof: By assumption, there exists a sequence $v_n \in D(A_n)$, $n \in I$, with $w_n = A_n v_n \to Au$ $(n \in I)$. Because of the inverse discrete convergence, $v_n \to u$ $(n \in I)$ so that (v_n) is a consistency sequence for u. □

In analogy to Theorem 6.15, the following interesting result allows us to infer from the inverse stability of (A_n) a corresponding property for A via the consistency of A, A_n, $n \in I$.

Lemma 6.19. If A, A_n, $n \in I$, is consistent and if (A_n) is inversely stable at a consistency sequence for each $u \in D(A)$, then A is injective.

Proof: Let $Au = Av$ for arbitrary $u, v \in D(A)$, and let (u_n'), (v_n') be associated consistency sequences such that (A_n) is inversely stable at these sequences. It thus follows from

$$||A_n u_n' - A_n v_n'||_n \to ||Au - Av|| = 0 \ (n \in I),$$

that $||u_n' - v_n'||_n \to 0$ $(n \in I)$. Hence, $u = v$. □

We now proceed to prove a theorem analogous to Theorem 6.14, which will provide equivalent conditions for the inverse discrete convergence of equicontinuously equidifferentiable mappings.

Theorem 6.20. Let (A_n) be locally bijective at (u_n^0) with the sequence $(\tilde{A}_n^{-1})$ equicontinuously equidifferentiable at $(w_n^0) = (A_n u_n^0)$. Suppose $w_n^0 \to Au^0$ $(n \in I)$ for a $u^0 \in D(A)$. Then the following statements are equivalent:

(17a) The sequence (A_n) is inversely discretely convergent to A at u^0 and $\tilde{A}_n^{-1}$ is continuous at w_n^0 for each $n \in I$.

(17b) A, A_n, $n \in I$, is consistent at u^0 with consistency sequence (u_n^0) and $(\tilde{A}_n^{-1})$ is equicontinuous at (w_n^0).

(17c) A, A_n, $n \in I$, is consistent at u^0 and the sequence $||(\tilde{A}_n^{-1})'(w_n^0)||$, $n \in I$, is bounded.

If one of the statements (17a) to (17c) is valid, then, for every $\beta \in (0, 1/\alpha')$ with $\alpha' \equiv \sup_n ||(\tilde{A}_n^{-1})'(w_n^0)|| < \infty$, there is a $\sigma > 0$ so that the error estimate

$$||u_n - u_n^1||_n \leq \frac{1}{\beta} ||A_n u_n - w_n^0||_n + ||u_n^0 - u_n^1||_n \tag{17d}$$

is valid for all $n \in I$, $u_n \in D(A_n)$ with $A_n u_n \in B_\sigma(w_n^0)$, and every sequence $u_n^1 \to u^0$ $(n \in I)$.

Proof: (17a) ➡ (17b) ➡ (17c). From $A_n u_n^0 \to Au^0$ $(n \in I)$, and the inverse discrete convergence at u^0, we have immediately that $u_n^0 \to u^0$ $(n \in I)$, so that A, A_n, $n \in I$, is consistent at u^0. From Theorem 6.17, (A_n) is inversely stable at (u_n^0), which is equivalent to the stability of $(\tilde{A}_n^{-1})$ at $(w_n^0) = (A_n u_n^0)$. Now Theorem 6.8 applied to $(\tilde{A}_n^{-1})$ furnishes the equicontinuity of $(\tilde{A}_n^{-1})$ at (w_n^0), which in its own right is equivalent to the boundedness of the sequence $||(\tilde{A}_n^{-1})'(w_n^0)||$, $n \in I$.

(17c) ➡ (17a). Under (17c), again from Theorem 6.8, we know that (A_n) is inversely stable at (u_n^0). For any consistency sequence (u_n'), we thus deduce from

$$||A_n u_n^0 - A_n u_n'||_n \to 0 \quad \text{that} \quad ||u_n^0 - u_n'||_n \to 0 \ (n \in I).$$

This shows that $u_n^0 \to u^0$ $(n \in I)$ and that (u_n^0) is also a consistency sequence. Theorem 6.17 now yields the inverse discrete convergence; the continuity of each $\tilde{A}_n^{-1}$ at w_n^0 is trivial.

(17c) ➡ (17d). Condition (8d) in Theorem 6.8 for the sequence $(\tilde{A}_n^{-1})'(w_n^0)$, $n \in I$, provides for every $1/\beta$ $(= \alpha) > \alpha' \equiv \sup_n ||(\tilde{A}_n^{-1})'(w_n^0)||$, a $\sigma > 0$ such that

$$||\tilde{A}_n^{-1} w_n - \tilde{A}_n^{-1} z_n||_n \leq \beta^{-1} ||w_n - z_n||_n, \quad w_n, z_n \in B_\sigma(w_n^0), \quad n \in I.$$

For $u_n = \tilde{A}_n^{-1} w_n \in D(A_n)$ with $w_n = A_n u_n \in B_\sigma(w_n^0)$ and $z_n = w_n^0$, we see that

$$||u_n - u_n^0||_n \leq \beta^{-1} ||A_n u_n - w_n^0||_n, \quad n \in I.$$

The triangle inequality finally yields the estimate in (17d). □

As we have observed in the proof of the preceding theorem, we do not need that $B_{\rho_0}^0((u_n^0)) \subset D((A_n))$ (cf. the definition of local injectivity in Section 6.1), but only that a local inverse $(\tilde{A}_n^{-1})$ exists with $B_{\sigma_0}^0((w_n^0)) \subset D((\tilde{A}_n^{-1})) \subset R((A_n))$ for some $\sigma_0 > 0$. In any case, however, we must assume in Theorem 6.20 that the approximate equations are solvable in a neighborhood of (w_n^0). At this point, we would like to emphasize that the following two theorems will additionally prove this local solvability.

In Theorem 6.12, we have characterized the bistability of equicontinuous equidifferentiable mappings. Together with consistency, bistability is equivalent to the so-called biconvergence, as we shall see in the following Theorem 6.21. A sequence (A_n) is called *discretely biconvergent to* A *at* $u \in D(A)$ if the relation

$$u_n \to u \Longleftrightarrow A_n u_n \to Au \ (n \in I) \tag{18}$$

holds for every sequence $u_n \in D(A_n)$, $n \in I$.

<u>Theorem 6.21</u>. Suppose the sequence (A_n) is equicontinuously equidifferentiable at (u_n^0) and suppose $u_n^0 \to u^0$ $(n \in I)$ for some $u^0 \in D(A)$. Let, moreover, w_n^1, $n \in I$, be any sequence converging discretely to Au^0. Then the following statements are equivalent:

(19a) (A_n) is locally bijective at (u_n^0); the associated restrictions $\tilde{A}_n$ are bijective and bicontinuous at u_n^0 for every $n \in I$; the sequence of inverses $(\tilde{A}_n^{-1})$ is equicontinuously equidifferentiable at $(w_n^0) = (A_n u_n^0)$; and $(\tilde{A}_n)$ is discretely biconvergent to A at u^0.

(19b) A, A_n, $n \in I$, is consistent at u^0 with consistency sequence (u_n^0), and one of the statements (13a)-(13d) is valid.

(19c) Each of the statements (13a)-(13d) is valid, and the biconvergence relation (18) at $u = u^0$ is true with the associated two-sided error estimate

$$\begin{aligned}\frac{1}{\alpha}\|A_n u_n - w_n^1 - d_n^0\|_n &\le \|u_n - u_n^0\|_n \\ &\le \frac{1}{\beta}\|A_n u_n - w_n^1 - d_n^0\|_n\end{aligned}$$

for all $u_n \in B_\rho(u_n^0)$, $n \in I$, where $d_n^0 \equiv A_n u_n^0 - w_n^1$.

<u>Proof:</u> (19a) $\Rightarrow$ (19b). Because of the biconvergence and $u_n^0 \to u^0$ $(n \in I)$. $A_n u_n^0$, $n \in I$, converges discretely to Au^0. Hence A, A_n, $n \in I$, is consistent at u^0, u_n^0, $n \in I$. Because of $u_n^0 \to u^0$ and $w_n^1 \to Au^0$ $(n \in I)$, we see that (18) (for $u = u^0$) is equivalent to

$$\|u_n - u_n^0\|_n \to 0 \Longleftrightarrow \|A_n u_n - w_n^1\|_n \to 0 \ (n \in I), \tag{20}$$

for all $u_n \in D(A_n)$, $n \in I$. Further, due to $w_n^1 \to Au^0$ and $A_n u_n^0 \to Au^0$ $(n \in I)$, we have immediately the equivalence of (20) with the bistability at (u_n^0), and (19b) will follow by using Theorem 6.12.

(19b) $\Rightarrow$ (19c). If one of the statements (13a)-(13d) is true, then each of them will be true by Theorem 6.12. If we use the following relation

$$A_n u_n - A_n u_n^0 = A_n u_n - w_n^1 - d_n^0$$

in the inequality of (13d), then we get immediately the two-sided error estimate of (19c). From consistency, we see that $||d_n^0||_n \to 0$ $(n \in I)$. This result together with the two-sided error estimate, proves the biconvergence relation (20).

(19c) ⇒ (19a). Statement (19a) follows immediately from (19c) used in connection with Theorem 6.12. □

In applications of the biconvergence theorem, one is interested above all in deducing the solvability and convergence of solutions of $A_n u_n = w_n$, $n \in I$, where the w_n lie in a neighborhood of w_n^1 which may be regarded as initially given, e.g., by restrictions $w_n^1 = r_n A u^0$. In particular, (13c) together with consistency gives all the desired results. The following theorem is a corollary of Theorem 6.21.

<u>Theorem 6.22</u>. Under the assumptions of Theorem 6.21, let A, A_n, $n \in I$, be consistent at u^0 with consistency sequence (u_n^0), and let (A_n) satisfy one (and therefore all) of the statements (13a)-(13d). Then there are numbers $\rho, \sigma > 0$ and an index $\nu \in I$ such that the equation $A_n u_n = w_n$ is uniquely solvable for every $w_n \in B_\sigma(w_n^1)$, $n \geq \nu$, with the solution $u_n \in B_\rho(u_n^0)$ depending continuously on w_n.

<u>Proof</u>: From $w_n^1 \to Au^0$, $w_n^0 = A_n u_n^0 \to Au^0$ $(n \in I)$, we know that $B_\sigma(w_n^1) \subset A_n B_\rho(u_n^0)$, $n \geq \nu$, with certain $\sigma > 0$, $\nu \in \mathbb{N}$ and ρ from (13d). The assertion then follows from Theorem 6.21. □

For linear mappings, the statements (13a)-(13d) are equivalent to the bijectivity of A_n for every $n \in I$ and to the uniform boundedness of the sequences $||A_n||$, $||A_n^{-1}||$, $n \in I$.

We can further apply Theorem 6.21 to sequences (A_n) whose derivatives are bijective with $(A_n'(u_n^0)^{-1})$ uniformly bounded. (No such assumption is made on the boundedness of sequence $(A_n'(u_n^0))$ itself.) For this case, we have up to now Theorem 6.20 at hand, which does require though the local solvability of the equations $A_n u_n = w_n$, $n \in I$, If the $A_n'(u_n^0)$, $n \in I$, are bijective with $||A_n'(u_n^0)^{-1}||$, $n \in I$, uniformly bounded - the letter equivalent to the existence of a $\beta' > 0$ such that

$$\beta' ||u_n||_n \leq ||A_n'(u_n^0)u_n||_n, \quad u_n \in E_n, \quad n \in I - \tag{21}$$

then we have the equibicontinuity of the derivatives with respect to the following norms,

$$|||u_n|||_n \equiv \max(||u_n||_n, ||A_n'(u_n^0)u_n||_n), \quad u_n \in E_n, \quad n \in I. \tag{22}$$

The norms $|||\cdot|||_n$ are chosen analogously to *graph-norms*. The equicontinuity can be readily observed from the following inequality,

$$\tilde{\beta}|||u_n|||_n \le ||A_n'(u_n^0)u_n||_n \le |||u_n|||_n, \quad u_n \in E_n, \; n \in I, \tag{23}$$

with $\tilde{\beta} = \min(1,\beta')$. Conversely, (21) follows from (23) with $\beta' = \tilde{\beta}$. If we now apply Theorem 6.21 to the sequence of spaces E_n equipped with the norms $|||\cdot|||_n$, then we obtain the following result.

Theorem 6.23. In addition to the assumptions of Theorem 6.21 let A, A_n, $n \in I$, be consistent at u^0 with consistency sequence (u_n^0). Furthermore, suppose that $A_n'(u_n^0)$ is bijective for every $n \in I$, and that (21) is satisfied. Then, to each $\beta \in (0,\beta')$, there are positive numbers ρ, σ and an index $\nu \in I$ such that $A_n u_n = w_n$ has a unique solution u_n in $|||u_n - u_n^0|||_n \le \rho$ for all $w_n \in B_\sigma(w_n^1)$, $n \ge \nu$. Moreover, the inverse convergence relation (16) holds along with the error estimate

$$||u_n - u_n^0||_n \le |||u_n - u_n^0|||_n \le \frac{1}{\beta}||A_n u_n - w_n^1 - d_n^0||_n, \tag{24}$$

for every u_n such that $A_n u_n \in B_\sigma(w_n^1)$, $n \ge \nu$.

Proof: We apply Theorems 6.21 and 6.22 with $|||\cdot|||_n$ as the underlying norm in E_n. Statement (13c) is true due to the equivalence of (21) and (23). With $\alpha' = 1$ and $\beta' = \beta$, statement (13d) results, and the asserted inequality is obtained from the inequality in (13d) with $u_n^0 = u_n$, $v_n = u_n$, and $A_n u_n \in B_\sigma(w_n^1) \subset A_n B_\rho(u_n^0)$, $n \ge \nu$. The solution u_n of $A_n u_n = w_n$ satisfies $|||u_n - u_n^0|||_n \le \rho$ for every $w_n \in B_\sigma(w_n^1)$, $n \ge \nu$. Whenever $A_n u_n \to Au^0$ $(n \in I)$, $u_n \in D(A_n)$, the convergence

$$||u_n - u_n^0||_n \le |||u_n - u_n^0|||_n \to 0 \; (n \in I)$$

will result from (24) in conjunction with $||d_n^0||_n \to 0$ $(n \in I)$. The inverse convergence relation (16) is thereby proven. □

REFERENCES

Anselone & Ansorge (1981)*, Aubin (1967a)*, Chartres & Stepleman (1972)*, Dieudonné (1969), Grigorieff (1973b), Grigorieff (1975)*, John (1968)*, Kantorovich & Akilov (1964), Ortega & Rheinboldt (1970),

---

*Article(s)

Petryshyn (1967b,1968a)*, Reinhardt (1975a)*, Rinow (1961), Stummel (1970, 1973a),1976b)*, Stummel (1973b), Stummel & Reinhardt (1973)*.

---

*Article(s)

Chapter 7
Compactness Criteria for Discrete Convergence

We begin this chapter by defining the concept of a discretely compact sequence of elements, and use this notion to introduce the concepts of a-regular, regularly convergent, and discretely compact operator sequences. These properties provide criteria for inverse stability (respectively, bistability) which, as we know from the theory developed in the preceding chapter, are essential for deducing the inverse discrete convergence (respectively, biconvergence) of a sequence of mappings.

In Section 7.1, we present the concept of discrete compactness, provide concrete examples, and point out some theoretical implications of this concept. Section 7.2 examines both the conditions of a-regularity and of regular convergence for sequences of linear operators and shows their relationship. The former concept essentially provides inverse stability and the latter furnishes the bistability of an operator sequence (cf. Theorem 7.5, respectively 7.9). If we apply either criterion to sequences of derivatives and combine it with the results of Chapter 6, then we have compactness criteria for both the convergence of solutions and their biconvergence (see Theorem 7.10 and 7.11). We introduce in Section 7.3 the concept of discretely compact sequences of operators and then illustrate this notion with several important examples. For fixed point equations, the discrete compactness shows the existence of solutions as limits of sequences of fixed points (cf. Theorem 7.13). For equations of the second kind, which represent a generalization of fixed point equations, the mappings $I - K$, $I - K_n$, $n \in \mathbb{N}$, associated with these equations and their approximates, respectively, are regularly convergent in the linear case, if K, K_n, $n \in \mathbb{N}$, are consistent and the sequence (K_n) is discretely compact (cf. Lemma 7.15). Applying this fact to derivatives of differentiable mappings, we obtain compactness criteria for the biconvergence of solutions of nonlinear equations of the second kind. Such criteria are applied in Section 7.4 to projection methods for approximating solutions of nonlinear fixed point equations. In this application, the discrete compactness is a consequence of compactness properties of the underlying, unperturbed mappings or of their derivatives.

The concept of a-regularity has been introduced for the linear case by Grigorieff (1973a) as a generalization of Petryshyn's concept of "a-proper mappings" (1968a, 1968b). The notion of regular convergence goes back to Vainikko (cf. Vainikko (1976) and the cited references therein). The theorems we formulate in Section 7.2, giving equivalent statements about a-regularity, are due to Grigorieff (1973a). From these theorems, we easily get the relationship of this concept to that of regular convergence. Grigorieff (1975) later extended the concept of a-regularity to nonlinear mappings. We shall not follow Grigorieff's approach as far as nonlinear mappings are concerned, but we shall need these properties only for linear mappings (i.e., to apply to Fréchet-derivatives of nonlinear mappings) and proceed to examine nonlinear problems with the use of the results from Chapter 6. A similar approach can be found in the book by Vainikko (1976), whose results are improved here by establishing equivalence theorems. The presentation in Section 7.4 is similar to that of Stummel (1976b), 3.3, who does not, however, utilize the concept of regular convergence.

7.1. DISCRETE COMPACT SEQUENCES OF ELEMENTS

In this section, we introduce the concept of a discretely compact sequence of elements and, moreover, discuss several properties and give examples.

Let $A(E,\pi_n E_n,\lim^E)$ be a discrete approximation of normed spaces E,E_n, $n \in I$ $(= \mathbb{N})$. We shall always assume in the following that discrete convergences possess the property of asymptotic linearity and, moreover, that the norms of discretely convergent sequences converge. In order to define discrete compactness, we first need the definition of a discretely convergent subsequence. Let I' be a subsequence of I and $u_n \in E_n$, $n \in I'$. Then the *subsequence* $(u_n)_{n\in I'}$ *is discretely convergent* to $u \in E$ (in symbols, $u_n \to u$ $(n \in I')$) if there is a discretely convergent sequence $u'_n \to u$ $(n \in I)$ with

$$||u_n' - u_n||_n \to 0 \ (n \in I'). \qquad (1)$$

The convergence of subsequences is clearly independent of the specially selected sequence u_n', $n \in I$ (cf. Property 5.1). Convergent subsequences (and, indeed, entire sequences) can be characterized as follows.

Lemma 7.1. The sequence u_n, $n \in I' \subset I$, converges discretely to u if, and only if, every subsequence $I'' \subset I'$ has a subsequence $I''' \subset I''$ such that u_n, $n \in I'''$, converges to u.

Proof: From the convergence of u_n, $n \in I'$, to u, the convergence of u_n, $n \in I''$, to u follows immediately for every subsequence $I'' \subset I$. Conversely, let the above criterion be satisfied, but suppose that u_n, $n \in I'$, is not discretely convergent to u. Then there is a sequence (u_n') with $u_n' \to u$ $(n \in I)$, and a subsequence $I'' \subset I'$, with $||u_n - u_n'||_n \geq m > 0$, $n \in I''$. By assumption, there is a subsequence $I''' \subset I''$, such that u_n, $n \in I'''$, converges to u. This has as a consequence the convergence of $||u_n - u_n'||_n$, $n \in I'''$, to zero, which provides a contradiction to the above inequality. □

The above definition of discretely convergent subsequences now enables us to explain the concept of discrete compactness of sequences of vectors. A sequence of elements $u_n \in E_n$, $n \in I$, is called *discretely compact* if every subsequence u_n, $n \in I' \subset I$ contains a discretely convergent subsequence, i.e., to every subsequence $I' \subset I$, there exist a $u \in E$ and another subsequence $I'' \subset I'$ so that $u_n \to u$ $(n \in I'')$. A discretely compact subsequence $(u_n)_{n \in I'}$ is analogously defined.

The next lemma is of use in our analysis.

Lemma 7.2. The following statements are true:

(2a) Every discretely convergent sequence is discretely compact.

(2b) Every discretely compact sequence is bounded.

(2c) If (u_n), (u_n') are discretely compact, then $(\alpha u_n + \beta u_n')$ is also discretely compact, where $\alpha, \beta \in \mathbb{K}$.

Proof: Statements (2a) and (2c) are obvious. The boundedness of discretely compact sequences will now be shown by a contradiction argument. Let us assume that $||u_n||_n$, $n \in I$, is not bounded; then there is a subsequence $I' \subset I$ with $||u_n||_n \to \infty$ $(n \in I')$. On account of the discrete compactness of (u_n), there exist a $u \in E$ and a subsequence $I'' \subset I'$ such that $u_n \to u$ $(n \in I'')$. The sequence u_n, $n \in I''$, can be "filled out" to be a convergent sequence v_n, $n \in I$, with $v_n \to u$ $(n \in I)$ because of the assumed surjectivity of $\lim^E$. Because of the discretely convergent norms, we have

$$\lim_{n \in I''} ||u_n||_n = \lim_{n \in I} ||v_n||_n = ||u||,$$

in contradiction to $||u_n||_n \to \infty$ $(n \in I'')$. □

As a consequence of a result of Wolf (1974), one can obtain the precompactness of the set of limits of subsequences of a discretely compact sequence under the additional assumption that the space E - or the set of limits itself - is separable (see, also Grigorieff (1975), 3.(8), Reinhardt (1975a), 2.(6), Vainikko (1976), 1.(26)).

We shall now provide some trivial and nontrivial *examples* of discretely compact sequences.

Example 1. We consider a discrete approximation, which is given by a stable sequence of restriction operators $R_n: E \to E_n$, $n \in I$, and a relatively compact sequence $u^{(n)} \in E$, $n \in I$. On account of the stability of (R_n) (cf. 5.(14)), we have immediately the discrete compactness of $(R_n u^{(n)})$. Because of property 5.(9), we can easily convince ourselves that the set of accumulation points of $\{u^{(n)}\}$ agrees with the set of discrete limits of subsequences of $(R_n u^{(n)})$ (cf. Anselone & Ansorge (1981), Theorem 6.4). These considerations lead to a useful criterion for showing the discrete compactness of a sequence (u_n). Namely, if there is a compact sequence $u^{(n)} \in E$, $n \in I$, so that $||u_n - R_n u^{(n)}||_n \to 0$ $(n \in I)$, then both $(R_n u^{(n)})$ and the sequence (u_n) are discretely compact. □

Example 2. The well-known Arzela-Ascoli Theorem gives necessary and sufficient conditions for the compactness of a sequence $u^{(n)} \in C(G)$, $n \in I$ (see Lemma 3.3), where $C(G)$ is the Banach space of all continuous, real- or complex-valued functions on a compact metric space G. This important result can be generalized to discrete approximations with essentially no change in its formulation. We shall only quote this generalization by Reinhardt (1975a), and refer the reader to the original literature for the rather extensive proof. But in the following we shall make precise the terminology and concepts used. We begin by assuming that G, G_n, $n \in I$, are closed subsets of a compact metric space M with $\text{Lim}\, G_n = G$. Then there exists, according to the results in Chapter 5, a discrete approximation $A(C(G), \pi_n C(G_n), \lim)$ having the discrete uniform convergence lim (cf. 5.(22)) for which the maximum norms also converge. With an arbitrary but fixed extension $\hat{u} \in C(M)$ of each $u \in C(G)$, we can define a sequence of linear restriction operators $R_n\colon C(G) \to C(G_n)$, $n \in I$, by $R_n u = \hat{u}|G_n$, $n \in I$, which are uniformly bounded. A sequence $u_n \in C(G_n)$, $n \in I$, is called *equicontinuous*, if, for every sequence $x_n \in G_n$, $n \in I$, the following is true:

$$\forall \varepsilon > 0 \quad \exists \delta > 0 \quad \ni: \forall n \in I, \quad \forall x_n' \in G_n: \qquad |x_n, x_n'| \le \delta \Rightarrow |u_n(x_n) - u_n(x')| \le \varepsilon. \tag{3}$$

The number δ can depend on (x_n) in addition to ε. From the discrete compactness of every sequence in $\pi_n G_n$, it follows that (u_n) is uniformly equicontinuous, i.e., the number δ in (3) is independent of $(x_n) \in \pi_n G_n$. In the case $G_n = G$, we have the usual condition of equicontinuity of functions on compact sets (see 3.(3b)).

With these preparatory remarks, we are now in the position to present the following generalization of the Arzela-Ascoli Theorem.

Theorem 7.3. Let G, G_n, $n \in I$, be closed subsets of a compact metric space M for which $\operatorname{Lim} G_n = G$. Then the following two conditions are necessary and sufficient for the discrete compactness of a sequence $u_n \in C(G_n)$, $n \in I$:

(4a) (u_n) is equicontinuous, i.e., (3) holds.

(4b) (u_n) is uniformly bounded, i.e.,

$$\exists C \ge 0 \quad \ni: \forall n \in I, \quad \forall x_n \in G_n \Rightarrow |u_n(x_n)| \le C. \qquad \square$$

The proof is similar to the classical proof insofar as the discrete compactness can be characterized by ε-nets. These concepts must, however, be defined in a suitable way for the setting of discrete approximations. The result is also valid for vector-valued functions, and has even been proved in the cited work for continuous functions having values in arbitrary Banach spaces. On close observation, we see that Theorem 7.3 is a consequence of Theorem 4.(6) in Reinhardt (1975a) because the uniform boundedness (4b) is equivalent to the relative compactness of the sequence $(u_n(x_n))$ in $\mathbb{K}$ for every sequence $(x_n) \in \pi_n G_n$ due to the compactness of each G_n and the associated Bolzano-Weierstrass property. $\square$

7.2. A-REGULAR AND REGULARLY CONVERGENT MAPPINGS

In Theorem 6.23 from Section 6.4, we have seen that, in addition to consistency, the uniform boundedness of the inverses of the associated derivatives is the essential condition for the discrete convergence of solutions. The a-regularity and regular convergence of a sequence of mappings have special compactness properties and turn out to be sufficient criteria - and in several cases also necessary ones - for deducing their inverse stability and bistability. Using the results from Chapter 6, we obtain compactness criteria ensuring inverse convergence and biconvergence.

As before, let $A(E,\pi_n E_n,\lim^E)$ and $A(F,\pi_n F_n,\lim^F)$ be discrete approximations of normed spaces E,F,E_n,F_n, $n \in I$, which we assume to be Banach spaces for simplicity. The properties of asymptotic linearity and Property 5.3 of convergent norms are to be satisfied for the discrete convergences $\lim^E$ and $\lim^F$. The aforementioned assumptions should be taken as valid for the entire section. In addition, let $L: D(L) \subset E \to F$, $L_n: D(L_n) \subset E_n \to F_n$, $n \in I$, be linear mappings. Then it is well known that their domains of definition form a discrete approximation $A(D(L), \pi_n D(L_n),\lim^E)$ if L,L_n, $n \in I$, is consistent.

The sequence L,L_n, $n \in I$, is called *a-regular* (or, *approximation regular*), if, for every bounded subsequence $u_n \in D(L_n)$, $n \in I' \subset I$, for which $L_n u_n \to w \in F$ $(n \in I')$, there is an element $u \in D(L)$ and a subsequence $I'' \subset I'$ such that

$$u_n \to u \ (n \in I'') \quad \textit{and} \quad Lu = w. \tag{5}$$

The interrelationship between this concept and that of discrete compactness becomes clear if we additionally introduce the notion of discrete closedness. The sequence L,L_n, $n \in I$, of linear operators is said to be *discretely closed* if, for every given $u \in E$, $w \in F$, and every subsequence $u_n \in D(L_n)$, $n \in I' \subset I$, for which

$$u_n \to u, \quad L_n u_n \to w \ (n \in I'), \quad \textit{then } u \in D(L),\ Lu = w. \tag{6}$$

With this concept, we can present the following characterization of a-regular mappings.

<u>Lemma 7.4</u>. The sequence L,L_n, $n \in I$, is a-regular if, and only if, it is discretely closed, and, for every bounded subsequence $u_n \in D(L_n)$, $n \in I' \subset I$, we have the following relation,

$$(L_n u_n)_{n\in I'} \text{ discretely convergent} \Rightarrow (u_n)_{n\in I'} \text{ discretely compact.} \tag{7}$$

Proof: Let L, L_n, $n \in I$, be a-regular. The uniqueness of the discrete limits immediately implies the discrete closedness; and the discrete convergence of a sequence $(L_n u_n)_{n\in I'}$ implies the discrete compactness of $(u_n)_{n\in I'}$. Conversely, let $I' \subset I$ and let $u_n \in D(L_n)$, $n \in I'$, be a bounded sequence with $L_n u_n \to w$ $(n \in I')$ for some $w \in F$. Then, according to (7), $(u_n)_{n\in I'}$ is discretely compact. In particular, there is a subsequence $I'' \subset I$ and a $u \in E$ such that $u_n \to u$ $(n \in I'')$. The discrete closedness hypothesis, with now I'' in place of I', guarantees that $u \in D(L)$ as well as $Lu = w$. □

The following result is basic for our purposes and enables us to conclude the inverse stability of a sequence of operators from its a-regularity and from the injectivity of L. We recall that the inverse stability of (L_n) is equivalent to inequality 6.(12), where now each E_n is replaced by $D(L_n)$.

Theorem 7.5. If L is injective and the sequence L, L_n, $n \in I$, is a-regular, then (L_n) is inversely stable.

Proof: We assume that (L_n) is not inversely stable. Then there are a subsequence $I' \subset I$ and elements $u_n \in D(L_n)$, $n \in I'$, for which

$$||u_n||_n = 1, \quad n \in I' \quad \text{and} \quad ||L_n u_n||_n \to 0 \quad (n \in I').$$

Because of the a-regularity, there exist a $u \in D(L)$ and a subsequence $I'' \subset I$ such that $u_n \to u$ $(n \in I'')$ and $Lu = 0$. Because $||u_n||_n = 1$, $u \neq 0$ due to the convergence of the norms, thereby contradicting the hypothesized injectivity of L. □

We note that injectivity of L is necessarily present in case L, L_n, $n \in I$, is consistent and (L_n) is inversely stable (see Lemma 6.19).

The following condition on the sequence of mappings L_n, $n \in I$, will be needed in order to provide the desired characterization of a-regularity. To express this condition precisely, we say that a mathematical property holds *for almost all* $n \in I$ if, and only if, it holds for all but a finite subset of I.

Property 7.6. For almost all $n \in I$, $N(L_n) = \{0\}$ implies $R(L_n) = F_n$.

Property 7.6 is satisfied if, for example, $L_n: E_n \to F_n$ is Fredholm with index zero for almost all $n \in I$ (see Kato (1966) for the definition of "Fredholm" and "index"). In particular, this property is present in many applications where the L_n are mappings between spaces of finite and equal dimensions. The injectivity of the L_n for almost all $n \in I$ (expressed by $N(L_n) = \{0\}$) is a consequence of the inverse stability (cf. inequality 6.(12)), which will follow from the injectivity of L itself and the a-regularity (see Theorem 7.5, above). A theorem providing equivalent conditions for the a-regularity of linear mappings can now be formulated as follows.

Theorem 7.7. The following statements are equivalent:

(8a) The operator L is injective; the sequence L_n, $n \in I$, obeys Property 7.6; and L, L_n, $n \in I$, is a-regular.

(8b) The operator L is surjective; the sequence (L_n) is inversely stable and satisfies Property 7.6; and L, L_n, $n \in I$, is consistent.

(8c) The inverse L^{-1} exists and belongs to $B(F,E)$; and, for almost all $n \in I$, $L_n^{-1} \in B(F_n, E_n)$ with $L_n^{-1} \to L^{-1}$ $(n \in I)$.

Proof: (8a) ➡ (8b). The inverse stability follows from Theorem 7.5. As remarked above, L_n is necessarily injective, and, by Property 7.6, is also surjective for almost all $n \in I$. For such $n \geq \nu_0$, denote $u_n' \equiv L_n^{-1}(w_n)$. To prove the surjectivity of L, let $w \in F$ and $w_n \in F_n$, $n \in I$, with $w_n \to w$ $(n \in I)$. The boundedness of u_n', $n \geq \nu_0$, follows from the inverse stability and the boundedness of the sequence (w_n). The a-regularity provides, for every infinite $I' \subset I$, a subsequence $I'' \subset I'$ and a $u \in D(L)$ such that $u_n \to u$ $(n \in I'')$ with $Lu = w$. Because of the injectivity of L, u must be the same limit for every subsequence indexed by arbitrary $I' \subset I$; appealing to Lemma 7.1, we see that the entire sequence converges to the common limit u. This proves the surjectivity of L. At the same time, we have generated a consistency sequence (u_n') for $u \in D(L)$.

(8b) ➡ (8c). According to Lemma 6.19, L is injective, and thus $L: D(L) \to F$ will be bijective. From the inverse stability and Property 7.6, we can deduce the existence of bounded inverses L_n^{-1} of F_n onto $D(L_n) \subset E_n$ for almost all $n \in I$ (cf. 6.(12)). Theorem 6.17 implies the discrete convergence of (L_n^{-1}) to L^{-1} for the special case when L and L_n, $n \in I$, are bijective and linear. Finally, Theorem 6.15 provides the continuity - and thus the boundedness - of L^{-1}.

(8c) ➡ (8a). Since the $L_n: D(L_n) \to F_n$ are bijective for almost all $n \in I$, Property 7.6 is particularly true. The injectivity of L is trivially present. It remains now to deduce the a-regularity from the discrete convergence of the inverses. To this end, let $I' \subset I$ be an arbitrary subsequence and $u_n \in D(L_n)$, $n \in I'$, be a bounded sequence with $L_n u_n \to w$ $(n \in I')$ for some $w \in F$. Then it follows that

$$u_n = L_n^{-1} w_n \to L^{-1} w = u \ (n \in I')$$

for the solution $u \in D(L)$ of $Lu = w$, which shows the a-regularity. □

Several *examples* of a-regular operators can be constructed from the examples of discretely compact sequences of operators presented in Section 7.3 in conjunction with Lemma 7.15, below.

Closely related to a-regularity is the concept of regular convergence, which is a fundamental tool in the investigations of Vainikko (1976). A sequence L_n: $D(L_n) \subset E_n \to F_n$, $n \in I$, *converges regularly to* L: $D(L) \subset E \to F$ if it is discretely convergent to L and, moreover, if the following property holds:

Property 7.8. (u_n) bounded and $(L_n u_n)$ discretely compact imply that (u_n) is discretely compact.

For denoting regular convergence, we write for brevity $L_n \overset{r}{\to} L$ $(n \in I)$. Regular convergence is stronger than a-regularity; conversely, Property 7.8 can be deduced from a-regularity together with additional assumptions. We make these remarks more precise in the following theorem.

Theorem 7.9. The regular convergence $L_n \overset{r}{\to} L$ $(n \in I)$, together with the existence of the discrete approximation $A(D(L), \pi_n D(L_n), \lim^E)$, is equivalent to the stability of (L_n) along with a-regularity and consistency of L, L_n, $n \in I$.

Proof: (i) Stability and consistency follow from the discrete convergence of (L_n) to L provided that the domains of definition form a discrete approximation (cf. Theorem 6.13). To prove a-regularity, let $I' \subset I$ be an arbitrary subsequence and let $u_n \in D(L_n)$, $n \in I'$, a bounded sequence with $L_n u_n \to w$ $(n \in I')$. We set $u_n = 0$, $n \in I - I'$; then, in particular, we see that $(L_n u_n)_{n \in I}$ is discretely compact. By assumption of regular convergence, (u_n) is discretely compact (cf. Property 7.8), so that there is a subsequence $I'' \subset I$ and a $u \in D(L)$ for which $u_n \to u$ $(n \in I'')$. Because of discrete convergence, we have $L_n u_n \to Lu$ $(n \in I'')$, and thus $Lu = w$.

(ii) Conversely, the discrete convergence $L_n \to L$ $(n \in I)$ and the existence of $A(D(L), \pi_n D(L_n), \lim^E)$ follow from stability and consistency. With every bounded sequence (u_n) for which $(L_n u_n)$ is discretely compact, there is associated to every subsequence $I' \subset I$ another subsequence $I'' \subset I'$ and some $w \in F$ for which $L_n u_n \to w$ $(n \in I'')$. The a-regularity further provides a $u \in D(L)$ and a subsequence $I''' \subset I''$ such that $u_n \to u$ $(n \in I''')$, and we have Property 7.8. □

A combination of Theorems 7.7 and 7.9 shows that we must add only the stability of (L_n) to condition (8a), in order to obtain the regular convergence $L_n \overset{r}{\to} L$ $(n \in I)$ along with the convergence of the inverses $L_n^{-1} \to L^{-1}$ $(n \in I)$. We can easily convince ourselves that the latter convergence is also a regular one. The concept of regular convergence is suitable for proving biconvergence, whereas a-regularity is useful for describing convergence of approximate solutions. This observation applies to both linear and nonlinear problems, as the following two theorems will show. These theorems essentially depend on replacing the stability - and especially the inverse stability - by compactness properties which pertain only to the derivatives. This is possible by using the equivalence theorems of this section in conjunction with those in Chapter 6.

As in Section 6.3, we now consider (not necessarily linear) mappings $A\colon D(A) \subset E \to F$, $A_n\colon D(A_n) \subset E_n \to F_n$, $n \in I$, and we further assume the existence of the discrete approximations as before.

Theorem 7.10. In addition to the hypothesis of the Biconvergence Theorem 6.21, let A be differentiable at u^0 and let A, A_n, $n \in I$, be consistent at u^0 with consistency sequence (u_n^0). Then the following conditions are equivalent:

(9a) The derivative $A'(u^0)$ is injective; the sequence of the derivatives $A_n'(u_n^0)$, $n \in I$, satisfies Property 7.6, and $A_n'(u_n^0) \overset{r}{\to} A'(u^0)$ $(n \in I)$.

(9b) The inverse $A'(u^0)^{-1} \in B(F,E)$ and, for almost all $n \in I$, the inverses $A_n'(u_n^0)^{-1} \in B(F_n,E_n)$ exist satisfying the regular convergences

$$A_n'(u_n^0) \overset{r}{\to} A'(u^0), \quad A_n'(u_n^0)^{-1} \overset{r}{\to} A'(u^0)^{-1} \quad (n \in I).$$

(9c) $A'(u^0)$ is bijective; $A'(u^0)$, $A_n'(u_n^0)$, $n \in I$, is consistent; and each of the conditions 6.(19a-c) holds with I replaced by $I_1 \equiv \{n \geq \nu\}$ for some $\nu \in I$.

Proof: (9a) ➡ (9b). From Theorem 7.9, $A'(u^0)$, $A_n'(u_n^0)$, $n \in I$, is necessarily a-regular, and the existence of $A'(u^0)^{-1} \in B(F,E)$, $A_n'(u_n^0)^{-1} \in B(F_n,E_n)$ for $n \in I_1 \equiv \{n \geq \nu\}$, along with the discrete convergence $A_n'(u_n^0)^{-1} \to A'(u^0)^{-1}$ $(n \in I)$, follows by virtue of Theorem 7.7. Together with the discrete convergence $A_n'(u_n^0) \to A'(u^0)$ $(n \in I)$, the preceding conclusions easily yield Property 7.8 for $L_n = A_n'(u_n^0)^{-1}$, $n \in I$, and therefore ensure the regular convergence $A'(u_n^0)^{-1} \overset{r}{\to} A'(u^0)^{-1}$ $(n \in I)$.

(9b) ➡ (9c). The consistency of the derivatives is trivial, and that of A, A_n, $n \in I$, is assumed. It suffices, therefore, to show only the statement 6.(13c) (cf. 6.(19b)). This follows, however, for a subset of I of the form $I_1 \equiv \{n \geq \nu\}$, from the discrete convergences $A_n'(u_n^0) \to A'(u^0)$, $A_n'(u_n^0)^{-1} \to A'(u^0)^{-1}$ $(n \in I)$ (cf. Theorem 6.11 and 6.13).

(9c) ➡ (9a). The implication (8b) ➡ (8a) from Theorem 7.7 for $L = A'(u^0)$, $L_n = A_n'(u_n^0)$, $n \in I_1$, yields Property 7.6 for the derivatives, along with the a-regularity of $A'(u^0)$, $A_n'(u_n^0)$, $n \in I$. Using Theorem 7.9, we obtain the desired regular convergence $A_n'(u_n^0) \overset{r}{\to} A'(u^0)$ $(n \in I)$. □

The preceding theorem yields the Biconvergence Theorem 3.(14) of Vainikko (1976) containing a series of sufficient conditions guaranteeing biconvergence. For proving biconvergence, the use of condition (9a) is preferable, where the regular convergence of the sequence of derivatives can be obtained by its a-regularity and stability (cf. Theorem 7.9). The consistency of the derivatives is then a necessary consequence of the other conditions in (9a) (cf. Theorem 7.7). The final result of this section, Lemma 7.12 below, is a useful tool for proving the regular convergence in (9a) directly.

But before proving Lemma 7.12, we investigate the situation where the stability of the derivatives may not hold. Then convergence of the approximations can still be shown by Theorem 6.23, where now the a-regularity is invoked to prove the inverse stability of $A_n'(u_n^0)$, $n \in I$.

Theorem 7.11. Let the following conditions be satisfied:

(10a) The assumptions of Theorem 6.21 are valid, and A is differentiable at u^0.

(10b) $A'(u^0)$ is injective.

(10c) A, A_n, $n \in I$, is consistent at u^0 with consistency sequence (u_n^0).

(10d) $A_n'(u_n^0)$, $n \in I$, satisfies Property 7.6.

(10e) $A'(u^0)$, $A_n'(u_n^0)$, $n \in I$, is a-regular.

Then the statements in Theorem 6.23 hold, where the error estimate 6.(24) is valid only for all $n \in I_1 \equiv \{n \geq \nu\}$ with some $\nu \in I$.

Proof: According to Theorem 6.23, it suffices to show the bijectivity of the $A_n'(u_n^0)$, $n \in I_1 = \{n \in \nu\}$, along with their inverse stability. However, these statements follow immediately from Theorem 7.7. □

As announced before, we shall finally show a result which serves to prove the regular convergence of $(A_n'(u_n^0))$ to $A'(u^0)$. By definition, we must check stability and consistency, along with Property 7.8 for the derivatives. In particular, the question arises of whether the consistency of the derivatives follows from that of the mappings themselves. An answer has been provided by Theorem 3.(28) in Vainikko (1976). We improve this result by using the fact that the stability of (A_n) at (u_n^0) is equivalent to the stability of $(A_n'(u_n^0))$. Moreover, our result is locally formulated so that the assumptions in Vainikko (1976), 3.3, on the regions of definition follow naturally from the equicontinuous equidifferentiability. We should remark that we could have proved the following result in Chapter 6 since the compactness properties of this section are found neither in the statements of the theorem nor in its proof. Because of the aforementioned connection with condition (9a) of Theorem 7.10, this result will be formulated here.

Lemma 7.12. Suppose that (A_n) is equicontinuously equidifferentiable at (u_n^0), that A is differentiable at u^0, and that $u_n^0 \to u^0$ $(n \in I)$. Further assume that the $A_n'(u_n^0)$, $n \in I$, are uniformly bounded and that A, A_n, $n \in I$, is consistent at all $v \in B_\rho(u^0)$ for some $\rho > 0$.[1)] Then the derivatives converge discretely, i.e.,

$$A_n'(u_n^0) \to A'(u^0) \quad (n \in I). \tag{11}$$

Proof: We have, for arbitrary $\varepsilon > 0$, a $\delta > 0$ such that

$$||A(u^0+h) - Au^0 - A'(u^0)h|| \leq \tfrac{\varepsilon}{4}||h||,$$

$$||A_n(u_n^0+h_n) - A_nu_n^0 - A_n'(u_n^0)_n||_n \leq \tfrac{\varepsilon}{4}||h_n||_n,$$

for all $n \in I$ and all $||h|| \leq \delta$, $||h_n||_n \leq \delta$. We first show that $A_n \to A$ $(n \in I)$ at v for all v in a neighborhood of u^0. From the uniform boundedness of $A_n'(u_n^0)$, $n \in I$, and the equicontinuous equidiffer-

[1)] Because of the hypothesis of differentiability, there exists a $\rho > 0$ such that $B_\rho(u^0) \subset D(A)$.

entiability at (u_n^0), we know that there exists a δ' in $0 < \delta' \le \min(\delta,\rho,1)$ such that $A_n'(v_n)$, $n \in I$, is uniformly bounded for all $v_n \in B_{\delta'}(u_n^0)$, $n \in I$. Now, let $v \in B_{\delta'/2}(u^0)$ and let (v_n^0) be an associated consistency sequence. Then there is a $\nu_0 \in I$ such that $v_n^0 \in B_{\delta'}(u_n^0)$, $n \ge \nu_0$, and the $A_n'(v_n^0)$, $n \ge \nu_0$, are uniformly bounded. By means of Theorem 6.8, (A_n) is stable at (v_n^0), and by Theorem 6.13 we can conclude that A_n, $n \in I$, is discretely convergent to A at each $v \in B_{\delta'/2}(u^0)$. We also have the discrete convergence of (A_n) to A at u^0.

To prove the assertion in the statement of the theorem, let $0 \neq v \in E$, $v_n \in E_n$, $z_n \in F_n$, $n \in I$, be such that $v_n \to v$, $z_n \to A'(u^0)v$ $(n \in I)$. Let us define

$$h \equiv \delta'' \frac{v}{||v||}, \quad h_n \equiv \delta'' \frac{v_n}{||v||}, \; n \in I, \text{ with } \delta'' \equiv \frac{\delta'}{2}.$$

Then $h_n \to h$ $(n \in I)$, and together with $u_n^0 \to u^0$ $(n \in I)$ we have $A_n(u_n^0 + h_n) \to A(u^0+h)$ and $A_n u_n^0 \to Au^0$ $(n \in I)$. Moreover, $\delta'' z_n/||v|| \to A'(u^0)h$ $(n \in I)$, and, from the first estimate in this proof, we obtain

$$||A_n(u_n^0 + h_n) - A_n u_n^0 - \frac{\delta''}{||v||} z_n||_n \le \frac{\varepsilon}{2}\delta'', \quad n \ge \nu_1,$$

for some $\nu_1 \in I$. In addition, with some $\nu_2 \ge \nu_1$ we know that $||h_n||_n \le \delta'$ $(= 2\delta'') < \delta, n \ge \nu_2$, and thus

$$||A_n(u_n^0 + h_n) - A_n u_n^0 - A_n'(u_n^0)h_n||_n \le \frac{\varepsilon}{4}||h_n||_n, \quad n \ge \nu_2.$$

The triangle inequality finally yields

$$\begin{aligned} \frac{\delta''}{||v||}||z_n - A_n'(u_n^0)v_n||_n &\le ||\frac{\delta''}{||v||} z_n - (A_n(u_n^0+h_n) - A_n u_n^0)||_n \\ &\quad + ||A_n(u_n^0 + h_n) - A_n u_n^0 - A_n'(u_n^0)h_n||_n \\ &\le \frac{\varepsilon}{2}\delta'' + \frac{\varepsilon}{4}||h_n||_n \le \frac{\varepsilon}{4}\delta' + \frac{\varepsilon}{4}\delta' = \frac{\varepsilon}{2}\delta', \quad n \ge \nu_2, \end{aligned}$$

which shows that

$$||z_n - A_n'(u_n^0)v_n||_n \le \varepsilon\delta'||v||, \quad n \ge \nu_2,$$

and that $A_n'(u_n^0)v_n \to A'(u^0)v$ $(n \in I)$. In the case $v = 0$, it suffices to prove that $A_n'(u_n^0)h_n \to 0$ for all $h_n \to 0$ $(n \in I)$. But this will follow immediately from the uniform boundedness of the derivatives. □

7.3. DISCRETE COMPACT SEQUENCES OF MAPPINGS AND BICONVERGENCE FOR EQUATIONS OF THE SECOND KIND

This section presents the definition of discrete compactness of operator sequences. Immediate and important consequences of this concept are the existence of fixed points as limits of sequences of fixed points and the stability of discretely compact linear mappings. Fixed point equations are special cases of equations of the second kind. In the linear case, it turns out that the associated operators $I-K$, $I-K_n$, $n \in \mathbb{N}$, are both a-regular and regularly convergent, provided that K, K_n, $n \in \mathbb{N}$, is consistent, and (K_n) discretely compact (see Lemma 7.15). With such properties, we can obtain results about the biconvergence of solutions of equations of the second kind by appealing to the theorems of the previous section. In analogy with the treatment of linear problems, we must require - and indeed check - that the corresponding properties hold for the derivatives in the nonlinear setting.

Let us adopt the assumptions of Section 7.2 on the spaces E, F, E_n, F_n. Let $K\colon D(K) \subset E \to F$, $K_n\colon D(K_n) \subset E_n \to F_n$, $n \in I$ $(= \mathbb{N})$, be (not necessarily linear) mappings whose domains of definition form a discrete approximation $A(D(K), \pi_n D(K_n), \lim^E)$. As we have remarked before, this requirement is especially satisfied whenever K, K_n, $n \in I$, is consistent. The sequence (K_n) is called *discretely compact* if, for every bounded sequence $u_n \in D(K_n)$, $n \in I$, the sequence $K_n u_n$, $n \in I$, is discretely compact. We now turn to present several *examples* of discretely compact operators.

Example 1. Let $E_n = E$, $F_n = F$, $K_n = K$, $n \in I$, and suppose that $K\colon D(K) \subset E \to F$ is compact. Then the sequence K_n, $n \in I$, is discretely compact. Indeed, the image sequence $K_n u_n = K u_n$, $n \in I$, is relatively compact in F whenever $u_n \in E$, $n \in I$, is a bounded sequence. □

Example 2. Let $E_n \subset E$, $F_n \subset F$, $n \in I$, be subspaces of normed spaces with $P_n\colon F \to F_n$, $n \in I$, a sequence of bounded linear projection operators with the properties

$$P_n w \to w \ (n \in I), \quad w \in F, \quad \|P_n\| \leq \delta, \quad n \in I. \tag{12}$$

Also, let $K\colon D(K) \subset E \to F$ be compact and let each $K_n \equiv P_n K|E_n$, $n \in I$, possess a nonempty domain of definition $D(K_n) = E_n \cap D(K)$, $n \in I$. Then (K_n) is discretely compact. To see this, we note that if $u_n \in D(K_n) \subset E$, $n \in I$, is bounded, then $(Ku_n)_{n\in I}$ is relatively compact. The sequence $K_n u_n$, $n \in I$, is also relatively compact because of the properties of (P_n).

Indeed, from $Ku_n \to w \in F$, $n \in I' \subset I$, we obtain that $K_n u_n$, $n \in I'$, also converges to w as the following estimates show,

$$||K_n u_n - w|| \le ||P_n K u_n - P_n w|| + ||P_n w - w||$$

$$\le \gamma ||Ku_n - w|| + ||P_n w - w||. \quad \square$$

Example 3. The class of collectively compact operator sequences, introduced first by Anselone (see e.g. (1971)) can be viewed as a special case of discretely compact operator sequences. Let E and F be normed spaces. Then a sequence $K_n: X \subset E \to F$, $n \in I$, of (not necessarily linear) mappings with common domain of definition X is called *collectively compact,* if the set

$$\bigcup_{n \in I} K_n M$$

is relatively compact in F for every bounded set $M \subset X$. This definition could also be given verbatim for the case when E and F are only metric spaces. The rather simple setting in Example 1 clearly provides an example of a collectively compact operator sequences. The reader can easily convince himself that the following equivalence is true:

$$(K_n) \text{ collectively compact} \Longleftrightarrow (K_n) \text{ discretely compact and } K_n \text{ compact } \forall n \in I.$$

The right-hand side of the equivalence can be taken as the definition of a collectively compact operator sequence; the above equivalence then provides the immediate generalization of this concept to the discrete approximation setting. $\square$

Example 4. Our final example is a generalization of the preceding ones. As in Example 2, we consider subspaces $E_n \subset E$, $F_n \subset F$, $n \in I$, and projection operators $P_n: F \to F_n$, $n \in I$, with property (12). If $K^{(n)}$: $X \subset E \to F$, $n \in I$, is a collectively compact sequence of operators, then clearly the mappings $K_n \equiv P_n K^{(n)}|E_n: X_n \to F_n$, $n \in I$, (with nontrivial domains of definition $X_n = E_n \cap X$, $n \in I$) are compact for every $n \in I$, and the sequence is discretely compact. The latter can be seen because projection operators having property (12) provide an example of a stable sequence of restriction operators. Thus, the sequence $(P_n K^{(n)} u_n)_{n \in I}$ will be discretely compact for every compact sequence $K^{(n)} u_n$, $n \in I$, in F. $\square$

The following theorem points out the significance of the discrete compactness concept for solving fixed point equations.

Theorem 7.13. Let $K: D(K) \subset E \to E$, $K_n: D(K_n) \subset E_n \to E_n$, $n \in I$, be (not necessarily linear) operators with (K_n) discretely compact and discretely convergent to K. For every $n \in I$, suppose there is a fixed point u_n of

$$u_n = K_n u_n,$$

and, moreover, suppose that the sequence (u_n) is bounded. Then (u_n) is discretely compact, and each limit u of discretely convergent subsequences u_n, $n \in I' \subset I$, is a fixed point of the equation

$$u = Ku.$$

If the latter equation has at most one solution, then there exists precisely one solution u, and the entire sequence u_n, $n \in I$, converges discretely to u.

Proof: Because of the discrete compactness of (K_n) and the boundedness of the sequence of fixed points, there is, to every subsequence $I' \subset I$, another subsequence $I'' \subset I$ and an element $u \in E$ such that

$$u_n = K_n u_n \to u \ (n \in I'').$$

Since $K_n \to K$ $(n \in I)$, then $u_n = K_n u_n \to Ku$ $(n \in I'')$, and therefore Ku must equal u. To see the latter convergence, we note that there is a sequence $u_n' \in D(K_n)$ such that $u_n' \to u$ $(n \in I)$, since $A(D(K), \pi_n D(K_n), \lim^E)$ represents a discrete approximation. If we define $v_n \equiv u_n$, $n \in I''$, $v_n \equiv u_n'$, $n \in I-I''$, then we see that $K_n v_n \to Ku$ $(n \in I)$, and hence the subsequence $K_n u_n = K_n v_n$, $n \in I''$, is also discretely convergent to Ku. If $u = Ku$ has at most one solution, then there exists exactly one solution u (the existence having been shown by the above arguments). There is then to every subsequence $I' \subset I$, a subsequence $I'' \subset I'$ such that $u_n \to u$ $(n \in I'')$. From Lemma 7.1 (applied to all of I), we see that u_n, $n \in I$, converges discretely to u. □

The following theorem indicates that for linear operators the discrete convergence of K_n, $n \in I$, to K hypothesized in the preceding theorem can be replaced by the consistency of K, K_n, $n \in I$.

Theorem 7.14. A discretely compact sequence of linear operators $K_n: E_n \to F_n$, $n \in I$, is necessarily stable.

Proof: Let us assume that the assertion is false, i.e., $||K_n|| \to \infty$ $(n \in I')$ for a subsequence $I' \subset I$ (see Lemma 6.9). Then there are elements $v_n \in E_n$, $n \in I'$, such that

$$||K_n v_n||_n / ||v_n||_n \to \infty \quad (n \in I').$$

If we define $u_n \equiv v_n/||v_n||_n$, $n \in I'$, then $||u_n||_n = 1$, $n \in I'$, and $||K_n u_n||_n \to \infty$ $(n \in I')$. We complete the subsequence u_n, $n \in I'$, to a sequence parametrized by all $n \in I$, by letting $u_n = 0$, $n \in I-I'$. Then $(K_n u_n)_{n\in I}$ is discretely compact and, according to Lemma 7.2, it is bounded - a contradiction to $||K_n u_n||_n \to \infty$ $(n \in I')$. □

The fact that linear mappings $I-K$, $I-K_n$, $n \in I$, are a-regular, if (K_n) is discretely compact and K, K_n, $n \in I$, is consistent, plays a fundamental role in the ensuing analysis. (In addition to representing the set of positive integers, the symbol I will also be used to indicate the identity mapping in all the spaces E, E_n, $n \in I$, but in any case its use should be clear from the context.) The following lemma discusses still other important consequences of discrete compactness. Used in combination with the example presented at the beginning of this section, this lemma will also provide examples of a-regular operators.

Lemma 7.15. Let (K_n) be a discretely compact sequence of linear operators $K_n: E_n \to E_n$, $n \in I$, and let K, K_n, $n \in I$, be consistent where K is a mapping of E into itself. Then K is linear and bounded; the regular convergence

$$I - K_n \overset{r}{\to} I - K \quad (n \in I) \tag{13}$$

holds; and in particular the sequence $I-K, I-K_n$, $n \in I$ is a-regular. Under the above assumptions, $I-K$ is injective if, and only if, $I-K_n$, $n \in I$, is inversely stable.

Proof: (i) We first show that $K_n \to K$ $(n \in I)$ and $K \in B(E,E)$. The stability of (K_n) follows from its discrete compactness, and, together with the consistency of K, K_n, $n \in I$, yields the discrete convergence of (K_n) to K. From Property 5.2 of a discrete convergence, we can conclude the linearity of K as well as its boundedness from Theorem 6.15.

(ii) Next, we show a-regularity. Let u_n, $n \in I'$, be bounded and suppose $(I-K_n)u_n \to v$ $(n \in I')$ for some $v \in E$ and an arbitrary subsequence $I' \subset I$. Due to the discrete compactness of (K_n), there is a subsequence $I'' \subset I'$ and a $w \in E$ such that $K_n u_n \to w$ $(n \in I'')$. This implies that the sequence $(u_n)_{n\in I''}$ converges directly to $v + w$. It is shown in Part (i) of this proof that (K_n) converges discretely to K. Therefore $K_n u_n \to K(v+w) = w$ $(n \in I'')$, and hence $(I-K)(v+w) = v$. The a-regularity is thus shown.

(iii) Regular convergence follows immediately from Theorem 7.9, since $K_n \to K$, and hence $(I-K_n) \to (I-K)$ $(n \in I)$.

(iv) If $I-K$ is injective, then Theorem 7.5 gives the inverse stability of $I-K_n$, $n \in I$. Conversely, Lemma 6.19 allows us to conclude the injectivity of $I-K$ from the consistency of $I-K$, $I-K_n$, $n \in I$, and the inverse stability of $I-K_n$, $n \in I$. □

If E is a separable Banach space, then we can additionally show from the hypotheses of Lemma 7.15 that K is a compact operator (cf. Wolf (1974), 2.(4)); K then is completely continuous. Moreover, using Theorem 3.1, we see that the inverse stability of $(I-K_n)$, in conjunction with the hypothesized consistency and discrete compactness, yields a bijective, bicontinuous operator $I-K$. We can thus conclude from properties of K_n, $n \in I$, the unique and continuous solvability of the linear equation of the second kind

$$(I-K)u = w.$$

The bijectivity of $I-K$ can be obtained without E being separable, if the approximate mappings K_n are assumed completely continuous for every $n \in I$. Then we obtain from inverse stability the unique and continuous solvability of the approximate equations

$$(I-K_n)u_n = w_n$$

for almost all $n \in I$. The following equivalence theorem summarizes the above remarks, and, in addition, proves the biconvergence.

Theorem 7.16. Let (K_n) be a discretely compact sequence of linear, completely continuous operators $K_n: E_n \to E_n$, $n \in I$, and let K be a mapping of E into itself. Then the following two statements are equivalent:

(14a) $I-K$ is injective and K, K_n, $n \in I$, is consistent.

(14b) K is linear and bounded; $(I-K)$ as well as $(I-K_n)$ for almost all $n \in I$ are bijective and bicontinuous, and

$$I-K_n \overset{r}{\to} I-K, \quad (I-K_n)^{-1} \overset{r}{\to} (I-K)^{-1} \quad (n \in I).$$

Proof: (14a) ⇒ (14b). Lemma 7.15 shows that K is linear and bounded, that $I-K_n \overset{r}{\to} I-K$ $(n \in I)$, and that $(I-K_n)$ is inversely stable. The latter conclusion yields the injectivity of almost all $I-K_n$, $n \in I$. Since K_n has also been assumed completely continuous, $I-K_n$ is bijective and bicontinuous for almost all $n \in I$ (see Theorem 3.1). Applying Theorem 7.7 with $L = I-K$, $L_n = I-K_n$, $n \in I$, we deduce the bijectivity and bicontinuity of $I-K$, along with the discrete convergence $(I-K_n)^{-1} \to (I-K)^{-1}$ $(n \in I)$. The latter convergence is also a regular one.

(14b) $\Rightarrow$ (14a). To prove the converse, we see that Lemma 6.18 yields consistency of $I-K$, $I-K_n$, $n \in I$, which clearly implies the consistency of K, K_n, $n \in I$. The injectivity of $I-K$ follows from the inverse stability of $I-K_n$, $n \in I$, via Lemma 7.15. □

Fixed point equations and the linear equations that we have considered in the preceding paragraphs are special cases of *equations of the second kind*, which we write in the general form as

$$Su - Tu = w, \tag{15a}$$

$$S_n u_n - T_n u_n = w_n, \quad n \in I. \tag{15b}$$

Here, S,T (respectively, S_n, T_n, $n \in I$) are (not necessarily) linear mappings from E into F (respectively, E_n into F_n, $n \in I$), whose associated domains of definition are assumed to have the properties that $D(S) \cap D(T) \neq \emptyset$ $(D(S_n) \cap D(T_n) \neq \emptyset$, $n \in I)$. If we define mappings $A = S-T$ and $A_n = S_n - T_n$, $n \in I$, then we see for the corresponding domains of definition that $D(A) = D(S) \cap D(T)$ and $D(A_n) = D(S_n) \cap D(T_n)$, $n \in I$, respectively. We shall restrict our discussion to the case where $S: E \to F$ and $S_n: E_n \to F_n$, $n \in I$, are *linear mappings*. But the following analysis could just as well be extended to differentiable mappings S, S_n, $n \in I$ (see also the following Section 7.4). The setting considered here in case of compact mappings T, T_n, $n \in I$, corresponds to "nonlinear equations with a linear principal part" in Vainikko (1976), 3.5. We shall later also assume compactness properties, but only for the derivatives of T_n, $n \in I$.

Now, A is obviously (continuously) differentiable at a $u^0 \in E$ if, and only if, the same is true for T; (A_n) is equidifferentiable (equicontinuously equidifferentiable) at a sequence (u_n^0), if, and only if, (T_n) has the corresponding property. We have the following representation for the derivatives,

$$A'(u^0) = S - T'(u^0), \quad A_n'(u_n^0) = S_n - T_n'(u_n^0), \quad n \in I.$$

We shall rely on Theorem 7.10 for results concerning biconvergence of solutions of (15a,b). It is incumbent upon us to show that its assumptions and conditions (in particular (9a)) are satisfied for the problem at hand.

The following result addresses the question of regular convergence of sums of linear operators.

Lemma 7.17. Let $B, C\colon E \to F$ and $B_n, C_n\colon E_n \to F_n$, $n \in I$, be linear mappings and suppose

$$B_n \to B, \quad C_n \to C \ (n \in I).$$

If, in addition, B is surjective, (B_n) is inversely stable, and (C_n) discretely compact, then $(B_n+C_n)_{n\in I}$ converges regularly to $B+C$.

Proof: See Vainikko (1976), Theorem 2.(55). □

We maintain that the mapping B is even bijective according to Lemma 6.19 and bicontinuous according to Theorem 6.15. Moreover, a certain converse to the preceding lemma has been shown by Vainikko (1976), 2.(57), who proved a representation of regularly convergent mappings as sums of mappings.

If we check the other conditions from (9a) in Theorem 7.10, then we observe that Property 7.6 is satisfied for B_n+C_n, $n \in I$, whenever, for example, $B_n^{-1} \in B(F_n,E_n)$ and C_n is completely continuous for almost all n. Then, for the appropriate n, we can write

$$B_n + C_n = B_n(I-K_n) \quad \text{where} \quad K_n = -B_n^{-1}C_n, \quad n \geq \nu.$$

Each K_n, $n \geq \nu$, is then completely continuous, and Property 7.6 follows from Theorem 3.1.

We now consider again the special mappings in (15a & b), i.e., we set

$$B = S, \quad B_n = S_n, \quad n \in I, \quad \text{and} \quad C = T'(u^0), \quad C_n = T_n'(u_n^0), \quad n \in I.$$

The preceding lemma and the remarks following provide sufficient conditions for satisfying statement (9a) in Theorem 7.10. The next result, concerning biconvergence of solutions of (15a,b), is a consequence of this theorem.

Theorem 7.18. Suppose the following statements are true:

(16a) The sequence (T_n) is equicontinuously equidifferentiable at (u_n^0); T is differentiable at u^0; and $u_n^0 \to u^0$, $w_n^1 \to Au^0$ $(n \in I)$.

(16b) $A'(u^0)$ is injective.

(16c) S, S_n, $n \in I$, is consistent, and T, T_n, $n \in I$, is consistent at all $v^0 \in B_{\rho_0}(u^0)$ for some $\rho_0 > 0$.

(16d) S is surjective; S_n is bijective for almost all $n \in I$; and the sequence (S_n) is bistable.

(16e) $T_n'(u_n^0)$ is completely continuous for almost every $n \in I$, and the sequence $(T_n'(u_n^0))$ is discretely compact.

Then S, along with $A'(u^0)$, is bijective and bicontinuous and

$$S_n \to S, \quad S_n^{-1} \to S^{-1}, \quad A_n'(u_n^0) \overset{r}{\to} A'(u^0) \ (n \in I).$$

Moreover, each of the statements (9a-c) from Theorem 7.10 hold; and, in particular, all the statements in Theorem 6.21 on biconvergence are valid though only for n belonging to a subset of I of the form $I_1 = \{n \geq \nu\}$ for some positive integer ν.

Proof: From (16a), we see that the differentiability assumptions are satisfied in Theorem 7.10 and, hence, in the Biconvergence Theorem 6.21. The consistency of S, S_n, $n \in I$, the bistability of (S_n), and the surjectivity of S imply the discrete convergence $S_n \to S$ $(n \in I)$, the bijectivity and bicontinuity of S, along with the discrete convergence $S_n^{-1} \to S^{-1}$ $(n \in I)$ (cf. Lemma 6.19, Theorems 6.15 and 6.17). The discrete compactness of $T_n'(u_n^0)$, $n \in I$, implies its stability (see Theorem 7.14). Together with the consistency assumption on T, T_n, $n \in I$, the stability of $T_n'(u_n^0)$, $n \in I$, yields the discrete convergence of $(T_n'(u_n^0))$ to $T'(u^0)$ by Lemma 7.12. Moreover, Theorem 6.14 allows us to conclude that $T_n \to T$ $(n \in I)$ at u^0. Consequently,

$$A_n \to A \ (n \in I) \ \text{ at } \ u^0 \quad \text{and} \quad A_n'(u_n^0) = S_n - T'(u_n^0) \to S - T'(u^0) \ (n \in I).$$

If we apply the discrete compactness of $(T_n'(u_n^0))$ again, and use the inverse stability of (S_n) along with the surjectivity of S, we get from Lemma 7.17 the regular convergence

$$A_n'(u_n^0) = S_n - T_n'(u_n^0) \overset{r}{\to} S - T'(u^0) = A'(u^0) \ (n \in I).$$

This shows statement (9a) of Theorem 7.10 if we note further that Property 7.6 is satisfied because of the complete continuity of almost all $T_n'(u_n^0)$, $n \in I$ (see also the remarks following Lemma 7.17). The result of Theorem 7.10 assures us that the statements (9b) and (9c) are also true. Therefore we can conclude the validity of all the statements 6.(19a-c) of the Biconvergence Theorem 6.21. □

If we consider again the special case of *fixed point equations* defined by setting $w = 0$ and $w_n = 0$ in (15a) and (15b), respectively, then Theorem 7.18, under the assumptions (16a-e), implies the following: From the discrete convergence of (A_n) to A at the fixed point u^0, we see that $w_n^0 = A_n u_n^0 \to Au^0 = 0$ $(n \in I)$; to a pair of positive numbers ρ,σ from Theorem 6.22, there is a $\nu \in I$ so that $||w_n^0||_n \leq \sigma$, $n \geq \nu$; there is a unique fixed point $u_n \in E_n$ of the equation

$$S_n u_n = T_n u_n \Longleftrightarrow A_n u_n = 0$$

in the set $||u_n - u_n^0||_n \leq \rho$ for all $n \geq \nu$; the sequence of fixed points converges discretely to u^0 with the two-sided error estimate

$$\frac{1}{\alpha}||(S_n - T_n)u_n^0||_n \leq ||u_n - u_n^0||_n \leq \frac{1}{\beta}||(S_n - T_n)u_n^0||_n, \quad n \geq \nu. \tag{17}$$

7.4. PROJECTION METHODS FOR THE APPROXIMATE SOLUTION OF NONLINEAR FIXED POINT EQUATIONS

We now proceed to apply the theory of the previous section in order to investigate the existence and convergence of fixed points of nonlinear differentiable mappings, which arise when we approximate a given fixed point equation by a projection method.

Suppose we have a sequence of subspaces $E_n \subset E$, $n \in I$, with the property (cf. 5.(3) and Example 2 in Section 7.3) that

$$|u,E_n| \to 0 \ (n \in I), \quad u \in E. \tag{18}$$

Further, let F_n, $n \in I$, be a sequence of subspaces in F, and $P_n\colon F \to F_n$, $n \in I$, be an associated sequence of bounded, linear projection operators which are pointwise convergent,

$$||P_n w - w|| \to 0 \ (n \in I), \quad w \in F. \tag{19}$$

We assume that E and F are Banach spaces and designate their norms by the common notation $||\cdot||$. Using the uniform boundedness principle, we see from (19) that the P_n, $n \in I$, constitute a uniformly bounded sequence, i.e.,

$$||P_n|| \leq \gamma, \quad n \in I. \tag{20}$$

Next, let S,T be (not necessarily linear) mappings from E into F and let A be given by $A = S-T$, where $D(A) = D(S) \cap D(T)$ is supposed nonempty. We assume that the fixed point equation,

$$Su = Tu \Longleftrightarrow Au = 0, \tag{21}$$

possesses a fixed point $u^0 \in D(A)$. Further, let S and T - and thereby also A - be uniformly differentiable in a neighborhood $B_{\rho_0}(u^0)$ of u^0, with the derivatives S', T', A' continuous at each $v \in B_{\rho_0}(u^0)$, i.e., (written down for A)

$$\forall \varepsilon > 0, \quad \exists \delta > 0 \quad \ni: \forall u \in B_{\rho_0}(u^0), \quad \forall h: \|h\| \leq \delta$$
$$\Rightarrow u+h \in D(A) \quad \text{and} \quad \|A(u+h) - Au - A'(u)h\| \leq \varepsilon\|h\|, \tag{22a}$$

and

$$\forall u \in B_{\rho_0}(u^0), \quad \forall \varepsilon > 0, \quad \exists \eta > 0 \quad \ni: \forall v: \|u - v\| \leq \eta, \forall h \in E$$
$$\Rightarrow \|(A'(u) - A'(v))h\| \leq \varepsilon\|h\|. \tag{22b}$$

We note that η can depend on u in (22b); thus (22a,b) represent weaker conditions than that of uniform continuous differentiability of A in $B_{\rho_0}(u^0)$.

Projection methods for approximating the fixed point equation (21) have the following features in common. For given $w_n \in F_n$, solutions $u_n \in E_n$ are sought for the equations

$$P_nSu_n = P_nTu_n + w_n \quad \text{or} \quad P_nAu_n = w_n, \quad n \in I. \tag{23}$$

It is convenient to define

$$S_n \equiv P_nS|E_n, \quad T_n \equiv P_nT|E_n, \quad A_n \equiv P_nA|E_n, \quad n \in I,$$

with the domains of definitions for these operators naturally given by $D(S_n) = E_n \cap D(S)$, $D(T_n) = E_n \cap D(T)$, and $D(A_n) = E_n \cap D(A)$. Because of assumption (18), we know that to every $u \in E$, there exists a sequence $u_n \in E_n$, $n \in I$, such that $\|u_n - u\| \to 0$ $(n \in I)$. If u is an interior point of $D(A)$, then u_n lie in $D(A)$ and in $D(A_n)$ for almost every $n \in I$. We shall denote an arbitrary but fixed sequence converging to u^0 by (u_n^0), $\|u^0 - u_n^0\| \to 0$ $(n \in I)$. For simplicity, we assume that $u_n^0 \in E_n \cap B_{\rho_0/2}(u^0)$ for all $n \in I$. (Otherwise we must restrict the parameter n to lie in a subset I_1 of I defined by $I_1 \equiv \{n \geq \nu\}$ for some ν). We now have the following result on the equicontinuous equidifferentiability of A_n, $n \in I$.

Lemma 7.19. Under the differentiability assumptions (22a,b) and the above assumptions on (u_n^0), the sequence (A_n) is equicontinuously equidifferentiable at (u_n^0) with the derivative

$$A_n'(u_n) = P_nA'(u_n)|E_n, \quad u_n \in E_n \cap B_{\rho_0}(u^0), \quad n \in I.$$

Proof: (i) We show first that (A_n) is equidifferentiable at all $(v_n) \in B_{\rho_0/2}((u_n^0))$. Let $\varepsilon > 0$ be arbitrary; let $\delta > 0$ be determined from condition (22a); and let $(v_n) \in B_{\rho_0/2}((u_n^0))$. Then $||u^0-v_n|| \leq \rho_0$, $n \in I$, and, from (22a), we know that $v_n + h_n \in D(A)$ and

$$||A_n(v_n+h_n) - A_nv_n - P_nA'(v_n)h_n||$$
$$\leq \gamma||A(v_n+h_n) - Av_n - A'(v_n)h_n|| \leq \gamma\varepsilon||h_n||$$

for all $n \in I$, and all $h_n \in E_n$ with $||h_n|| \leq \delta$.

(ii) It remains to show that (A_n') is equicontinuous at (u_n^0). To this end, let $\varepsilon > 0$ be arbitrary and let $\eta > 0$ be determined by the continuity of A' at u^0. Then there exists a $\nu_1 \in I$, so that $||u^0 - u_n^0|| \leq \eta/2$, $n \geq \nu_1$; and, for every $u_n \in E_n \cap B_{\eta/2}(u_n^0)$, we have $||u_n - u^0|| \leq \eta$, $n \geq \nu_1$, along with

$$||(A'(u_n) - A'(u_n^0))h|| \leq ||(A'(u_n) - A'(u^0))h||$$
$$+ ||(A'(u^0) - A'(u_n^0))h|| \leq 2\varepsilon||h||, \quad n \geq \nu_1, \; h \in E.$$

Moreover, let $\hat{\eta} > 0$ be the common number determined by continuity of A' at u_n^0, $n = 1,\ldots,\nu_1-1$ - i.e., let $\hat{\eta}$ be such that

$$||(A'(v) - A'(u_n^0))h|| \leq 2\varepsilon||h||, \quad n = 1,2,\ldots,\nu_1-1,$$

for all $h \in E$ and all v in $||v - u_n^0|| \leq \hat{\eta}$. If we now define $\tilde{\eta} \equiv \min(\eta/2,\hat{\eta})$, we have

$$||P_n(A'(u_n^0) - A'(u_n^0))h_n|| \leq 2\gamma\varepsilon||h_n||, \quad n \in I,$$

for all sequences $(u_n) \in B_{\tilde{\eta}}((u_n^0))$ and all $h_n \in E_n$, $n \in I$. □

Concerning the problems (15a,b), we also wish to give conditions guaranteeing that the hypotheses and statement (9a) of Theorem 7.10 are satisfied. The following lemma provides such conditions. For brevity of presentation, we write

$$B \equiv S'(u^0), \quad C \equiv T'(u^0), \quad B_n \equiv P_nB|E_n, \quad C_n \equiv P_nC|E_n, \quad n \in I.$$

Lemma 7.20. Let $A'(u^0)$ be injective, B be surjective and continuous, and let C be completely continuous. Further, let each B_n be surjective for almost all $n \in I$ with the sequence (B_n) inversely stable. Then B is bijective and bicontinuous; B_n is bijective and bicontinuous for almost all $n \in I$; C_n is completely continuous for all $n \in I$; the sequence (C_n) is discretely compact; and $B_n \to B$, $B_n^{-1} \to B^{-1}$, $C_n \to C$ $(n \in I)$. In addition, $A'(u^0)$ is bijective and bicontinuous; $A_n'(u_n^0)$, $n \in I$, satisfies Property 7.6, and

$$A_n'(u_n^0) \xrightarrow{r} A'(u^0) \quad (n \in I).$$

Proof: Since B, C are continuous, $B_n \to B$, $C_n \to C$ $(n \in I)$ (cf. Example 4, Section 6.3). Because of the inverse stability of (B_n), B and B_n, (for almost all) $n \in I$, are injective (cf. Lemma 6.19). Also, $B_n^{-1} \to B^{-1}$ $(n \in I)$ (cf. Theorem 6.17) and B^{-1} is continuous (cf. Theorem 6.15). Because C is completely continuous and the P_n are bounded, $C_n = P_nC|E_n$ is completely continuous for all $n \in I$, and the sequence (C_n) is discretely compact (cf. Example 2 in Section 7.3). The derivative

$$A'(u^0) = B - C = B(I - K)$$

is injective, where $K \equiv B^{-1}C$ is completely continuous. From Theorem 3.1, we see that $A'(u^0)$ is bijective and bicontinuous. An analogous argument shows that each $A_n'(u_n^0) = B_n(I-K_n)$, $K_n \equiv B_n^{-1}C_n$, $n \in I$, is also bijective and bicontinuous, with the result that Property 7.6 is satisfied. Finally, the regular convergence of $(A_n'(u_n^0))$ to $A'(u^0)$ follows from Lemma 7.17. □

At this point, we still need to show the consistency of A, A_n, $n \in I$, at u^0, u_n^0, $n \in I$, in order to be able to apply Theorem 7.10 to show biconvergence of the solutions of (23) to the fixed point u^0 of $Su^0 = Tu^0$. Under the assumptions of Lemma 7.20, $A'(u^0)$ is, in particular, continuous, and also A is continuous at u^0 by Theorem 6.1. As in Example 4 from Section 6.3, $A_n \to A$ $(n \in I)$ at u^0; and, in particular, the sequence A, A_n, $n \in I$, is consistent at u^0, u_n^0, $n \in I$, i.e., $||P_nAu_n^0|| \to 0$ $(n \in I)$. If we finally substitute $w_n^1 = 0$, $n \in I$, in Theorem 6.21 and in Theorem 7.10, then obviously $w_n^1 \to Au^0 = 0$ $(n \in I)$. Theorem 7.10, together with Lemma 7.20, thus yields the following result.

Theorem 7.21. Under the differentiability condition (22a,b) and under the hypotheses of Lemma 7.20, each of the statements (19a-c) of the Biconvergence Theorem 6.21 holds for possibly a subset I_1 of I defined by $I_1 \equiv \{n \geq \nu\}$ for some $\nu \in I$, and with $w_n^1 = 0$, $n \in I_1$. □

In the concluding paragraph of this chapter, we consider a *special projection method* for approximating the fixed point equation (21) (cf. Krasnoselskii et al. (1972), Chapter 19, with $S = I$, $w_n = 0$),

$$Su_n = P_n Tu_n + w_n, \quad n \in I, \tag{24}$$

with solutions $u_n \in E$ and perturbations $w_n \in F$ of the null vector. P_n, $n \in I$, are to be projections of F into itself satisfying (19). We choose $E_n = E$, $F_n = F$, and define $S_n = S$, $T_n = P_n T$, and $A_n = S-P_n T$, $n \in I$. In order to obtain a result analogous to Lemma 7.10 about the equicontinuous equidifferentiability of (A_n) at the constant sequence $u_n^0 = u^0$, $n \in I$, we need the differentiability condition (22a) for S, T at the points u^0, and - in place of (22b) - it is enough to assume the continuity of the derivatives S', T' at u^0. We would like to remark here that the continuous differentiability of A at u^0 is not sufficient to show the equicontinuous equidifferentiability of (A_n) at u^0. The assumptions of Lemma 7.20 mean in this special case that $A'(u^0)$ is injective, $S'(u^0)$ bijective and bicontinuous, and $T'(u^0)$ completely continuous. Here $B_n = B = S'(u^0)$, $n \in I$. The mappings $C_n = P_n T'(u^0)$, $n \in I$, are then collectively compact (cf. the definition in Example 3 of Section 7.3), and hence the results from Theorem 7.21 are valid.

REFERENCES

Anselone (1971), Anselone & Ansorge (1979,1981)*, Grigorieff (1972, 1973a,1975)*, Kato (1966), Krasnoselskii et al. (1972), Petryshyn (1968a, 1968b)*, Reinhardt (1975a)*, Stummel (1970,1973a,1976b)*, Vainikko (1969)*, Vainikko (1976), Wolf (1974)*.

---

*Article(s)

Part III
Convergence Analysis for Approximate Solutions to Boundary-Value Problems and Integral Equations

In this part of the book, we analyze the stability, consistency, and convergence properties of approximation methods for boundary-value problems and integral equations by applying the convergence theory from Part II.

On the one hand, the general concepts in our convergence theory will be further elucidated by reformulating the examples already discussed in Part II in a more concrete setting. On the other hand, we demonstrate the efficiency of our theory by obtaining results, including two-sided error estimates, for the most diverse classes of problems and approximation methods.

Compactness arguments are applicable to the theoretical analysis of approximation methods for both boundary-value problems and integral equations, and account for treating these two different problem areas in this portion of the text. The methods for approximating boundary-value problems are further subdivided into finite-difference methods and projection methods (for the associated variational formulation). Besides compactness properties, maximum principles will be used to analyze finite-difference methods. An analysis of projection methods does not require compactness arguments but techniques are applied which we summarize as variational principles.

The applications point out that a powerful convergence theory in a suitable general framework constitutes an essential basis for a successful analysis of specific problems, but that a hard analysis is often required to verify stability, inverse stability, discrete compactness, and consistency. Thus, we present only a limited number of examples to show the applicability of our theory as well as to provide analytical techniques.

To understand the analysis in each of the chapters of Part III, the reader must have some knowledge of the material presented in the chapters of Part I according to the following diagram:

Part I	Part III	Problem
Chapter: 1	8	Boundary-value problems
2	9	Variational equations
3	10	Integral equations

The theoretical basis for our analysis, however, is the convergence theory from Part II. Note that, in Sections 5.3 and 5.4, we have provided the discrete approximations underlying the analysis of Chapters 8 and 10 together with important properties of sequences of measures needed for the integral equations.

Chapter 8
Convergence of Finite-Difference Methods for Boundary-Value Problems

In this chapter, we obtain results on the convergence of solutions to finite-difference approximations of boundary-value problems. More specifically, we study the convergence of finite-difference approximations to both linear and nonlinear ordinary differential equations of second order and to Poisson's equation on a rectangle. In Chapter 1, appropriate finite-difference approximations were introduced, and both the exact and the approximate equations were expressed as operator equations. In addition, consistency of these methods (in the sense of Section 6.3) has been shown by an analysis of the truncation errors derived in Chapter 1.

The main goal of this chapter, then, is to show the inverse stability of the difference operators. This will be accomplished by using maximum principles in both Section 8.1 and 8.3 for finite-difference approximations of, respectively, linear ordinary differential equations and Poisson's equation in two independent variables. In Section 8.2, we shall establish the inverse stability from a compactness property, i.e., from the property of a-regularity which has already been studied in Chapter 7. The extension to nonlinear ordinary differential equations is possible by the approach described in Chapters 6 and 7. Such an approach means here that we essentially show maximum principles and a-regularity for the associated Fréchet-derivatives of the finite-difference operators. In a concrete application presented in Section 8.1, it will turn out that no Fréchet-derivatives exist however, but that a maximum principle can be directly applied to the nonlinear operators. Applying the results from Part II along with the consistency properties discussed in Chapter 1 will yield the convergence along with error estimates. We note the interesting fact that compactness arguments in Section 8.2 ensure the convergence in norms stronger than those used in Section 8.1.

The presentation in Section 8.1 parallels an unpublished work of Grigorieff (1973b). Section 8.2 extends the investigations of Vainikko (1976), §6, and is an obvious consequence of the convergence theory from Section 7.2. Section 8.3 presents a maximum principle for the discrete Laplace operator as is found, e.g., in Isaacson & Keller (1966). There are similar maximum principles for finite-difference approximations of general elliptic differential operators, and these also lead to existence and convergence results inclusive of error estimates (cf. e.g., Ciarlet (1970)). Compactness arguments can likewise play a crucial role in the convergence analysis of finite-difference approximations for elliptic differential equations, which is outlined in, e.g., Vainikko (1970), §7.

We do not treat here the finite-difference approximation for the beam bending problem presented in Section 1.3. Problems of fourth order (or, more generally, n-th order) can be expressed as first order systems of equations, and such systems can be approximated by finite-differences. We refer the reader to Keller (1968, 1976) for a discussion of such problems. An appropriate and convergent finite-difference scheme can be obtained for the beam problem by employing Ritz-Galerkin techniques with piecewise linear trial functions. We refer the reader to Chapters 2 and 9 for a detailed treatment of these methods.

8.1. CONVERGENCE OF DIFFERENCE METHODS FOR ORDINARY DIFFERENTIAL EQUATIONS VIA MAXIMUM PRINCIPLES

We begin our discussion by considering linear, second-order ordinary differential equations of the form

$$(Lu)(x) \equiv u''(x) + p(x)u'(x) + q(x)u(x) = f(x), \quad x \in [a,b], \tag{1a}$$

with boundary conditions

$$\begin{aligned} \ell_0(u) &\equiv \alpha_0 u(a) - \alpha_1 u'(a) = \gamma_0, \\ \ell_1(u) &\equiv \beta_0 u(b) + \beta_1 u'(b) = \gamma_1, \end{aligned} \tag{1b}$$

where $[a,b]$ is a compact interval in $\mathbb{R}$. With the terminology and notation from Section 1.2, we approximate the boundary-value problem (1a,b) by the finite-difference equations,

$$(L_h u_h)(x) \equiv D_h^2 u_h(x) + p_h(x) D_h u_h(x) + q_h(x) u_h(x) = f_h(x), \quad x \in I_h', \tag{2a}$$

$$\begin{aligned} \ell_{0,h}(u_h) &\equiv \alpha_{0,h} u_h(a_h) - \alpha_{1,h} D_h^+ u_h(a_h) = \gamma_{0,h}, \\ \ell_{1,h}(u_h) &\equiv \beta_{0,h} u_h(b_h) + \beta_{1,h} D_h^- u_h(b_h) = \gamma_{1,h}, \end{aligned} \tag{2b}$$

where h is a positive mesh width. We have already expressed these problems in the form of operator equations in Section 1.2,

$$(1a,b) \iff Au = w,$$

$$(2a,b) \iff A_h u_h = w_h,$$

where $Au \equiv (Lu, \ell_0(u), \ell_1(u))$, $A_h u_h \equiv (L_h u_h, \ell_{0,h}(u_h), \ell_{1,h}(u_h))$, $w = (f, \gamma_0, \gamma_1) \in F$, $w_h = (f_h, \gamma_{0,h}, \gamma_{1,h}) \in F_h$. The appropriate spaces, $E = C^2[a,b]$, $F = C[a,b] \times \mathbb{K}^2$, $E_h = C(I_h)$, $F_h = C(I_h') \times \mathbb{K}^2$, are equipped with the norms defined in 1.(6) and 1.(7), respectively, which in essence are maximum norms. We assume that all functions are real valued, i.e., $\mathbb{K} = \mathbb{R}$.

From Theorem 5.11, we see that discrete approximations $A(E, \pi_h E_h, \lim^E)$ and $A(F, \pi_h F_h, \lim^F)$ will exist, in case the mesh widths constitute a null sequence $\Lambda = (h_1, h_2, \ldots)$ and if $a_h \to a$, $b_h \to b$ as $h \to 0$, $h \in \Lambda$. For brevity, we shall always write $(h \in \Lambda)$ instead of $(h \to 0$, $h \in \Lambda)$. The discrete (uniform) convergences are given by

$$\lim{}^E u_h = u \iff \exists \hat{u} \in C(M): \hat{u}|[a,b] = u, \max_{x \in I_h} |u_h(x) - \hat{u}(x)| \to 0 \ (h \in \Lambda), \tag{3}$$

$$\begin{aligned} \lim{}^F w_h = w \iff & \exists \hat{f} \in C(M): \hat{f}|[a,b] = f, \\ & \max_{x \in I_h'} |f_h(x) - \hat{f}(x)| \to 0, \quad \gamma_{0,h} \to \gamma_0, \ \gamma_{1,h} \to \gamma_1 \quad (h \in \Lambda), \end{aligned} \tag{4}$$

for $u \in E$, $u_h \in E_h$, $w = (f, \gamma_0, \gamma_1) \in F$, $w_h = (f_h, \gamma_{0,h}, \gamma_{1,h}) \in F_h$. Moreover, we let $M = [a_0, b_0]$ be a compact interval in $\mathbb{R}$ such that $[a,b]$, $I_h \subset M$, $h \in \Lambda$. In addition, we see that the discrete maximum norms will satisfy Property 5.3 under the above mentioned hypotheses. Conversely, the results from Section 5.3 show that the convergences $a_h \to a$ and $b_h \to b$ $(h \in \Lambda)$ are even necessary for the existence of the above discrete approximations with norms satisfying the Convergence Property 5.3.

The discrete convergences (3) and (4) can also be characterized in terms of restriction operators, which in this case are defined by

$$(r_h u)(x) = \hat{u}(x), \quad x \in I_h, \quad u \in C[a,b], \quad h \in \Lambda,$$

where $\hat{u} \in C(M)$ denotes a fixed extension of $u \in C[a,b]$. We denote the restrictions to I_h and to I_h' by the common notation r_h. The discrete convergences (3) and (4) can then be expressed by

$$\lim^E u_h = u \Longleftrightarrow \|u_h - r_h u\|_h \to 0 \quad (h \in \Lambda)$$

and

$$\lim^F w_h = w \Longleftrightarrow \|f_h - r_h f\|_h + |\gamma_{0,h} - \gamma_0| + |\gamma_{1,h} - \gamma_1| \to 0 \quad (h \in \Lambda),$$

respectively. The maximum norms in $C(I_h)$ and $C(I_h')$ are denoted by the same notation $\|\cdot\|_h$.

The consistency of the mappings A, A_h, $h \in \Lambda$, has already been established in Section 1.2 for the special case of $a_h = a$, $b_h = b$, $p_h = r_h p$, $q_h = r_h q$, $\alpha_{i,h} = \alpha_i$, $\beta_{i,h} = \beta_i$, $i = 0,1$, by examining the truncation errors. The consistency sequence is thus directly given by the sequence of restrictions of the solution u of (1). For the general case, we can easily convince ourselves that, for every function $u \in C^2[a,b]$, the *truncation errors at* u

$$\tau_h(u) = (L_h r_h u - r_h Lu, \ell_{0,h}(r_h u) - \ell_0(u), \ell_{1,h}(r_h u) - \ell_1(u))$$

tend to zero, and therefore consistency at the sequence of restrictions $(r_h u)$ is present, provided that

$$\|p_h - r_h p\|_h \to 0, \quad \|q_h - r_h q\|_h \to 0 \quad (h \in \Lambda), \tag{5a}$$

$$\alpha_{i,h} \to \alpha_i, \quad \beta_{i,h} \to \beta_i \quad (h \in \Lambda), \quad i = 0,1. \tag{5b}$$

In the notation from Chapter 1, the first component of $\tau_h(u)$ represents, for the solution u of the boundary-value problem (1), the truncation error associated with the difference equation, whereas both $\tau_h^{(1)}(u) \equiv \ell_{0,h}(r_h u) - \ell_0(u)$ and $\tau_h^{(2)}(u) \equiv \ell_{1,h}(r_h u) - \ell_1(u)$ represent the truncation errors associated with the approximate boundary conditions.

We now turn to showing the inverse stability of the mappings A_h, $h \in \Lambda$. We restrict our discussion in the following to Dirichlet boundary conditions, i.e., we set

$$\alpha_0 = \beta_0 = \alpha_{0,h} = \beta_{0,h} = 1, \quad \alpha_1 = \beta_1 = \alpha_{1,h} = \beta_{1,h} = 0.$$

We recall that, in Section 1.2, we have also expressed the finite-difference operators as

$$(L_h v_h)(x) = a_{-1,h}(x) v_h(x-h) + a_{0,h}(x) v_h(x) + a_{1,h}(x) v_h(x+h), \qquad x \in I_h', \; v_h \in E_h, \; h \in \Lambda \tag{6}$$

(cf. 1.(3)). At this point, it is useful to examine several properties of linear difference operators of this form. A difference operator L_h expressed as in (6) is said to be of *positive type*, if

$$a_{-1,h} > 0, \quad a_{1,h} > 0, \quad a_{-1,h} + a_{0,h} + a_{1,h} \leq 0 \quad \text{in} \quad I_h'. \tag{7}$$

This "positivity" property for finite-difference operators is very closely intertwined to the "M-matrix" property of the associated system of equations. We shall not, however, explore this relationship in the text. The following maximum principle is valid for the class of finite-difference operators of positive type.

Theorem 8.1. Suppose that the finite-difference operator L_h is of positive type and that $v_h \in C(I_h)$ satisfies $L_h v_h \geq 0$, $x \in I_h'$. Then

$$\max_{x \in I_h} v_h(x) \leq \max(0, v_h(a_h), v_h(b_h)). \tag{8}$$

If v_h assumes a nonnegative maximum value at a point in I_h', then it is constant on I_h.

Proof: We first prove the second assertion by contradiction. We assume that, for $v\ (= v_h) \in C(I_h)$ with $L_h v \geq 0$, there is an $x \in I_h'$ such that

$$v(x-h) \leq v(x), \quad v(x+h) \leq v(x), \quad v(x) \geq 0,$$

where at least one of the first two inequalities is a strict one. Using (7), we get

$$\begin{aligned} 0 \leq L_h v(x) &= a_{-1,h}(x) v(x-h) + a_{0,h}(x) v(x) + a_{1,h}(x) v(x+h) \\ &< (a_{-1,h}(x) + a_{0,h}(x) + a_{1,h}(x)) v(x) \leq 0, \end{aligned}$$

thereby leading to a contradiction. Inequality (8) follows directly from the proved assertion, for, if v assumes its nonnegative maximum at a point $x \in I_h$, then $0 \leq v(x) = v(a_h)$ or $0 \leq v(x) = v(b_h)$. Trivially, $v(x) \leq 0$ if v assumes a negative maximum in I_h. □

One of the immediate consequences of Theorem 8.1 is that the homogeneous finite-difference equation,

$$L_h v_h = 0, \quad v_h(a_h) = v_h(b_h) = 0,$$

possesses only the trivial solution $v_h = 0$. Indeed, we see that $v_h(x) \leq 0$, $x \in I_h$, for every solution v_h, and application of Theorem 8.1 to $-v_h$ yields $v_h(x) \geq 0$, $x \in I_h$.

If, in addition, $a_{-1,h} + a_{0,h} + a_{1,h} = 0$, then the hypotheses of

Theorem 8.1 imply that

$$\max_{x\in I_h} v_h(x) \leq \max(v_h(a_h), v_h(b_h)).$$

We can then reach a contradiction as in the proof of Theorem 8.1 without having to use $v_h(x) \geq 0$ (in the case of a nonnegative maximum).

Finally, we see that the first assertion of the theorem immediately yields the following *monotonicity property:*

$$L_h v_h(x) \leq L_h u_h(x), \quad x \in I_h', \quad v_h(a_h) \geq u_h(a_h), \quad v_h(b_h) \geq u_h(b_h) \\ \Rightarrow v_h(x) \geq u_h(x), \quad x \in I_h. \tag{9}$$

The coefficients occurring in the special finite-difference approximations (2a),

$$a_{\pm 1,h}(x) = h^{-2}(1 \pm \tfrac{h}{2}p_h(x)), \quad a_{0,h}(x) = -h^{-2}(2-h^2 q_h(x)), \quad x \in I_h', \tag{10}$$

clearly satisfy the positivity condition (7) in case that

$$q_h(x) \leq 0, \quad |p_h(x)| \leq c_1, \quad x \in I_h', \tag{11a}$$

for some $c_1 \geq 0$, and that the mesh widths satisfy $h \in (0,h_0]$ for some h_0 restricted by

$$0 < h_0 < 2/c_1. \tag{11b}$$

Under these conditions, we are now able to show the inverse stability of the mappings A_h, $h \in \Lambda$, by making essential use of the monotonicity property (9). To this end, we tacitly assume that $[a,b]$, $I_h \subset [a_0,b_0]$ for all $h \in (0,h_0]$.

<u>Theorem 8.2</u>. Suppose the condition (11a) is satisfied for all $h \in (0,h_0]$ with h_0 restricted by (11b). Then there are numbers $\beta > 0$, $\kappa_0 \in (0,h_0]$ such that the finite-difference operators defined by (6) and (10) satisfy the following inequality

$$\beta \max_{x\in I_h} |v_h(x)| \leq |v_h(a_h)| + |v_h(b_h)| + \max_{x\in I_h'} |(L_h v_h)(x)| \tag{12}$$

for all $h \in (0,\kappa_0]$, $v_h \in C(I_h)$.

<u>Proof:</u> It is clear that (11a) implies (7) for all $h \in (0,h_0]$. For an arbitrary but fixed $m' > 0$, choose λ so that $\lambda^2 - \lambda c_1 \geq 2m'$. Letting $d > \exp(\lambda(a-a_0))$, we define $v(x) \equiv d-\exp(-\lambda(x-a))$, $x \in \mathbb{R}$. Then

$$v(x) \geq d - \exp(\lambda(a-a_0)) > 0, \quad x \in I_h,$$
$$\lambda^2 - \lambda p_h(x) \geq 2m', \qquad x \in I_h', \quad h \in (0,h_0].$$

With respect to the maximum norm, we have further that

$$||D_h^2 v - v''||_h \leq \frac{m'}{2} \exp(-\lambda\ell), \quad c_1||D_h v - v'||_h \leq \frac{m'}{2} \exp(-\lambda\ell)$$

for sufficiently small h where $\ell \equiv b_0 - a_0$. (For notational purposes, we let v also denote its restriction to either of the meshes I_h or I_h'.) Consequently, we have the following inequalities for our finite-difference operator,

$$\begin{aligned}(L_h v)(x) &= (v'' + p_h v' + q_h v)(x) + (D_h^2 v - v'')(x) + p_h(D_h v - v')(x) \\ &\leq -(\lambda^2 - \lambda p_h(x))\exp(-\lambda(x-a)) + (q_h v)(x) + m'\exp(-\lambda\ell) \\ &\leq -2m'\exp(-\lambda\ell) + m'\exp(-\lambda\ell) = -m' \exp(-\lambda\ell), \; x \in I_h' \;\; h \in (0,\kappa_0],\end{aligned}$$

where κ_0 is some number in $(0,h_0]$. If we now define the constant c by

$$c \equiv \max\left\{-\frac{1}{m} \min_{x\in I_h'} g_h(x), \; v_h(a_h)/m_0, \; v_h(b_h)/m_0, 0\right\}$$

for $v_h \in C(I_h)$ and $g_h = L_h v_h$, $h \in (0,\kappa_0]$, with $m \equiv m'\exp(-\lambda\ell)$, $m_0 \equiv d - \exp(\lambda(a-a_0))$, then $g_h(x) \geq -mc$, $x \in I_h'$, and hence

$$L_h(cv)(x) = cL_h v(x) \leq -mc \leq g_h(x), \quad x \in I_h', \quad h \in (0,\kappa_0].$$

Since, moreover, $cv(a_h) \geq cm_0 \geq v_h(a_h)$ (and likewise when a_h is replaced by b_h), the monotonicity (9) implies that $v_h(x) \leq cv(x)$, $x \in I_h$, $h \in (0,\kappa_0]$. For $(-v_h)$, we analogously obtain $v_h(x) \geq c'v(x)$, where

$$c' \equiv \max\left\{\frac{-1}{m} \min_{x\in I_h'} (-g_h(x)), \; -v_h(a_h)/m_0, \; -v_h(b_h)/m_0, 0\right\},$$

so that

$$|v_h(x)| \leq \max(|c|,|c'|)\max_{z\in I_h} |v(z)|, \quad x \in I_h, \quad h \in (0,\kappa_0],$$

thereby proving (12). □

If we make the stronger assumption (as found in Keller (1968), for example) that q_h is uniformly bounded away from zero, then we can show (12) without using the maximum principle. Maximum principles and monotonicity properties are also present for problems with boundary conditions of the third kind. Under the same assumption (11a,b), together with some conditions on the coefficients $\alpha_{i,h}$, $\beta_{i,h}$, we can again prove an inverse stability inequality of the form (12) where now the functionals $\ell_{0,h}$ and $\ell_{1,h}$ occur in the right-hand side. For reasons of brevity, we shall not carry out details of the proof.

Let us again focus our attention on Dirichlet boundary conditions, for which we are able to prove the following result.

Theorem 8.3. Suppose the boundary-value problem (1) with $\alpha_0 = \beta_0 = 1$, $\alpha_1 = \beta_1 = 0$ satisfies the hypotheses that $p, q \in C[a,b]$, and $q \leq 0$ in $[a,b]$. Suppose, moreover, that the finite-difference approximation (2) with $\alpha_{0,h} = \beta_{0,h} = 1$, $\alpha_{1,h} = \beta_{1,h} = 0$ satisfies (5a) and that $a_h \to a$, $b_n \to b$ $(h \in \Lambda)$. Then there exists a number $\kappa_1 > 0$ so that the finite-difference equation (2) possesses a unique solution $u_h \in E_h$ for every $h \in \Lambda_1 \equiv \{h \in \Lambda\colon h \leq \kappa_1\}$ and every $(f_h, \gamma_{0,h}, \gamma_{1,h}) \in F_h$. Furthermore, there exists a unique solution $u \in C^2[a,b]$ of (1) for every $f \in C[a,b]$ and $\gamma_0, \gamma_1 \in \mathbb{R}$. If, in addition,

$$\gamma_{0,h} \to \gamma_0,\ \gamma_{1,h} \to \gamma_1 \quad \text{and} \quad \max_{x \in I_h'} |f_h(x) - \hat{f}(x)| \to 0 \ (h \in \Lambda)$$

then

$$\max_{x \in I_h} |(u_h - r_h u)(x)| \to 0 \ (h \in \Lambda),$$

where $\hat{f} \in C(M)$ is an arbitrary extension of f. Moreover, for some $\beta > 0$ independent of h, the following error estimates hold,

$$\beta \max_{x \in I_h} |(u_h - r_h u)(x)| \leq \max_{x \in I_h'} |(f_h - L_h r_h u)(x)| + |\gamma_{0,h} - \gamma_0| + |\gamma_{1,h} - \gamma_1|, \quad h \in \Lambda_1. \tag{13}$$

Proof: (i) We first apply Theorem 8.2 to a finite-difference operator $\tilde{L}_h$ defined as in (2a), but with q_h replaced by $\tilde{q}_h(x) \equiv \min(0, q_h(x))$, $x \in I_h'$. From (5a), $||p_h - r_h p||_h \to 0$ $(h \in \Lambda)$ so that the p_h, $h \in \Lambda$, are uniformly bounded. Hence, $\tilde{q}_h$ and p_h satisfy condition (11a), and Theorem 8.2 yields the estimate

$$\beta ||v_h||_h \leq ||\tilde{L}_h v_h||_h + |v_h(a_h)| + |v_h(b_h)|, \qquad v_h \in E_h, \quad h \in \Lambda \cap (0, \kappa_0'],$$

for some $\kappa_0' > 0$. This implies a corresponding inequality for L_h,

$$\beta' ||v_h||_h \leq ||L_h v_h||_h + |v_h(a_h)| + |v_h(b_h)|, \quad v_h \in E_h,\ h \in \Lambda_1, \tag{14}$$

with $\beta' = \beta/2$ and $0 < \kappa_1 \leq \kappa_0'$. In order to assure the latter inequality we first note that

$$||(L_h - \tilde{L}_h)v_h||_h \leq ||q_h - \tilde{q}_h||_h ||v_h||_h, \quad v_h \in E_h, \ h \in \Lambda.$$

Because $q \leq 0$ and $||q_h - r_h q||_h \to 0 \ (h \in \Lambda)$, $||q_h - \tilde{q}_h||_h \to 0 \ (h \in \Lambda)$ thereby showing the desired inequality with β' in place of β. (Indeed, the convergence of $||q_h - \tilde{q}_h||_h$ to zero can be shown by a contradiction argument which uses the fact that $q_h - \tilde{q}_h \geq 0$.)

(ii) We claim now that the other assertions in the theorem follow from (14). First, (14) shows the inverse stability of the mappings A_h, $h \in \Lambda$ (cf. 6.(12)), from which follows, in particular, the injectivity of each A_h, $h \in \Lambda_1$. Since solving the finite-difference equation (2) is tantamount to solving $A_h u_h = w_h$ - and the latter system has as many equations as unknowns - we can easily see that $A_h u_h = w_h$ is uniquely solvable for every $w_h \in F_h$, $h \in \Lambda_1$. We can deduce the consistency of A, A_h, $h \in \Lambda$, from (5a) and the existence of discrete approximations (since $a_h \to a$, $b_h \to b$ $(h \in \Lambda)$). The discrete uniform convergence of the solutions u_h to u (in the sense of (3)) then follows from the discrete convergence of the problem data w_h to w (cf. Theorem 6.17 in the linear case). Further, the injectivity of A follows from Lemma 6.19, and its surjectivity is a consequence of well-known results about boundary-value problems. Substituting the error $u_h - r_h u$ into the inverse stability inequality (14) yields the error estimate (13). □

The first part of the right-hand side of (13) can be estimated in a meaningful way by

$$||f_h - L_h r_h u||_h \leq ||f_h - r_h f||_h + ||\tau_h^{(0)}(u)||_h, \quad h \in \Lambda_1,$$

where $\tau_h^{(0)}(u) = L_h r_h u - r_h f$ is the truncation error associated with the difference equation (see 1.(9a)). In the special case that $a_h = a$, $b_h = b$, $p_h = p|I_h'$, $q_h = q|I_h'$, $f_h = f|I_h'$, we have seen in Section 1.2 that

$$||\tau_h^{(0)}(u)||_h = ||L_h r_h u - f_h||_h = O(h^2) \quad (h \in \Lambda)$$

whenever $u \in C^4[a,b]$. If, moreover, $|\gamma_{0,h} - \gamma_0| = O(h^2)$, $|\gamma_{1,h} - \gamma_1| = O(h^2)$, then $||u_h - r_h u||_h$ converges to zero as $O(h^2)$, too.

At this particular point we should mention the so-called *upwind-differencing*. Here, instead of the central difference quotient of first order, the forward or backward ones are used depending on the sign of

$p_h(x)$. More precisely, $p(x)u'(x)$ is approximated by

$$p_h(x)D_h^+u_h(x), \text{ if } p_h(x) \geq 0; \text{ and by } p_h(x)D_h^-u_h(x), \text{ if } p_h(x) < 0.$$

The *upwind-scheme* for approximating the Dirichlet boundary-value problem (1) is thus given by

$$D_h^2u_h(x) + p_h^+(x)D_h^+u_h(x) + p_h^-(x)D_h^-u_h(x) + q_h(x)u_h(x) = f_h(x), \quad x \in I_h',$$
$$u_h(a_h) = \gamma_{0,h}, \quad u_h(b_h) = \gamma_{1,h},$$

where the grid functions p_h^+, p_h^- are respectively defined by

$$p_h^+(x) \equiv \max(0, p_h(x)), \quad p_h^-(x) \equiv \min(0, p_h(x)), \quad x \in I_h'.$$

If we write the associated finite-difference operator in the form (6), the coefficients of the upwind-scheme are given by

$$a_{-1,h}(x) = h^{-2}(1 - hp_h^-(x)), \quad a_{1,h}(x) = h^{-2}(1 + hp_h^+(x)),$$
$$a_{0,h}(x) = -h^{-2}(2 - h^2q_h(x) + h|p_h(x)|), \quad x \in I_h'.$$

The advantage of this scheme is immediate, in that it is of positive type under the same assumptions as in (11a) but without any restriction on the mesh width. This is particularly important for problems where $|p_h(x)|$ exhibits large values. For such problems, the previous scheme (10) may produce useless results for mesh sizes violating (11b) whereas the upwind-scheme can be efficiently applied for any h (cf. Bohl (1981), VI.).

The disadvantage of the upwind-scheme becomes apparent if the truncation errors are analyzed. In the special case $a_h = a$, $b_h = b$, $p_h = p|I_h'$, $q_h = q|I_h'$, we see that the truncation errors in general do not behave better than

$$||\tau_h^{(0)}(u)||_h = O(h) \quad (h \in \Lambda),$$

and thus, for $\gamma_{0,h} = \gamma_0$, $\gamma_{1,h} = \gamma_1$, the order of convergence is also $O(h)$.

Using the results from Chapter 6, we are able to extend our investigations to *nonlinear problems*. We again treat only Dirichlet boundary-value problems which we state as follows: Given a continuous function $f\colon [a,b] \times \mathbb{R}^2 \to \mathbb{R}$ and $\gamma_0, \gamma_1 \in \mathbb{R}$, we seek a real-valued solution $u \in C^2[a,b]$ of the boundary-value problem

$$(Tu)(x) \equiv u''(x) + f(x,u(x),u'(x)) = 0, \quad x \in [a,b], \tag{15a}$$

$$u(a) = \gamma_0, \quad u(b) = \gamma_1. \tag{15b}$$

A finite-difference approximation analogous to (2) is to find a solution $u_h \in C(I_h)$ to the equations

$$(T_h u_h)(x) \equiv D_h^2 u_h(x) + f_h(x,u_h(x),D_h u_h(x)) = 0, \quad x \in I_h', \tag{16a}$$

$$u_h(a_h) = \gamma_{0,h}, \quad u_h(b_h) = \gamma_{1,h}, \tag{16b}$$

where, for every $h \in \Lambda$, f_h is a continuous, real-valued function with domain $I_h \times \mathbb{R}^2$, and $\gamma_{0,h}$ and $\gamma_{1,h}$ are real numbers approximating γ_0 and γ_1, respectively.

The associated mappings A: $C^2[a,b] \to C[a,b] \times \mathbb{R}^2$, A_h: $C(I_h) \to C(I_h') \times \mathbb{R}^2$, $h \in \Lambda$, are given by

$$Av \equiv (Tv, v(a), v(b)), \quad A_h v_h \equiv (T_h v_h, v_h(a_h), v_h(b_h)), \quad h \in \Lambda. \tag{17}$$

We regard A as a mapping with domain of definition $D(A) = \{v \in C^2[a,b]: Av = w^0\}$ with $w^0 = (0,\gamma_0,\gamma_1)$ and assume that $D(A)$ is nonempty. In order that discrete approximations exist, we again assume $a_h \to a$, $b_h \to b$ $(h \in \Lambda)$. The consistency of A, A_h, $h \in \Lambda$, at some $v \in D(A)$ with the restrictions $r_h v$, $h \in \Lambda$, as the consistency sequence is tantamount to

$$\max_{x \in I_h'} |\tau_h^{(0)}(v)(x)| \to 0 \quad (h \in \Lambda) \tag{18}$$

where $\tau_h^{(0)}(v)$, $h \in \Lambda$, are the *truncation errors* given by

$$\tau_h^{(0)}(v)(x) \equiv (T_h r_h v - r_h Tv)(x) = D_h^2 r_h v(x) + f_h(x, r_h v(x), D_h r_h v(x)),$$
$$x \in I_h', \quad h \in \Lambda.$$

Here no truncation error associated with the boundary conditions are present, so that $\tau_h^{(0)}(v)$ is precisely what we have earlier termed "the truncation error associated with the difference equation." A simple choice of f_h which will lead to a consistent approximation is

$$f_h(x,y,z) = \hat{f}(x,y,z), \quad (x,y,z) \in I_h' \times \mathbb{R}^2,$$

where $\hat{f}$ is a continuous extension of f to $\mathbb{R}^3$. Such an extension will, of course, only be required in case there are points of I_h' lying outside of $[a,b]$.

We now provide a condition guaranteeing the inverse stability of the difference approximations defined in (16). For brevity, for numbers $\kappa_i > 0$, $i = 0,1,2,\ldots$, we write $\Lambda_i \equiv \Lambda \cap (0,\kappa_i]$ in the following.

Theorem 8.4. For each $h \in \Lambda$, let the function $f_h(\cdot,\cdot,\cdot)$ be differentiable with respect to the second and third arguments in the region

$$U_h \equiv \{(x,y,z) \in I_h' \times \mathbb{R}^2 : |y - r_h v(x)| \leq \rho_0,\ |z - r_h v'(x)| \leq \rho_0\} \tag{19}$$

for some $\rho_0 > 0$ where $v \in C^1[a,b]$. Moreover, suppose that the partial derivatives $\partial f_h/\partial y$, $\partial f_h/\partial z$ are continuous in U_h in their second and third arguments, with

$$\frac{\partial f_h}{\partial y}(x,y,z) \leq 0, \quad \left|\frac{\partial f_h}{\partial z}(x,y,z)\right| \leq c_1, \quad (x,y,z) \in U_h, \quad h \in \Lambda_0, \tag{20}$$

for some $c_1 > 0$ and $\kappa_0 > 0$. Then there are a $\kappa_1 \in (0,\kappa_0]$ and a $\beta > 0$ so that for all $h \in \Lambda_1$ and u_h, v_h in

$$\begin{aligned} X_h \equiv \{\phi_h \in E_h :\ & |\phi_h(x) - r_h v(x)| \leq \rho_0, \\ & |D_h\phi_h(x) - r_h v'(x)| \leq \rho_0,\ x \in I_h'\}, \end{aligned} \tag{21}$$

the following inequality holds

$$\begin{aligned} \beta \max_{x \in I_h} |(u_h - v_h)(x)| \leq\ & |(u_h - v_h)(a_h)| + |(u_h - v_h)(b_h)| \\ & + \max_{x \in I_h'} |(T_h u_h - T_h v_h)(x)|. \end{aligned} \tag{22}$$

Proof: From the Mean Value Theorem, we get for $u_h, v_h \in X_h$, $h \in \Lambda$, that

$$(T_h u_h - T_h v_h)(x) = D_h^2 z_h + \frac{\partial \tilde{f}_h}{\partial z} D_h z_h(x) + \frac{\partial \tilde{f}_h}{\partial y} z_h(x), \quad x \in I_h',$$

where $z_h \equiv u_h - v_h$, and the tilde denotes evaluation of the appropriate derivative at a certain intermediate value of the respective argument. The right-hand side can be expressed as $\tilde{L}_h(u_h,v_h) z_h$ with $\tilde{L}_h(u_h,v_h)$ a linear finite-difference operator which is of positive type by assumption (20) for all $u_h, v_h \in X_h$ and sufficiently small $h \in \Lambda$ (cf. (7), (11)). Hence, Theorem 8.1 implies the monotonicity property

$$(\tilde{L}_h(u_h,v_h)\phi_h)(x) \geq 0,\ \ x \in I_h',\ \ \phi_h(a_h) \leq 0,\ \ \phi_h(b_h) \leq 0 \Rightarrow \phi_h(x) \leq 0, \quad x \in I_h,$$

for every $u_h, v_h \in X_h$, $\phi_h \in E_h$, $h \in \Lambda_1$ $(\subset \Lambda_0)$. As in the proof of Theorem

8.2, we can thereby deduce the existence of a number $\beta > 0$ and a $\kappa_2 \in (0,\kappa_1]$ such that

$$\beta||\phi_h||_h \leq |\phi_h(a_h)| + |\phi_h(b_h)| + ||L_h(u_h,v_h)\phi_h||_h,$$
$$u_h,v_h \in X_h, \quad \phi_h \in E_h, \quad h \in \Lambda_2.$$

The asserted inequality (22) follows with κ_1 replaced by κ_2 and with u_h-v_h substituted for ϕ_h. □

It is remarkable that the proof of Theorem 8.4 directly uses - through the Mean Value Theorem - a monotonicity property of the associated nonlinear mappings themselves. The inverse stability is *not* proved by exploiting a positivity property of Fréchet-derivatives in conjunction with results from Section 6.2. We use this approach here, since the equi-(Fréchet-) differentiability of the nonlinear mappings is not present for the underlying maximum norms. This will be made more explicit at the end of this section by an example whose finite-difference approximations are shown to be inversely stable by use of the results of Section 6.2 in contrast to the arguments above.

We can now show the following convergence theorem. Let us recall that $D(A)$ is supposed to be nonempty which means that the existence of a solution u of (15) is assumed.

Theorem 8.5. With a solution u of (15), suppose the consistency condition (18) and the assumptions of Theorem 8.4 hold (for $v = u$). Further, suppose there is a solution $u_h \in X_h$ of the finite-difference approximation (16) for every $h \in \Lambda$, and that $\gamma_{0,h} \to \gamma_0$, $\gamma_{1,h} \to \gamma_1$ $(h \in \Lambda)$. Then, for almost all $h \in \Lambda$, u_h is the unique solution of (16) and

$$\max_{x\in I_h} |(u_h - r_h u)(x)| \to 0 \quad (h \in \Lambda)$$

with the following error estimates

$$\beta \max_{x\in I_h} |(u_h-r_h u)(x)| \leq |\gamma_{0,h}-\gamma_0| + |\gamma_{1,h}-\gamma_1|$$
$$+ \max_{x\in I_h'} |\tau_h^{(0)}(u)(x)|, \quad h \in \Lambda_1.$$

Proof: The boundary-value problems (15) and (16) define mappings A and $A_h\colon X_h \to C(I_h') \times \mathbb{R}^2$, where A should be viewed as a mapping of $D(A) = \{v \in C^2[a,b]\colon Av = w^0\} \subset E$ onto $R(A) = \{w^0\} \subset F$. By assumption, A, A_h,

$h \in \Lambda$, is consistent at u. Theorem 8.4 shows that (A_h) is inversely stable at every sequence $v_h \in X_h$, $h \in \Lambda_1$. Because $\gamma_{0,h} \to \gamma_0$, $\gamma_{1,h} \to \gamma_1$ $(h \in \Lambda)$, $||u_h - r_h u||_h \to 0$ $(h \in \Lambda)$ by Theorem 6.17. Finally, the inverse stability inequality (22) shown in Theorem 8.4 yields the injectivity of A_h for almost all $h \in \Lambda$ along with the asserted error estimate. □

Let us remark that the uniqueness of the solution of (15) can be ensured by Lemma 6.19 if the assumptions of Theorem 8.4 hold globally and the consistency condition (18) is satisfied at every $v \in D(A)$.

To conclude this section, we consider the nonlinear example from Section 1.4 (cf. 1.(13)),

$$-(pu')'(x) + q(x)u(x) = f(x,u(x)), \quad x \in [0,1], \quad u(0) = u(1) = 0.$$

We can view this example in the setting of the above considered class of nonlinear boundary-value problems if we assume that p is continuously differentiable. The boundary-value problem can then be formulated in the manner described at the end of Section 1.2,

$$pu'' + p'u' - qu + f(x,u) = 0 \quad \text{in} \quad [0,1], \quad u(0) = u(1) = 0.$$

For the simplest case where p, p', q and f are merely restricted to the mesh points, the finite-difference approximations will have the form

$$(T_h u_h)(x) = 0, \quad x \in I_h', \quad u_h(0) = u_h(1) = 0, \tag{23}$$

where

$$T_h u_h \equiv L_h u_h + f(\cdot, u_h) \quad \text{and}$$
$$(L_h u_h)(x_j) \equiv p(x_j) D_h^2 u_h(x_j) + p'(x_j) D_h v_h(x_j) - q(x_j) u_h(x_j),$$
$$1 \le j \le J_h - 1.$$

If $p \not\equiv 0$, then the finite-difference approximation can also be expressed in the form (16) with

$$f_h(x,y,z) = \frac{1}{p(x)} (p'(x)z - q(x)y + f(x,y)), \quad x \in I_h', \quad y,z \in \mathbb{R}.$$

The hypotheses of Theorem 8.4 are satisfied in case f is continuously differentiable with respect to the second argument, $q \in C[0,1]$, $p \in C^1[0,1]$, and

$$p(x) \ge c_0 > 0, \quad \frac{\partial f}{\partial y}(x,y) - q(x) \le 0, \quad x \in [0,1], \quad y \in \mathbb{R}. \tag{24}$$

Indeed, we note that

$$\frac{\partial f_h}{\partial y} = \frac{1}{p}\left(\frac{\partial f}{\partial y} - q\right), \quad \frac{\partial f_h}{\partial z} = \frac{p'}{p}.$$

Thus we have an inverse stability inequality of the form (22); and, since the truncation errors are $O(h^2)$ for sufficiently smooth solutions, the results of Theorem 8.5 hold with the error estimate $||u_h - r_h u||_h = O(h^2)$. We note that the hypotheses in (24) follow from those of Theorem 1.1 (cf. 1.(16) and 1.(17)). As in Theorem 1.1, the conditions $\frac{\partial f}{\partial y} \leq 0$, $q \geq 0$ ensure solvability of the finite-difference equations, and the solutions can, for example, be computed by Newton's method.

In 1.(15), we gave another finite-difference approximation with the same order of accuracy for the truncation errors. This scheme can be expressed as in (23), where now

$$(L_h u_h)(x_j) \equiv \frac{1}{h^2}\{p_{j+1/2}u_h(x_{j+1}) - (p_{j+1/2} + p_{j-1/2} + q_j h^2)u_h(x_j)$$
$$+ p_{j-1/2}u_h(x_{j-1})\}, \quad 1 \leq j \leq J_h-1, \quad u_h \in C(I_h),$$

and $T_h u_h \equiv L_h u_h + f(\cdot,u_h)$. (We note that the finite-difference equations presented here differ from those of Section 1.4 by a multiplicative factor of (-1).) In a corresponding manner, we define

$$Tu \equiv Lu + f(\cdot,u), \quad (Lu)(x) \equiv (pu')' - qu, \quad u \in C^2[0,1].$$

If f is differentiable with respect to the second argument with $\partial f/\partial y$ continuous, then the associated mappings $Au = (Tu,u(0),u(1))$, $u \in C^2[0,1]$, $A_h u_h = (T_h u_h, u_h(0), u_h(1))$, $u_h \in C(I_h)$, are differentiable, with derivatives given by

$$A'(u)v = (Lv + \frac{\partial f}{\partial y}(\cdot,u)v, v(0), v(1)), \quad u,v \in C^2[0,1],$$

$$A_h'(u_h)v_h = (L_h v_h + \frac{\partial f}{\partial y}(\cdot,u_h)v_h, v_h(0), v_h(1)), \quad u_h, v_h \in C(I_h).$$

It is clear that A_h, $h \in \Lambda$, is equicontinuously equidifferentiable at $u_h = u|I_h$, $h \in \Lambda$, for every $u \in C^1[0,1]$. If we express the derivative as $A_h'(u_h)v_h = (T_h'(u_h)v_h, v_h(0), v_h(1))$, with

$$(T_h'(u_h)v_h)(x) = a_{-1,h}(x)v_h(x-h) + a_{0,h}(x)v_h(x) + a_{1,h}(x)v_h(x+h),$$
$$x \in I_h',$$

and

$$a_{\pm 1,h}(x_j) = \frac{1}{h^2}\, p_{j\pm 1/2},$$

$$a_{0,h}(x_j) = -\frac{1}{h^2}(p_{j+1/2} + p_{j-1/2}) - q_j + \frac{\partial f}{\partial y}(x_j, u_h(x_j)), \quad 1 \leq j \leq J_h - 1,$$

then we can easily see that the finite-difference operator $T_h'(u_h)$ is of positive type for every $u_h \in C(I_h)$ in case $p,q \in C[0,1]$, and (24) holds. If, in addition, $p \in C^1[0,1]$, then we can show as in Theorem 8.2 that

$$\beta \max_{x \in I_h} |v_h(x)| \leq |v_h(0)| + |v_h(1)|$$
$$+ \max_{x \in I_h'} |(T_h'(u_h)v_h)(x)|, \quad u_h, v_h \in C(I_h), \quad h \in \Lambda_1.$$

This inequality is tantamount to the uniform boundedness of $A_h'(u_h)^{-1}$, $h \in \Lambda_1$; bijectivity of the derivatives $A_h'(u_h)$ follows from their injectivity as mappings of the spaces E_h to F_h which are of finite and equal dimension. If, in addition, we take the consistency 1.(21) into account, then, uniqueness of the solution u of the given boundary value problem follows from Lemma 6.19. Application of Theorem 6.23 with $u_h = u|I_h$, $w = w_h^1 = 0$, yields the convergence of the approximate solutions and the error estimate

$$\max_{x \in I_h} |(u_h - u)(x)| \leq \frac{1}{\beta} \max_{x \in I_h'} |\tau_h^{(0)}(u)(x)| = O(h^2) \quad (h \to 0).$$

8.2 CONVERGENCE OF DIFFERENCE METHODS FOR ORDINARY DIFFERENTIAL EQUATION VIA COMPACTNESS ARGUMENTS

In this section, we continue investigating the same classes of boundary-value problems and finite-difference approximations as in the previous section. The convergence - or more precisely the biconvergence - will now be essentially assured by showing the regular convergence and by applying the results from Chapter 7. For our analysis, it is crucial that the norms for the solution spaces be chosen differently from those in the preceding section.

We begin this discussion by again considering the linear differential equation (1), and we restrict our analysis to the case of homogeneous Dirichlet boundary conditions, i.e.,

$$Lu = f \quad \text{in} \quad [a,b], \quad u(a) = u(b) = 0. \tag{25}$$

The finite-difference approximation is expressed by

$$L_h u_h = f_h \quad \text{in} \quad I_h', \quad u_h(a) = u_h(b) = 0, \tag{26}$$

with L_h defined in (2a). The meshes are again given by

$$I_h = \{x \in [a,b]: x = x_j \equiv a+jh, \quad j = 0,\dots,J_h\}, \quad I_h' = I_h - (\{a\}\cup\{b\}).$$

We select the following spaces for our analysis

$$E = \{u \in C^2[a,b]: u(a) = u(b) = 0\}, \quad F = C[a,b],$$

$$E_h = \{u_h \in C(I_h): u_h(a) = u_h(b) = 0\}, \quad F_h = C(I_h').$$

We thus have incorporated the (homogeneous) boundary conditions in the spaces and shall not need the additional mappings A, A_h. Unlike the previous section, but similar to 1.(8), we include respectively maxima of the first and second derivatives and of first- and second-order difference quotients in the norms of E and E_h, i.e.,

$$||u|| = \max_{0\le i\le 2} \max_{a\le x\le b} |u^{(i)}(x)|, \quad u \in E,$$

$$||u_h||_h \equiv \max(\max_{x\in I_h} |u_h(x)|, \max_{x\in I_h^0} |D_h^+ u_h(x)|, \max_{x\in I_h'} |D_h^2 u_h(x)|), \quad u_h \in E_h,$$

where $I_h^0 \equiv I_h - \{b\}$. As norms in F and F_h we set

$$||f|| = \max_{a\le x\le b} |f(x)|, \quad f \in F, \text{ and } \quad ||f_h||_h = \max_{x\in I_h'} |f_h(x)|, \quad f_h \in F_h,$$

respectively.

We should remark at this point that the finite-difference operator L_h is Fredholm with index zero because of the equal dimensions of E_h and F_h.

Let Λ again be a null sequence of positive mesh widths. Then $A(F, \pi_h F_h, \lim^F)$ is a discrete approximation with the discrete convergence

$$\lim{}^F f_h = f \iff \max_{x\in I_h'} |(f_h - f)(x)| \to 0 \quad (h \in \Lambda) \tag{27}$$

and the norms satisfy Property 5.3 (cf. Lemma 5.7 and Theorem 5.11). The pointwise restrictions to the meshes I_h, I_h^0, I_h', $h \in \Lambda$, are equally denoted by $r_h u$, $h \in \Lambda$. For elements $u \in E$, $u_h \in E_h$, $h \in \Lambda$, we define a discrete convergence by

$$\lim{}^E u_h = u \iff ||u_h - r_h u||_h \to 0 \quad (h \in \Lambda) \tag{28}$$

$$\iff \max_{x\in I_h} |(u_h - u)(x)| \to 0, \quad \max_{x\in I_h^0} |D_h^+(u_h - r_h u)(x)| \to 0,$$

$$\max_{x\in I_h'} |D_h^2(u_h - r_h u)(x)| \to 0 \quad (h \in \Lambda).$$

Indeed, it is easy to see that by (28) a discrete convergence is defined in the sense of Section 5.1. It is not difficult to see that the convergence of the norms of discretely convergent sequences is necessary and sufficient for the convergences

$$\max_{x\in I_h} |u(x)| \to \max_{a\le x\le b} |u(x)|, \quad \max_{x\in I_h^0} |D_h^+ u(x)| \to \max_{a\le x\le b} |u'(x)|,$$

$$\max_{x\in I_h'} |D_h^2 u(x)| \to \max_{a\le x\le b} |u''(x)| \quad (h \in \Lambda), \quad u \in C^2[a,b].$$

We have the same results on the consistency of the operators L, L_h, $h \in \Lambda$, as in the preceding section. Indeed, if we have (5a) - i.e., $||p_h - p||_h \to 0$, $||q_h - r_h q||_h \to 0$ $(h \in \Lambda)$ with respect to the norms in F_h - then the sequence of restrictions $(r_h u)$ is also a consistency sequence. In the special case that $p_h = r_h p$, $q_h = r_h q$, the truncation errors satisfy

$$\max_{x\in I_h'} |(L_h r_h \phi - L\phi)(x)| = O(h^2) \quad (h \in \Lambda), \tag{29}$$

for every function $\phi \in C^4[a,b]$.

With the above choice of norms, the sequence of finite-difference operators L_h, $h \in \Lambda$, is easily seen to be stable for uniformly bounded mesh functions $p_h, q_h, h \in \Lambda$. If we again assume (5a), then we obtain the estimate

$$||L_h|| \le 2(1 + \max_{a\le x\le b} |p(x)| + \max_{a\le x\le b} |q(x)|), \quad h \in \Lambda_0. \tag{30}$$

Here, and in several places in the following text, we use that

$$\max_{x\in I_h'} |D_h v_h(x)| \le \max_{x\in I_h^0} |D_h^+ v_h(x)|, \quad v_h \in E_h.$$

We now wish to apply the results of Section 7.2 - in particular, Theorem 7.7 - in order to show the convergence of solutions. Toward this end, we begin by demonstrating the a-regularity of (L_n), which, together with the consistency (29) and the stability (30), will ensure the regular convergence (cf. Theorem 7.9).

Lemma 8.6. Let $p,q \in C[a,b]$ and suppose that (5a) holds. Then (L_h) is a-regular and is regularly convergent to L.

Proof: Since the consistency and stability are already guaranteed, we need only to check Property 7.8 for regular convergence; then, according to Theorem 7.9, (L_h) is a-regular. Let $u_h \in E_h$, $h \in \Lambda$, be a sequence for which $||u_h||_h \leq 1$, $h \in \Lambda$, and for which $(L_h u_h)$ is discretely compact. If we define $u^h \in C[a,b]$, $h \in \Lambda$, as continuous, piecewise linear interpolants of u_h, i.e., $u^h(x) = u_h(x)$, $x \in I_h$, then $||u^h||_\infty \leq 1$, $h \in \Lambda$, where $||\cdot||_\infty$ is an abbreviation for the maximum norm in $C[a,b]$. We know that

$$|u^h(x)-u^h(x')| \leq |x-x'| \max_{z\in I_h^0} |D_h^+ u_h(z)|, \quad x,x' \in [a,b], \quad h \in \Lambda,$$

since u^h is piecewise continuously differentiable and

$$|(u^h)'(x)| \leq \max_{x\in I_h^0} |D_h^+ u_h(z)|, \quad x \in [x_{i-1},x_i], \quad i = 1,\ldots,J_h.$$

Since we have assumed $||u_h||_h \leq 1$, $h \in \Lambda$, (u^h) is equicontinuous and uniformly bounded. Thus the sequence (u^h) is compact in $C[0,1]$ according to the Arzela-Ascoli Theorem. With respect to the maximum norm on $C(I_h)$, the sequence of restrictions $(r_h u^h) = (u_h)$ is therefore discretely compact (see the example in the discussion following Lemma 7.2). Now let $v^h \in C[0,1]$, $h \in \Lambda$, be the continuous piecewise linear functions defined by $v^h(x) = D_h^+ u_h(x)$, $x \in I_h^0$, $v^h(b) = D_h^+ u_h(b-h)$, $h \in \Lambda$. Then $||v^h||_\infty \leq 1$, $h \in \Lambda$, and, as above, we know that

$$|v^h(x) - v^h(x')| \leq |x-x'| \max_{z\in I_h^0} |D_h^+ v^h(z)|, \quad x,x' \in [a,b], \quad h \in \Lambda.$$

Here,

$$\begin{aligned} D_h^+ v^h(x_j) &= \frac{1}{h}(v^h(x_{j+1}) - v^h(x_j)) \\ &= \frac{1}{h^2}(u_h(x_{j+2}) - 2u_h(x_{j+1}) + u_h(x_j)) = D_h^2 u_h(x_{j+1}), \\ &\qquad\qquad j = 0,\ldots,J_h-2, \end{aligned}$$

so that

$$|v^h(x)-v^h(x')| \leq |x-x'| \max_{z\in I_h'} |D_h^2 u_h(z)| \leq |x-x'|, \quad x,x' \in [a,b], \quad h \in \Lambda,$$

since $||u_h||_h \leq 1$, $h \in \Lambda$. According to the Arzela-Ascoli Theorem, this sequence (v^h) is also compact in $C[0,1]$ and hence $(D_h^+ u_h) = (r_h v^h)$ is discretely compact (with respect to the maximum norms in $C(I_h^0)$). Consequently, also $(D_h^- u_h)$ is discretely compact (with respect to the maximum norm in $C(I_h-\{a\})$). It is easy to see that $D_h u_h = \frac{1}{2}(D_h^+ + D_h^-)u_h$, $h \in \Lambda$, is also directly compact (with respect to the maximum norm in

$C(I_h'))$. Since we have assumed that $(L_h u_h)$ is discretely compact - and since

$$D_h^2 u_h = L_h u_h - p_h(D_h u_h) - q_h u_h, \quad h \in \Lambda,$$

with (p_h), (q_h) uniformly bounded - $(D_h^2 u_h)$ is discretely compact. Altogether, (u_h), $(D_h u_h)$, and $(D_h^2 u_h)$ are discretely compact in their respective discrete maximum norms. From this fact, we can claim that (u_h) is discretely compact with respect to $||\cdot||_h$. In order to see this, we must show that $v = u'$, $w = v'$ whenever, for an arbitrary subsequence $\Lambda' \subset \Lambda$,

$$\max_{x\in I_h} |(u_h-u)(x)| \to 0, \quad \max_{x\in I_h^0} |(D_h^+ u_h - v)(x)| \to 0,$$

$$\max_{x\in I_h'} |(D_h^2 u_h - w)(x)| \to 0 \quad (h \in \Lambda').$$

We show only that $u' = v$. Let $x > a$ and $x_h \to x$ $(h \in \Lambda)$. Then

$$|u_h(x_h) - u(x)| \leq \max_{z\in I_h} |(u_h-u)(z)| + |u(x_h)-u(x)| \to 0 \quad (h \in \Lambda').$$

Using the relations

$$u_h(x_h) = u^h(x_h) = u^h(a) + \int_a^{x_h} (u^h)'(t)dt = u_h(a) + h \sum_{\substack{a\leq y<x_h \\ x\in I_h^0}} D_h^+ u_h(y)$$

and passing to the limit $h \to 0$, $h \in \Lambda'$, we obtain

$$u(x) = u(a) + \int_a^x v(t)dt.$$

This proves $v = u'$. A similar analysis shows $v' = w$, if we note that $D_h^2 u_h(x) = D_h^+ v_h(x-h)$, $x \in I_h'$, where $v_h \equiv D_h^+ u_h$. □

If, moreover, we stipulate that $L: E \to F$ is injective - i.e. that the associated homogeneous boundary-value problem

$$u'' + pu' + qu = 0 \quad \text{in} \quad [a,b], \quad u(a) = u(b) = 0,$$

has only the trivial solutions - then inverse stability of (L_h) follows from Theorem 7.5. Also, Theorem 7.7 yields the unique solvability of the given boundary-value problem and of the finite-difference equations for h small enough along with the biconvergence of their solutions. We summarize the preceding remarks in the following theorem.

Theorem 8.7. Let the assumptions of Lemma 8.6 hold, and let the associated homogeneous boundary-value problem possess only the trivial solution. Then, the sequence (L_h) is inversely stable, i.e., there are $\kappa_0 > 0$ and $\beta > 0$ such that

$$\beta||u_h||_h \leq ||L_h u_h||_h, \quad u_h \in E_h, \quad h \in \Lambda_0.$$

Moreover, the exact boundary-value problem (25), and the finite-difference approximations (26) for all $h \in \Lambda_0$ are uniquely solvable for arbitrary functions f and f_h, respectively. Also,

$$\max_{x \in I_h'} |(f-f_h)(x)| \to 0 \iff ||u_h - r_h u||_h \to 0 \quad (h \in \Lambda)$$

with the two-sided error estimate

$$\frac{1}{\alpha}||\tau_h^{(0)}(u) - (f_h - r_h f)||_h \leq ||u_h - r_h u||_h$$
$$\leq \frac{1}{\beta}||\tau_h^{(0)}(u) - (f_h - r_h f)||_h, \quad h \in \Lambda_0,$$

where α is the constant on the right-hand side of (30) and $\tau_h^{(0)}(u) = L_h r_h u - r_h f$, $h \in \Lambda$, denote the truncation errors. □

If we compare the results of the previous theorem with those of Theorem 8.3 from the preceding section, then we observe the following. The results in Theorem 8.7 essentially follow from the injectivity of L; in Theorem 8.3, the condition that $q \leq 0$ is essential in obtaining the injectivity and the other results. It is also remarkable that the compactness methods of this section can be utilized to obtain an error estimate in a stronger norm which additionally guarantees the convergence of the difference quotients of first and second order.

We now proceed to extend the above analysis to nonlinear boundary-value problems of the form of (15) which are approximated by finite-difference approximations of the form (16). Without loss of generality, we consider homogeneous Dirichlet boundary conditions with the result that the problem

$$Tu = 0 \quad \text{in} \quad [a,b], \quad u(a) = u(b) = 0, \tag{31}$$

is approximated by

$$T_h u_h = 0 \quad \text{in} \quad I_h', \quad u_h(a) = u_h(b) = 0, \tag{32}$$

where $T\colon E \to F$ and $T_h\colon E_h \to F_h$ are defined in (15a) and (16a), respectively. Again, let $\Lambda = (h_1, h_2, \ldots)$ be a denumerable null sequence of positive mesh widths. The statements on consistency in the preceding section apply. Since the underlying norms in E now include derivatives, we have the continuous differentiability of T at $u \in E$, if the following condition is satisfied for some $\rho_0 > 0$:

(33) The function $f(.,.,.)$ possesses continuous, partial derivatives $\partial f/\partial y$, $\partial f/\partial z$ in the region

$$U = \{(x,y,z) \in [a,b] \times \mathbb{R}^2 \colon |y-u(x)| \le \rho_0, \ |z-u'(x)| \le \rho_0\}.$$

The derivative of T is then given by

$$(T'(v)\phi)(x) = \phi''(x) + \frac{\partial f}{\partial z}(x,v(x),v'(x))\phi'(x) + \frac{\partial f}{\partial y}(x,v(x),v'(x))\phi(x),$$

$$x \in [a,b], \quad v \in B^0_{\rho_0}(u), \quad \phi \in C^2[a,b].$$

The associated *linearized problem* to (31) consists of finding a solution v in E to $T'(u^0)v = w$, where w is an arbitrary function in F and u^0 is a solution of (31) which again is assumed to exist.

The equicontinuous equidifferentiability of T_h, $h \in \Lambda$, at $r_h u$, $h \in \Lambda$, will be present whenever the following condition is met:

(34) For some $\rho_1 > 0$, the functions $f_h(.,.,.)$ are equidifferentiable with respect to the second and third arguments uniformly in

$$U_h = \{(x,y,z) \in I_h' \times \mathbb{R}^2 \colon |y-u(x)| \le \rho_1,$$

$$|z-D_h^+ r_h u(x)| \le \rho_1\}, \quad h \in \Lambda,$$

and the partial derivatives $\partial f_h/\partial y$, $\partial f_h/\partial z$ are uniformly equicontinuous in U_h, $h \in \Lambda$.

The derivatives then have the form

$$(T_h'(v_h)\phi_h)(x) = D_h^2\phi_h(x) + \frac{\partial f_h}{\partial z}(x,v_h(x),D_h v_h(x))D_h\phi_h(x)$$

$$+ \frac{\partial f_h}{\partial y}(x,v_h(x),D_h v_h(x))\phi_h(x), \quad x \in I_h', \quad \phi_h \in C(I_h), \quad h \in \Lambda,$$

where $v_h \in B^0_{\rho_1}(r_h u)$. If $f_h(x,y,z) = f(x,y,z)$, $x \in I_h'$, $y,z \in \mathbb{R}$, then (33) implies (34), where here Λ is possibly replaced by a subset $\Lambda_1 = \{h \in \Lambda, h \le \kappa_1\}$ and $\rho_1 = \rho_0/2$. The proof of these assertions is left to the reader. Further, the consistency of the derivatives $T'(u)$,

$T_h'(r_hu)$, $h \in \Lambda$, at all $v \in E$ is guaranteed by the following conditions (cf. (5a)),

$$\max_{x\in I_h'} \left|\frac{\partial f_h}{\partial y}(x,u(x),D_hr_hu(x)) - \frac{\partial f}{\partial y}(x,u(x),u'(x))\right| \to 0,$$
$$\max_{x\in I_h'} \left|\frac{\partial f_h}{\partial z}(x,u(x),D_hr_hu(x)) - \frac{\partial f}{\partial z}(x,u(x),u'(x))\right| \to 0 \quad (h \in \Lambda). \tag{35}$$

Using the Convergence Theorem 7.10, we can now prove the following result.

Theorem 8.8. Suppose (31) has a solution $u^0 \in C^2[a,b]$ and assumptions (33) and (34) are met along with consistency conditions (18) and (35) for $u = u^0$. Further, suppose the homogeneous linearized problem $T'(u^0)v = 0$ has only the trivial solution. Then there are positive numbers κ_0, ρ, α, β such that each of the finite-difference equations (32) possesses a unique solution u_h^0 in $||u_h^0 - r_hu^0||_h \le \rho$, $h \in \Lambda_0$, such that the convergences

$$e_h^{(0)} \equiv \max_{x\in I_h} |(u_h^0-u^0)(x)| \to 0, \quad e_h^{(1)} \equiv \max_{x\in I_h^0} |(D_h^+u_h^0-(u^0)')(x)| \to 0,$$
$$e^{(2)} \equiv \max_{x\in I_h'} |(D_h^2u_h^0-(u^0)'')(x)| \to 0 \quad (h \in \Lambda),$$

hold along with the two-sided error estimates

$$\frac{1}{\alpha}||T_hr_hu^0||_h \le ||u_h^0 - r_hu^0||_h \le \frac{1}{\beta}||T_hr_hu^0||_h, \quad h \in \Lambda_0.$$

If, moreover, $f_h = f$ and $u^0 \in C^4[a,b]$, then $||u_h^0 - r_hu^0||_h = O(h^2)$ $(h \in \Lambda)$.

Proof: We apply Theorem 7.10, whose hypotheses are clearly satisfied under the conditions presented above. To show statement 7.(9a), we initially recognize that $(A'(u^0)=)\ T'(u^0)$ is injective by hypothesis and next observe that each of the mappings $T_h'(r_hu^0)\colon E_h \to F_h$, $h \in \Lambda$, is Fredholm with index zero. Thus Property 7.6 is satisfied. From (35), we know that the sequence of derivatives $T'(u^0)$, $T_h'(r_hu^0)$, $h \in \Lambda$, is consistent at all $v \in E$, and $(T_h'(r_hu^0))$ is stable by virtue of the uniform boundedness of $\partial f_h/\partial y$, $\partial f_h/\partial z$ (with argument $(x,u^0(x), D_hr_hu^0(x))$). Application of Lemma 8.6 to $L = T'(u^0)$, $L_h = T_h'(r_hu^0)$, $h \in \Lambda$, shows that $T_h'(r_hu^0) \overset{r}{\to} T'(u^0)$ $(h \in \Lambda)$. Hence, all the conditions of 7.(9a) are met and an application of Theorem 7.10 yields 7.(9c). Combined with 6.(13d),

7.(9c) (for $w_h^0 = 0$) then implies all the assertions for unique solvability of (32) as well as the two-sided error estimate (since $\tau_h^{(0)}(u^0) = T_h r_h u^0$). Convergence of the $e_h^{(i)}$, $i = 0,1,2$, follows from the convergence of $||u_h^0 - r_h u^0||_h$, $h \in \Lambda$, to zero, since

$$\max_{x\in I_h^0} |(D_h^+ r_h u^0 - (u^0)')(x)| \to 0, \quad \max_{x\in I_h'} |(D_h^2 r_h u^0 - (u^0)'')(x)| \to 0 \quad (h \in \Lambda).$$

For the special case that $f_h = f$, $u^0 \in C^4[a,b]$, $||\tau_h^{(0)}(u^0)||_h = O(h^2)$ $(h \in \Lambda)$, thereby showing the last assertion of the theorem. □

8.3. CONVERGENCE OF THE FIVE-POINT DIFFERENCE APPROXIMATION FOR POISSON'S EQUATION

In this section, we restrict our analysis of finite-difference approximations of elliptic boundary-value problems to the 5-point finite difference approximation of Poisson's equation given in 1.(25). We provide a maximum principle from which inverse stability follows as in Section 8.1. The convergence of the approximate solutions with error estimates will then be a direct result of the behavior of the truncation errors already examined in Section 1.5. The purpose of the analysis in this section is nothing else than to show how well-known results on the particular problem we consider can be viewed from the perspective of the convergence theory in Chapter 6. We shall make crucial use of the well-known maximum principle for the discrete Laplace operator and of its consequences discussed at length in Isaacson & Keller (1966).

We consider Poisson's equation,

$$-\Delta u = f \text{ in } G, \quad u|\partial G = g$$

on a rectangle $G = \{(x_1,x_2): 0 < x_1 < a,\ 0 < x_2 < b\}$ and its associated 5-point finite-difference approximation Δ_h defined in 1.(23),

$$-\Delta_h u_h = f_h \text{ in } G_h, \quad u_h|\partial G_h = g_h,$$

where $h_1 = a/N_h$, $h_2 = b/M_h$, $h \equiv \max(h_1,h_2)$, $N_h, M_h \in \mathbb{N}$, and

$$\overline{G}_h = \{(x_1,x_2): x_1 = nh_1,\ 0 \le n \le N_h,\ x_2 = mh_2,\ 0 \le m \le M_h\},$$

$$\partial G_h = \partial G \cap \overline{G}_h, \quad G_h = \overline{G}_h - \partial G_h.$$

We let $E \equiv C^2(G) \cap C(\overline{G})$, $F \equiv C(G) \times C(\partial G)$, both supplied with the supremum-norm, and define $A: E \to F$ by

$$Au = (-\Delta u, u|\partial G), \quad u \in E.$$

Further, we define E_h, F_h and $A_h\colon E_h \to F_h$ respectively by $E_h \equiv C(\overline{G}_h)$, $F_h \equiv C(G_h) \times C(\partial G_h)$ and

$$A_h u_h \equiv (-\Delta_h u_h, u_h|\partial G_h), \quad u_h \in E_h,$$

and let $r_h v$ denote any of the following restrictions of a function v defined on either $\overline{G}, G$ or ∂G to either $\overline{G}_h$, G_h or ∂G_h.

Let Λ be a null sequence of maximal mesh widths. Then we see as in Theorem 5.11 that

$$\operatorname{Lim} G_h = \operatorname{Lim} \overline{G}_h = \overline{G}, \quad \operatorname{Lim} \partial G_h = \partial G.$$

From Theorem 5.10, a discrete convergence, along with an associated discrete approximation $A(E, \pi_h E_h, \lim^E)$, is then given by

$$u_h \to u \ (h \in \Lambda) \iff \max_{x \in \overline{G}_h} |(u_h - u)(x)| \to 0 \quad (h \in \Lambda)$$

for $u \in E$, $u_h \in E_h$, $h \in \Lambda$. In a corresponding manner, a discrete convergence and the associated discrete approximation $A(F, \pi_h F_h, \lim^F)$ will be defined by

$$w_h \to w \ (h \in \Lambda) \iff \max_{x \in G_h} |(f_h - f)(x)| \to 0, \quad \max_{x \in \partial G_h} |(g_h - g)(x)| \to 0 \quad (h \in \Lambda)$$

for $w = (f,g) \in F$, $w_h = (f_h, g_h) \in F_h$, $h \in \Lambda$. It is worth noting here that, since $G_h \subset G$, $h \in \Lambda$, we have surjectivity of the discrete convergence, and the closedness of G is not needed (recall that G was assumed open).

The inverse stability of the sequence A_h, $h \in \Lambda$, results from a maximum principle for the discrete Laplace operator which can be stated as follows:

Theorem 8.9. For every mesh function $v_h \in C(\overline{G}_h)$ satisfying

$$\Delta_h v_h(x) \geq 0 \quad \text{for all} \quad x \in G_h,$$

we have

$$\max_{x \in G_h} v_h(x) \leq \max_{x \in \partial G_h} v_h(x).$$

The proof can be found in Isaacson & Keller (1966), Section 9.1, Theorem 1.

From the maximum principle, we can deduce that the associated homogeneous system of equations has the null solution as its unique solution.

Well-known results from linear algebra therefore guarantee the unique solvability of the discrete Poisson equation for every $w_h = (f_h, g_h) \in F_h$.

By choosing an appropriate mesh function, we can derive, via the maximum principle (as in Section 8.1 for the case of ordinary differential equations), the following inequality.

Theorem 8.10. For every mesh function $u_h \in C(\overline{G}_h)$,

$$\max_{x\in\overline{G}_h} |u_h(x)| \le \max_{x\in\partial G_h} |u_h(x)| + \frac{1}{2}\min(a^2,b^2)\max_{x\in G_h} |\Delta_h u_h(x)|, \quad h \in \Lambda. \tag{36}$$

The proof can be found in Isaacson & Keller (1966), Section 9.1, Theorem 2.

Inequality (36) shows the inverse stability of A_h, $h \in \Lambda$, for this inequality is equivalent to

$$\beta||u_h||_h \le ||A_h u_h||_h, \quad u_h \in E_h, \quad h \in \Lambda,$$

if we set $\beta = \min(1,\max(2/a^2,2/b^2))$ and define the norms in E_h and F_h respectively by

$$||u_h||_h = \max_{x\in\overline{G}_h} |u_h(x)|, \quad u_h \in E_h,$$

and

$$||w_h||_h = \max_{x\in G_h} |f_h(x)| + \max_{x\in\partial G_h} |g_h(x)|, \quad w_h = (f_h,g_h) \in F_h.$$

From the following representation of the truncation errors

$$\tau_h(u) = (\tau_h^{(0)}(u),\tau_h^{(1)}(u)),$$

$$\tau_h^{(0)}(u) = r_h\Delta u - \Delta_h r_h u, \quad \tau_h^{(1)}(u) = (r_h u-u)|\partial G_h = 0, \quad h \in \Lambda,$$

in conjunction with 1.(24) (where $\tau_h^{(0)}$ is replaced by $-\tau_h^{(0)}(u)$) we see that

$$||\tau_h(u)||_h \to 0 \ (h \in \Lambda), \text{ if } u \in C^2(G),$$

$$||\tau_h(u)||_h = O(h^2), \text{ if } u \in C^4(G).$$

With the inverse stability inequality (36), we can finally obtain the following convergence result.

Theorem 8.11. Let $u \in C^2(G) \cap C(\overline{G})$ be the solution of Poisson's equation for $w = (f,g) \in F$ given, and suppose

$$\max_{x\in G_h} |(f-f_h)(x)| \to 0, \quad \max_{x\in\partial G_h} |(g-g_h)(x)| \to 0 \quad (h \in \Lambda).$$

Then the solutions u_h, $h \in \Lambda$, of the 5-point finite-difference approximations converge to u with the errors estimated by

$$\beta \max_{x\in\bar{G}_h} |(u_h-u)(x)| \le \max_{x\in G_h} \{|(f_h-f)(x)| + |\tau_h^{(0)}(u)(x)|\}$$
$$+ \max_{x\in\partial G_h} |(g-g_h)(x)|, \quad h \in \Lambda. \qquad \square$$

For general elliptic differential operators 1.(26), there are also maximum principles for the associated finite-difference approximations which satisfy a positively property and a certain diagonal dominance property. Such conditions are summarized in Ciarlet (1970) under the concept of "generalized nonnegative type". Elliptic differential operators themselves satisfy maximum principles which are essential tools in the classical study of solutions to the corresponding boundary-value problems (cf. Courant & Hilbert (1966), Garabedian (1964)).

REFERENCES (cf. also References in Chapter 1)

Bohl (1981), Ciarlet (1970)*, Courant & Hilbert (1966), Dorr (1970)*, Garabedian (1964), Grigorieff (1973b), Isaacson & Keller (1966), Keller (1968,1976), Mitchell (1969), Mitchell & Griffiths (1980), Vainikko (1976).

*Article

Chapter 9
Biconvergence for Projection Methods via Variational Principles

In Chapter 2, we formulated linear operator equations in variational form which we then approximated by projection methods. In an entirely analogous manner, we were able to apply projection methods to nonlinear problems. The prototype examples for illustrating our methods were examples of boundary-value problems in ordinary and partial differential equations already introduced in Chapter 1; the convergence analysis for the finite-difference approximations of these examples was the topic of the preceding chapter.

This chapter is devoted to an analysis of the underlying biconvergence of projection methods for both linear and nonlinear operator equations where the crucial requirements of stability and inverse stability are characterized by properties of the associated variational equations. This approach motivates the terminology "variational principles". The preparatory Section 9.1 lists the conditions guaranteeing the existence of discrete approximations and the consistency of mappings for projection methods; these conditions already appeared in the examples illustrating the general abstract setting in Chapters 6 and 7. In Section 9.2, we shall characterize stability and inverse stability for linear problems by utilizing the aforementioned variational principles. A biconvergence theorem including two-sided error estimates will then be immediate. These results are additionally specialized in a Hilbert space setting. They are then used to deduce the convergence of the Ritz-Galerkin approximations of two of the examples introduced in Chapter 2. Section 9.3 treats projection methods for nonlinear problems where now stability and inverse stability are characterized by variational principles for the sequence of Fréchet-derivatives. The results of Chapter 6 then provide a biconvergence theorem for approximating nonlinear problems via projection methods. In the final part of this section, we show biconvergence for the nonlinear boundary-value problem from Section 2.2, which is approximated in a manner analogous to the Ritz-Galerkin procedure for the linear examples.

The approximability condition (3a) in Section 9.1, ensuring the existence of a discrete approximation associated with a sequence of subspaces, is the starting point of the theory of finite elements. Indeed, the theory of finite elements, in particular, addresses the question of just what decompositions of the (usually, multidimensional) regions guarantee the approximability condition. Interpolation estimates provide sufficient conditions for (3a) to hold from which, moreover, the accuracy of the approximations themselves can be discerned. We are only able to discuss the theory of finite elements in the setting of our examples where here the finite elements are given by subintervals or rectangles. The biconvergence theorem for linear problems yields for the Hilbert space setting the basic result proved by Babuska and Azia (1972) in their treatise of the finite element method. The results in Section 9.3 can therefore be considered as an extension of this fundamental result to the setting of nonlinear problems.

9.1. APPROXIMABILITY

In this rather brief section, we restate the general form of projection methods for linear and nonlinear operator equations as was done in Section 7.4. The approximability conditions (3a,b) will ensure (in the terminology of Chapter 5 to 7) the existence of discrete approximations by means of subspaces. In addition to stating the approximability condition, we give conditions guaranteeing the consistency of the sequence of associated mappings.

A *projection method* for approximating the solution of

$$Au = w, \tag{1}$$

where $A: D(A) \subset E \to F$ is a mapping between normed spaces E and F, has the following form. Let $E_n \subset E$, $F_n \subset F$ be closed subspaces and $P_n^F: F \to F_n$, $n \in I$ (= N), be bounded, linear projection operators. For approximating (1), we seek, for given $w_n \in F_n$, $n \in I$, solutions $u_n \in D_n \equiv E_n \cap D(A)$ of

$$P_n^F Au_n = w_n, \quad n \in I. \tag{2}$$

The sequence of approximating operators is thus given by $A_n = P_n^F A|E_n$, $n \in I$. As a first *assumption*, we require that

$$|v,E_n| \equiv \inf_{\phi \in E_n} ||v-\phi||_E \to 0 \quad (n \in I), \quad v \in E, \tag{3a}$$

and

$$||P_n^F z - z||_F \to 0 \ (n \in I), \quad z \in F, \quad ||P_n^F|| \leq \gamma, \quad n \in I. \tag{3b}$$

The first condition in (3b) can also be replaced by $|z,F_n| \to 0$ $(n \in I)$, $z \in F$, as we have already seen in Example 3 of Section 5.1. If F is complete, then it is well known that the second condition in (3b) follows from the first one because of the Uniform Boundedness Principle. By virtue of (3a,b), discrete approximations $A(E,\pi_n E_n,\lim^E)$ and $A(F,\pi_n F_n,\lim^F)$ exist, where the discrete convergences are given by convergences in the norm of E and F, respectively.

We can easily convince ourselves that consistency of A, A_n, $n \in I$, at $u \in D(A)$ occurs if, and only if, there is a sequence $u_n' \in D_n$, $n \in I$, such that

$$u_n' \to u \quad \text{and} \quad P_n^F A u_n' - P_n^F A u \to 0 \quad (n \in I).$$

In the parlance of Chapter 1, the quantities $d_n^A(u) \equiv P_n^F A u_n' - P_n^F A u$ represent the *truncation errors* at u. These measure how well Au can be approximated by elements from AE_n relative to projection onto F_n. From Example 4 in Section 6.3, we know that (A_n) is discretely convergent to A at u whenever A is continuous in a neighborhood of u; and hence, in particular, that A, A_n, $n \in I$, is consistent at u. We recall that the continuity of a differentiable mapping A at u is a consequence of the boundedness of the derivative $A'(u)$ (cf. Section 6.1).

9.2. STABILITY, INVERSE STABILITY, AND BICONVERGENCE FOR LINEAR PROBLEMS

The principal goal of this section is to characterize stability and inverse stability of linear mappings by properties of associated linear forms which, in the Hilbert space setting, can be formulated in terms of properties of the associated variational equations themselves. Such characterizations will immediately imply a biconvergence theorem which ensures quasi-optimal convergence in the Hilbert space setting. In the final part of this section, we shall show the convergence of the Ritz-Galerkin method for the one- and two-dimensional elliptic boundary-value problems of second order from Section 2.2. We leave to the reader a similar analysis of the example of a fourth-order elliptic boundary-value problem (the "cantilevered beam problem").

We first make a few preparatory remarks concerning stability and inverse stability of linear projection methods. The norm of a vector u of a normed space E can be represented via functionals of the dual space E^* as

$$||u||_E = \sup_{0 \neq \Phi \in E^*} \frac{|\langle u, \Phi\rangle|}{||\Phi||_{E^*}} . \tag{4}$$

This is not only valid for real, but also for complex normed spaces. The proof of (4) relies crucially on the Hahn-Banach Theorem (cf. e.g., Kantorovich & Akilov (1964), IV.2).

The representations of the norm of u given by (4) enables us to characterize stability and inverse stability of a sequence of linear operators $L_n: E_n \to F_n$, $n \in I$, by the following theorem.

Theorem 9.1. (L_n) is stable, if, and only if,

$$|\langle L_n v_n, \Psi\rangle| \leq \mu_1 ||\Psi||_{F_n^*} ||v_n||_E, \quad \Psi \in F_n^*, \quad v_n \in E_n, \quad n \geq \nu. \tag{5}$$

(L_n) is inversely stable, if, and only if,

$$\mu_0 ||v_n||_E \leq \sup_{0 \neq \Psi \in F_n^*} \frac{|\langle L_n v_n, \Psi\rangle|}{||\Psi||_{F_n^*}}, \quad v_n \in E_n, \quad n \geq \nu. \tag{6}$$

Proof: (i) Using (4), we have

$$||L_n v_n||_F = \sup_{0 \neq \Psi \in F_n^*} \frac{|\langle L_n v_n, \Psi\rangle|}{||\Psi||_{F_n^*}} .$$

It follows directly from (5) that $||L_n v_n||_F \leq \mu_1 ||v_n||_E$, $v_n \in E_n$, $n \geq \nu$, and stability of (L_n) is thereby proved (cf. Lemma 6.9). Conversely using the stability in conjunction with 6.(9), we see that

$$|\langle L_n v_n, \Psi\rangle| \leq ||\Psi||_{F_n^*} ||L_n v_n||_F \leq \mu_1 ||\Psi||_{F_n^*} ||v_n||_E, \quad v_n \in E_n, \quad \Psi \in F_n^*, \quad n \geq \nu.$$

(ii) It is now quite easy to see via (4) that inequality (6) is equivalent to $||L_n v_n||_F \geq \mu_0 ||v_n||_E$, $v_n \in E_n$, $n \geq \nu$, i.e., that (L_n) is inversely stable (cf. Section 6.2). □

In particular, the inverse stability guarantees the injectivity of almost all approximating operators L_n, $n \geq \nu$. Surjectivity is a consequence of injectivity whenever, for example, $\dim E_n = \dim F_n < \infty$. If the finite dimensionality is not present, then Theorem 2.8 (i.e., the generalization of the Lax-Milgram Lemma), states that 2.(17b) (for L_n in place of A) provides a sufficient criterion for the surjectivity of L_n whenever E_n and F_n are Hilbert spaces. The following lemma generalizes this statement to the arbitrary Banach space setting.

Lemma 9.2. Let E_n, F_n be Banach spaces and $L_n: E_n \to F_n$ a bounded linear, continuously invertible mapping. Then L_n is surjective whenever

$$\sup_{v_n \in E_n} |\langle L_n v_n, \Psi\rangle| > 0 \quad \forall \Psi \in F_n^*, \quad \Psi \neq 0. \tag{7}$$

Proof: Since L_n is continuously invertible and E_n is complete, we know that $R(L_n)$ is closed. It thus suffices to show that $R(L_n)$ is dense in F_n. Let us assume that this is not the case; or, that there is an element $w_n^0 \in F_n$ such that $|w_n^0, R(L_n)| > 0$. The Hahn-Banach Theorem then provides a $\Psi^0 \in F_n^*$ which vanishes on $R(L_n)$ and for which $\langle w_n^0, \Psi^0\rangle = |w_n^0, R(L_n)| > 0$ (cf. Kantorovich & Akilov (1964), IV. 2.3 & 2.4). This, however, contradicts assumption (7). □

We have immediately the following biconvergence theorem for the projection method.

Theorem 9.3. Let $L: E \to F$ be a bounded linear mapping between Banach spaces E and F, and suppose $E_n \subset E$, $F_n \subset F$, $n \in I$, are closed subspaces satisfying assumptions (3a,b). The sequence $L_n \equiv P_n^F L|E_n$, $n \in I$, is then stable and discretely convergent to L. Furthermore, if (6) is valid and (7) holds for almost all $n \in I$, then (L_n) is inversely stable and, for the appropriate $n \in I$, L_n is both bijective and bicontinuous. The operator L itself is injective and continuously invertible. For arbitrary $w \in R(L)$, $w_n \in F_n$, $n \in I$, the following biconvergence relation holds,

$$w_n - P_n^F w \to 0 \iff u_n \to u \quad (n \in I),$$

where u and u_n are solutions of (1) and (2), respectively. Moreover, the following two-sided error estimates hold,

$$\mu_1^{-1}||w_n - P_n^F w - d_n^L(u)||_F \leq ||u_n - u_n'||_E \leq \mu_0^{-1}||w_n - P_n^F w - d_n^L(u)||_F, \quad n \geq \nu, \tag{8}$$

where u_n', $n \in I$, is an arbitrary sequence converging to u and the $d_n^L(u)$ are the associated truncation errors given by

$$d_n^L(u) = P_n^F L u_n' - P_n^F L u, \quad n \in I.$$

Proof: Stability of (L_n) and the discrete convergence $L_n \to L$ $(n \in I)$ follow from continuity of L and assumptions (3a,b), with $\mu_1 = \gamma||L||$ in (5). (L_n) is inversely stable by Theorem 9.2 and assumption (7) - also bijective. The injectivity of L has been shown in Lemma 6.19 and its continuous invertibility is a consequence of the discrete convergence of the inverses (cf. Thm. 6.15). The two-sided estimates in (8) follow immediately from the stability and inverse stability when we substitute the errors $u_n - u_n'$. Consequently, the biconvergence relation holds since $d_n^L(u) \to 0$ $(n \in I)$. □

We now specialize the biconvergence results to the case when E and F are Hilbert spaces. Let the operators P_n^F and P_n^E, respectively, denote orthogonal projections of F onto F_n and of E onto E_n, $n \in I$. From the well-known relations

$$||v - P_n^E v||_E = |v,E_n|, \quad v \in E, \quad ||z - P_n^F z||_F = |z,F_n|, \quad z \in F,\ n \in I,$$

we can deduce via the Pythagorean Theorem that

$$||v_n - v||_E^2 = ||v_n - P_n^E v||_E^2 + |v,E_n|^2, \quad v \in E, \quad v_n \in E_n, \quad n \in I. \tag{9}$$

In the Hilbert space setting, then problems (1) and (2) are respectively equivalent to the variational equations,

$$u \in E\colon (Lu,z)_F = (w,z)_F, \quad \forall z \in F, \tag{10}$$

and

$$u_n \in E_n\colon \quad (Lu_n,z_n)_F = (w_n,z_n)_F, \quad \forall z_n \in F_n, \quad n \in I. \tag{11}$$

The stability and inverse stability of $L_n = P_n^F L|E_n$, $n \in I$, can be equivalently expressed as

$$|(Lv_n,z_n)_F| \le \mu_1||v_n||_E||z_n||_F, \quad v_n \in E_n,\ z_n \in F_n,\ n \ge \nu, \tag{12}$$

and

$$\mu_0||v_n||_E \le \sup_{0 \ne z_n \in F_n} \frac{|(Lv_n,z_n)_F|}{||z_n||_F}, \quad v_n \in E_n, \quad n \ge \nu, \tag{13}$$

respectively. Here the μ_1 in (12) can be chosen to be equal to $||L||$, since $||P_n^F|| \le 1$, $n \in I$. Surjectivity of L_n follows, for example, from the injectivity whenever $\dim E_n = \dim F_n < \infty$, or from (cf. (7))

$$\sup_{v_n \in E_n} |(Lv_n, z_n)_F| > 0, \quad \forall\, 0 \neq z_n \in F_n. \tag{14}$$

With the above formulations reflecting our Hilbert space setting, we see that the results of Thoerem 9.3 hold, where an obvious choice is $u_n' = P_n^E u$. Using (9), we can obtain yet another two-sided error estimate (again, $u = L^{-1}w$)

$$\begin{aligned} |u,E_n|^2 + \mu_1^{-2}||w_n - P_n^F w - d_n^L(u)||_F^2 &\leq ||u_n - u||_E^2 \\ &\leq |u,E_n|^2 + \mu_0^{-2}(||d_n^L(u)||_F + ||w_n - P_n^F w||_F)^2, \quad n \geq \nu. \end{aligned} \tag{15}$$

The truncation errors $d_n^L(u) = P_n^F L(P_n^E u - u)$ can be estimated from above via (with $\mu_1 \equiv ||L||$)

$$||d_n^L(u)||_F \leq \mu_1 |u,E_n|, \quad n \in I.$$

In particular, if $w_n = P_n^F w$, $n \in I$, then we get from (15),

$$||u_n - u||_E^2 \leq (1 + \mu_1^2/\mu_0^2)|u,E_n|^2, \quad n \geq \nu. \tag{16}$$

The convergence is therefore *quasi-optimal*, i.e., the errors approach zero with the same rate of convergence as $|u,E_n|$. For this special case, we have reestablished the fundamental Theorem 6.2.1 of Babuska and Aziz (1972), if we note that $1 + \mu_1^2/\mu_0^2 \leq (1 + \mu_1/\mu_0)^2$.

In the final portion of this section, we analyze the convergence of the Ritz-Galerkin methods for Examples 1 and 3 from Section 2.2. For such methods, we have that $E = F$, $E_n = F_n$, and $\dim E_n < \infty$. The ellipticity of L implies condition (13) of inverse stability with $\mu_0 = \alpha_0$ (cf. Section 2.1), and condition (14) is superfluous in light of the finite dimensionality of $E_n = F_n$, $n \in I$. Combined with the approximability condition (3a), the ellipticity of L thus implies for the Ritz-Galerkin method all the assertions of Theorem 9.3, and, in particular, yields the error estimates in (15) and (16). In Examples 1 and 3, the underlying complete spaces corresponding to V in Section 2.2 are commonly denoted by $H_0^1(a,b)$ and $H_0^1(G)$, respectively.

Example 1 from Section 2.2 (cf. also Section 2.3): The variational equation

$$a(u,v) \equiv (pu',v')_0 + (qu,v)_0 = (w,v)_0, \quad v \in H_0^1(a,b),$$

for determining $u \in H_0^1(a,b)$ for given $w \in L^2(a,b)$ is approximated by the Ritz-Galerkin method with piecewise polynomial functions. As in

Fairweather (1978), we let

$$M_j^s(a,b) = \{v \in C^j[a,b]: v|I_k \in P_s(I_k), \quad k = 1,\dots,n+1\}$$

be the space of j-times continuously differentiable functions which are polynomials of at most degree s on each subinterval $I_k = [x_{k-1}, x_k]$, $k = 1,\dots,n+1$, where $a = x_0 < x_1 < \dots < x_{n+1} = b$. In Section 2.3, we approximated the solution of the variational equation by continuous, piecewise linear functions which vanish at $x = a$ and $x = b$, i.e., $E_n = M_0^1(a,b) \cap H_0^1(a,b)$. From "interpolation estimates", we know that with $h_n \equiv \max|x_k - x_{k-1}|$,

$$|v,E_n|_m \equiv \inf_{\phi \in E_n} ||v - \phi||_m \leq Ch_n^{j-m}|v|_j$$

for all $v \in W^{j,2}(a,b)$, $m = 0,1$, $m \leq j \leq 2$. Here, $W^{j,2}(a,b)$ is the space of all functions possessing generalized derivatives of j-th order in $L^2(a,b)$ and $|v|_j = ||v^{(j)}||_0$ denotes the associated seminorm (cf. e.g., Ciarlet (1978), Fairweather (1978)).

Under the hypotheses of Lemma 2.1 on the coefficients p and q, the associated bilinear form $a(.,.)$ is elliptic with respect to $||\cdot||_1$, and the given variational equation thus has a uniquely determined solution $u \in E \equiv H_0^1(a,b)$ for every $w \in L^2(a,b)$. The associated mapping $L: E \to E$ is defined by

$$(Lv,\phi)_1 = a(v,\phi), \quad v,\phi \in E.$$

To every $w \in L^2(a,b)$ and to the functional $f(\phi) = (w,\phi)_0$ on E, there corresponds an element $\hat{w} \in E$ uniquely determined by

$$(\hat{w},\phi)_1 = (w,\phi)_0, \quad \phi \in E.$$

The given variational equation is then clearly equivalent to $Lu = \hat{w}$. The equations of the Ritz-Galerkin method,

$$(pu_n', v_n')_0 + (qu_n, v_n)_0 = (w, v_n)_0, \quad v_n \in E_n, \quad n \in I,$$

can be expressed as

$$P_n^E L u_n = P_n^E \hat{w}, \quad n \in I,$$

where P_n^E, $n \in I$, denote the H_0^1-projections on E_n. Together with the L^2-projections P_n^0, $n \in I$, the latter satisfy the relations $(P_n^0 w, v_n)_0 = (P_n^E \hat{w}, v_n)_1$, $v_n \in E_n$. For the continuous, piecewise linear solutions of the Ritz-Galerkin method, (16) together with the interpolation estimates yields

$$||u_n - u||_1 \leq C|u,E_n|_1 \leq Ch_n|u|_2, \quad n \in I,$$

provided $u \in W^{2,2}(a,b)$. We can also deduce an error estimate with a higher power of h_n but with respect to a weaker norm by a procedure which is referred to as the "Aubin-Nitsche trick" in the literature. We do not provide a general treatment here but give the corresponding result for our concrete example. For the L^2-norm of the error, we obtain

$$||u_n - u||_0 \leq Ch_n^2|u|_2, \quad n \in I,$$

provided, again, $u \in W^{2,2}(a,b)$ (cf., e.g., Ciarlet (1978), Sec. 3.2). □

Example 3 from Section 2.2 (cf. also Section 2.4): For the associated variational formulation of Poisson's Equation,

$$u \in H_0^1(G): \ [u,v]_1 = (w,v)_0, \quad v \in H_0^1(G),$$

we let G be a rectangular region with a decomposition into rectangles as indicated in Section 2.4. For this simple example of a finite element approximation, we obtain from the interpolation estimates

$$||v - \pi_h v||_1 = O(h),$$

whenever $v \in H^2(G)$ (cf. Ciarlet (1978), Section 3.1). Here $\pi_h v$ denotes the interpolant of v, which is continuous in G, linear on each rectangle and agrees with v at the corners of the rectangles. Letting $\Lambda = (h_1,h_2,\ldots)$ denote a null sequence of rectangular diameters, we obtain the following estimate for solutions u_h, $h \in \Lambda$, of the Ritz-Galerkin method 2.(47),

$$||u - u_h||_1 = O(h) \quad (h \in \Lambda),$$

whenever the solution u of the above variational equation belongs to $W^{2,2}(G)$. □

9.3. BICONVERGENCE FOR NONLINEAR PROBLEMS

Our investigations in Chapter 6 now allow us to extend the results of the previous section to nonlinear problems. Indeed, we recall that in Section 7.4 we have established results on the equidifferentiability of the approximating operators generated by projection methods when used to solve fixed point equations. It mainly remains to characterize the bistability of the sequence of derivatives; and, for this purpose, we can exploit the results on linear mappings in the previous section. In the final part of this section, we analyze the nonlinear example from Sections 2.2 and 2.5 taken from the work of Ciarlet et al. (1967). Our approach enables us to obtain the biconvergence with two-sided error estimates with respect to appropriately chosen norms. A few helpful results will be cited (without proof) from the original literature.

The general problem setting is thus the same as in Section 9.1, with A not necessarily linear and with conditions (3a,b) satisfied. The spaces E and F are assumed to be complete with E_n and F_n closed subspaces of E and F, respectively. We further assume that $Au = 0$ has a solution u^0, where (u_n^0) is to be an arbitrary, but fixed, sequence of elements $u_n^0 \in E_n$, $n \in I$, converging to u^0. The operator A is to satisfy the differentiability conditions 7.(22a,b) which we know imply the equicontinuous equidifferentiability of the mappings $A_n \equiv P_n^F A | E_n$, $n \in I$, at (u_n^0) with derivatives,

$$A_n'(v_n) = P_n^F A'(v_n) | E_n, \quad v_n \in E_n \cap B_{\rho_0}(u^0), \quad n \in I.$$

(We should bear in mind that the above result could be true only for almost every $n \in I$ (cf. Section 7.4, remark before Lemma 7.19).)

From the analysis of Section 6.2, we see that the central conditions of stability and inverse stability of (A_n) at (u_n^0) are characterized by the stability and the inverse stability, respectively, of $A_n'(u_n^0)$, $n \in I$. These can be further characterized by

$$|\langle A_n'(u_n^0)v_n, \Psi\rangle| \le \mu_1 ||\Psi||_{F_n^*} ||v_n||_E, \quad \Psi \in F_n^*, \quad v_n \in E_n, \quad n \ge \nu, \tag{17}$$

and

$$\mu_0 ||v_n||_E \le \sup_{0 \ne \Psi \in F_n^*} \frac{|\langle A_n'(u_n^0)v_n, \Psi\rangle|}{||\Psi||_{F_n^*}}, \quad v_n \in E_n, \quad n \ge \nu, \tag{18}$$

respectively (cf. Theorem 9.1). The stability (17) results, whenever $A'(u^0)$ is a bounded, linear mapping or whenever A itself is continuous at u^0. In this case, the derivatives $A'(u)$ are uniformly bounded for all u in a neighborhood of u^0 because of 7.(22a,b) (see the proof of Theorem 6.1). This neighborhood also contains u_n^0 for almost all $n \in I$. For showing the surjectivity of $A_n'(u_n^0)$, we have at our disposal the results from Section 9.2, which point out that surjectivity follows either from the injectivity of the respective $A_n'(u_n^0)$ whenever $\dim E_n = \dim F_n < \infty$, or from (17) and (18) combined with (cf. Lemma 9.2)

$$\sup_{v_n \in E_n} |\langle A_n'(u_n^0)v_n, \Psi\rangle| > 0, \qquad \forall\, 0 \neq \Psi \in F_n^*. \tag{19}$$

With these considerations, we are now in a position to state the following biconvergence theorem for projection methods for approximating nonlinear problems. We set $w_n^0 \equiv A_n u_n^0$, $n \in I$, where $u_n^0 \in E_n$, $n \in I$, is the aforementioned arbitrary, but fixed, sequence converging to u^0, whose elements are assumed (without loss of generality) to lie in $D(A)$ for all $n \in I$.

Theorem 9.4. Assume that

(a) assumptions (3a,b) are satisfied with $E_n \subset E$, $F_n \subset F$ Banach spaces;

(b) the differentiability condition 7.(22a,b) is satisfied with a solution u^0 of $Au^0 = 0$;

(c) $A'(u^0)$ is a bounded mapping; and

(d) condition (18) holds, and (19) is satisfied for almost all $n \in I$.

Then (17) holds; and, for every $\alpha > \mu_1$ and $\beta \in (0,\mu_0)$, there are positive numbers ρ,σ such that the approximating equations (2) are uniquely solvable for all $n \geq \nu$ and all $w_n \in B_\sigma(w_n^0)$; the solutions u_n lie in each $B_\rho(u_n^0)$, and the sequence (A_n) is locally bijective and equibicontinuous at (u_n^0). Further, the following biconvergence relation holds,

$$||u_n - u^0||_E \to 0 \Longleftrightarrow ||w_n||_F \to 0 \quad (n \in I),$$

along with the two-sided error estimate

$$\alpha^{-1}||w_n - d_n^A(u^0)||_F \leq ||u_n - u_n^0||_E \leq \beta^{-1}||w_n - d_n^A(u^0)||_F, \quad n \geq \nu.$$

Proof: As remarked above, the uniform boundedness of $A_n'(u_n^0)$, $n \geq \nu$, is a consequence of the boundedness of $A'(u^0)$ used in conjunction with the differentiability hypotheses 7.(22a,b). Hence (17) follows. From (18) and (19), we obtain the inverse stability of $(A_n'(u_n^0))$ as well as the bijectivity of almost all of the derivatives. The remaining assertions follow from the Theorems 6.21 and 6.22. □

We conclude this section by applying our theory to the nonlinear boundary-value problem from Section 2.2 and to its Ritz-Galerkin approximation formulated in Section 2.5.

Example 4 from Section 2.2 (cf. also Section 2.5): The associated variational equation has the form (cf. 2.(24))

$$a(u,v) \equiv (pu',v')_0 + (qu,v)_0 = \int_0^1 f(x,u(x))v(x)dx, \quad v \in H_0^1(0,1).$$

In Section 2.5, we expressed the associated mappings in the form $A = \hat{A}-T$, where

$$(\hat{A}u,v)_1 = a(u,v), \quad u,v \in H_0^1(0,1),$$
$$(Tu,v)_1 = (f(.,u),v)_0, \quad u,v \in H_0^1(0,1).$$

Under the hypotheses on p, q, and f found in Section 2.2, we have seen that the derivatives are given by $A'(u)v = \hat{A}v-T'(u)v$, where

$$(T'(u)v,\phi)_1 = (f_y(.,u)v,\phi)_0, \quad u,v,\phi \in H_0^1(0,1).$$

A is, moreover, continuously differentiable at every $u \in H_0^1(0,1)$. We can easily convince ourselves that the uniform differentiability condition 7.(22a) holds for every $u \in H_0^1(0,1)$ and every closed ball $B_{\rho_0}(u)$ with center u (with respect to the norm $||\cdot||_1$).

As preparation for the proof of stability and of inverse stability, we define

$$[u]_\beta \equiv \{a(u,u) - \beta||u||_0^2\}^{1/2}, \quad u \in H_0^1(0,1), \quad \beta \in \mathbb{R}. \tag{20}$$

As in Lemma 2 of Ciarlet et al. (1967), we see that this expression defines a norm on $H_0^1(0,1)$ for every $\beta < \Lambda$; moreover, for $\beta,\beta' < \Lambda$, we have

$$c_1[u]_{\beta'}^2 \leq [u]_\beta^2 \leq c_2[u]_{\beta'}^2, \quad u \in H_0^1(0,1), \tag{21}$$

with

$$c_1 = (1 + \frac{\max(\beta-\beta',0)}{\Lambda-\beta})^{-1}, \quad c_2 = 1 + \frac{\max(\beta'-\beta,0)}{\Lambda-\beta'} .$$

Hence, a scalar product will also be defined on $H_0^1(0,1)$ by $[u,v]_\beta \equiv a(u,v) - \beta(u,v)_0$, $\beta \in (-\infty,\Lambda)$. The definiteness property follows from the properties of the norm $[u]_\beta$; other properties of the scalar product are trivial to verify.

Clearly, $\beta = 0$ lies in $(-\infty,\Lambda)$; and, moreover, the norm $[\cdot]_0$ is equivalent to $||\cdot||_1$ since $a(.,.)$ is bounded and elliptic. Also we know that the supremum norm $||\cdot||_{0,\infty}$ can be estimated by the norm

$||\cdot||_1$; this is nothing other than one of the well-known Sobolev Inequalities in the one-dimensional case (see also the proof of Lemma 2.1). For arbitrary $\beta \in (-\infty,\Lambda)$, we can thus conclude from (21) that

$$||u||_{0,\infty} \leq K_\beta[u]_\beta, \quad u \in H_0^1(0,1). \tag{22}$$

We remark that, with any λ satisfying $f_y \leq \lambda < \Lambda$ (cf. 2.(23)), $[\cdot]_\lambda$ also represents a norm; and, moreover, that the relation $f_y \leq 0$ required in 1.(17) yields, in particular, the condition that $\lambda < \Lambda$ (required here) for $\lambda = 0$.

We now consider finite-dimensional subspaces E_n, $n \in I$, of $H_0^1(0,1)$ - e.g., piecewise polynomial functions defined on meshes in $[0,1]$. Also, we let $P_n^E: H_0^1(0,1) \to E_n$, $n \in I$, be orthogonal projections with respect to $(.,.)_1$. We now have the following result on stability and inverse stability with respect to the norm $[\cdot]_\lambda$.

<u>Theorem 9.5</u>. Under the above hypotheses, each $A_n'(u_n)$ satisfies

$$[v_n]_\lambda^2 \leq |(A_n'(u_n)v_n,v_n)_1|, \quad v_n \in E_n, \quad n \in I, \tag{23}$$

for every $u_n \in E_n$, $n \in I$. For every sequence $(u_n) \in \pi_n E_n$ bounded in the norm $||\cdot||_{0,\infty}$, there is a $C > 0$ such that

$$|(A_n'(u_n)v_n,\psi_n)_1| \leq C[v_n]_\lambda[\psi_n]_\lambda, \quad v_n,\psi_n \in E_n, \quad n \in I. \tag{24}$$

<u>Proof</u>: From the representation of the derivative $A'(u)$ and by definition of orthogonal projections, we have

$$(A_n'(u_n)v_n,\psi_n)_1 = a(v_n,\psi_n) - (f_y(\cdot,u_n)v_n,\psi_n)_0, \quad u_n,v_n,\psi_n \in E_n, \quad n \in I.$$

The hypotheses thus imply the estimate (23), for

$$|(A_n'(u_n)v_n,v_n)_1| \geq a(v_n,v_n) - \lambda||v_n||_0^2 = [v_n]_\lambda^2, \quad v_n \in E_n, \quad n \in I.$$

If, moreoever, $u_n \in E_n$, $n \in I$, is a sequence uniformly bounded in the sup-norm, i.e., $||u_n||_{0,\infty} \leq c$, $n \in I$, then we get from the above representation of $A_n'(u_n)$ that

$$(A_n'(u_n)v_n,\psi_n)_1 \leq a(v_n,\psi_n) + \Gamma(v_n,\psi_n)_0, \quad v_n,\psi_n \in E_n, \quad n \in I,$$

where

$$\Gamma \equiv \max\{-f_y(x,y): 0 \leq x \leq 1, \ |y| \leq c\}.$$

Because $f_y \leq \lambda < \Lambda$, it follows that $-\Gamma \leq \lambda < \Lambda$. The right-hand side of the above inequality is precisely the expression for the scalar product $[v_n,\psi_n]_{-\Gamma}$; and the Hölder Inequality, used in conjunction with (21) where now $\beta = -\Gamma$, $\beta' = \lambda$, yields

$$|(A_n'(u_n)v_n,\psi_n)_1| \leq [v_n]_{-\Gamma}[\psi_n]_{-\Gamma}$$

$$\leq \left[1 + \frac{\max(\Gamma+\lambda,0)}{\Lambda-\lambda}\right][v_n]_\lambda[\psi_n]_\lambda, \quad v_n,\psi_n \in E_n, \quad n \in I. \quad \square$$

We consider now the special case when $u_n^0 = P_n^E u^0$, $n \in I$, with u^0 a solution of the given variational equation. Our analysis in Section 2.2 has shown that such a u^0 is unique if it exists, since such a u^0 minimizes the strictly convex function $J(\cdot)$ defined in 2.(25). Moreover, we have that $J(u^0) \leq J(0) = 0$. From Ciarlet et al. (1967) it is known that every function $w \in H_0^1(a,b)$, with $J(w) \leq 0$, satisfies $[w]_\lambda \leq 2M/\sqrt{\Lambda-\lambda}$ where

$$M \equiv \max\{|f(x,0)|: 0 \leq x \leq 1\}.$$

Using (22), we get that $||u^0||_{0,\infty} \leq 2MK_\lambda/\sqrt{\Lambda-\lambda}$. If, furthermore, we assume the approximability condition (3a) in the $||\cdot||_1$-norm, then $||u_n^0 - u^0||_{0,\infty} \to 0$ $(n \in I)$, and hence there is a $\nu \in I$ for which

$$||u_n^0||_{0,\infty} \leq 3MK_\lambda/\sqrt{\Lambda-\lambda}, \quad n \geq \nu.$$

A perusal of the proof of Theorem 9.5 shows that (24) is satisfied for almost all $n \in I$, and can thus be expressed as

$$|(A_n'(u_n^0)v_n,\psi_n)_1| \leq \left[1 + \frac{\max(\Gamma_0+\lambda,0)}{\Lambda-\lambda}\right][v_n]_\lambda[\psi_n]_\lambda, \quad v_n,\psi_n \in E_n, \quad n \geq \nu,$$

where $\Gamma_0 \equiv \max\{-f_y(x,y): 0 \leq x \leq 1, |y| \leq 3MK_\lambda/\sqrt{\Lambda-\lambda}\}$.

From (23) and (24), we therefore see that $(A_n'(u_n^0))$ is both stable and inversely stable. For almost all $n \in I$, the derivatives are injective and bijective as mappings of a finite-dimensional space into itself. Because of (21), (23), (24), and the equivalence of the norms $[\cdot]_0$ and $||\cdot||_1$, we have the following uniform boundedness properties for the operator norms defined with respect to $||\cdot||_1$,

$$\alpha' \equiv \sup_{n\geq\nu} ||A_n'(u_n^0)|| < \infty, \quad \frac{1}{\beta'} \equiv \sup_{n\in I} ||A_n'(u_n^0)^{-1}|| < \infty.$$

We now apply the results of Theorem 9.4 (with $w_n = 0 \in E_n$, $n \in I$) for the solution u_n of the resulting approximating equations derived in a manner analogous to the Ritz-Galerkin method for linear problems; $||\cdot||_1$

is the underlying norm. We obtain the two-sided error estimates,

$$\alpha^{-1}||d_n^A(u^0)||_1 \leq ||u_n - P_n^E u^0||_1 \leq \beta^{-1}||d_n^A(u^0)||_1, \quad n \geq \nu,$$

and

$$|u^0,E_n|_1^2 + \alpha^{-2}||d_n^A(u^0)||_1^2 \leq ||u_n - u^0||_1^2$$
$$\leq |u^0,E_n|_1^2 + \beta^{-2}||d_n^A(u^0)||_1^2, \quad n \geq \nu,$$

where $d_n^A(u^0) = P_n^E(AP_n^E u^0 - Au^0) = P_n^E A P_n^E u^0$, $n \in I$, are the truncation errors and both α and β are arbitrary with $\alpha > \alpha'$, $0 < \beta < \beta'$. Because the derivative $A'(u^0)$ is bounded, the truncation errors can, moreover, be estimated by $||d_n^A(u^0)||_1 \leq C||u^0 - P_n^E u||_1$, $n \in I$. This estimate, together with (22), leads to the following estimates for the errors in our approximations,

$$||u_n - u^0||_{0,\infty} \leq C_1||u_n - u^0||_1 \leq C_2|u^0,E_n|_1, \quad n \geq \nu,$$

with C_i, $i = 1,2$, denoting generic constants. The convergence is thus seen to be quasi-optimal. □

REFERENCES (cf. also References in Chapter 2)

Aubin (1972,1979), Babuska & Aziz (1972)*, Ciarlet (1978), Ciarlet, Schultz & Varga (1967)*, Douglas & Dupont (1974)*, Fairweather (1978), Kantorovich & Akilov (1964), Krasnoselskii, Vainikko et al. (1972), Lions & Magenes (1972), Mitchell & Wait (1977), Oden & Reddy (1976), Stummel (1970,1972,1976a,1976b,1977)*.

*Article(s)

Chapter 10
Convergence of Perturbations of Integral Equations of the Second Kind

In Chapter 3, we presented several methods for approximating solutions of linear and nonlinear integral equations of the second kind. The approximation schemes that we considered can be divided into two classes, namely, those where the approximate equations can be also expressed in the form of integral equations with perturbed kernels and perturbed regions of integration, and those which represent projection methods.

The convergence theory of projection methods for approximating solutions of equations of the second kind has already been studied in Section 7.4. The derivation of the associated algebraic systems of equation was done in Chapter 3. Application of the convergence results in Section 7.4 is straightforward, so that there is no need to delve further into projection methods for integral equations of the second kind. The required assumptions on the projection operators have already been investigated in Chapter 9 for Galerkin methods. And for the convergence analysis of particular collocation methods, we refer the reader to Atkinson (1976), Baker (1977), Ikebe (1972), Atkinson & Graham & Sloan (1983), and Sloan (1980a), and the literature cited therein.

In this chapter, we shall apply the results from Section 7.3 to show convergence of approximations defined by integral equations with perturbed kernels and regions. We saw in Chapter 3 that Nyström methods and product integration methods can be studied in such a setting. This presentation closely parallels that of Stummel (1974b) except that here our analysis requires weaker assumptions. For the sake of clarity, we shall restrict our study to closed, bounded regions in $\mathbb{R}^d$; we have in Section 3.1 indicated how our analysis is generalizable to the setting of arbitrary compact metric spaces.

10.1. STATEMENT OF THE PROBLEM AND CONSISTENCY

We begin our study by giving a brief description of the problems being approximated and refer the reader to Chapter 3 for a more detailed presentation. The bulk of this section discusses conditions on the measures, regions, and kernels in order that the associated integral operators are consistent in the sense of Section 6.3.

With the notation of Section 3.4, we consider the following nonlinear integral equations of the second kind in $C(G)$,

$$u(x) - \int_G k(x,y,u(y))dy = w(x), \quad x \in G, \tag{1}$$

along with the approximate equations, which we can also express in the form of integral equations, albeit with perturbed regions and kernels,

$$u_n(x) - \int_{G_n} k_n(x,y,u_n(y))d\mu_n(y) = w_n(x), \quad x \in G_n,\ n \in I\ (= \mathbb{N}). \tag{2}$$

The sets G, G_n, $n \in I$, are assumed to be closed subsets of a compact subset M of $\mathbb{R}^d$, $d \in \mathbb{N}$. We shall use the notation $|\cdot|$ to designate the underlying norm in $\mathbb{R}^d$, and denote the supremum norms in $C(G)$ and $C(G_n)$ by $||\cdot||$ and $||\cdot||_n$, respectively.

The integral in (1) is to be understood in the Lebesgue sense. The μ_n, $n \in I$, are to be arbitrary nonnegative measures on G_n, which, as we know from Chapter 3, generate bounded linear functionals on $C(G_n)$. Note that the Lebesgue integral also specifies a nonnegative measure on G which we denote by $\mu(v) = \int_G v\,dx$, $v \in C(G)$. In Section 5.4, we have studied in detail the case where the μ_n, $n \in I$, are also given by Lebesgue integrals over perturbed regions (cf., in particular, Theorem 5.15), and the case where the μ_n, $n \in I$, are specified by quadrature formulas with finite sets of nodes (cf. Theorems 5.17 and 5.18).

We further assume that the integral operator associated with (1),

$$(Ku)(x) \equiv \int_G k(x,y,u(y))dy,$$

represents a mapping of $C(G)$ into itself (cf. Section 3.4). This will be true, for example, whenever the kernel $k(.,.,.)$ is continuous in all its arguments. We shall assume for the approximating equations (2) that the kernels $k_n(.,.,.)$ are continuous with respect to all arguments. This requirement reduces to continuity with respect to the third argument whenever G_n are discrete point sets. Discrete point sets are present with Nyström methods or product integration methods which we have already shown in Chapter 3 to be expressible in the form (2). The associated approximate integral operators,

$$(K_n u_n)(x) \equiv \int_{G_n} k_n(x,y,u_n(y))d\mu_n(y), \quad u_n \in C(G_n), \quad x \in G_n, \quad n \in I,$$

then map $C(G_n)$ into itself. Further assumptions on the measures and kernels will subsequently be specified in the text. In particular, the special case of linear integral equations and their approximations can be cast in the forms (1) and (2) by setting $k(x,y,z) = k(x,y)z$, and $k_n(x,y,z) = k_n(x,y)z$, $n \in I$.

We next investigate the consistency of the integral operator. In order to ensure the existence of an underlying discrete approximation $A(C(G),\pi_n C(G_n),\lim)$ in the sense of Section 5.3, we shall assume that

$$\operatorname{Lim\,sup} G_n = G. \tag{3}$$

For any continuous kernel $k(.,.,.)$, we know that there is a continuous extension $\hat{k}(.,.,.)$ to $M \times M \times \mathbb{R}$ by the Tietze-Urysohn extension theorem. If $k(.,.,.)$ is not necessarily continuous, an extension may also exist. In any case, we assume that the extended kernel $\hat{k}$: $M \times M \times \mathbb{R} \to \mathbb{R}$, if it exists, is such that

$$(\hat{K}v)(x) \equiv \int_G \hat{k}(x,y,v(y))dy, \quad x \in M, \quad v \in C(M), \tag{4}$$

is well defined and maps $C(M)$ into itself. In case that $G_n \subset G$, $n \in I$, the consideration of extensions is superfluous, and we can then set $M = G$, $\hat{k} = k$.

We have the following result guaranteeing consistency at a special consistency sequence.

<u>Lemma 10.1</u>. The sequence of integral operators K,K_n, $n \in I$, is consistent at $u \in C(G)$, if extensions $\hat{u} \in C(M)$ of u and $\hat{k}(.,.,.)$ of $k(.,.,.)$ exist such that

$$||d_n^K(u)||_n \equiv \sup_{x\in G_n} |d_n^K(u)(x)| \to 0 \quad (n \in I) \tag{5}$$

where

$$d_n^K(u)(x) \equiv \int_{G_n} k_n(x,y,\hat{u}(y))d\mu_n(y) - \int_G \hat{k}(x,y,u(y))dy, \quad x \in G_n, \quad n \in I.$$

Proof: We set $u'_n = \hat{u}|G_n$, $z'_n = \hat{z}|G_n$, $n \in I$, where $\hat{z} = \hat{K}\hat{u}$ with $\hat{K}$ defined in (4). Then we have the convergences

$$u'_n \to u, \quad z'_n \to Ku \quad (n \in I),$$

and $K_n u'_n \to Ku$ $(n \in I)$, since, according to (5),

$$||K_n u'_n - z'_n||_n = ||d_n^K(u)||_n \to 0 \quad (n \in I). \qquad \square$$

The mappings $A = I-K$, $A_n = I-K_n$, $n \in I$, are likewise consistent at u by assumption (5), with the consistency sequence $(u'_n) = (\hat{u}|G_n)$. We define further $w_n^1 \equiv (\hat{u}-\hat{K}\hat{u})|G_n$, $n \in I$, with $\hat{K}$ defined in (4). Then

$$d_n^A(u) \equiv A_n u'_n - w_n^1 = -d_n^K(u), \quad n \in I,$$

and (5) is equivalent to

$$||d_n^A(u)||_n \to 0 \quad (n \in I). \tag{6}$$

We also call $d_n^A(u)$, $n \in I$, the sequence of *truncation errors* at u.

For the *Nyström method* with a continuous kernel k and its continuous extension $\hat{k}$, we have (cf. 3.(35))

$$\begin{aligned} d_n^K(u)(x) &= \sum_{y \in G_n} \alpha_n(y)\hat{k}(x,y,\hat{u}(y)) - \int_G \hat{k}(x,y,u(y))dy \\ &= (\hat{\mu}_n - \mu)\tilde{k}(x,\cdot), \quad x \in G_n, \quad n \in I, \end{aligned}$$

where $\tilde{k}(x,y) = \hat{k}(x,y,\hat{u}(y))$ represents a continuous function in y for every $x \in M$. Here, $\hat{\mu}$ is the natural extension of the measure on G given by the Lebesgue integral (see 3.(5) for the definition of a natural extension of a measure), and $\hat{\mu}_n$, $n \in I$, are the natural extensions of the associated approximating measures defined in 3.(35).

Whenever the kernel has product form, i.e., $k(x,y,z) = h(x,y)r(x,y,z)$, with h not necessarily continuous, we have

$$\begin{aligned} d_n^K(u)(x) = \sum_{i=1}^n \int_a^b h(x,s)\eta_i(s)dy \; r(x,y_i,u(y_i)) \\ - \int_a^b h(x,y)r(x,y,u(y))dy \end{aligned}$$

for the *product integration method* over $G = [a,b] \subset \mathbb{R}$ with discrete point sets $G_n = \{y_1,\ldots,y_n\}$, $n \in I$, in G (see 3.(17) and 3.(36)). For a continuous factor $r(.,.,.)$, we can easily convince ourselves that consistency of the product integration method will follow in case

$$\max_{x\in G_n} \left| \sum_{i=1}^{n} g(y_i)\int_a^b h(x,s)\eta_i(s)ds - \int_a^b h(x,y)g(y)dy\right| \to 0 \quad (n \in I)$$

for every $g \in C(G)$.

Continuous kernels allow, in general, a splitting of the truncation errors into two parts, $d_n^K(u) = d_n^0(u) + d_n^1(u)$, where

$$d_n^0(u)(x) \equiv \int_{G_n} \hat{k}(x,y,\hat{u}(y))d\mu_n(y) - \int_G \hat{k}(x,y,u(y))dy,$$

and

$$d_n^1(u)(x) \equiv \int_{G_n} (k_n(x,y,\hat{u}(y)) - \hat{k}(x,y,\hat{u}(y))d\mu_n(y), \quad x \in G_n, \quad n \in I.$$

The first part can also be written as

$$d_n^0(u)(x) = (\hat{\mu}_n-\hat{\mu})\tilde{k}(x,\cdot), \quad x \in G_n, \quad n \in I,$$

and the second part as

$$d_n^1(u)(x) = [(K_n-\hat{K}_n)u_n'](x), \quad x \in G_n, \quad n \in I,$$

where $\hat{\mu},\hat{\mu}_n$ denote the natural extensions of μ,μ_n, respectively, where $\tilde{k}(x,y) = \hat{k}(x,y,u(y))$, $x,y \in M$, with $\hat{k}$ the continuous extension of k, and where the $\hat{K}_n$: $C(G_n) \to C(G_n)$, $n \in I$, are defined by

$$(\hat{K}_n u_n)(x) \equiv \int_{G_n} \hat{k}(x,y,u_n(y))d\mu_n(y), \quad x \in G_n, \quad n \in I. \tag{7}$$

Note that, for Nyström's method, $d_n^1(u) = 0$.

For continuous kernels $k(.,.,.)$ and weakly convergent measures, we shall show in the following theorem that $d_n^0(u) \to 0$ $(n \in I)$ always, and hence $d_n^1(u) \to 0$ must only be proved in order to show consistency.

Theorem 10.2. Let $\hat{k}$ be a continuous extension of the continuous kernel k, and let the weak convergence $\hat{\mu}_n \rightharpoonup \hat{\mu}$ $(n \in I)$ be satisfied. Then

$$||d_n^0(u)||_n = \sup_{x\in G_n} |(\hat{\mu}_n-\hat{\mu})\tilde{k}(x,\cdot)| \to 0 \quad (n \in I).$$

The sequence of integral operators K,K_n, $n \in I$ - and also $A = I-K$, $A_n = I-K_n$, $n \in I$ - is consistent at u, if

$$||d_n^1(u)||_n = ||(K_n-\hat{K}_n)u_n'||_n \to 0 \quad (n \in I),$$

where $\hat{u} \in C(M)$ is an extension of u, and $u_n' = \hat{u}|G_n$, $n \in I$.

Proof: We show that the extensions

$$\hat{d}_n^0(u)(x) = (\hat{\mu}_n-\mu)\tilde{k}(x,\cdot), \quad x \in M, \quad n \in I,$$

of $d_n^0(u)$, $n \in I$, are equicontinuous and converge pointwise to zero. According to a well-known theorem (see, e.g., Dieudonné (1969), Theorem (7.5.6)), $d_n^0(u)$, $n \in I$, then converge uniformly to zero due to the compactness of M, and hence,

$$||d_n^0(u)||_n \le \sup_{x \in M} |\hat{d}_n^0(u)(x)| \to 0 \quad (n \in I).$$

Under the assumptions given above, $||\hat{\mu}_n||$, $n \in I$, is bounded, with the result that

$$||\hat{\mu}|| + ||\hat{\mu}_n|| \le M_1, \quad n \in I,$$

for some number $M_1 > 0$. The function $\tilde{k}(.,.)$ defined above is continuous on $M \times M$ and hence is uniformly continuous there. Then, for every $\varepsilon > 0$, there is a $\delta > 0$ such that $|\tilde{k}(x,y) - \tilde{k}(x',y)| \le \varepsilon/M_1$ for all $|x-x'| \le \delta$, $x,x' \in M$, and all $y \in M$. By definition of $\hat{d}_n^0(u)$,

$$|\hat{d}_n^0(u)(x) - \hat{d}_n^0(u)(x')| = |(\hat{\mu}_n-\hat{\mu})(\tilde{k}(x,\cdot) - \tilde{k}(x',\cdot))|$$

$$\le M_1 \sup_{y \in M} |\tilde{k}(x,y) - \tilde{k}(x',y)| \le \varepsilon$$

for all $|x-x'| \le \delta$, $n \in I$, Therefore $\hat{d}_n^0(u)$, $n \in I$, is equicontinuous. By assumption, $\hat{\mu}_n(w) \to \hat{\mu}(w)$ $(n \in I)$ for every $w \in C(M)$. In particular, for $w = \tilde{k}(x,\cdot)$ and with arbitrary $x \in M$, we get

$$\hat{d}_n^0(u)(x) = (\hat{\mu}_n-\hat{\mu})\tilde{k}(x,\cdot) \to 0 \quad (n \in I). \qquad \square$$

We would like to point out that the convergence $||d_n^0(u)||_n \to 0$ $(n \in I)$ has been proven for every $u \in C(G)$. The convergence of $||d_n^1(u)||_n$, $n \in I$, to zero is nevertheless required only for the particular function u under consideration.

In addition, we remark that the weak convergence of the extended measures, assumed in the preceding theorem, along with (3) ensure $\text{Lim}\, G_n = G$ provided, moreover, the Lebesgue integral specifies a positive measure on G; the latter requirement is met whenever G possesses nonempty interior (cf. Sec. 5.4). The convergence of the underlying regions then guarantees the convergence of the maximum norms of discretely convergent sequences (cf. Sec. 5.3).

We finally note that the weak convergence $\hat{\mu}_n \rightharpoonup \hat{\mu}$ $(n \in I)$ is ensured in Section 5.4 for the case of approximating Lebesgue measures on perturbed regions and for quadrature formulas.

10.2. EQUIDIFFERENTIABILITY

The primary task of this section is to show the equicontinuous equidifferentiability of the integral operators K_n, $n \in I$, defined in the previous section. Toward this end, we shall need a series of assumptions, and subsequently we shall state criteria which ensure that these assumptions are met. For the sake of clarity, we shall first assemble all the assumptions in this section which will additionally ensure the consistency and - in the following section - will imply the regular convergence along with the biconvergence of the underlying operator sequence.

In Section 3.4, we gave conditions on the kernels which guarantee that the integral operators K, K_n are continuously differentiable. The explicit representation of the derivative was already presented in 3.(38) and 3.(41). In order to be able to apply the results from Chapter 7, we need even stronger assumptions ensuring the equicontinuous equidifferentiability of (K_n). The property will be obtained by a certain uniformity requirement in condition 3.(37). Moreover, we shall give a somewhat stronger condition than 3.(40) which yields a more verifiable criterion for the continuous differentiability of K itself.

In the following, we let u be an arbitrary but fixed function from $C(G)$, and we set $w \equiv Au = u - Ku$. For clarity of presentation, we now collect all our needed assumptions, which include conditions ensuring consistency of the associated sequence of operators.

(8) $\operatorname{Lim} G_n = G$ and the sequence of measures μ_n, $n \in I$, is uniformly bounded.

(9) Extensions $\hat{u} \in C(M)$ of u and $\hat{k}(.,.,.): M \times M \times \mathbb{R} \to \mathbb{R}$ of $k(.,.,.)$ exist for which $||d_n^K(u)||_n \to 0$ $(n \in I)$, where $d_n^K(u)$ is defined as in Lemma 10.1.

(10) There is a positive ρ_0 such that the extension $\hat{k}$ of k is differentiable with respect to the third argument in

$$W \equiv \{(x,y,z) \in M \times M \times \mathbb{R}: |z-\hat{u}(y)| \le \rho_0\}.$$

The partial derivative $\partial\hat{k}/\partial z(x,y,\cdot)$ is continuous in $\{z \in \mathbb{R}: |z-\hat{u}(y)| \le \rho_0\}$ uniformly for all $x,y \in M$ and moreover satisfies

$$\sup_{x\in G} \int_G \left|\frac{\partial\hat{k}}{\partial z}(x,y,\phi(y))\right| dy < \infty, \quad \phi \in B^0_{\rho_0}(u), \tag{10a}$$

$$\lim_{x'\to x} \int_G \left|\frac{\partial\hat{k}}{\partial z}(x',y,\phi(y)) - \frac{\partial\hat{k}}{\partial z}(x,y,\phi(y))\right| dy = 0, \quad x \in G, \quad \phi \in B^0_{\rho_0}(u). \tag{10b}$$

(11) The associated homogeneous linearized integral equation

$$h(x) - \int_G \frac{\partial k}{\partial z}(x,y,u(y))h(y)\,dy = 0, \quad x \in G,$$

possesses only the trivial solution.

(12) There exists a positive ρ_1 such that the sequence of kernels k_n, $n \in I$, is equidifferentiable with respect to the third argument in the region

$$W_n \equiv \{(x,y,z) \in G_n \times G_n \times \mathbb{R}: |z-\hat{u}(y)| \le \rho_1\}.$$

For every $n \in I$, the partial derivative $\partial k_n/\partial z$ is continuous with respect to all three arguments in W_n and $\partial k_n/\partial z(x,y,\cdot)$ is equicontinuous in $\{z \in \mathbb{R}: |z-\hat{u}(y)| \le \rho_1\}$ uniformly for all $x,y \in G_n$, $n \in I$.

(13) $$C_1 \equiv \sup_{n\in I} \sup_{x\in G_n} \int_{G_n} \left|\frac{\partial k_n}{\partial z}(x,y,\hat{u}(y))\right| d\mu_n(y) < \infty.$$

(14) $\omega(\hat{u};\delta) \to 0$ $(\delta \to 0)$ where

$$\omega(v;\delta) \equiv \sup\left\{\int_{G_n} \left|\frac{\partial k_n}{\partial z}(x,y,v(y)) - \frac{\partial k_n}{\partial z}(x',y,v(y))\right| d\mu_n(y) : |x-x'| \le \delta,\ n \in I\right\}.$$

(15) To every $h \in C(G)$, there is an extension $\hat{h} \in C(M)$ for which

$$\sup_{x\in G_n}\left|\int_{G_n}\frac{\partial k_n}{\partial z}(x,y,\hat{u}(y))\hat{h}(y)d\mu_n(y) - \int_G \frac{\partial \hat{k}}{\partial z}(x,y,u(y))h(y)dy\right| \to 0 \ (n \in I).$$

Before we show the equicontinuous equidifferentiability of K_n, $n \in I$, we mention several obvious conclusions from these assumptions. Condition (8) is satisfied whenever $\operatorname{Lim\,sup} G_n \subset G$, whenever the measures $\hat{\mu}_n$, $n \in I$, are weakly convergent to $\hat{\mu}$, and whenever μ is itself positive (cf. Theorem 5.16). Condition (9) assures the consistency of K, K_n, $n \in I$, at u, for which still other criteria have been given in the preceding section. Lemma 3.6 from Section 3.4 implies that (10) will guarantee in particular the differentiability of K at all functions in a neighborhood of u, with the derivatives

$$(K'(\phi)h)(x) = \int_G \frac{\partial k}{\partial z}(x,y,\phi(y))h(y)dy, \quad \phi \in B^0_{\rho_0}(u), \quad x \in G,$$

representing completely continuous mappings of $C(G)$ onto itself. Condition (11) means that $A'(u) = I-K'(u)$ is injective, from which we easily conclude that $A'(u)$ is bijective as $K'(u)$ is completely continuous. Condition (12) is stronger than 3.(37). The derivatives K_n', $n \in I$, are hence given by

$$(K_n'(v_n)h_n)(x) = \int_{G_n} \frac{\partial k_n}{\partial z}(x,y,v_n(y))h_n(y)d\mu_n(y), \quad x \in G_n.$$

The following theorem shows that (12) will moreover ensure the equicontinuous equidifferentiability of the sequence K_n, $n \in I$.

<u>Theorem 10.3</u>. If μ_n, $n \in I$, is uniformly bounded and (12) holds, then K_n, $n \in I$, is equicontinuously equidifferentiable at $u_n^0 = \hat{u}|G_n$, $n \in I$, with the derivatives $K_n'(v_n)$, $v_n \in B_{\rho_1/2}(u_n^0)$, $n \in I$, completely continuous mappings of $C(G_n)$ into itself.

<u>Proof</u>: By assumption, there is a $M_0 > 0$ such that $|\mu_n(\phi)| \le M_0||\phi||_n$, $\phi \in C(G_n)$, $n \in I$. Using assumption (12) and the Mean Value Theorem, we see that the remainder terms

$$r_n(x,y,z;\zeta) \equiv k_n(x,y,z+\zeta) - k_n(x,y,z) - \frac{\partial k_n}{\partial z}(x,y,z)\zeta$$

of the kernels will satisfy the following uniform differentiability property (cf. also the proof of Lemma 3.5),

$$\forall \varepsilon > 0, \quad \exists \delta > 0 \quad \ni: |r_n(x,y,z;z'-z)| \le (\varepsilon/M_0)|z-z'|$$

for every (x,y,z), $(x,y,z') \in W_n$ with $|z-z'| \le \delta$ and for every $n \in I$.

For arbitrary $n \in I$, $v_n \in B_{\rho_1/2}(u_n^0)$, and $h_n \in C(G_n)$ such that $||h_n||_n \leq \min(\delta,\rho_1/2)$, we see that the quantity

$$R_n(v_n;h_n)(x) \equiv \int_{G_n} r_n(x,y,v_n(y);h_n(y))d\mu_n(y), \quad x \in G_n,$$

can be estimated by

$$|R_n(v_n;h_n)(x)| \leq (\varepsilon/M_0)||h_n||_n \; ||\mu_n|| \leq \varepsilon||h_n||_n, \quad x \in G_n.$$

This shows the equidifferentiability of the mappings K_n, $n \in I$, at each sequence of elements $v_n \in B_{\rho_1/2}(u_n^0)$, $n \in I$, with derivatives expressed as above. Since each $\partial k_n/\partial z$ is continuous, the Fréchet-derivatives will clearly be completely continuous mappings of $C(G_n)$ into itself for every $n \in I$. The equicontinuity of K_n', $n \in I$, at u_n^0, $n \in I$, easily follows from the hypothesized uniform equicontinuity property of the $\partial k_n/\partial z$, $n \in I$. □

We now turn to discuss the remaining assumptions in some detail and subsequently specify criteria ensuring their validity. Condition (13), in connection with the uniform boundedness of the measures μ_n, $n \in I$, will furnish the uniform boundedness - or equivalently the stability - of the sequence of derivatives $K_n'(u_n^0)$, $n \in I$. Namely, we have

$$|(K_n'(u_n^0)h_n)(x)| \leq C_1||h_n||_n, \quad x \in G_n, \quad h_n \in C(G_n), \quad n \in I. \tag{16}$$

In Theorem 10.6 below, we show that (13) and (14) imply the discrete compactness of $K_n'(u_n^0)$, $n \in I$. The last requirement (15) guarantees the consistency of the derivatives $K'(u)$, $K_n'(u_n^0)$, $n \in I$, at all $h \in C(G)$, with $h_n' = \hat{h}|G_n$, $n \in I$, constituting a consistency sequence. The result of Lemma 7.12 shows that consistency of the derivatives follows from the other conditions if we require in lieu of (9) the consistency of K, K_n, $n \in I$, at all functions in a neighborhood of u.

In the following lemmas, we provide other sufficient conditions which guarantee assumptions (8) to (15) - especially (13) and (14). The continuity of $\partial k/\partial z$ or the equicontinuity of $\partial k_n/\partial z$ is always taken to be with respect to all three arguments, unless stated otherwise in the text.

Lemma 10.4. Let μ_n, $n \in I$, be uniformly bounded, suppose that (12) holds and $\partial k_n/\partial z$ is equicontinuous in W_n, $n \in I$. Then (14) is satisfied.

Proof: The assertion is immediate since $\omega(u;\delta)$ can be estimated by

$$||\mu_n|| \sup\left\{\left|\frac{\partial k_n}{\partial z}(x,y,\hat{u}(y)) - \frac{\partial k_n}{\partial z}(x',y,\hat{u}(y))\right| : \ x,x',\ y \in G_n, \right.$$
$$\left. |x-x'| \le \delta, \ n \in I\right\}. \qquad \square$$

Lemma 10.5. Let $\partial\hat{k}/\partial z$ be continuous in W; let $\partial k_n/\partial z$ be continuous in W_n for each $n \in I$; and let μ_n, $n \in I$, be uniformly bounded. Then the following convergence condition

$$\sup_{x\in G_n} \int_{G_n} \left|\frac{\partial k_n}{\partial z}(x,y,\hat{u}(y)) - \frac{\partial \hat{k}}{\partial z}(x,y,\hat{u}(y))\right| d\mu_n(y) \to 0 \quad (n \in I) \tag{17}$$

implies (13) and (14). Moreover, the following condition

$$\sup_{x,y\in G_n} \left|\frac{\partial k_n}{\partial z}(x,y,\hat{u}(y)) - \frac{\partial \hat{k}}{\partial z}(x,y,\hat{u}(y))\right| \to 0 \quad (n \in I) \tag{18}$$

is sufficient for (17). If, additionally, $\hat{\mu}_n$, $n \in I$, converges weakly to $\hat{\mu}$, then condition (15) follows also from (17) or (18).

Proof: (i) The continuity of $\partial\hat{k}/\partial z$ in the compact set W implies the uniform continuity as well as the boundedness of $\partial\hat{k}/\partial z$ in W. From (17),

$$\int_{G_n} \left|\frac{\partial k_n}{\partial z}(x,y,\hat{u}(y))\right| d\mu_n(y)$$
$$\le \int_{G_n} \left|\frac{\partial k_n}{\partial z}(x,y,\hat{u}(y)) - \frac{\partial \hat{k}}{\partial z}(x,y,\hat{u}(y))\right| d\mu_n(y) + C_2M_0 \le 2C_2M_0$$

for all $x \in G_n$, $n \ge \nu_0$, where ν_0 is some element of I, C_2 a bound for $\partial\hat{k}/\partial z$, and $M_0 > 0$ a bound for $||\mu_n||$, $n \in I$. For those indices $n < \nu_0$, we can produce a bound C_3 because of the continuity of each $\partial k_n/\partial z$ and the compactness of W_n. Hence (13) is valid with $C_1 = \max(2C_2M_0, C_3)$.

(ii) In particular, the uniform continuity of $\partial\hat{k}/\partial z$ will yield, for arbitrary $\varepsilon > 0$, a $\delta > 0$ such that

$$\left|\frac{\partial \hat{k}}{\partial z}(x,y,u(y)) - \frac{\partial \hat{k}}{\partial z}(x',y,u(y))\right| \le \varepsilon/(3M_0)$$

for all $n \in I$, x,x', $y \in G_n$ with $|x-x'| \le \delta$. Now (17) provides an $\nu_0 \in I$ such that

$$\sup_{x\in G_n} \int_{G_n} \left|\frac{\partial k_n}{\partial z}(x,y,\hat{u}(y)) - \frac{\partial \hat{k}}{\partial z}(x,y,\hat{u}(y))\right| d\mu_n(y) \le \frac{\varepsilon}{3}, \quad n \ge \nu_0.$$

We finally get

$$\int_{G_n} \left|\frac{\partial k_n}{\partial z}(x,y,\hat{u}(y)) - \frac{\partial k_n}{\partial z}(x',y,\hat{u}(y))\right| d\mu_n(y) \le \varepsilon$$

for all $n \ge \nu_0$ and $x,x', y \in G_n$ with $|x-x'| \le \delta$. An analogous estimate is obtained for the case where $n < \nu_0$ since $\partial k_n/\partial z$ is continuous in $\hat{W}_n$. Thus $\omega(\hat{u};\delta) \to 0$ $(\delta \to 0)$.

(iii) The implication (18) ⇒ (17) is clear, since the μ_n, $n \in I$, are assumed to be uniformly bounded.

(iv) In order to show consistency of the sequence of derivatives, we observe, as in Theorem 10.2, that

$$\sup_{x\in G_n} |(\hat{\mu}_n - \hat{\mu})\tilde{k}'(x,\cdot)| \to 0 \quad (n \in I)$$

because of the continuity of $k'(x,y) \equiv \partial\hat{k}/\partial z(x,y,u(y))$. It remains to show that, for arbitrary $h \in C(G)$ with an extension $\hat{h} \in C(M)$,

$$\sup_{x\in G_n} \left|\int_{G_n} \left(\frac{\partial k_n}{\partial z}(x,y,\hat{u}(y)) - \frac{\partial \hat{k}}{\partial z}(x,y,\hat{u}(y))\right)\hat{h}(y)\, d\mu_n(y)\right| \to 0 \quad (n \in I).$$

For every continuation $\hat{h}$ of h, such a result follows directly from (17). □

In closing, we remark that (17) implies the norm convergence $||K_n'(u_n^0) - \hat{K}_n'(u_n^0)|| \to 0$ $(n \in I)$, where $\hat{K}_n'$, $n \in I$, denote the derivatives of the operators $\hat{K}_n$ defined in (7). In the linear case, this property has the consequence that $||K_n - \hat{K}_n|| \to 0$ $(n \in I)$, which clearly is a very strong requirement. We shall see in the following section that we can proceed with our analysis using only conditions (8) to (15). We note that (14) is a condition on the kernels which ensures the equicontinuity of $K_n'(u_n^0)$, $n \in I$.

10.3. BICONVERGENCE

Tho preliminary results and remarks in the preceding sections enable us to show the bistability of the sequence $A_n = I-K_n$, $n \in I$, and its biconvergence to $A = I-K$ by means of Theorem 7.18. For this task, we now prove the following result, which above all will guarantee the discrete compactness of the sequence of derivatives $K_n'(u_n^0)$, $n \in I$, and the regular convergence of the derivatives

$$A_n'(u_n^0) = I - K_n'(u_n^0), \quad n \in I, \quad A'(u) = I - K'(u),$$

and their inverses, where $u_n^0 = \hat{u}|G_n$, $n \in I$, with u an arbitrary but fixed element of $C(G)$.

Theorem 10.6. Let assumptions (8) to (15) be satisfied. Then $A'(u)$ is bijective and bicontinuous; the sequence $K_n'(u_n^0)$, $n \in I$, is discretely compact and discretely convergent to $K'(u)$; and there exists an index $\nu \in I$ such that the derivatives $A_n'(u_n^0)$ are bijective and bicontinuous for every $n \geq \nu$. Moreover, the sequences $||A_n'(u_n^0)||$, $n \in I$, and $||A_n'(u_n^0)^{-1}||$, $n \geq \nu$, are bounded, and the following regular convergences hold,

$$A_n'(u_n^0) \overset{r}{\to} A'(u^0), \quad A_n'(u_n^0)^{-1} \overset{r}{\to} A'(u^0)^{-1} \quad (n \in I).$$

Proof: We can easily conclude the bijectivity and bicontinuity of $A'(u)$ from the remarks to condition (11) (cf. Theorem 3.1). The integral operators $K_n'(u_n^0)$ are completely continuous for each $n \in I$ as seen by Theorem 10.3. In order to apply Theorem 7.16 - for $K'(u)$, $K_n'(u_n^0)$ in place of K, K_n - we must show both the discrete compactness of $(K_n'(u_n^0))$ and the consistency of $K'(u)$, $K_n'(u_n^0)$, $n \in I$. To prove the discrete compactness, we let $h_n \in C(G_n)$, $n \in I$, be a bounded sequence. Then conditions (13) and (14) provide the uniform boundedness and equicontinuity of $g_n \equiv K_n'(u_n^0)h_n$, $n \in I$. We have namely the estimates

$$|g_n(x)| \leq ||h_n||_n \int_{G_n} \left|\frac{\partial k_n}{\partial z}(x,y,\hat{u}(y))\right| d\mu_n(y),$$

and

$$|g_n(x)-g_n(x')| \leq ||h_n||_n \int_{G_n} \left|\frac{\partial k_n}{\partial z}(x,y,u(y)) - \frac{\partial k_n}{\partial z}(x',y,u(y))\right| d\mu_n(y),$$

for all $x, x' \in G_n$, $n \in I$. Theorem 7.3, which generalizes the Arzela-Ascoli Theorem, provides the discrete compactness of (g_n). The remaining assertions follow from statement 7.(14b) of Theorem 7.16. □

The following biconvergence theorem is a direct consequence of the preceding theorem and the Theorems 6.21 and 6.22.

Theorem 10.7. Let u be a solution of the integral equation (1). Then, under assumptions (8-15), there is an index $\nu \in I$ such that all three equivalent statements of the Biconvergence Theorem 6.21 are valid for n belonging to the subset $I_1 = \{n \geq \nu\}$ of I. Hence, there are positive numbers ρ, σ such that the integral equation (2) is uniquely solvable for all $w_n \in B_\sigma^0(w_n^0)$, $n \geq \nu$, where $w_n^0 = A_n u_n^0$; the solutions u_n lie in $B_\rho^0(u_n^0)$; and (A_n) is locally bijective and equibicontinuous at $(u_n^0) = (\hat{u}|G_n)$. Moreover, the following biconvergence relation holds

$$u_n \to u \Longleftrightarrow w_n \to w \ (n \in I),$$

along with the associated two-sided error estimates,

$$\frac{1}{\alpha}||A_n v_n - w_n^1 - d_n^A(u)||_n \leq ||v_n - u_n^0||_n \leq \frac{1}{\beta}||A_n v_n - w_n^1 - d_n^A(u)||_n,$$

for all $v_n \in B_\rho(u_n^0)$, $n \geq \nu$, where $w_n^1 = (I-\hat{K}_n)u_n^0$, $n \in I$. □

With $u_n = v_n$ in the error estimate, $d_n = A_n u_n - w_n^1 - d_n^A(u)$ can be split into the parts $d_n = d_n^0(u) + d_n^1(u) + d_n^2(u)$ where $d_n^0(u)$ and $d_n^1(u)$ are defined as in Section 10.1, and

$$d_n^2(u) \equiv w_n - w_n^1, \quad n \in I,$$

describes the error in approximating w by the sequence (w_n^1).

With our Convergence Theorem 10.7, we treat as a special case *linear integral equations of the second kind*. The differentiability assumptions in (10) and (12) become trivial, and conditions (10a) and (10b) reduce to 3.(4a) and 3.(4b). In addition to (8), we need the following conditions:

(19) There exist extensions $\hat{u} \in C(M)$ of u and $\hat{k}(.,.)\colon M \times M \to \mathbb{R}$ of $k(.,.)$ such that $||d_n^K(u)||_n \to 0$ $(n \in I)$ where

$$d_n^K(u) = \int_{G_n} k_n(x,y)\hat{u}(y)d\mu_n(y) - \int_G \hat{k}(x,y)u(y)dy, \quad x \in G_n, \ n \in I.$$

(20) The homogeneous integral equation

$$v(x) - \int_G k(x,y)v(y)dy = 0, \quad x \in G,$$

possesses only the trivial solution.

(21) $$\alpha \equiv \sup_{n\in I} \sup_{x\in G_n} \int_{G_n} |k_n(x,y)|d\mu_n(y) < \infty.$$

(22) $$\omega(\delta) \equiv \sup\left\{\int_{G_n} |k_n(x,y)-k_n(x',y)|d\mu_n(y)\colon x,x' \in G_n,\ |x-x'| \leq \delta,\ n \in I\right\} \to 0 \quad (\delta \to 0).$$

Sufficient conditions, analogous to those formulated in Theorem 10.2 and

Lemma 10.5, are rather apparent for equations with continuous kernels. As an immediate consequence of the preceding theorem we now have the following convergence theorem for the solutions of linear integral equations.

Theorem 10.8. Under the hypotheses expressed in 3.(4a), 3.(4b), (8), and (19) - (22), there is an index $\nu \in I$ such that the linear integral equations

$$u_n(x) - \int_{G_n} k_n(x,y)u_n(y)d\mu_n(y) = w_n(x), \quad x \in G_n,$$

are uniquely solvable for all $w_n \in C(G_n)$, $n \geq \nu$, with every $I-K_n$, $n \geq \nu$, bicontinuous. In addition, we have the biconvergence relation

$$u_n \to u \Longleftrightarrow w_n \to w \quad (n \in I)$$

along with the two-sided error estimates

$$\frac{1}{\alpha}||w_n - w_n^1 + d_n^k(u)||_n \leq ||u_n - \hat{u}|G_n||_n \leq \frac{1}{\beta}||w_n - w_n^1 + d_n^K(u)||_n, \quad n \geq \nu,$$

where $u \in C(G)$ is the solution of the linear integral equation

$$u(x) - \int_G k(x,y)u(y)dy = w(x), \quad x \in G. \qquad \square$$

REFERENCES (cf. also References in Chapter 3)

Anselone (1965)*, Anselone (1971), Anselone & Moore (1964)*, Anselone & Palmer (1968)*, Atkinson (1967)*, Atkinson (1976), Baker (1977), Brakhage (1960)*, Dieudonné (1969), Ikebe (1972)*, Kantorovich & Akilov (1964), Reinhardt (1975a)*, Sloan (1980a,1981)*, Stummel (1974b,1975)*, Atkinson & Graham & Sloan (1983)*.

*Article(s)

Part IV
Inverse Stability, Consistency and Convergence for Initial Value Problems in Partial Differential Equations

In this part, we first establish in a general framework a convergence theory for numerical methods approximating initial value problems. This analysis is made possible by incorporating the given problem and its approximations into the setting of Chapter 6. First of all, however, we have to specify the discrete approximations and discrete convergences underlying the specific problem area considered. They have already been provided, together with the verification of the corresponding properties, by the treatment of corresponding examples in Section 5.3 of Part II. We shall often refer to Chapter 4, in which a series of examples of initial value problems and appropriate numerical methods have been introduced and, moreover, represented in suitable operator notation.

The underlying concepts are again inverse stability and consistency which ensure discrete convergence in a sense appropriate for the present special framework. We shall in Chapter 11 develop the corresponding convergence theory. According to our investigations in Part II, a basic requirement is the equicontinuous equidifferentiability of the approximating mappings. This will be verified at the end of Chapter 11 for several classes of examples. It is worth noticing that here - as well as later in Chapter 12 - the choice of norms in the underlying spaces is of paramount importance.

All of Chapter 12 is dedicated to a study of inverse stability. We establish and apply several special criteria for inverse stability. Positivity properties, for example, guarantee inverse stability with respect to the supremum norm (in spaces of grid functions), Fourier methods provide a tool for proving inverse stability with respect to discrete L^2-

norms. With respect to such norms, the von Neumann stability criterion, basic for the classical Lax-Richtmyer theory, will at times follow from our general concept of inverse stability in a very special situation. The above mentioned special criteria, along with appropriate techniques for verifying them, again rely strongly on the choice of underlying norms.

Applying the general convergence results from Chapter 11, along with the stability criteria of Chapter 12, we are finally able to carry out in Chapter 13 a convergence analysis of special methods by investigating the behavior of the associated truncation errors.

Chapter 11
Inverse Stability and Convergence for General Discrete-Time Approximations of Linear and Nonlinear Initial Value Problems

In this chapter, we develop a convergence theory for discrete-time approximations to both linear and nonlinear initial value problems. We shall assume that such problems are pure initial value problems, and allow that the mappings occurring in both the exact formulation of the problem, and in the associated approximations, depend on time. Our convergence theory established for the problems in this chapter will essentially consist of a rather concrete description and characterization of the concepts of inverse stability, consistency, and discrete convergence. These concepts were discussed at length in the development of our general convergence theory in Part II.

The central results of this chapter rely crucially on a characterization of inverse stability in terms of an inequality for the Fréchet-derivatives of the operators occurring in each time step of the approximate problems (cf. Theorem 11.4). Results on consistency and convergence are then easily obtained by interpreting the appropriate results in Chapter 6 in the context of the setting of this chapter.

By appropriately exploiting the special form of the mappings occurring in our problems, we further develop a concept of convergence appropriate for the associated semihomogeneous approximations. Such a concept turns out to be necessary and sufficient for the convergence of approximate solutions to totally inhomogeneous problems.

As a first step in applying the convergence results from Chapter 6, we must ensure the equidifferentiability of the appropriate mappings occurring in the approximating equations. For this step, we give in Section 11.1 suitable conditions and show that these are satisfied under appropriate assumptions for three classes of examples. In our discussion, it becomes rather clear that we must make a judicious choice of norms for the underlying Banach spaces, which are not yet specified in the general theorems of this chapter. Indeed, it is remarkable that the equidifferentiability property in Examples 1 and 2 of Section 11.1 can be shown only for norms incorporating a negative power of the time step widths.

In Section 11.2, we prove the afore-mentioned and central Theorem 11.4 for characterizing inverse stability and apply this theorem to the special case of linear problems. In our discussion, we see that inverse stability for linear problems is preserved under suitably small perturbations of the associated mappings. This result is well known for the case of time independent mappings occurring in the approximating equations and, is also valid for the time dependent case. Moreover, this result is also applicable to nonlinear methods, for the reason (cf. Chapter 6) that inverse stability is essentially ensured for nonlinear mappings if it is present for the associated (linear) Fréchet-derivatives.

We apply the convergence theorems from Section 6.4 in the concluding Section 11.3, but beforehand we give a less abstract characterization of the concept of consistency for the present problems. The "discrete convergence of semihomogeneous methods uniformly with respect to all initial times" will be seen to be the appropriate conceptual basis which will be necessary and sufficient for the convergence of solutions to the associated, totally inhomogeneous problems. Convergence concepts for semihomogeneous problems form the central core in the treatise by Ansorge (1978), of which that of von Dein (1976) essentially agrees with ours. The results of our Theorem 11.10 therefore show that such a chosen convergence concept for semihomogeneous problems represents nothing other than the inverse discrete convergence in the sense of Chapter 6.

The material in this chapter is based on the works (1975a, 1975b, 1977) of the author, which have been extended in several aspects in the text and, in particular, enlarged by a thorough treatment of the examples also in the following chapters.

11.1 STATEMENT OF THE PROBLEM AND DIFFERENTIABILITY

In this section, we study a general form of discrete-time approximations for nonlinear (pure) initial value problems, in which all the examples of finite-difference and (discrete-time) Galerkin methods considered in Chapter 4 can be expressed. A general representation of discrete-time approximations has already been presented in Section 4.5 and is thus repeated here only briefly. In order to be able to apply our convergence theory from Part II, we must first make sure that the associated approximating mappings are equicontinuously equidifferentiable. Theorem 11.1 ensures such a property via conditions on the mappings $C_n^{(\ell)}(t)$ occurring in the approximating equations in each time step. At the end of this section, we shall give the appropriate Fréchet-derivatives and check the equidifferentiability for three classes of examples. In our discussion of the examples, we shall see that such a property depends essentially on the choice of norms, which even can contain a negative power of the time step width, as in the case of the first example.

Suppose we have a Banach space F equipped with norm $|\cdot|_F$ and a subspace E of F with norm $|\cdot|_E$ which is continuously embedded in F (i.e., $|u|_F \le c|u|_E$ for all $u \in E$, and some $c > 0$). For every $t \in [0,T]$ $(T > 0)$, we let $A(t)$ be an operator (not necessarily linear) from E into F, and let $u_0 \in E$, $w(t) \in F$, $t \in [0,T]$, specify the problem data. By a pure initial value problem (abbreviation: IVP = initial value problem) we mean the following,

$$u(0) = u_0, \quad \frac{du}{dt}(t) = A(t)u(t) + w(t), \quad t \in [0,T], \tag{1}$$

where a solution $u \in C([0,T],E) \cap C^1([0,T],F)$ is sought. For nonlinear problems, we can incorporate $w(t)$ in $A(t)$, i.e., we can assume $w = 0$ without loss of generality. We shall assume, for the sake of our convergence analysis, that a solution u^0 will exist to the nonlinear, semi-homogeneous IVP,

$$u^0(0) = u_0, \quad \frac{du^0}{dt}(t) = A(t)u^0(t), \quad t \in [0,T]. \tag{1$_0$}$$

With the spaces $\hat{X}$, X, and Y, given respectively by

$$\hat{X} \equiv C([0,T],E), \quad X \equiv \hat{X} \cap C^1([0,T],F), \text{ and } \quad Y \equiv E \times C([0,T],F),$$

a mapping T from X into Y can be defined via

$$Tv = (v(0), \frac{dv}{dt} - A(\cdot)v),$$

where the domain of definition $D(T)$ of T is determined by that of $A(t)$, $t \in [0,T]$. The IVP $(1)_0$ is then equivalent to $Tu^0 = (u_0,0)$.

We now describe a general discrete-time approximation of $(1)_0$ in the following manner: Suppose we are given Banach spaces E_n,F_n, $n \in I$ $(= \mathbb{N})$, and a null sequence of positive step sizes (in the time direction) τ_n, $n \in I$, by which the following meshes can be defined,

$$[0,T]_n \equiv \{t \in [0,T]: t = t_k \equiv k\tau_n, \quad k = 0,\ldots,N\},$$

$$[0,T]_n' \equiv [0,T]_n - \{0\},$$

where $N\tau_n = T$. Further, suppose we have mappings $C_n^{(\ell)}(t): D_n^{(\ell)} \subset E_n \to F_n$, $t \in [0,T]_n'$, $\ell = 0,1$, $n \in I$, whose domains of definition do not depend on t, for the sake of simplicity. With elements $u_{0,n} \in E_n$, $w_n(t) \in F_n$, $t \in [0,T]_n'$, $n \in I$, we seek as approximations to the solution of (1) (or more precisely: of $(1)_0$) solutions $u_n(t) \in E_n$, $t \in [0,T]_n$, of

$$u_n(0) = u_{n,0}, \quad C_n^{(0)}(t)u_n(t) = C_n^{(1)}(t)u_n(t-\tau_n) + \tau_n w_n(t), \qquad t \in [0,T]_n', \quad n \in I. \tag{2}$$

We define $X_n \equiv C([0,T]_n,E_n)$, $Y_n \equiv E_n \times C([0,T]_n',F_n)$ and

$$\begin{aligned} T_n v_n &\equiv (v_n(0), T_n^1 v_n), \\ (T_n^1 v_n)(t) &\equiv \frac{1}{\tau_n}\{C_n^{(0)}(t)v_n(t) - C_n^{(1)}(t)v_n(t')\}, \quad t \in [0,T]_n', \ n \in I, \end{aligned} \tag{3}$$

where $t' \equiv t - \tau_n$ denotes the predecessor of $t \in [0,T]_n'$. The domain of definition $D(T_n)$ of T_n includes all elements $v_n \in X_n$ for which $v_n(t) \in D_n^{(0)} \cap D_n^{(1)}$, $t \in [0,T]_n$. With this notation, we see that (2) is equivalent to

$$u_n(0) = u_{n,0}, \quad (T_n^1 u_n)(t) = w_n(t), \text{ or } T_n u_n = (u_{n,0},w_n).$$

We define norms in $\hat{X}$ and Y, respectively, by

$$||v||_\infty \equiv \max_{0 \le t \le T} |v(t)|_E, \quad v \in \hat{X}, \tag{4a}$$

$$\left\{ \begin{aligned} ||y||_p &\equiv |y^0|_E + \left(\int_0^T |y^1(t)|_F^p dt\right)^{1/p}, \quad 1 \le p < \infty, \\ ||y||_\infty &\equiv |y^0|_E + \max_{0 \le t \le T} |y^1(t)|_F, \quad y = (y^0,y^1) \in Y. \end{aligned} \right. \tag{4b}$$

The space X is equipped with the norm induced by $\hat{X}$. The norms in E_n and F_n can depend on n and are expressed by $|\cdot|_{E_n}$ and $|\cdot|_{F_n}$, $n \in I$, respectively. We are then able to define norms in X_n and Y_n by

$$||v_n||_{\infty,n} \equiv \max_{t\in[0,T]_n} |v_n(t)|_{E_n}, \quad v_n \in X_n, \tag{5a}$$

$$\begin{cases} ||y_n||_{p,n} \equiv |y_n^0|_{E_n} + \left(\tau_n \sum_{t\in[0,T]_n'} |y_n^1(t)|_{F_n}^p\right)^{1/p}, & 1 \le p < \infty, \\ ||y_n||_{\infty,n} \equiv |y_n^0|_{E_n} + \max_{t\in[0,T]_n'} |y_n^1(t)|_{F_n}, & y_n = (y_n^0,y_n^1) \in Y_n,\ n \in I. \end{cases} \tag{5b}$$

Because $\Sigma_{t\in[0,T]_n'} \tau_n = T$, we see that the following inequalities are valid for the p-norms in Y_n,

$$||y_n||_{p,n} \le \max(1,T^{1/p-1/p'})||y_n||_{p',n}, \quad y_n \in Y_n, \quad 1 \le p \le p' \le \infty. \tag{6}$$

We assume that a discrete approximation $A(E,\pi_nE_n,\lim^E)$ exists by virtue of a stable sequence of linear restriction operators $R_n^E: E \to E_n$, $n \in I$. Then, from Theorem 5.12, a sequence of restrictions $R_n^X: \hat{X} \to X_n$, $n \in I$, is given by

$$(R_n^X v)(t) = R_n^E v(t), \quad v \in \hat{X}, \quad t \in [0,T]_n, \quad n \in I,$$

and $A(\hat{X},\pi_nX_n,\lim^X)$ is a discrete approximation where the discrete convergence is given in the usual manner by

$$\lim{}^X v_n = v \Longleftrightarrow ||v_n - R_n^X v||_{\infty,n} \to 0 \ (n \in I).$$

If the norms $|\cdot|_E$, $|\cdot|_{E_n}$, $n \in I$, satisfy the Convergence Property 5.3 (i.e., cf. also 5.(13), $|R_n^E g|_{E_n} \to |g|_E$, $g \in E$) - which we in the following shall also assume - then this property also holds for the norms $||\cdot||_\infty$, $||\cdot||_{\infty,n}$, $n \in I$ (which can be seen by the techniques used in the proof of Theorem 5.12).

With a stable sequence of linear restriction operators $R_n^F: F \to F_n$, $n \in I$ - which also satisfies the Convergence Property 5.(13) for the norms - we define

$$R_n^Y y \equiv (R_n^E y^0, y_n^1), \quad y_n^1(t) \equiv R_n^F y^1(t), \quad t \in [0,T]_n', \quad y = (y^0,y^1) \in Y, \quad n \in I.$$

It is not difficult to see that a sequence of restriction operators is given by R_n^Y, $n \in I$, and that $A(Y,\pi_nY_n,\lim^Y)$ represents a discrete approximation with respect to all p-norms; the associated discrete convergence is defined for every p by

$$\lim^Y y_n = y \Longleftrightarrow \|y_n - R_n^Y y\|_{p,n} \to 0 \quad (n \in I)$$

and also satisfies Property 5.3 of the discrete convergence of the norms.

For the examples considered in Chapter 4, we have that $E = F$, or, more generally, that E is dense and continuously embedded in F. Further, the approximating spaces E_n and F_n coincide as vector spaces, but possibly are equipped with different norms. If we do assume the same norm in E_n and F_n, and that E is dense in F, then a discrete approximation $A(F,\pi_n E_n,\lim^F)$ exists provided $A(E,\pi_n E_n,\lim^E)$ exists, where $\lim^F$ can be characterized as in 5.(17). For this special case, requiring the existence of a second sequence of restrictions (R_n^F) is not at all necessary, for its existence is ensured by the existence of (R_n^E).

If, for some $n \in I$ and all $t \in [0,T]_n'$, the mappings $C_n^{(0)}(t)$ and $C_n^{(1)}(t)$ are differentiable at $u_n(t)$ and $u_n(t')$, respectively, then T_n is also differentiable at u_n and has its Fréchet-derivative given by

$$T_n'(u_n)x_n = (x_n(0),(T_n^1)'(u_n)x_n)$$
$$[(T_n^1)'(u_n)x_n](t) = \frac{1}{\tau_n}\{C_n^{(0)}(t)'(u_n(t))x_n(t) \qquad (7)$$
$$- C_n^{(1)}(t)'(u_n(t'))x_n(t')\}, \quad t \in [0,T]_n', \ x_n \in X_n.$$

The analysis in Chapter 6 requires as an essential assumption that equicontinuous equidifferentiability of the sequence T_n, $n \in I$, be present. The following theorem gives conditions on $C_n^{(\ell)}(t)$ for this assumption to be satisfied for the mappings defined in (3).

Theorem 11.1. Let $u_n^0 \in X_n$, $n \in I$, and let $\rho > 0$ be such that for every $\ell = 0,1$, $n \in I$ and every $(t,g_n) \in U_n^{(\ell)}$, where

$$U_n^{(\ell)} \equiv \{(t,g_n) \in [0,T]_n' \times E_n: |g_n - u_n^0(t-\ell\tau_n)|_{E_n} \le \rho\},$$

the derivative $C_n^{(\ell)}(t)'(g_n)$ exists and has the following properties:

(8a) $\forall \varepsilon > 0, \ \exists \delta > 0 \quad \forall n \in I, \ (t,g_n) \in U_n^{(\ell)}, \quad e_n \in E_n: |e_n|_{E_n} \le \delta$

$$\Rightarrow g_n+e_n \in D_n^{(\ell)}, \ \frac{1}{\tau_n}|C_n^{(\ell)}(t)(g_n+e_n) - C_n^{(\ell)}(t)g_n$$
$$- C_n^{(\ell)}(t)'(g_n)e_n|_{F_n} \le \varepsilon|e_n|_{E_n};$$

(8b) $\forall\varepsilon > 0,\ \exists\eta > 0,\ \forall n \in I,\ e_n \in E_n,\ (t,g_n) \in U_n^{(\ell)}:\ |u_n^0(t-\ell\tau_n)-g_n|_{E_n} \leq \eta$

$$\Rightarrow \frac{1}{\tau_n}|(C_n^{(\ell)}(t)'(u_n^0(t-\ell\tau_n)) - C_n^{(\ell)}(t)'(g_n))e_n|_{F_n} \leq \varepsilon|e_n|_{E_n}.$$

Then T_n, $n \in I$, is equicontinuously equidifferentiable at (u_n^0).

Proof: For arbitrary $\varepsilon > 0$, let $\delta > 0$ be the number determined by (8a). Let $v_n \in B_\rho(u_n^0)$ and $x_n \in X_n$ be such that $||x_n||_{\infty,n} \leq \delta$, $n \in I$. Then $(t,v_n(t)) \in U_n^{(0)}$, $t \in [0,T]_n'$, $n \in I$; and, with $g_n = v_n(t)$, $e_n = x_n(t)$, we get from (8a) that $(v_n + x_n)(t) \in D_n^{(0)}$ and that

$$\frac{1}{\tau_n}|C_n^{(0)}(t)(v_n+x_n)(t) - C_n^{(0)}(t)v_n(t) - C_n^{(0)}(t)'(v_n(t))x_n(t)|_{F_n} \leq \varepsilon|x_n(t)|_{E_n},\quad t \in [0,T]_n'.$$

Correspondingly, we have $(v_n+x_n)(t') \in D_n^{(1)}$ and

$$\frac{1}{\tau_n}|C_n^{(1)}(t)(v_n+x_n)(t') - C_n^{(1)}(t)v_n(t') - C_n^{(1)}(t)'(v_n(t'))x_n(t')|_{F_n} \leq \varepsilon|x_n(t')|_{E_n},\quad t \in [0,T]_n'.$$

For the norm $||\cdot||_{\infty,n}$ in Y_n, we therefore get

$$\begin{aligned}||T_n(v_n+x_n) - T_nv_n - T_n'(v_n)x_n||_{\infty,n} &= \frac{1}{\tau_n}\max_{t\in[0,T]_n'}|C_n^{(0)}(t)(v_n+x_n)(t) - C_n^{(0)}(t)v_n(t)\\ &\qquad - C_n^{(0)}(t)'(v_n(t))x_n(t)\\ &\quad - [C_n^{(1)}(t)(v_n+x_n)(t') - C_n^{(1)}(t)v_n(t') - C_n^{(1)}(t)'(v_n(t'))x_n(t')]|_{F_n}\\ &\leq \varepsilon\Big\{\max_{t\in[0,T]_n'}|x_n(t)|_{E_n} + \max_{t\in[0,T]_n'}|x_n(t')|_{E_n}\Big\} \leq 2\varepsilon||x_n||_{\infty,n}.\end{aligned}$$

Because of the relations in (6), we have a corresponding estimate for all p-norms in Y_n, which shows the equidifferentiability of (T_n) at each sequence (v_n) of elements $v_n \in B_\rho(u_n^0)$, $n \in I$. The equicontinuity of the derivatives at (u_n^0) follows from (8b) in an analogous manner. □

We now check the differentiability conditions (8a,b) for the relevant approximations used for three classes of initial value problems.

Example 1. We consider the following *semilinear parabolic initial value problem* (cf. also 4.(47a,b) in Section 4.4),

$$
\begin{aligned}
&u_t = a(x,t)u_{xx} + F(x,t,u,u_x) \quad \text{in} \quad [0,1] \times [0,T], \\
&u(x,0) = u_0(x) \quad \text{in} \quad [0,1], \quad u(0,t) = u(1,t) = 0, \quad t \in [0,T].
\end{aligned} \tag{9}
$$

As a compatibility condition, we require $u_0(0) = u_0(1) = 0$. We now assume that a solution u^0 of (9) exists and that the following conditions are satisfied:

(10a) a is continuous with respect to both arguments and there exist $\alpha_0, \alpha_1 > 0$ such that

$$0 < \alpha_0 \le a(x,t) \le \alpha_1, \quad x \in [0,1], \quad t \in [0,T].$$

(10b) The first and second partial derivatives of $F(x,t,y,z)$ with respect to y and z exist in

$$U \equiv \{(x,t,y,z): |y - u^0(x,t)| \le \rho_0, \ |z - u_x^0(x,t)| \le \rho_0\},$$

for some $\rho_0 > 0$, and are moreover continuous.

We first investigate the *explicit finite-difference method*, which is obtained by central differencing in the x-variable on the meshes

$$[0,1]_n \equiv \{x_j \equiv jh_n, \ j = 0,\dots,J\}, \quad [0,1]_n' \equiv [0,1]_n - (\{0\} \cup \{1\})$$

in $[0,1]$, $Jh_n = 1$; the right-hand side of the above differential equations is approximated at t_k. Thus the following scheme results,

$$
\begin{aligned}
v_j^{k+1} &= (1-2ra_j^k)v_j^k + ra_j^k(v_{j-1}^k+v_{j+1}^k) \\
&\quad + \tau_n F(x_j,t_k,v_j^k, \tfrac{1}{2h_n}(v_{j+1}^k-v_{j-1}^k)), \quad j = 1,\dots,J-1, \ k = 0,1,\dots,N-1, \\
v_j^0 &= u_0(x_j), \ 0 \le j \le J, \quad v_0^{k+1} = v_J^{k+1} = 0, \quad k = 0,1,\dots,N-1.
\end{aligned} \tag{11}
$$

Here, $r = \tau_n/h_n^2$ denotes the associated *mesh ratio*, which is assumed constant for all $n \in I$. For brevity, we set $a_j^k = a(x_j,t_k)$ and $v_j^k = u_n(x_j,t_k)$ for $u_n \in C([0,1]_n)$.

We obtain a corresponding *implicit finite-difference* method when we replace t_k by t_{k+1} in approximating the right-hand side of the differential equation in (9),

$$
\begin{aligned}
&(1 + 2ra_j^{k+1})v_j^{k+1} - ra_j^{k+1}(v_{j-1}^{k+1} + v_{j+1}^{k+1}) \\
&\quad - \tau_n F(x_j,t_{k+1},v_j^{k+1}, \tfrac{1}{2h_n}(v_{j+1}^{k+1}-v_{j-1}^{k+1})) = v_j^k, \\
&\qquad\qquad j = 1,\dots,J-1, \quad k = 0,1,\dots,N-1,
\end{aligned} \tag{12}
$$

$$v_j^0 = u_0(x_j), \quad 0 \le j \le J, \quad v_0^{k+1} = v_J^{k+1} = 0, \quad k = 0,1,\ldots,N-1. \tag{12}$$

For the methods (11) and (12), the operators $C_n^{(\ell)}(t)$, mapping $E_n = \{g_n \in C([0,1]_n) : g_n(x_0) = g_n(x_J) = 0\}$ into itself in the general representation (2), have the following forms, respectively,

$$C_n^{(0)}(t) = I,$$

$$(C_n^{(1)}(t)g_n)(x_j) = (1-2ra(x_j,t'))g_j + ra(x_j,t')(g_{j-1}+g_{j+1}) \tag{13}$$

$$+ \tau_n F(x_j,t',g_j,\frac{1}{2h_n}(g_{j+1}-g_{j-1})), \quad j = 1,\ldots,J-1, \quad t \in [0,T]_n',$$

and

$$(C_n^{(0)}(t)g_n)(x_j) = (1+2ra(x_j,t))g_j - ra(x_j,t)(g_{j-1}+g_{j+1})$$

$$- \tau_n F(x_j,t,g_j,\frac{1}{2h_n}(g_{j+1}-g_{j-1})), \quad j = 1,\ldots,J-1, \quad g_n \in E_n, \tag{14}$$

$$C_n^{(1)}(t) = I, \quad t \in [0,T]_n',$$

where $g_n(x_j)$ is set equal to g_j for brevity. At the boundary points, both $C_n^{(1)}(t)g_n$ in (13) and $C_n^{(0)}(t)g_n$ in (14) are set equal to g_n which are zero there.

Because of the occurrence in the last argument of F of central first-order difference quotients (also denoted by $D_h v$) in the x-direction, the following result shows the differentiability conditions (8a,b) only in a weak form, i.e., with respect to the norms

$$|g_n|_{E_n} \equiv \frac{1}{\tau_n} \max_{0 \le j \le J} |g_n(x_j)|, \quad g_n \in E_n, \quad n \in I. \tag{15}$$

<u>Lemma 11.2</u>. Under the assumption (10b), the finite-difference operators $C_n^{(\ell)}(t)$, $\ell = 0,1$, defined in (13) and (14), satisfy the differentiability conditions (8a,b) at $u_n^0 = u^0|[0,1]_n \times [0,T]_n$, $n \in I$, with respect to the norms defined in (15) where I is possibly replaced by $I_1 = \{n \in \nu_1\}$, for some $\nu_1 \in I$.

<u>Proof</u>: (i) (for the explicit method (11)). In the proof, we shall also write u^0 to denote the restriction of u^0 to the mesh, and we shall let $|\cdot|_{0,\infty}$ denote the maximum norm in both E_n and $C([0,1]_n')$. Clearly, it suffices to check (8a,b) for $C_n^{(1)}(t)$. First, we know that

$$|D_h u^0(t) - u_x^0(t)|_{0,\infty} \le \rho_0/2, \quad t \in [0,T]_n, \quad n \ge \nu_0,$$

for some $\nu_0 \in I$. We define

$$W_n \equiv \{(t,g_n) \in [0,T]_n' \times E_n : |g_n(x) - u^0(x,t')| \le \rho_0,$$
$$|D_h g_n(x) - u_x^0(x,t')| \le \rho_0, \quad x \in [0,1]_n'\}, \quad n \in I,$$

and claim that, with $|\cdot|_{E_n}$ from (15) and for some $\nu_1 \ge \nu_0$, we have

$$U_n^{(1)} \equiv \{(t,g_n): |g_n - u_n^0(t')|_{E_n} \le \rho_0/2\} \subset W_n, \quad n \ge \nu_1.$$

To see this, we observe that $|D_h g_n|_{0,\infty} \le (1/h_n)|g_n|_{0,\infty}$ and that

$$|D_h g_n - u_x^0(t)|_{0,\infty} \le |D_h(g_n - u^0)|_{0,\infty} + |D_h u^0 - u_x^0|_{0,\infty}$$
$$\le \frac{1}{h_n}|g_n - u^0|_{0,\infty} + \frac{\rho_0}{2} \le \frac{\rho_0}{2}\left(1 + \frac{\tau_n}{h_n}\right) = \frac{\rho_0}{2}(1 + rh_n), \quad n \ge \nu_0,$$

for $(t,g_n) \in U_n^{(1)}$. For some $\nu_1 \ge \nu_0$, we have $rh_n \le 1$, $n \ge \nu_1$, so that the inclusion $U_n^{(1)} \subset W_n$, $n \ge \nu_1$, is shown. We now select ν_1 so large that $h_n \le 1$ for all $n \ge \nu_1$. Then $\tau_n = rh_n^2 \le 1$, $n \ge \nu_1$.

For $(t,g_n) \in U_n^{(1)}$, $n \ge \nu_1$, it is clear that $(x,t',g_n(x),D_h g_n(x)) \in U$, $x \in [0,1]_n'$ (with U defined in (10b)). Hence, the Fréchet-derivative of $C_n^{(1)}(t)$ exists at g_n and has the representation

$$(C_n^{(1)}(t)'(g_n)\phi_n)(x_j) = (1-2ra(x_j,t'))\phi_j + ra(x_j,t')(\phi_{j-1}+\phi_{j+1})$$
$$+ \tau_n(\phi_j F_y + D_h\phi_n(x_j)F_z), \quad 1 \le j \le J-1, \tag{16}$$

where F_y and F_z, respectively, denote $\partial F/\partial y$ and $\partial F/\partial z$ with argument $(x_j,t',g_n(x_j), D_h g_n(x_j))$ where $\phi_j = \phi_n(x_j)$, $\phi_n \in E_n$. The remainder term in the differentiation of $C_n^{(1)}(t)$ can be represented via the Mean Value Theorem as

$$\frac{1}{\tau_n} R_n^{(1)}(t,g_n;\phi_n)(x_j) = \phi_n(x_j)(\tilde{F}_y - F_y) + D_h\phi_n(x_j)(\tilde{F}_z - F_z),$$

where

$$\tilde{F}_y = F_y(x_j,t',g_n(x_j) + \xi, D_h g_n(x_j)), \text{ with some } |\xi| \le |\phi_n(x_j)|,$$
$$\tilde{F}_z = F_z(x_j,t',(g_n+\phi_n)(x_j), D_h g_n(x_j) + \zeta), \text{ with some } |\zeta| \le |D_h\phi_n(x_j)|.$$

Applying the Mean Value Theorem once again, we see that

$$\tilde{F}_z - F_z = \phi_n(x_j)\tilde{F}_{yz} + \zeta\tilde{F}_{zz}$$

with the second partial derivatives evaluated at certain intermediate points. By assumption, the partial derivatives of F are continuous in U and hence are uniformly continuous and bounded. Let $C_{yz} \geq 0$ and $C_{zz} \geq 0$ be bounds for F_{yz} and F_{zz}, respectively, in the set U. For arbitrary $\varepsilon > 0$, let δ be the number determined by the property of uniform continuity of F_y in U; without loss of generality, let $\delta \leq \varepsilon/r$. Then

$$|\tilde{F}_y - F_y| \leq \varepsilon \quad \text{for every} \quad 1 \leq j \leq J-1, \quad t \in [0,T]_n', \quad n \in I,$$

in case $|\phi_n|_{0,\infty} \leq \delta$. Now let $n \geq \nu_1$ and $\phi_n \in E_n$ with $|\phi_n|_{E_n} \leq \delta$ be arbitrary. Then, $|\phi_n|_{0,\infty} \leq \delta\tau_n \leq \delta$, $|D_h\phi_n|_{0,\infty} \leq |\phi_n|_{0,\infty}/h_n \leq \delta r h_n$, and, for the above remainder terms, we have the estimates

$$\begin{aligned}\frac{1}{\tau_n}|R_n^{(1)}(t,g_n;\phi_n)(x_j)| &\leq \varepsilon|\phi_n(x_j)| + |D_h\phi_n(x_j)|\{C_{yz}|\phi_n(x_j)| \\ &\qquad + C_{zz}|D_h\phi_n(x_j)|\} \\ &\leq \varepsilon|\phi_n|_{0,\infty} + C_{yz}\delta r h_n|\phi_n|_{0,\infty} + C_{zz}\frac{1}{h_n}|\phi_n|_{0,\infty}\delta r h_n \\ &\leq \varepsilon|\phi_n|_{0,\infty}(1 + h_n C_{yz} + C_{zz}), \quad j = 1,\ldots,J-1.\end{aligned}$$

We multiply both sides by τ_n^{-1} and obtain (8a) for $C_n^{(1)}(t)$, with $I_1 = \{n \geq \nu_1\}$ in place of I and with the norms $|\cdot|_{E_n}$ defined in (15).

The second condition (8b), which ensures the equicontinuity of the derivatives, will follow in an analogous manner from the following estimates,

$$\begin{aligned}&\frac{1}{\tau_n}|([C_n^{(1)}(t)'(g_n) - C_n^{(1)}(t)'(f_n)]\phi_n)(x_j)| \\ &\quad = |(F_y(x_j,t',g_n(x_j),D_hg_n(x_j)) - F_y(x_j,t',f_n(x_j),D_hf_n(x_j)))\phi_n(x_j) \\ &\quad + (F_z(x_j,t',g_n(x_j),D_hg_n(x_j)) - F_z(x_j,t',f_n(x_j),D_hf_n(x_j)))D_h\phi_n(x_j)| \\ &\quad \leq \varepsilon|\phi_n(x_j)| + |D_h\phi_n(x_j)|\{C_{yz}|g_n-f_n|_{0,\infty} + C_{zz}|D_h(g_n-f_n)|_{0,\infty}\} \\ &\quad \leq \varepsilon|\phi_n|_{0,\infty}(1 + h_nC_{yz} + C_{zz}), \quad 1 \leq j \leq J-1,\end{aligned}$$

for all (t,g_n), $(t,f_n) \in U_n^{(1)}$, $n \geq \nu_1$, with $|g_n-f_n|_{E_n} \leq \delta$.

(ii) (for the implicit method (12)). The proof of conditions (8a,b) for $C_n^{(0)}(t)$ proceeds analogously to that of part (i) and is left to the reader. We give here only the Fréchet-derivatives,

$$(C_n^{(0)}(t)'(g_n)\phi_n)(x_j) = (1 + 2ra(x_j,t))\phi_j - ra(x_j,t)(\phi_{j-1}+\phi_{j+1}) \\ - \tau_n(\phi_j F_y + D_h\phi_n(x_j)F_z), \quad 1 \le j \le J-1, \tag{17}$$

where

$$(t,g_n) \in U_n^{(0)} \equiv \{(t,g_n): |g_n - u_n^0(t)|_{E_n} \le \rho_0/2\}, \quad n \ge \nu_1,$$

and the arguments of F_y and F_z are precisely $(x_j,t,g_n(x_j),D_h g_n(x_j))$.
□

At this point, we remark that the requirement $n \ge \nu_1$ needed for ensuring the validity of conditions (8a,b) stems from restrictions on h_n, namely that $h_n \le \min(1,1/r)$ and that the central difference quotients approximate u_x^0 sufficiently well. If we have global existence and boundedness of the partial derivatives of F, then such a restriction is not needed.

We further remark that the differentiability conditions (8a,b) also hold with respect to other norms which include another negative power of τ_n or even no such factor. In these cases, however, we have to consider the $C_n^{(\ell)}(t)$ as mappings from E_n into F_n $(= E_n)$ which are equipped with different norms, e.g.,

$$|g_n|_{E_n} = \tau_n^{-1/2}|g_n|_{0,\infty}, \quad |f_n|_{F_n} = |f_n|_{0,\infty}$$

or

$$|g_n|_{E_n} = |g_n|_{0,\infty} + |D_h g_n|_{0,\infty}, \quad |f_n|_{F_n} = |f_n|_{0,\infty}.$$

But in view of our stability analysis in Section 12.4 below, we choose the norms in E_n and $F_n = E_n$ equally (as in (15)). Otherwise an inverse stability property with respect to the differently chosen norms has to be provided which is not immediately available from the maximum principles used in Section 12.4. □

Example 2. Another type of nonlinearity than present in Example 1 occurs in the following *quasilinear parabolic initial value problem* (cf. 4.(52) in Sec. 4.5):

$$u_t = (a(x,u)u_x)_x + F(x,t,u) \quad \text{in} \quad [0,1] \times [0,T], \\ u(x,0) = u_0(x) \quad \text{in} \quad [0,1], \quad u(0,t) = u(1,t) = 0 \quad \text{in} \quad [0,T]. \tag{18}$$

We assume at the outset that $u_0(0) = u_0(1) = 0$, and that a solution u^0 of (18) exists, such that the following holds (with some $\rho_0 > 0$, $\alpha_0 > 0$):

(19a) $a(x,y)$ is continuous with respect to both arguments, is twice differentiable with respect to the second argument in

$$V \equiv \{(x,y) \in [0,1] \times \mathbb{R}:\ \exists t \in [0,T]: |y-u^0(x,t)| \leq \rho_0\},$$

with continuous partial derivatives a_y, a_{yy}, and satisfies the following inequality,

$$0 < \alpha_0 \leq a(x,y), \quad (x,y) \in V.$$

(19b) $F(x,t,y)$ is continuous with respect to all three arguments and twice differentiable with respect to the third argument in

$$W \equiv \{(x,t,y) \in [0,1] \times [0,T] \times \mathbb{R}: |y-u^0(x,t)| \leq \rho_0\}$$

with continuous partial derivatives F_y, F_{yy}.

We investigate here the differentiability properties of the following *nonlinear Crank-Nicolson-Galerkin method* (cf. 4.(55) in finite-dimensional subspaces $E_n \subset H_0^1(0,1)$, $n \in I$,

$$\begin{aligned}(D_\tau^+ v^k,\phi_n)_0 &+ \tfrac{1}{2}(\alpha(v^{k+1})v_x^{k+1} + \alpha(v^k)v_x^k,\phi_n')_0 \\ &= \tfrac{1}{2}(\Phi(t_{k+1},v^{k+1}) + \Phi(t_k,v^k),\phi_n)_0, \quad \phi_n \in E_n.\end{aligned} \tag{20}$$

Here $v^k \in E_n$ denote approximations at time t_k; $\Phi(t,v)(x) \equiv F(x,t,v(x))$, $\alpha(v)(x) \equiv a(x,v(x))$, $v \in H_0^1(0,1)$; and $D_\tau^+ v^k \equiv (v^{k+1}-v^k)/\tau_n$ approximates the partial derivative with respect to t.

Using the L^2-scalar product $(.,.)_0$ (L^2-norm: $|\cdot|_0$), we can define, via the Poincaré-Friedrichs inequality, a scalar product in $E = H_0^1(0,1)$ by

$$(\psi,\phi)_1 = (\psi',\phi')_0, \quad \phi,\psi \in E.$$

The associated norm is written as $|\cdot|_1 = (.,.)_1^{1/2}$. In order to express method (20) in the form (2), we define the following mappings $\hat{C}_n^{(\ell)}(t)$ from E into itself:

$$\begin{aligned}(\hat{C}_n^{(0)}(t)g,\phi)_1 &= (g,\phi)_0 + \tfrac{1}{2}\tau_n(\alpha(g)g',\phi')_0 - \tfrac{1}{2}\tau_n(\Phi(t,g),\phi)_0, \\ (\hat{C}_n^{(1)}(t)g,\phi)_1 &= (g,\phi)_0 - \tfrac{1}{2}\tau_n(\alpha(g)g',\phi')_0 + \tfrac{1}{2}\tau_n(\Phi(t',g),\phi)_0,\end{aligned} \tag{21a}$$

$$\phi \in E, \quad t \in [0,T]_n', \quad n \in I.$$

With the H^1-projection onto E_n,

$$P_n g \in E_n\colon\ (P_n g,\phi_n)_1 = (g,\phi_n)_1,\quad g \in E,\quad \phi_n \in E_n,\quad n \in I,$$

method (20) can be expressed in the general form (2) by setting

$$C_n^{(\ell)}(t) \equiv P_n \hat{C}_n^{(\ell)}(t)|E_n,\quad \ell = 0,1,\quad t \in [0,T]_n',\quad n \in I. \tag{21b}$$

We now investigate the two nonlinear parts of $C_n^{(\ell)}(t)$, $\ell = 0,1$, separately and, for this purpose, define mappings $Q,F(t)$ from E into itself via

$$(Qg,\phi)_1 = (\alpha(g)g',\phi')_0,\quad \phi \in E, \tag{22a}$$

$$(F(t)g,\phi)_1 = (\Phi(t,g),\phi)_0,\quad \phi \in E, \tag{22b}$$

for appropriate t,g specified below. Analogous to Section 2.5 (cf. 2.(59)), we can express the Fréchet-derivatives of $F(t)$ by

$$(F(t)'(g)\psi,\phi)_1 = (F_y(\cdot,t,g)\psi,\phi)_0,\quad \phi,\psi \in E, \tag{23a}$$

where $(t,g) \in U$, with U defined by

$$U \equiv \{(t,g) \in [0,T] \times E\colon |g - u^0(t)|_1 \le \rho_0/2\},$$

and $F(t)$ is, moreoever, continuously differentiable in U. In the following lemma, we shall show that the derivative of Q is given by

$$(Q'(g)\psi,\phi)_1 = (\alpha_y(g)g'\psi,\phi')_0 + (\alpha(g)\psi',\phi')_0,\quad \psi,\phi \in E, \tag{23b}$$

where $\alpha_y(g)(x) = a_y(x,g(x))$ and g is such that $(t,g) \in U$ for some $t \in [0,T]$. Note that the statements concerning the differentiability of $F(t)$ and Q hold with respect to $|\cdot|_1$ as the underlying norm in E.

As a consequence of (23a,b), it is obvious that the derivatives of the mappings $\hat{C}_n^{(\ell)}(t)$ defined in (21a) are given by

$$\begin{aligned}(\hat{C}_n^{(\ell)}(t)'(g)\psi,\phi)_1 &= (\psi,\phi)_0\\ &+/- \tfrac{1}{2}\tau_n\{(\alpha_y(g)g'\psi + \alpha(g)\psi',\phi')_0 - (F_y(\cdot,t-\ell\tau_n,g)\psi,\phi)_0\},\\ &\qquad (t,g) \in U,\quad n \in I,\end{aligned} \tag{24a}$$

where $+/-$ holds for $\ell = 0/\ell = 1$. When we "multiply" these mappings (from the left) by P_n and then restrict them to E_n - compare the similar formula for $C_n(t)$ in (21b) - we obtain mappings from E_n into itself which we denote by

$$C_n^{(\ell)}(t)'(g) \equiv P_n\hat{C}_n^{(\ell)}(t)'(g)|E_n, \quad (t,g) \in U, \quad \ell = 0,1, \quad n \in I. \tag{24b}$$

These are the derivatives of $C_n^{(\ell)}(t)$ in the usual sense if we additionally restrict (t,g) to $U \cap ([0,T]_n' \times E_n)$.

It is another aim of the following lemma to prove the uniform differentiability properties (8a,b). To this end, we have to choose suitable norms in E_n, $n \in I$. We shall see that the desired properties hold in a weak form expressible by negative powers of τ_n in the definition of the norms. This has some similarities to the analysis of Example 1 in this section but, due to the Galerkin methods considered here, the norms are completely different. Indeed, we consider $C_n^{(\ell)}(t)$ and $C_n^{(\ell)}(t)'(g)$ given in (21b) and (24b), respectively, as mappings from E_n into $F_n = E_n$ supplied with the norms

$$\begin{aligned} |g_n|_{E_n} &\equiv \tau_n^{-1/2}|g_n|_1, \quad g_n \in E_n, \\ |f_n|_{F_n} &\equiv \tau_n^{-1/2}[f_n]_n, \quad f_n \in F_n \quad (= E_n), \quad n \in I, \end{aligned} \tag{25a}$$

respectively, where

$$[f_n]_n \equiv \sup_{0 \neq \phi_n \in E_n} |(f_n,\phi_n)_1|/|\phi_n|_0.$$

Note that, by the Poincaré-Friedrichs inequality, the norm $[\cdot]_n$ is stronger than $|\cdot|_1$ on E_n. (At this point, it is appropriate to mention that the supremum norm will be denoted by $|\cdot|_{0,\infty}$ in the following.)

To verify conditions (8a,b), we need an *inverse assumption* for the finite dimensional spaces E_n, namely

$$|\phi_n|_1 \leq C_0\tau_n^{-1/2}|\phi_n|_0, \quad \phi_n \in E_n, \quad n \in I, \tag{25b}$$

with $C_0 > 0$ independent of n. It is satisfied, for example, in case E_n are spaces of piecewise polynomial functions associated with quasi-uniform meshes (with maximal mesh widths h_n) and the mesh ratio $r = \tau_n/h_n^2$ remains constant in n (cf. Ciarlet (1978), Sec. 3.2). We are now in the position to prove the following lemma.

<u>Lemma 11.3</u>. Under the assumption (19a), Q is continuously differentiable at each g such that $(t,g) \in U$ for some $t \in [0,T]$, and the derivative is given by (23b). Moreover, under (19a,b), the inverse assumption (25b), and $\tau_n \leq 1$, $n \in I$, the differentiability properties (8a,b) are satisfied where $U_n^{(\ell)}$ is replaced by U, where u_n^0 in (8b) is replaced by an

arbitrary v with $(t,v(t)) \in U$, $t \in [0,T]$, and where the corresponding norms are specified in (25a).

Proof: (i). Let $(t,g) \in U$ for some $t \in [0,T]$. Then we have, via the Mean Value Theorem, the relations

$$
\begin{aligned}
&(\alpha(g+\psi)(g'+\psi'),\phi')_0 - (\alpha(g)g',\phi')_0 \\
&\quad = ((\alpha(g+\psi) - \alpha(g))(g'+\psi'),\phi')_0 \\
&\qquad + (\alpha(g)(g'+\psi'),\phi')_0 - (\alpha(g)g',\phi')_0 \\
&\quad = (\tilde{\alpha}_y\psi(g'+\psi'),\phi')_0 + (\alpha(g)\psi',\phi')_0 \\
&\quad = ((\tilde{\alpha}_y - \alpha_y(g))g'\psi,\phi')_0 + (\tilde{\alpha}_y\psi\psi',\phi')_0 \\
&\qquad + (\alpha_y(g)g'\psi,\phi')_0 + (\alpha(g)\psi',\phi')_0, \quad \psi,\phi \in E,
\end{aligned}
$$

where

$$\tilde{\alpha}_y(x) = a_y(x,g(x) + \xi) \quad \text{with} \quad |\xi| \le |\psi(x)|, \quad x \in [0,1].$$

Here, we must assume that $|\psi|_{0,\infty} \le \rho_0/2$ in order that $(x,(g+\psi)(x)) \in V$. With $C_1 > 0$ and $C_2 > 0$ bounds for a_y and a_{yy}, respectively, in V, we can obtain the following estimates,

$$
\begin{aligned}
&|(\tilde{\alpha}_y - \alpha_y(g))g'\psi + \tilde{\alpha}_y\psi\psi',\phi')_0| \\
&\quad \le (C_2|\psi|_{0,\infty}^2|g|_1 + C_1|\psi|_{0,\infty}|\psi|_1)|\phi|_1 \\
&\quad \le \max(C_1,C_2)(|g|_1 + 1)|\psi|_1^2|\phi|_1
\end{aligned}
$$

for all $\psi,\phi \in E$ with $|\psi|_{0,\infty} \le \rho_0/2$, where the last inequality uses the fact that the supremum norm can be bounded by the H^1-norm $|\cdot|_1$ in E. We have thus shown the existence of the Fréchet-derivative $Q'(g)$ and the representation thereof in (23b) with $|\cdot|_1$ as the underlying norm.

Continuity of the Fréchet-derivatives follows from the relations

$$
\begin{aligned}
&|(Q'(g)\psi - Q'(f)\psi,\phi)_1| \\
&\quad = |([\alpha_y(g)g' - \alpha_y(f)f']\psi + [\alpha(g) - \alpha(f)]\psi',\phi')_0| \\
&\quad \le |(\alpha_y(g)(g' - f')\psi,\phi')_0| + |([\alpha_y(g) - \alpha_y(f)]f'\psi,\phi')_0| \\
&\qquad + |([\alpha(g) - \alpha(f)]\psi',\phi')_0| \\
&\quad \le \{C_1|g-f|_1|\psi|_{0,\infty} + C_2|g-f|_{0,\infty}|\psi|_{0,\infty}|f|_1 + C_1|g-f|_{0,\infty}|\psi|_1\}|\phi|_1 \\
&\quad \le \max(C_1,C_2)(2 + |f|_1)|g - f|_1|\psi|_1|\phi|_1
\end{aligned}
$$

for all $\psi,\phi \in E$ and $g,f \in E$ with (t,g), $(t,f) \in U$ for some $t \in [0,T]$. Here, again, C_i denote bounds for $\partial^i a/\partial y^i$ in V, $i = 1,2$.

(ii). The derivatives of $F(t)$, Q and $C_n^{(\ell)}(t)$ are given in (23a), (23b) and (24b), respectively. Part (i) of this proof and the proof of 2.(59) - using additionally the existence and boundedness of $F_{yy}(x,t,y)$ in W - show that, with some generic constant C only depending on u^0, V, W,

$$\frac{1}{\tau_n}|(C_n^{(0)}(t)(g + \psi_n) - C_n^{(0)}(t)g - C_n^{(0)}(t)'(g)\psi_n,\phi_n)_1|$$

$$= \frac{1}{2}|([\tilde{\alpha}_y - \alpha_y(g)]g'\psi_n + \tilde{\alpha}_y\psi_n\psi_n',\phi_n')_0$$

$$- (\Phi(t,g+\psi_n) - \Phi(t,g) - F_y(\cdot,t,g)\psi_n,\phi_n)_0|$$

$$\leq C\{(1 + |g|_1)|\psi_n|_1^2|\phi_n|_1 + |\psi_n|_{0,\infty}^2|\phi_n|_0\}$$

for all $\psi_n,\phi_n \in E_n$, $n \in I$, and all $(t,g) \in U$. We again estimate $|\cdot|_{0,\infty}$ by $|\cdot|_1$, use the inverse assumption (25b) and the fact that $|g|_1$ is bounded for $(t,g) \in U$, and obtain a further estimation of the last expression by

$$C(\tau_n^{-1/2} + 1)|\psi_n|_1^2|\phi_n|_0$$

with another generic constant $C > 0$. If we now take the definition of $|\cdot|_{E_n}$ into consideration (cf. (25a)) and choose, for an arbitrary $\varepsilon > 0$, $\delta = \min(\varepsilon,\rho_0/2)$ then, for every $(t,g) \in U$, $n \in I$ and $\psi_n \in E_n$ with $|\psi_n|_{E_n} \leq \delta$ (i.e. $|\psi_n|_1 \leq \delta\tau_n^{1/2}$), we obtain

$$\frac{1}{\tau_n}|(C_n^{(0)}(t)(g + \psi_n) - C_n^{(0)}(t)g - C_n^{(0)}(t)'(g)\psi_n,\phi_n)_1|$$

$$\leq C\varepsilon(1 + \tau_n^{1/2})|\psi_n|_1|\phi_n|_0, \quad \psi_n,\phi_n \in E_n, \quad n \in I.$$

Multiplying by $\tau_n^{-1/2}$ and using the definition of $|\cdot|_{F_n}$, the desired estimate in condition (8a) is proved (for $\ell = 0$). The estimate for $C_n^{(1)}(t)$ follows in a quite analogous manner.

To prove condition (8b), we see that, for every v such that $(t,v(t)) \in U$, $t \in [0,T]$,

$$\frac{1}{\tau_n}|(([C_n^{(0)}(t)'(v(t)) - C_n^{(0)}(t)'(g)]\psi_n,\phi_n)_1|$$

$$= \frac{1}{2}|(([\alpha_y(v(t))v(t)' - \alpha_y(g)g']\psi_n + [\alpha(v(t)) - \alpha(g)]\psi_n',\phi_n')_0$$

$$- ([F_y(\cdot,t,v(t)) - F_y(\cdot,t,g)]\psi_n,\phi_n)_0|$$

$$\le C\{(2 + |g|_1)|v(t) - g|_1|\psi_n|_1|\phi_n|_1 + |v(t) - g|_{0,\infty}|\psi_n|_0|\phi_n|_0\}$$

for all $(t,g) \in U$, ψ_n, $\phi_n \in E_n$, $n \in I$. Again, we use the inverse assumption (25b) and estimate $|\cdot|_{0,\infty}$ by $|\cdot|_1$; thus, for every $\varepsilon > 0$ and every $g_n \in E_n$ with $|v(t) - g_n|_{E_n} \le \delta \equiv \min(\varepsilon,\rho_0/2)$, we obtain

$$\frac{1}{\tau_n}|(([C_n^{(0)}(t)'(v(t)) - C_n^{(0)}(t)'(g_n)]\psi_n,\phi_n)_1|$$

$$\le C\varepsilon(|\psi_n|_1 + \tau_n^{1/2}|\psi_n|_0)|\phi_n|_0 \le C\varepsilon|\psi_n|_1|\phi_n|_0, \quad \psi_n,\phi_n \in E_n, \ n \in I.$$

Multiplying by $\tau_n^{-1/2}$ and using the definition of $|\cdot|_{F_n}$, we get the estimate in (8b). The proof for $C_n^{(1)}(t)$ is carried out quite analogously. □

The lemma just proven indicates that the conditions (8a,b) - in their original form - are satisfied for every sequence $u_n^0(t) \in E_n$, $t \in [0,T]_n'$, $n \in I$, for which $(t,u_n^0(t))$ remains in U; the sets $U_n^{(\ell)}$ can then be chosen with $\rho = \rho_0/2$. □

Example 3. As the third example, we consider *scalar, quasilinear, hyperbolic initial-value problems* of the form (cf. 4.(58) in Sec. 4.4)):

$$\begin{aligned} &u_t - c(x,t,u)u_x = s(x,t,u), \quad t > 0, \quad x \in \mathbb{R}, \\ &u(x,0) = u_0(x), \quad x \in \mathbb{R}. \end{aligned} \tag{26}$$

We refer the reader to John (1982), Törnig (1979), 17.3.1, and the literature cited therein for results on the existence and uniqueness of solutions. The finite-difference approximations cited in Section 4.4 can be expressed in the form

$$v_j^{k+1} = \sum_{\mu=-1}^{1} b_{\mu,j}^k(v_j^k)v_{j+\mu}^k + \tau_n s(x_j,t_k,v_j^k), \quad x_j \in G_n, \ k = 0,1,2,\ldots, \tag{27}$$

where $G_n = \{x_j \equiv jh_n, \ j = 0,\pm1,\pm2,\ldots\}$, $n \in I$, denotes an equally spaced mesh in $\mathbb{R}$;

$$b_{\mu,j}^k(y) \equiv b_\mu(x_j,t_k,y), \quad \mu = 0,\pm1,$$

and the functions b_μ can be determined from a given function b as follows,

$$\begin{aligned} b_0(x,t,y) &= 1 - \lambda b(x,t,y), \\ b_{\pm 1}(x,t,y) &= \frac{\lambda}{2}(\pm c(x,t,y) + b(x,t,y)). \end{aligned} \tag{28}$$

We assume that the associated *mesh ratio* $\lambda = \tau_n/h_n$ remains constant for all n. From the equations in (28) for the b_μ, we get immediately the following important relations,

$$\begin{aligned} &b(x,t,y) = (1 - b_0(x,t,y))/\lambda, \\ &\sum_{\mu=-1}^{1} b_\mu(x,t,y) = 1, \qquad \sum_{\mu=-1}^{1} \mu b_\mu(x,t,y) = \lambda c(x,t,y). \end{aligned} \tag{29}$$

As examples, we have the following finite-difference methods:

Friedrichs Method 4.(59): $b(x,t,y) = 1/\lambda$,

i.e. $b_0(x,t,y) = 0$

$$b_{\pm 1}(x,t,y) = \frac{1}{2}(1 \pm \lambda c(x,t,y));$$

Courant-Isaacson-Rees Method 4.(60): $b(x,t,y) = |c(x,t,y)|$,

i.e. $b_0(x,t,y) = 1 - \lambda|c(x,t,y)|$,

$$b_{\pm 1}(x,t,y) = \pm\lambda c^{\pm}(x,t,y),$$

where

$$c^+(x,t,y) \equiv \max(0,c(x,t,y)), \quad c^-(x,t,y) \equiv \min(0,c(x,t,y));$$

Lax-Wendroff Method 4.(61): $b(x,t,y) = \lambda c^2(x,t,y)$,

i.e. $b_0(x,t,y) = 1 - \lambda^2 c^2(x,t,y)$,

$$b_{\pm 1}(x,t,y) = \frac{1}{2}(\pm \lambda c(x,t,y) + \lambda^2 c^2(x,t,y)).$$

Methods of the form (27) are termed explicit and the associated finite-difference operators $C_n^{(\ell)}(t)$, $\ell = 0,1$, are given by

$$\begin{aligned} &C_n^{(0)}(t) = I, \\ &(C_n^{(1)}(t_{k+1})g_n)(x_j) = \sum_{\mu=-1}^{1} b_{\mu,j}^k(g_n(x_j))g_n(x_{j+\mu}) \\ &\qquad + \tau_n s(x_j,t_k,g_n(x_j)), \quad x_j \in G_n, \quad k = 0,1,\ldots, \quad n \in I. \end{aligned} \tag{30}$$

If the functions b,c, and s are differentiable with respect to the third argument y, then the Fréchet-derivative of $C_n^{(1)}(t)$ at $g_n \in C(G_n)$ is

$$(C_n^{(1)}(t_{k+1})'(g_n)\phi_n)(x_j) = \sum_{\mu=-1}^{1} b_\mu(x_j,t_k,g_n(x_j))\phi_n(x_{j+\mu})$$

$$+ \sum_{\mu=-1}^{1} \frac{\partial b_\mu}{\partial y}(x_j,t_k,g_n(x_j))g_n(x_{j+\mu})\phi_n(x_j) \tag{31}$$

$$+ \tau_n \frac{\partial s}{\partial y}(x_j,t_k,g_n(x_j))\phi_n(x_j), \quad x_j \in G_n, \quad \phi_n \in C(G_n),$$

$$k = 0,1,\dots,n \in I.$$

We can express the above in the following way with the help of forward and backward difference quotients in the x-direction (where, for brevity, we omit the argument $(x_j,t_k,g_n(x_j))$:

$$(C_n^{(1)}(t_{k+1})'(g_n)\phi_n)(x_j) = \sum_{\mu=-1}^{1} b_\mu\phi_n(x_{j+\mu})$$

$$+ \tau_n\left\{\frac{1}{\lambda}\left[\frac{\partial b_1}{\partial y} D_h^+ g_n(x_j) - \frac{\partial b_{-1}}{\partial y} D_h^- g_n(x_j)\right] + \frac{\partial s}{\partial y}\right\}\phi_n(x_j), \tag{31'}$$

$$x_j \in G_n, \quad \phi_n \in C(G_n), \quad k = 0,1,\dots,n \in I,$$

since

$$\sum_{\mu=-1}^{1} \frac{\partial b_\mu}{\partial y} = 0.$$

For the sake of brevity, we shall not verify the differentiability conditions (8a,b), which are here clearly taken in the supremum norm in $C(G_n)$, in case the sufficiently smooth functions b,c, and s are defined in a suitable compact neighborhood of $u(x,t)$. As a final remark, we note that our arguments are applicable to the Courant-Isaacson-Rees method only when c maintains the same sign. □

11.2. INVERSE STABILITY

By using an explicit representation of the inverses of $T_n'(u_n^0)$, $n \in I$, we are able to characterize their uniform boundedness by imposing suitable conditions on $C_n^{(\ell)}(t)$ (see Theorem 11.4). If we use the results from Section 6.2, then we get a criterion for the inverse stability of the mappings T_n, $n \in I$, themselves, which furnishes, in particular, two-sided estimates (see Theorem 11.5). We then specialize this characterization of inverse stability to the case of linear problems and, furthermore, establish a result which states that inverse stability is maintained under small perturbations. An analysis of inverse stability properties for various concrete methods will be done in detail in Chapter 12.

In this section, we shall assume that the assumptions of Theorem 11.1 are valid for a sequence $u_n^0 \in X_n$, $n \in I$, and that the $C_n^{(0)}(t)'(u_n^0(t))$, $t \in [0,T]_n'$, $n \in I$, are bijective. Then the linearized problems associated with (2) are

$$\begin{aligned} &x_n(0) = y_n^0, \\ &C_n^{(0)}(t)'(u_n^0(t))x_n(t) = C_n^{(1)}(t)'(u_n^0(t'))x_n(t') + \tau_n y_n^1(t), \qquad (32)\\ &\qquad t \in [0,T]_n', \quad n \in I, \end{aligned}$$

and are uniquely solvable for arbitrary $(y_n^0, y_n^1) \in Y_n$, $n \in I$. The solutions x_n, $n \in I$, can be determined explicitly by

$$\begin{aligned} &x_n(0) = y_n^0, \quad x_n(t) = D_n^{(0)}(t)^{-1}\{D_n^{(1)}(t)x_n(t') + \tau_n y_n^1(t)\}, \qquad (33)\\ &\qquad t \in [0,T]_n', \quad n \in I. \end{aligned}$$

Here, we write, for brevity,

$$D_n^{(\ell)}(t) \equiv C_n^{(\ell)}(t)'(u_n^0(t-\ell\tau_n)), \quad t \in [0,T]_n', \quad n \in I, \quad \ell = 0,1.$$

Also, the inverses of $T_n'(u_n^0)$, $n \in I$, exist and can be represented as

$$(T_n'(u_n^0)^{-1}y_n)(t_k) = \begin{cases} y_n^0, \quad k = 0, \\ \prod_{\nu=1}^{k} D_n(t_\nu)y_n^0 \\ \quad + \tau_n \sum_{\mu=1}^{k} \prod_{\nu=\mu+1}^{k} D_n(t_\nu)D_n^{(0)}(t_\nu)^{-1}y_n^1(t_\mu), \end{cases} \qquad (34)$$

$$1 \le k \le N, \quad y_n = (y_n^0, y_n^1) \in Y_n, \quad n \in I,$$

where $D_n(t) \equiv D_n^{(0)}(t)^{-1}D_n^{(1)}(t)$, and where the products are defined in the usual manner by

$$\prod_{\nu=m}^{k} D_n(t_\nu) \equiv I, k < m, \quad \prod_{\nu=m}^{k} D_n(t_\nu) \equiv D_n(t_k) \prod_{\nu=m}^{k-1} D_n(t_\nu), \quad k \ge m.$$

The representation (34) follows immediately from (33) by induction. Using (34), we see that, for each $n \in I$, $T_n'(u_n^0)^{-1}$ is a bounded, linear mapping from Y_n onto X_n (with respect to the norms $||\cdot||_{p,n}$, $1 \le p \le \infty$, in Y_n and $||\cdot||_{\infty,n}$ in X_n), in case the $D_n^{(0)}(t)$, $t \in [0,T]_n'$, are continuously invertible and the $D_n^{(1)}(t)$, $t \in [0,T]_n'$, are bounded.

We can now characterize in the following the uniform boundedness of the inverses $T_n(u_n^0)^{-1}$ by conditions on the $C_n^{(\ell)}(t)$ - or, more precisely, by conditions on their derivatives.

<u>Theorem 11.4</u>. The $T_n'(u_n^0)^{-1}$, $n \in I$, are uniformly bounded (with respect to the norms $||\cdot||_{\infty,n}$ in X_n and $||\cdot||_{1,n}$ in Y_n), if and only if, there exists a $\gamma \ge 0$ such that

$$\Big|\prod_{\nu=1}^{k} D_n(t_\nu) g_n\Big|_{E_n} \le \gamma |g_n|_{E_n}, \quad 1 \le k \le N, \quad g_n \in E_n, \quad n \in I, \tag{35a}$$

$$\Big|\prod_{\nu=m+1}^{k} D_n(t_\nu) D_n^{(0)}(t_m)^{-1} f_n\Big|_{E_n} \le \gamma |f_n|_{F_n}, \quad 1 \le m \le k \le N, \quad f_n \in F_n, \quad n \in I. \tag{35b}$$

<u>Proof</u>: The uniform boundedness of $T_n'(u_n^0)^{-1}$, $n \in I$, with respect to $||\cdot||_{\infty,n}$ in X_n and $||\cdot||_{1,n}$ in Y_n, is equivalent to the following estimates for the solutions x_n, $n \in I$, to equations (32),

$$\max_{t \in [0,T]_n} |x_n(t)|_{E_n} \le \gamma\Big(|y_n^0|_{E_n} + \tau_n \sum_{t \in [0,T]_n'} |y_n^1(t)|_{F_n}\Big), \quad n \in I. \tag{36}$$

Suppose (36) is true. For arbitrary $m \in \{1,\ldots,N\}$, $f_n \in F_n$, $n \in I$, let

$$y_n^{(0)} = 0, \quad y_n^1(t_k) = (\delta_{mk}/\tau_n) f_n, \quad k = 1,\ldots,N,$$

where δ_{mk} is the Kronecker delta. The solution of (32) then has the form

$$x_n(t_k) = 0, \quad k < m, \quad x_n(t_k) = \prod_{\nu=m+1}^{k} D_n(t_\nu) D_n^{(0)}(t_m)^{-1} f_n, \quad k \ge m.$$

Combining the above solution with (36), we get the inequality (35b). For proving (35a), we set $y_n^0 = g_n \in E_n$, $y_n^1(t_k) = 0$, $1 \le k \le N$; the solution of (32) then has the form

$$x_n(t_k) = \prod_{\nu=1}^{k} D_n(t_\nu) g_n, \quad 0 \le k \le N.$$

Expression (35a) then follows from (36). Conversely, from (35a,b), used in combination with the representation of the inverses in (34), we get immediately (36). □

Using (6), we see that we can obtain from (36) a corresponding inequality for all p-norms, $1 \leq p \leq \infty$,

$$\frac{1}{\gamma}||x_n||_{\infty,n} \leq ||T_n'(u_n^0)x_n||_{p,n}, \quad x_n \in X_n, \quad n \in I. \tag{37}$$

For $p = 1$, the above inequality (i.e., (35a,b)) is the strongest of all the inequalities of this type, this means that $p = 1$ implies (37) for all $1 \leq p \leq \infty$. Analogously, $p = \infty$ is the weakest estimate of the above form.

We now would like to remark that (35b) (with $m = k$) yields the uniform boundedness of $D_n^{(0)}(t)^{-1}$, $t \in [0,T]_n'$, $n \in I$.

In Section 6.2, we have seen that uniform boundedness of $T_n'(u_n^0)^{-1}$, $n \in I$, characterizes the inverse stability of (T_n) at (u_n^0). Here, we have the situation where the bijective derivatives satisfy a uniform inequality from below but are not necessarily uniformly bounded. Corresponding to the procedure used in the proof of Theorem 6.23, we define new norms in X_n by

$$|||x_n|||_{p,n} \equiv \max(||x_n||_{\infty,n}, ||T_n'(u_n^0)x_n||_{p,n}), \quad x_n \in X_n, \ n \in I \tag{38}$$

(cf. Definition 6.(22)). Using (38), we see that (37) is equivalent to the uniform, two-sided inequality

$$\frac{1}{\gamma}|||x_n|||_{p,n} \leq ||T_n'(u_n^0)x_n||_{p,n} \leq |||x_n|||_{p,n}, \quad x_n \in X_n, \ n \in I, \tag{39}$$

where the constants coincide for the case $\gamma \geq 1$. With the norm $|||\cdot|||_{p,n}$ in X_n and $||\cdot||_{p,n}$ in Y_n, we are able to directly apply the Bistability Theorem 6.12.

<u>Theorem 11.5</u>. The following statements are equivalent:

(40a) (T_n) is locally bijective at (u_n^0); the inverses $\tilde{T}_n^{-1}$ of the associated restrictions are continuous at $T_n u_n^0$ for every $n \in I$; $(\tilde{T}_n^{-1})$ is equicontinuously equidifferentiable at $(T_n u_n^0)$; and (T_n) is inversely stable at (u_n^0).

(40b) Inequality (37) is valid.

(40c) The quantity

$$\gamma' \equiv \sup_{n\in I} \sup_{||y_n||_{p,n}=1} ||T_n'(u_n^0)^{-1}y_n||_{\infty,n}$$

is finite; and, for every $\alpha > 1$ and $\beta \in (0,1/\gamma')$, there exist numbers $\rho,\sigma > 0$ such that $B_\sigma(T_n u_n^0) \subset T_n B_\rho(u_n^0)$ and

$$\frac{1}{\alpha}||T_n u_n - T_n v_n||_{p,n} \le |||u_n - v_n|||_{p,n} \le \frac{1}{\beta}||T_n u_n - T_n v_n||_{p,n},$$

$$u_n, v_n \in B_\rho(u_n^0), \quad n \in I.$$

Proof: Statement (40a) is equivalent to 6.(13a) in Theorem 6.12. Indeed, the equicontinuity of the sequence $(\tilde{T}_n)$ at (u_n^0) follows from the uniform boundedness of $T_n'(u_n^0)$, $n \in I$, with respect to the norms $|||\cdot|||_{p,n}$ in X_n and $||\cdot||_{p,n}$ in Y_n (see Theorem 6.8). From (40a), we therefore obtain the bistability of $(\tilde{T}_n)$ at (u_n^0); conversely, (40a) trivially follows from 6.(13a). Inequality (37) (cf. statement (40b)) is equivalent to (39) and hence to 6.(13c). Statement (40c) then follows from 6.(13d), and conversely yields (40b), in particular. □

For linear mappings $C_n^{(\ell)}(t)\colon E_n \to F_n$, $\ell = 0,1$, with $C_n^{(0)}(t)$ invertible, we obtain directly, via (34), an explicit representation of the solutions of the approximating problems (2), and, thereby, a representation of the inverses of T_n, namely,

$$u_n(t_k) = \prod_{\nu=1}^{k} C_n(t_\nu)u_{n,0} + \tau_n \sum_{\mu=1}^{k} \prod_{\nu=\mu+1}^{k} C_n(t_\nu)C_n^{(0)}(t_\mu)^{-1}w_n(t_\mu),$$

$$k = 0,\ldots,N, \quad n \in I, \tag{41}$$

where $C_n(t) \equiv C_n^{(0)}(t)^{-1}C_n^{(1)}(t)$. From Theorem 11.4, we can immediately get the following characterization of inverse stability in the linear case.

Theorem 11.6. Suppose that the mappings $C_n^{(\ell)}(t)$, $t \in [0,T]_n'$, $\ell = 0,1$, $n \in I$, are linear with $C_n^{(0)}(t)$, $t \in [0,T]_n'$, $n \in I$, bijective. Then the inverse stability of T_n (with respect to $||\cdot||_{\infty,n}$ in X_n and $||\cdot||_{1,n}$ in Y_n) is equivalent to

$$\Big|\prod_{\nu=1}^{k} C_n(t_\nu)g_n\Big|_{E_n} \le \gamma|g_n|_{E_n}, \quad k = 1,\ldots,N, \quad g_n \in E_n, \quad n \ge \nu_0, \tag{42a}$$

$$\Big|\prod_{\nu=m+1}^{k} C_n(t_\nu)C_n^{(0)}(t_m)^{-1}f_n\Big|_{E_n} \le \gamma|f_n|_{F_n}, \quad 1 \le m \le k \le N,$$

$$f_n \in F_n, \quad n \ge \nu_0. \quad \square \tag{42b}$$

If the mappings $C_n^{(\ell)}(t)$, $\ell = 0,1$, are linear and *constant* (i.e., independent of t), then the solutions of (2) can be represented by

$$u_n(t_k) = C_n^k u_{n,0} + \tau_n \sum_{\mu=1}^{k} C_n^{k-\mu} C_n^{(0)-1} w_n(t_\mu), \quad k = 0,\ldots,N,$$

and statements (42a,b) are equivalent to

$$|C_n^{(0)-1} f_n|_{E_n} \le \gamma |f_n|_{F_n}, \quad f_n \in F_n, \quad \text{and}$$
$$|C_n^k g_n|_{E_n} \le \gamma |g_n|_{E_n}, \quad k = 0,\dots,N, \quad g_n \in E_n, \quad n \ge \nu_0. \tag{43}$$

It is well-known that, for linear, time-independent mappings $C_n^{(\ell)}$, $\ell = 0,1$, $n \in I$, the stability property (43) is maintained under small perturbations of C_n, say, of order $O(\tau_n)$ (cf. the theorem due to Kreiss in Richtmyer & Morton (1967), 3.9, and in Meis & Marcowitz (1981), Thm. 5.13). This result is also true for mappings depending on time. The proof is very similar to that for the time-independent case.

Lemma 11.7. Suppose that the mappings $C_n^{(\ell)}(t)$, $\ell = 0,1$, $t \in [0,T]_n'$, $n \in I$, are linear and satisfy conditions (42a,b). Further, assume that $Q_n^{(1)}(t): E_n \to F_n$, $t \in [0,T]_n'$, $n \in I$, are linear mappings which are uniformly bounded,

$$|Q_n^{(1)}(t) g_n|_{F_n} \le \gamma_Q |g_n|_{E_n}, \quad t \in [0,T]_n', \quad g_n \in E_n, \quad n \in I. \tag{44}$$

Then, for the "perturbed method"

$$u_n(0) = u_{n,0}, \quad C_n^{(0)}(t) u_n(t) = [C_n^{(1)}(t) + \tau_n Q_n^{(1)}(t)] u_n(t') + \tau_n w_n(t),$$
$$t \in [0,T]_n', \quad n \in I, \tag{45}$$

inverse stability is present, i.e., (42a,b) is valid for $C_n^{(1)}(t) + \tau_n Q_n^{(1)}(t)$ in place of $C_n^{(1)}(t)$, $t \in [0,T]_n'$, $n \in I$.

Proof: We set

$$S_n(t) \equiv C_n^{(0)}(t)^{-1} C_n^{(1)}(t) + \tau_n C_n^{(0)}(t)^{-1} Q_n^{(1)}(t), \quad t \in [0,T]_n', \quad n \in I.$$

Then we obtain, for arbitrary m in $0 \le m \le N-1$, the representation

$$\prod_{\nu=1}^{\mu} S_n(t_{m+\nu}) = \sum_{\nu=0}^{\mu} \tau_n^\nu \sum_{\kappa=1}^{\binom{\mu}{\nu}} P_{\nu,\kappa}^{(\mu)}, \quad 1 \le m+\mu \le N,$$

where

$$P_{0,1}^{(1)} = C_n(t_{m+1}), \quad P_{0,1}^{(\mu)} = C_n(t_{m+\mu}) P_{0,1}^{(\mu-1)}, \quad \mu \ge 2,$$

$$P_{1,1}^{(1)} = Q_n(t_{m+1}), \quad P_{\mu,1}^{(\mu)} = Q_n(t_{m+\mu}) P_{\mu-1,1}^{(\mu-1)}, \quad \mu \ge 2,$$

$$P_{\nu,\kappa}^{(\mu)} = C_n(t_{m+\mu}) P_{\nu,\kappa}^{(\mu-1)}, \quad \kappa = 1,\dots,\binom{\mu-1}{\nu},$$

$$P_{\nu,z+\kappa}^{(\mu)} = Q_n(t_{m+\mu}) P_{\nu-1,\kappa}^{(\mu-1)}, \quad \kappa = 1,\dots,\binom{\mu-1}{\nu-1}, \quad (z \equiv \binom{\mu-1}{\nu}), \quad \nu = 1,\dots,\mu-1,$$

and

$$Q_n(t) \equiv C_n^{(0)}(t)^{-1}Q_n^{(1)}(t), \quad t \in [0,T]_n', \quad n \in I.$$

For $\nu = 0$ and $\nu = \mu$, the above sums $\sum_{\kappa=1}^{\binom{\mu}{\nu}} P_{\nu,\kappa}^{(\mu)}$ reduce to

$$P_{0,1}^{(\mu)} = \prod_{\ell=1}^{\mu} C_n(t_{\ell+m}) \quad \text{and} \quad P_{\mu,1}^{(\mu)} = \prod_{\ell=1}^{\mu} Q_n(t_{\ell+m}),$$

respectively. For every ν in $0 \le \nu \le \mu$, $P_{\nu,\kappa}^{(\mu)}$ consists of μ factors, each of which, in turn, is either $C_n(t_\ell)$ or $Q_n(t_\ell)$, where each t_ℓ, $\ell = m+1,\ldots,m+\mu$, appears exactly once and where a factor of the form $Q_n(t_\ell)$, $\ell \in \{m+1,\ldots,m+\mu\}$, occurs exactly ν times. The remaining $\mu-\nu$ factors of operators $C_n(t_\ell)$ are thus grouped into at most $\nu+1$ products of consecutive factors. Thus we have, for example, a term of the form

$$C_n(t_{m+\mu})Q_n(t_{m+\mu-1})Q_n(t_{m+\mu-2})\cdots C_n(t_{m+2})C_n(t_{m+1})$$

which we can also write as

$$C_n(t_{m+\mu})C_n^{(0)}(t_{m+\mu-1})^{-1}Q_n^{(1)}(t_{m+\mu-1})Q_n(t_{m+\mu-2})\cdots C_n(t_{m+1}).$$

By assumptions (42) and (44),

$$||Q_n(t)|| \le \gamma\gamma_Q, \quad t \in [0,T]_n', \quad n \in I.$$

If we assume, without loss of generality that $\gamma \ge 1$, then we get from (42) the following estimates for every $P_{\nu,\kappa}^{(\mu)}$,

$$m \ge 1: \quad |P_{\nu,\kappa}^{(\mu)}C_n^{(0)}(t_m)^{-1}f_n|_{E_n} \le (\gamma\gamma_Q)^{\nu}\gamma^{\nu+1}|f_n|_{F_n}, \quad f_n \in F_n, \ \mu = 1,\ldots,N-m,$$

$$m = 0: \quad |P_{\nu,\kappa}^{(\mu)}g_n|_{E_n} \le (\gamma\gamma_Q)^{\nu}\gamma^{\nu+1}|g_n|_{E_n}, \quad g_n \in E_n, \quad \mu = 1,\ldots,N.$$

Using the representation above, we conclude that

$$\left|\prod_{\nu=1}^{\mu} S_n(t_{m+\nu})C_n^{(0)}(t_m)^{-1}f_n\right|_{E_n} \le \sum_{\nu=0}^{\mu}\binom{\mu}{\nu}\tau_n^{\nu}\gamma^{\nu}\gamma_Q^{\nu}\gamma^{\nu+1}|f_n|_{F_n}$$

$$= \gamma(1 + \tau_n\gamma^2\gamma_Q)^{\mu}|f_n|_{F_n} \le \gamma\exp(T\gamma^2\gamma_Q)|f_n|_{F_n},$$

$$f_n \in F_n, \quad \mu = 1,\ldots,N-m,$$

and, correspondingly, that

$$\left|\prod_{\nu=1}^{\mu} S_n(t_\nu)g_n\right|_{E_n} \le \gamma\exp(T\gamma^2\gamma_Q)|g_n|_{E_n}, \quad \mu = 1,\ldots,N, \quad g_n \in E_n, \ n \in I. \quad \square$$

The last result is of great importance for proving inverse stability for concrete methods. It is, in addition, applicable to nonlinear problems, if indeed the associated linearized problems to the approximating equations (cf. (32)) are of the form (45).

11.3. CONSISTENCY AND CONVERGENCE

In this section, we present the main convergence results for discrete-time approximations of initial value problems. First, we investigate the meaning of consistency for the problems at hand. The basic Convergence Theorem 11.9 is a direct consequence of the results from Chapter 6. The second Convergence Theorem 11.10 answers the question of which convergence property of the associated semihomogeneous problems is necessary and sufficient for the convergence of the solutions to the totally inhomogeneous problems. In the concluding paragraphs of this section, we formulate the convergence theorems for the case of linear problems and, in particular, for linear operators independent of t. The convergence analysis of specific methods will be done in Chapter 13.

As in Section 11.2, we assume that the differentiability conditions (8a,b) are satisfied at a sequence (u_n^0) and that each $C_n^{(0)}(t)'(u_n^0(t))$ is bijective for every $t \in [0,T]_n'$, $n \in I$. In this section, we shall later specify the sequence (u_n^0) as $(R_n^X u^0)$.

We begin our analysis by discussing the consistency of T, T_n, $n \in I$. The *truncation errors* at $v \in D(T)$ are defined by $d_n^T(v) \equiv T_n R_n^X v - R_n^Y Tv$, i.e., $d_n^T(v) = (d_n^0(v), d_n^1(v))$, with

$$d_n^0(v) = 0,$$

$$d_n^1(v)(t) = \frac{1}{\tau_n}\{C_n^{(0)}(t)R_n^E v(t) - C_n^{(1)}(t)R_n^E v(t')\} \tag{46}$$

$$- R_n^F\left(\frac{dv}{dt}(t) - A(t)v(t)\right), \quad t \in [0,T]_n'.$$

This definition is, of course, only meaningful for those n for which $R_n^X v \in D(T_n)$. This is satisfied for all $n \in I$, for example, in case (T_n) is equidifferentiable at $(R_n^X v)$. The truncation error at the solution u^0 of $(1)_0$ is clearly given by

$$d_n^1(u^0)(t) = \frac{1}{\tau_n}\{C_n^{(0)}(t)R_n^E u^0(t) - C_n^{(1)}(t)R_n^E u^0(t')\}, \quad t \in [0,T]_n', \quad n \in I.$$

The sequence T, T_n, $n \in I$, is *consistent* at v (in the sense of Section 6.3) with *consistency sequence* $v_n' = R_n^X v$, $n \in I$ (with respect to the norms $||\cdot||_{p,n}$ in Y_n), if, and only if,

$$||d_n^T(v)||_{p,n} \to 0 \quad (n \in I). \tag{47}$$

This condition is satisfied for all $1 \leq p \leq \infty$, whenever

$$\max_{t \in [0,T]_n'} |d_n^1(v)(t)|_{F_n} \to 0 \quad (n \in I). \tag{48}$$

The latter condition is usually described in the literature as the consistency condition (with $v = u^0$) (cf., e.g., Richtmyer & Morton (1967), 3.2, 4.4). We label the number $q > 0$ as the *order of consistency* (relative to the $||\cdot||_{p,n}$-norm) if

$$||d_n^T(v)||_{p,n} = O(\tau_n^q) \quad (n \in I). \tag{49}$$

In the following, we shall require the consistency condition (47) at the solution u^0 of $(1')_0$. If the differentiability conditions (8a,b) are satisfied at (u_n^0) and, moreover, (u_n^0) and its associated sequence of restrictions $(R_n^X u^0)$ satisfy

$$|||u_n^0 - R_n^X u^0|||_{p,n} \to 0 \quad (n \in I), \tag{50}$$

then we can consider, without loss of generality, the sequence (u_n^0) as a consistency sequence. This is made clear in the following lemma.

Lemma 11.8. Suppose (50) is true. Then the consistency condition (47) at u^0 is equivalent to requiring that

$$||T_n u_n^0 - R_n^Y T u^0||_{p,n} \to 0 \quad (n \in I). \tag{51}$$

Proof: With the remainder terms R_n^T in the differentiation of T_n at u_n^0, $n \in I$, we have

$$d_n^T(u^0) = T_n'(u^0)(R_n^X u^0 - u_n^0) + R_n^T(u_n^0;\ R_n^X u^0 - u_n^0)$$
$$+ (T_n u_n^0 - R_n^Y T u^0), \quad n \in I.$$

Because of the equidifferentiability of T_n at u_n^0, $n \in I$, we see that (50) implies that

$$||R_n^T(u_n^0; R_n^X u^0 - u_n^0)||_{p,n} \to 0 \quad (n \in I)$$

and that

$$||T_n'(u_n^0)(R_n^X u^0 - u_n^0)||_{p,n} \le |||R_n^X u^0 - u_n^0|||_{p,n} \to 0 \quad (n \in I).$$

Hence the convergence in (51) follows from (47) and conversely. □

The last result suggests to take $u_n^0 = R_n^X u_0$, $n \in I$, for simplicity; this means that the differentiability conditions (8a,b) should be valid at the restriction sequence, which is also at the same time a consistency sequence.

The results from Section 6.3 now provide the following basic theorem on the existence and convergence of approximating solutions. Here, p denotes a number in $1 \le p < \infty$, or $p = \infty$, and u^0 is the solution of $(1)_0$.

Theorem 11.9. Suppose that the consistency condition (47) is valid at $v = u^0$, and that both the differentiability conditions (8a,b) and the inverse stability inequality (37) are satisfied at $(u_n^0) = (R_n^X u^0)$. Then there exist, for every β in $0 < \beta < 1/\gamma$, numbers $\sigma, \rho > 0$ and an index $\nu \in I$, such that equation (2) has a unique solution u_n in

$$|||u_n - R_n^X u^0|||_{p,n} \le \delta$$

for all $n \ge \nu$ and all $y_n = (u_{n,0}, w_n)$ in

$$||y_n - R_n^Y w^0||_{p,n} \le \sigma$$

(with $w^0 = (u_0, 0)$). Further, we have the inverse convergence relation

$$||y_n - R_n^Y w^0||_{p,n} \to 0 \Rightarrow ||u_n - R_n^X u^0||_{\infty,n} \to 0 \ (n \in I), \tag{52}$$

with the error estimates

$$||u_n - R_n^X u^0||_{\infty,n} \le |||u_n - R_n^X u^0|||_{p,n} \le \beta^{-1}||y_n - R_n^Y w^0 - d_n^T(u^0)||_{p,n}, \tag{53}$$

$$n \ge \nu. \quad □$$

The proof is immediate with the use of Theorem 6.23.

By substituting $u_n - R_n^X u^0$ in the two-sided estimate in statement (40c) from Theorem 11.5 (e.g. for $\alpha = 2$, $\beta = 1/(2\gamma')$), we can easily obtain the biconvergence relation

$$||y_n - R_n^Y w^0||_{p,n} \to 0 \Leftrightarrow |||u_n - R_n^X u^0|||_{p,n} \to 0 \quad (n \in I)$$

along with the following two-sided error estimates

$$\frac{1}{2}||y_n - R_n^Y w^0 - d_n^T(u^0)||_{p,n} \leq |||u_n - R_n^X u^0|||_{p,n}$$
$$\leq 2\gamma'||y_n - R_n^Y w^0 - d_n^T(u^0)||_{p,n}, \quad n \geq \nu.$$

In addition, we have, for $y_n = R_n^Y w^0$ (i.e., $u_{n,0} = R_n^E u_0$, $w_n = 0$), $n \in I$, the convergence of the solutions as $O(\tau_n^q)$ $(n \in I)$ where q is the order of consistency.

Also, the converse of the result of the last theorem holds, which yields, from the inverse convergence relation (52), the inverse stability at $(R_n^X u^0)$ in the sense of Definition 6.(11) from Section 6.2 (cf. Theorem 6.17); the latter is equivalent to inequality (37), if we assume local solvability properties of the approximating equations (more precisely, if the conditions of Theorem 11.10 below are satisfied).

We now discuss in detail the convergence results of the previous theorem for the different p-norms (in Y_n). In the previous section, we have remarked that the inverse stability inequality (37) is the strongest for $p = 1$ which is also equivalent to statements (35a,b) by Theorem 11.4. With regard to consistency, we observe that (47) is the weakest for $p = 1$, but the strongest for $p = \infty$ (i.e., (48)). Concerning the convergence of the problem data, we see correspondingly that the inverse convergence relation is the strongest for $p = \infty$, and the weakest for $p = 1$. The basis for these statements is relation (6) for the p-norms. We thus see that for the convergence of the solutions

$$||u_n - R_n^X u^0||_{\infty,n} \to 0 \quad (n \in I),$$

with a strong inverse stability inequality (i.e. for small p), we need only have a correspondingly weaker consistency condition and convergence of the data. Conversely, for inverse stability for large p $(p > 1)$, we must require that the consistency and convergence of the data be correspondingly stronger.

We now turn to the question of whether the study of semihomogeneous approximating problems is sufficient for the convergence analysis of discrete-time methods. The *semihomogeneous approximating problems* are given by

$$v_n^0(0) = u_{n,0}, \quad C_n^{(0)}(t)v_n^0(t) = C_n^{(1)}(t)v_n^0(t'), \quad t \in [0,T]_n'. \tag{54}$$

We note that we have already expressed the given IVP in semihomogeneous form $(1)_0$ without loss of any generality, but that, however, small perturbations $w_n(t)$ in each of the equations of (2) were allowed and considered in the sought-after inverse convergence relation (52). We know, of course, that the following convergence relation holds under the assumptions of Theorem 11.9,

$$|u_{n,0} - R_n^E u_0|_{E_n} \to 0 \Rightarrow ||v_n^0 - R_n^X u^0||_{\infty,n} \to 0 \quad (n \in I) \tag{55}$$

where the existence and uniqueness of the solutions of (54) is guaranteed for almost all $n \in I$. On the other hand, we may ask whether the inverse relation (52) (which takes perturbations $w_n(t)$ into account) can be inferred from the convergence of solutions of the semihomogeneous approximating problems.

Such an inference is not valid in general. We are able, however, to characterize the convergence of solutions to inhomogeneous problems by that of solutions to semihomogeneous problems if a stronger convergence property underlies the latter solutions. To formulate this property we first express the semihomogeneous problems (54) in explicit form,

$$v_n^0(0) = u_{n,0}, \quad v_n^0(t) = C_n(t)v_n^0(t'), \quad t \in [0,T]_n'. \tag{56}$$

We certainly know that under the differentiability conditions (8a,b), and under the invertibility of the $C_n^{(0)}(t)'(R_n^X u^0(t'))$, the inverses $T_n'(R_n^X u^0)^{-1}$ exist and that, if bounded, the given equations are locally solvable (cf. Chapter 6). Then, the equations of (54) are also solvable, assuming that $u_{n,0}$ lies sufficiently close to $R_n^E u_0$. The explicit equations in (56) are then easily derivable from (54), where $C_n(t) \equiv \tilde{C}_n^{(0)}(t)^{-1} C_n^{(1)}(t)$, with $\tilde{C}_n^{(0)}(t)^{-1}$ the local (i.e., in a neighborhood of $R_n^E u^0(t')$) inverse to $C_n^{(0)}(t)$. For simplicity, we shall henceforth consider the semihomogeneous approximating problems in explicit form, i.e., we set $C_n^{(0)}(t) = I$, $C_n^{(1)}(t) = C_n(t)$ and $E_n = F_n$. We can then write the solution of (56) as

$$v_n^0(t_k) = \prod_{\nu=1}^{k} C_n(t_\nu)u_{n,0}, \quad k = 1,\dots,N.$$

Under the assumptions discussed above, the $v_n^{(0)}(t')$ will always lie in the corresponding domains of definition of $C_n(t)$, $t \in [0,T]_n'$.

The solution of a semihomogeneous method corresponding to (56), where, the "initial value" is prescribed at some $t_m \in [0,T]_n$, i.e.,

$$v_n^0(t_m) = \phi_n, \quad v_n^0(t_k) = C_n(t_k)v_n^0(t_{k-1}), \quad k = m+1,\dots,N,$$

can be written as

$$v_n^0(t_k) = \prod_{\nu=m+1}^{k} C_n(t_\nu)\phi_n, \quad k = m,\dots,N.$$

This occasions the following definition. We call a semihomogeneous method of the form (56) *discretely convergent* to the semihomogeneous IVP $(1)_0$ *uniformly with respect to all initial times* if, for every $\varepsilon > 0$, there exist a $\delta > 0$, and an index $\nu \in I$, such that for all $n \geq \nu$, $\phi_n \in E_n$, $m,k \in \{0,\dots,N\}$, $k \geq m$, the relation

$$|\phi_n - R_n^E u^0(t_m)|_{E_n} \leq \delta \Rightarrow \Big|\prod_{\nu=m+1}^{k} C_n(t_\nu)\phi_n - R_n^E u^0(t_k)\Big|_{E_n} \leq \varepsilon \tag{57}$$

holds. The following result shows that this concept of convergence is the appropriate one for semihomogeneous methods for characterizing the inverse convergence relation (52) for the associated inhomogeneous equations. To state this result, we assume that the solvability conditions in statement (40a) hold along with the equicontinuous equidifferentiability of $(\tilde{T}_n^{-1})$ both of which will follow by Theorem 11.5 from the inverse stability inequality (37) used with the differentiability assumptions (8a,b). For brevity, we label the approximating method (2) as *regular* at (y_n), if (T_n) is locally bijective at (y_n) and the local inverses $\tilde{T}_n^{-1}$ themselves are continuous at y_n for every $n \in I$.

Theorem 11.10. Suppose that method (2) is regular at $(\hat{y}_n) \equiv (T_n R_n^X u^0)$, that $(\tilde{T}_n^{-1})$ is equicontinuously equidifferentiable at $(\hat{y}_n)$, and that the consistency condition (47) holds for some $p \in [1,\infty]$. Then the inverse convergence relation (52) for $p = 1$ is equivalent to the discrete convergence of the semihomogeneous method (56) to the IVP $(1)_0$ uniformly with respect to all initial times.

Proof: In the proof, we use quite essentially the fact that under the above assumptions, the inverse convergence relation (52) is equivalent to the inverse stability condition (37) for $p = 1$ (according to Theorems 6.17 and 11.5).

(i) We first show that the asserted convergence property of the semi-homogeneous method follows from (37). From Theorem 11.5, we have, with $\beta > 0$, $\sigma > 0$, a local inverse Lipschitz condition of the form

$$||\tilde{T}_n^{-1}z_n - R_n^X u^0||_{\infty,n} \le \frac{1}{\beta}||z_n - \hat{y}_n||_{1,n}$$

for all z_n in $||z_n - \hat{y}_n||_{1,n} \le \sigma$, $n \in I$. Specifically, for x_n^0 given by

$$x_n^0(0) = R_n^E u_0, \quad x_n^0(t) = C_n(t)x_n^0(t'), \quad t \in [0,T]_n', \quad n \in I,$$

we have that $T_n x_n^0 = R_n^Y w^0$, $n \in I$ (with $w^0 \equiv (u_0,0)$), and that there is a $\nu \in I$ such that

$$||x_n^0 - R_n^X u^0||_{\infty,n} \le \frac{1}{\beta}||R_n^Y w^0 - \hat{y}_n||_{1,n} = \frac{1}{\beta}||d_n^T(u^0)||_{1,n} \le \delta, \quad n \ge \nu.$$

For arbitrary $\varepsilon \in (0,\sigma/\beta)$, we set $\delta \equiv \varepsilon \min(1,\beta)/3$ and assert that (57) is valid. To see this, we let $n \ge \nu$, $m \in \{0,\ldots,N\}$, and $\phi_n \in E_n$ be such that $|\phi_n - R_n^E u^0(t_m)|_{E_n} \le \delta$. We define $z_n^0 = \phi_n$, $z_n^1(t) = 0$, $t \in [0,T]_n'$, in case $m = 0$, and

$$z_n^0 = R_n^E u_0, z_n^1(t_k) = \frac{\delta_{mk}}{\tau_n}\left(\phi_n - \prod_{\nu=1}^{m} C_n(t_\nu)R_n^E u^0\right), \quad k = 1,\ldots,N,$$

in case $m > 0$.

Then, since $x_n^0(t_k) = \prod_{\nu=1}^{k} C_n(t_\nu)R_n^E u^0$, $k = 0,\ldots,N$,

$$||z_n - R_n^Y w^0||_{1,n} = |\phi_n - x_n^0(t_m)|_{E_n}$$

$$\le |\phi_n - R_n^E u^0(t_m)|_{E_n} + |R_n^E u^0(t_m) - x_n^0(t_m)|_{E_n} \le 2\delta \le 2\sigma/3.$$

Observing that

$$(\tilde{T}_n^{-1}z_n)(t_k) = \prod_{\nu=m+1}^{k} C_n(t_\nu)\phi_n, \quad k = m,\ldots,N,$$

the asserted relation will therefore follow from the above Lipschitz condition, since

$$|\prod_{\nu=m+1}^{k} C_n(t_\nu)\phi_n - R_n^E u^0(t_k)|_{E_n} \le \frac{1}{\beta}||z_n - \hat{y}_n||_{1,n}$$

$$\le \frac{1}{\beta}(||z_n - R_n^Y w^0||_{1,n} + ||d_n^T(u^0)||_{1,n}) \le \frac{2\delta}{\beta} + \delta \le \varepsilon, \quad k = 1,\ldots,N.$$

(ii) The proof of the converse relies on the fact that the inverse stability condition (37) can be shown from the convergence property (57) for the semihomogeneous problems and the above assumptions. Because of the equidifferentiability of $(\tilde{T}_n^{-1})$ at $(\hat{y}_n)$, we have, for arbitrary $\varepsilon > 0$, the following estimate for the associated remainder terms

$$||\tilde{R}_n(\hat{y}_n;z_n - \hat{y}_n)||_{\infty,n} \leq \varepsilon||z_n - \hat{y}_n||_{1,n}, \quad n \in I,$$

whenever $||z_n - \hat{y}_n||_{1,n} \leq \eta$, $z_n \in Y_n$. Now, let μ $(\geq \nu)$ be an index such that $||d_n^T(u^0)||_{1,n} \leq \eta/3$, $\tau_n \leq 1$, and such that

$$\max_{1\leq k\leq N} |\prod_{\nu=1}^{k} C_n(t_\nu)R_n^E u_0 - R_n^E u^0(t_k)|_{E_n} \leq \eta/3, \quad n \geq \mu.$$

The latter inequality is a consequence of assumption (57), with $m = 0$ and $\phi_n = R_n^E u_0$. With δ from (57), let $\delta' \equiv \min(\eta/3,\delta,\varepsilon)$. For arbitrary $n \geq \mu$, $m \in \{0,\ldots,N\}$, $e_n \in E_n$, $|e_n|_{E_n} = 1$, we define $\phi_n \equiv R_n^E u^0(t_m)$, $\tilde{\phi}_n \equiv \phi_n + \delta' e_n$,

$$z_n^0 \equiv R_n^E u_0, \quad z_n^1(t_k) \equiv \frac{\delta_{mk}}{\tau_n}(\phi_n - \prod_{\nu=1}^{m} C_n(t_\nu)R_n^E u^0), \quad 1 \leq k \leq N,$$

$$\tilde{z}_n^0 \equiv R_n^E u_0 + \delta_{m0}\delta' e_n, \quad \tilde{z}_n^1(t_k) \equiv \frac{\delta_{mk}}{\tau_n}(\tilde{\phi}_n - \prod_{\nu=1}^{m} C_n(t_\nu)R_n^E u^0), \quad 1 \leq k \leq N.$$

We then have the relations,

$$||z_n - R_n^Y w^0||_{1,n} = |\phi_n - \prod_{\nu=1}^{m} C_n(t_\nu)R_n^E u^0|_{E_n},$$

$$||\tilde{z}_n - R_n^Y w^0||_{1,n} = |\tilde{\phi}_n - \prod_{\nu=1}^{m} C_n(t_\nu)R_n^E u^0|_{E_n},$$

and the estimates,

$$||z_n - \hat{y}_n||_{1,n} \leq ||z_n - R_n^Y w^0||_{1,n} + ||d_n^T(u^0)||_{1,n} \leq 2\eta/3,$$

$$||\tilde{z}_n - \hat{y}_n||_{1,n} \leq ||z_n - R_n^Y w^0||_{1,n} + \delta'|e_n|_{E_n} + ||d_n^T(u^0)||_{1,n} \leq \eta.$$

The associated solutions of the semihomogeneous approximating problems are given by

$$u_n(t_k) = (\tilde{T}_n^{-1} z_n)(t_k) = \prod_{\nu=m+1}^{k} C_n(t_\nu)\phi_n,$$

$$\tilde{u}_n(t_k) = (\tilde{T}_n^{-1}\tilde{z}_n)(t_k) = \prod_{\nu=m+1}^{k} C_n(t_\nu)\tilde{\phi}_n, \quad m \leq k \leq N;$$

and, by (57), the following estimates hold,

$$|u_n(t_k) - \tilde{u}_n(t_k)|_{E_n} \le |u_n(t_k) - R_n^E u^0(t_k)|_{E_n}$$
$$+ |R_n^E u^0(t_k) - \tilde{u}_n(t_k)|_{E_n} \le 2\varepsilon, \quad k = m,\dots,N.$$

When $m > 0$, $(u_n - \tilde{u}_n)(t_k) = 0$, $k = 0,\dots,m-1$. If we, in addition, make use of the representation (cf. Lemma 6.5 and (34))

$$(\tilde{T}_n^{-1})'(\hat{y}_n)(\tilde{z}_n - z_n)(t_k) = (T_n'(R_n^X u^0)^{-1}(\tilde{z}_n - z_n))(t_k)$$
$$= \delta' \prod_{\nu=m+1}^{k} C_n(t_k)'(R_n^E u^0(t_{k-1}))e_n$$

as well as

$$(\tilde{T}_n^{-1})'(\hat{y}_n)(\tilde{z}_n - z_n) = \tilde{T}_n^{-1}\tilde{z}_n - \tilde{T}_n^{-1}z_n - \tilde{R}_n(\hat{y}_n;\tilde{z}_n - \hat{y}_n) + \tilde{R}_n(\hat{y}_n;z_n - \hat{y}_n),$$

then we have

$$\left|\prod_{\nu=m+1}^{k} C_n(t_k)'(R_n^E u^0(t_{k-1}))e_n\right|_{E_n} \le \frac{1}{\delta'}||(\tilde{T}_n^{-1})'(\hat{y}_n)(\tilde{z}_n - z_n)||_{1,n}$$
$$\le \frac{1}{\delta'}\{||\tilde{u}_n - u_n||_{\infty,n} + \varepsilon||\tilde{z}_n - \hat{y}_n||_{1,n} + \varepsilon||z_n - \hat{y}_n||_{1,n}\} \le \frac{2\varepsilon}{\delta}(1+\eta),$$

which proves the asserted inequality (37) for $n \ge \nu$. For every $n < \nu$, we know that $(\tilde{T}_n^{-1})'(\hat{y}_n)$ is a bounded linear mapping, since $\tilde{T}_n^{-1}$ is differentiable and continuous at $\hat{y}_n$ (cf. Theorem 6.1). From (34), with $(\tilde{T}_n^{-1})'(\hat{y}_n) = T_n'(R_n^X u^0)^{-1}$, we get (37) for those $n < \nu$. □

In the concluding portion of this section, we apply our results to *linear problems*. Thus, we assume that we have linear mappings $A(t)$: $E \to F$, $t \in [0,T]$, $C_n^{(\ell)}(t)$: $E_n \to F_n$, $t \in [0,T]_n'$, $n \in I$, $\ell = 0,1$. We consider initial value problems in the form of (1), i.e., we allow an inhomogeneous term $w \in C([0,T],F)$. We demand of method (2) that the $C_n^{(0)}(t)$ are invertible in order that the approximating problems be solvable. On the question of convergence, we have the following theorem.

Theorem 11.11. Suppose the given IVP (1) is linear and solvable for every $u_0 \in E$, $w \in C([0,T],F)$. Further, assume the conditions of Theorem 11.6 are met for the approximation method (2) and that the consistency condition (47) is satisfied for every $v \in X$. Then the solutions of the approximating equations in (2) satisfy the inverse stability inequality

$$\max_{t\in[0,T]_n} |u_n(t)|_{E_n} \le \gamma\Big(|u_{n,0}|_{E_n} + (\tau_n \sum_{t\in[0,T]_n'} |w_n(t)|_{F_n}^p)^{1/p}\Big), \quad n \ge \nu, \tag{58}$$

if, and only if, the IVP (1) has a unique solution, the associated linear mapping $T: X \to Y$ is continuously invertible, and, for arbitrary $y = (u_0,w) \in Y$, $y_n = (u_{n,0},w_n) \in Y_n$, $n \in I$, with solutions u and u_n, $n \in I$, we have

$$||y_n - R_n^Y y||_{p,n} \to 0 \Rightarrow ||u_n - R_n^X u||_{\infty,n} \to 0 \quad (n \in I). \tag{59}$$

Proof: From (58), used in conjunction with Theorem 6.17 and Lemma 6.19, we know that the IVP (1) is always uniquely solvable and that $T_n^{-1} \to T^{-1}$ $(n \in I)$ in the sense of discrete convergence. So, T^{-1} is bounded according to Theorem 6.15. Conversely, from the inverse convergence relation (59), it follows that (T_n) is inversely stable at every sequence $(R_n^X u)$, $u \in X$, in the sense of definition 6.(11) (cf. Theorem 6.17); this is equivalent to inequality (58) (cf. 6.(12)). □

We know from Section 11.2 that the inverse stability inequality (58) follows from inequalities (42a,b) for all $p \in [1,\infty]$, and is moreover equivalent to these inequalities in case $p = 1$.

For the associated semihomogeneous problems (56), we have a result corresponding to that of Theorem 11.10, which characterizes the inverse convergence relation (59) (for $y = (u_0,0)$ and for arbitrary $y_n \in Y_n$, $n \in I$) by the convergence property (57) of the semihomogeneous approximation method. The regularity of the approximation method means in this context the continuous invertibility of T_n for every $n \in I$ (the existence of the inverses T_n^{-1} follows from that of $C_n^{(0)}(t)^{-1}$).

If we also know that E is dense in F, then we can obtain, under the assumptions of Theorem 11.11 (i.e., essentially under consistency and solvability of $(1)_0$ for every $u_0 \in E$), that the IVP $(1)_0$ is properly posed if the inverse stability inequality is satisfied (for a definition of "properly posed", see Section 4.5, also Richtmyer & Morton (1967), Chapter 3, and Meis & Marcowitz (1981), I.1).

For the case of (time) independent linear mappings A, $C_n^{(\ell)}$, it is sufficient to check the convergence (55) of solutions to the semihomogeneous problems (56) (with the solution of $(1)_0$ as the limit) in order to ensure the convergence of solutions to the approximating equations under perturbations. For this, we shall show that the inverse stability inequality (58) is necessarily present - even for $p = 1$ - if the solutions of the semihomogeneous problems converge. The inverse stability then ensures the desired results (for arbitrary $p \in [1,\infty]$) via Theorem 11.11.

Theorem 11.12. Suppose the operators occurring in (1) and (2) are linear and independent of t and that the $C_n^{(0)-1}$, $n \in I$, exist and are uniformly bounded. If, for arbitrary u_0, the semihomogeneous IVP $(1)_0$ is solvable and the convergence relation

$$|R_n^E u_0 - u_{n,0}|_{E_n} \to 0 \Rightarrow \max_{t \in [0,T]_n} |R_n^E u^0(t) - v_n^0(t)|_{E_n} \to 0 \quad (n \in I) \tag{60}$$

is satisfied, for arbitrary $u_{n,0} \in E_n$ and solutions u^0 and v_n^0 of $(1)_0$ and (56), respectively, then the inequalities in (43) hold - i.e., the inverse stability inequality (58) holds for $p = 1$.

Proof: We consider T and T_n, $n \in I$, as mappings onto $E \times \{0\}$ and $E_n \times \{0\}$, $n \in I$, respectively. Then (60) means the inverse discrete convergence of T_n to T at u^0. Since consistency is then trivially present (cf. Thm. 6.18), Theorem 6.17 yields the inverse stability of (T_n) at every sequence of solutions (v_n^0) of (56), for which

$$v_n^0(0) = u_{n,0} \to u_0 \quad (n \in I).$$

From Theorem 6.11 (or, more precisely, from inequality 6.(12)), we know that the inverse stability of the linear operators does not depend on any point, so that, with $\beta' > 0$, $\nu > 0$,

$$\beta' ||T_n^{-1} y_n||_{\infty,n} \le |y_n^0|_{E_n}, \quad y_n = (y_n^0, 0) \in E_n \times \{0\}, \quad \mu \ge \nu.$$

From the representation of the inverses of T_n, $n \in I$ (cf. (41)), it follows then (with $\gamma = 1/\beta'$) that

$$|C_n^{k-m} y_n^0|_{E_n} \le \gamma |y_n^0|_{E_n}, \quad 0 \le m \le k \le N, \quad y_n^0 \in E_n, \quad n \ge \nu,$$

which - along with the hypothesized uniform boundedness of $C_n^{(0)-1}$ - yields (43). □

We would further like to mention that we can prove the existence of generalized solutions to both linear and nonlinear initial value problems as the limit of solutions of approximating equations (cf., e.g., Meis & Marcowitz (1981), I.7, for special linear IVP's). Concerning the conditions for the nonlinear case, we refer the interested reader to the works of Kreth (1975) and the author (1975b).

REFERENCES (cf. also References in Chapter 4)

Ansorge (1978), von Dein (1976)*, John (1982), Kreth (1975)*, Meis & Marcowitz (1981), Reinhardt (1975a,1975b,1977)*, Richtmyer & Morton (1967), Stetter (1973), Törnig (1979).

*Article(s)

Chapter 12
Special Criteria for Inverse Stability

In this chapter, we analyze special criteria which guarantee for linear problems the inverse stability inequalities established in Chapter 11. These criteria strongly depend on the norms of the approximating spaces. The significance of the choice of norms was already made clear in Section 11.1 where we verified the differentiability requirements for several classes of examples. The analysis in this chapter, moreover, is applicable to nonlinear problems. Indeed, we know that the inverse stability of a nonlinear sequence of differentiable mappings is guaranteed whenever the associated sequence of Fréchet-derivatives is inversely stable.

The aim of Section 12.1 is to establish criteria for inverse stability with respect to the supremum norm of finite-difference methods with coefficients dependent on both space and time. The notion of a method of "positive type" was utilized in Forsythe & Wasow (1967), Chapter 14, to formulate a sufficient condition for inverse stability for explicit methods. A possible extension to implicit methods is suggested by Meis & Marcowitz (1981), §8. Their concept, however, only makes possible via Lemma 11.7 - and not directly - the treatment of the most simple (explicit) method for $u_t = u_{xx} + u$. Our extension of this treatment introduces the notion of a "positive method", and we prove that such methods are necessarily inversely stable whenever an additional growth condition on the coefficients is imposed. We shall check our conditions for a series of examples and, moreover, indicate how approximations of Neumann boundary conditions can affect the estimates of inverse stability.

In Section 12.2, we show how the well-known von Neumann stability criterion fits into our treatment of inverse stability. It turns out that this criterion is necessary - and in some situations also sufficient - for inverse stability in the special case of linear methods having constant coefficients with the approximating spaces consisting of periodic mesh functions equipped with the discrete L^2-norm. The von Neumann condition is basic for the general Lax-Richtmyer theory.

In Section 12.3, we establish inverse stability estimates for three different types of Galerkin methods having suitable Sobolev norms as underlying norms. For the case of continuous, piecewise linear basis functions, these methods will also be analyzed in the Sections 12.1 and 12.2 mentioned above. It becomes rather clear that one single method can exhibit different stability behaviors with respect to different norms, and, therefore, that different stability criteria may occur.

The last Section 12.4 is devoted to the inverse stability of concrete nonlinear approximation methods. For several finite-difference and Galerkin methods, we demonstrate how the results available for linear problems can be utilized to prove inverse stability for nonlinear methods. Indeed, this can be achieved by studying the associated linearized equations. From our investigations in Sections 12.1 and 12.3, we can provide criteria for inverse stability of nonlinear finite-difference and Galerkin methods with respect to supremum norms and Sobolev norms, respectively. The von Neumann condition from Section 12.2, however, is not applicable to nonlinear problems since the linearized equations contain coefficients which do depend, in general, on space and time.

12.1. LINEAR FINITE-DIFFERENCE METHODS WITH POSITIVITY PROPERTIES

In this section, we present criteria which guarantee the inverse stability with respect to the supremum norms (in the approximating spaces of grid functions) of linear, implicit finite-difference methods with coefficients depending on both space and time. The essential conditions needed are a positivity property (cf. (5a,b)) together with a growth condition (cf. (6)) on the coefficients. For a series of methods introduced in Chapter 4, we investigate whether the above mentioned criteria for inverse stability are satisfied. In most examples, a restriction on the associated mesh ratio must be required. This is the motivation for calling such methods "conditionally stable". As a consequence, inverse stability is only present when the time steps are proportional to h_n^2

(for parabolic problems) and to h_n (for hyperbolic problems) where h_n denotes a given mesh width in the spatial variable. We are therefore forced to compute approximations for a large number of time steps in order to reach a specific time when h_n is small.

Let us consider linear finite-difference methods of the following form,

$$\sum_{|\mu|\leq N_0} B^{(0)}_{n,\mu}(\cdot,t)T^{\mu}_n u_n(t) = \sum_{|\mu|\leq N_1} B^{(1)}_{n,\mu}(\cdot,t)T^{\mu}_n u_n(t') + \tau_n w_n(t), \tag{1}$$
$$t \in [0,T]'_n, \quad n \in I,$$

where, again, $t' = t-\tau_n$; T^{μ}_n denote the shift operators

$$(T^{\mu}_n f)(x) = f(x + \mu h_n), \quad x \in G_n, \quad f \in C(G_n);$$

G_n are uniform meshes in $\mathbb{R}$ with mesh widths h_n; $E_n = F_n = C(G_n)$ are the spaces of all bounded mesh functions (defined on G_n) with values in $\mathbb{R}^{\iota}$ ($\iota \in \mathbb{N}$); and $B^{(\ell)}_{n,\mu}(x,t)$, $|\mu| \leq N_1$, $\ell = 0,1$, $n \in I$, represent $\iota \times \iota$-matrices. We thus restrict our analysis to one spatial dimension, but allow the coefficients, i.e., the matrices $B^{(\ell)}_{n,\mu}(x,t)$, to depend on x and t. It is clear how the operators $C^{(\ell)}_n(t)$ appearing in the general form 11.(2) of an approximation method to an initial value problem are to be defined.

According to Forsythe & Wasow (1967), Sec. 14.1, a scalar explicit method (i.e., $r = 1$, $N_0 = 0$), given by

$$u_n(x,t) = \sum_{|\mu|\leq N_1} b^{(1)}_{n,\mu}(x,t)u_n(x+\mu h_n,t') + \tau_n w_n(x,t) \tag{2}$$
$$(x,t) \in G_n \times [0,T]'_n, \quad n \in I,$$

is said to be of *positive type* if $b^{(1)}_{n,\mu}(x,t) \geq 0$ for all $|\mu| \leq N_1$, $(x,t) \in G_n \times [0,T]'_n$, $n \geq \nu_0$, with some $\nu_0 \in \mathbb{N}$.

We can generalize this concept to implicit methods of the form (1). For this purpose, we assume that the matrices $B^{(\ell)}_{n,\mu}$ are *uniformly diagonalizable* which means that there exist regular matrices M_n, $n \in I$, independent of x and t which, together with their inverses, are uniformly bounded,

$$||M_n||_{\infty} \leq C_M, \quad ||M_n^{-1}||_{\infty} \leq C_M, \quad n \in I,$$

such that

$$D_{n,\mu}^{(\ell)}(x,t) \equiv M_n^{-1}B_{n,\mu}^{(\ell)}(x,t)M_n, \quad (x,t) \in G_n \times [0,T]_n', \quad |\mu| \leq N_1,$$
$$\ell = 0,1, \quad n \in I,$$

are diagonal matrices. (Here, $||\cdot||_\infty$ denotes the maximum absolute row sum of a matrix.) We can then rewrite the finite-difference method (1) in terms of the diagonal matrices and obtain the following equations for $v_n(t) = M_n^{-1}u_n(t)$,

$$\sum_{|\mu|\leq N_0} D_{n,\mu}^{(0)}(\cdot,t)T_n^\mu v_n(t) = \sum_{|\mu|\leq N_1} D_{n,\mu}^{(1)}(\cdot,t)T_n^\mu v_n(t') + \tau_n M_n^{-1} w_n(t),$$
$$t \in [0,T]_n', \quad n \in I. \tag{3}$$

The diagonal entries are denoted by $d_{n,\mu,j}^{(\ell)}(x,t)$, $j = 1,\ldots,\iota$. We further assume that the $B_{n,\mu}^{(0)}(x,t)$ are uniformly bounded, i.e.,

$$||B_{n,\mu}^{(0)}(x,t)||_\infty \leq C_B, \quad |\mu| \leq N_0, \quad (x,t) \in G_n \times [0,T]_n', \; n \in I, \tag{4}$$

and note that the diagonal matrices $D_{n,\mu}^{(0)}(x,t)$ are then uniformly bounded, too.

We now call a finite-difference method with uniformly diagonalizable coefficients *positive* if, for some numbers $\rho_0 > 0$, $n_0 \in I$, the following conditions are satisfied:

$$d_{n,\mu,j}^{(1)}(x,t) \geq 0, \quad |\mu| \leq N_1, \quad 1 \leq j \leq \iota, \quad (x,t) \in G_n \times [0,T]_n',$$
$$n \geq n_0, \tag{5a}$$

$$d_{n,0,j}^{(0)}(x,t) - \sum_{0\neq|\mu|\leq N_0} |d_{n,\mu,j}^{(0)}(x,t)| \geq \rho_0 > 0, \quad 1 \leq j \leq \iota,$$
$$(x,t) \in G_n \times [0,T]_n', \quad n \geq n_0. \tag{5b}$$

We would first like to mention that, in case of $d_{n,\mu,j}^{(0)} \leq 0$, $1 \leq j \leq \iota$, $\mu \neq 0$ (as required in Forsythe & Wasow (1967), (15.5)), condition (5b) is equivalent to

$$\sum_{|\mu|\leq N_0} d_{n,\mu,j}^{(0)}(x,t) \geq \rho_0 > 0, \quad 1 \leq j \leq \iota, \quad (x,t) \in G_n \times [0,T]_n', \; n \geq n_0.$$

As a second remark, (5a) obviously reduces to the above definition of "positive type" for a scalar, explicit method provided (5b), i.e., $d_{n,\mu}^{(0)} \geq \rho_0$, is satisfied (we then set $b_{n,\mu}^{(1)} \equiv d_{n,\mu}^{(1)}/d_{n,0}^{(0)}$ in (2)).

The property "positive" alone is not sufficient for ensuring the inverse stability with respect to the *supremum norms* defined by

$$|g|_{0,\infty} \equiv \max_{1\le j\le \iota} \sup_{x\in G_n} |g_j(x)|, \quad g = (g_1,\dots,g_\iota) \in C(G_n), \quad n \in I.$$

(The dependence on the index n is not explicitly expressed in the notation.) The desired stability behavior can be proved if we require an additional property on the coefficients, namely that

$$\frac{\sum_{|\mu|\le N_1} d^{(1)}_{n,\mu,j}(x,t)}{d^{(0)}_{n,0,j}(x,t) - \sum_{0\ne|\mu|\le N_0} |d_{n,\mu,j}(x,t)|} \le 1 + O(\tau_n) \quad (n \in I), \qquad j = 1,\dots,\iota. \tag{6}$$

The nominator is positive provided that (5b) holds. We can now state the central result of this section.

Theorem 12.1. Let the finite-difference method (1), with uniformly diagonalizable matrices, be positive and satisfy conditions (4) and (6). Then the associated difference operators $C^{(0)}_n(t)$, $t \in [0,T]'_n$, $n \ge n_0$, are bijective and continuously invertible, and, moreover, the inverse stability inequality 11.(42) is satisfied with respect to the supremum norms $|\cdot|_{0,\infty}$.

Proof: (i) The equation $C^{(0)}_n(t)g = f$ is equivalent to

$$\sum_{|\mu|\le N_0} D^{(0)}_{n,\mu}(x,t)M_n^{-1}g(x + \mu h_n) = M_n^{-1}f(x), \quad x \in G_n, \tag{7a}$$

or, with $\hat{g} \equiv M_n^{-1}g$, $\hat{g} = (\hat{g}_1,\dots,\hat{g}_\iota)$, and $\hat{f}$ given analogously, it is equivalent to

$$d^{(0)}_{n,0,j}\hat{g}_j(x) + \sum_{0\ne|\mu|\le N_0} d^{(0)}_{n,\mu,j}\hat{g}_j(x+\mu h_n) = \hat{f}_j(x), \qquad x \in G_n, \quad 1 \le j \le \iota, \tag{7b}$$

and to

$$\hat{g}_j(x) - \sum_{0\ne|\mu|\le N_0} \frac{-d^{(0)}_{n,\mu,j}}{d^{(0)}_{n,0,j}} \hat{g}_j(x+\mu h_n) = \frac{\hat{f}_j(x)}{d^{(0)}_{n,0,j}}, \quad x \in G_n, \qquad 1 \le j \le \iota. \tag{7c}$$

(The arguments (x,t) will be omitted in the following.) We define

$$(U_n(t)g)_j(x) \equiv \sum_{0\ne|\mu|\le N_0} \frac{-d^{(0)}_{n,\mu,j}}{d^{(0)}_{n,0,j}} g_j(x + \mu h_n), \quad g \in C(G_n),$$
$$(x,t) \in G_n \times [0,T]'_n, \quad 1 \le j \le \iota,$$

and proceed to prove that

$$|U_n(t)g|_{0,\infty} \le q|g|_{0,\infty}, \quad g \in C(G_n), \quad t \in [0,T]_n', \quad n \ge n_0, \tag{8}$$

with some constant $q < 1$ (independent of n,t). From (5b), it follows that

$$\sum_{0\neq|\mu|\le N_0} \frac{|d^{(0)}_{n,\mu,j}|}{d^{(0)}_{n,0,j}} \le \frac{\sum_{0\neq|\mu|\le N_0}|d^{(0)}_{n,\mu,j}|}{\rho_0 + \sum_{0\neq|\mu|\le N_0}|d^{(0)}_{n,\mu,j}|}, \quad 1 \le j \le \iota, \quad n \ge n_0.$$

The function $\alpha(z) \equiv z/(\rho_0+z)$ is monotonically increasing (for $z \ge 0$), and the coefficients $d^{(0)}_{n,\mu,j}$ are uniformly bounded because of (4) (with bound $C_D \equiv C_B C_M^2$), which implies that

$$\sum_{0\neq|\mu|\le N_0} |d^{(0)}_{n,\mu,j}|/\Big\{\rho_0 + \sum_{0\neq|\mu|\le N_0}|d^{(0)}_{n,\mu,j}|\Big\} \le \frac{2N_0C_D}{\rho_0+2N_0C_D} \equiv q < 1,$$
$$1 \le j \le \iota, \quad n \ge n_0.$$

We thus have the desired estimate (8) for $U_n(t)$, since

$$|(U_n(t)g)_j(x)| \le d^{(0)-1}_{n,0,j} \sum_{0\neq|\mu|\le N_0} |d^{(0)}_{n,\mu,j}||g_j(x+\mu h_n)|$$
$$\le q \sup_{y\in G_n} |g_j(y)|, \quad x \in G_n, \quad 1 \le j \le \iota, \quad n \ge n_0.$$

As a consequence of a well-known result, we now know that equation (7c) is uniquely solvable for every $\hat{f}$ and, together with $d^{(0)}_{n,0,j} \ge \rho_0 > 0$, we obtain the following estimate for its solution $\hat{g}$,

$$|\hat{g}|_{0,\infty} \le \frac{1}{\rho_0(1-q)}|\hat{f}|_{0,\infty}, \quad n \ge n_0.$$

Finally, the uniform boundedness of the sequences M_n and M_n^{-1}, $n \in I$, yields

$$|g|_{0,\infty} \le C_M|\hat{g}|_{0,\infty} \le \frac{C_M}{\rho_0(1-q)}|\hat{f}|_{0,\infty} \le \frac{C_M^2}{\rho_0(1-q)}|f|_{0,\infty}, \quad n \ge n_0,$$

for the solution $g = M_n\hat{g}$ of $C_n^{(0)}(t)g = f$, thereby proving the bijectivity and continuous invertibility of $C_n^{(0)}(t)$.

(ii) In order to prove the inverse stability estimate 11.(42), we define

$$(\hat{C}_n^{(\ell)}(t)g)(x) \equiv \sum_{|\mu|\leq N_\ell} D_{n,\mu}^{(\ell)}(x,t)g(x+\mu h_n), \quad g \in C(G_n), \tag{9}$$
$$\ell = 0,1, \quad n \in I.$$

As a result of Part (i) of this proof, $\hat{C}_n^{(0)}(t)^{-1} = (I - U_n(t))^{-1}$, $t \in [0,T]_n'$, $n \geq n_0$, exist and are uniformly bounded. We now derive an estimate for $\hat{C}_n(t) \equiv \hat{C}_n^{(0)}(t)^{-1}\hat{C}_n^{(1)}(t)$. The function $g = \hat{C}_n(t)f$ is a solution of

$$\hat{C}_n^{(0)}(t)g = \hat{C}_n^{(1)}(t)f,$$

or, equivalently, (cf. (7b)) of

$$d_{n,0,j}^{(0)}g_j(x) = -\sum_{0\neq|\mu|\leq N_0} d_{n,\mu,j}^{(0)}g_j(x+\mu h_n) + \sum_{|\mu|\leq N_1} d_{n,\mu,j}^{(1)}f_j(x+\mu h_n),$$
$$j = 1,\ldots,\iota, \quad (x,t) \in G_n \times [0,T]_n', \quad n \in I.$$

By means of (5a,b) and (6), we obtain the estimates

$$d_{n,0,j}^{(0)}|g_j(x)| \leq \sum_{0\neq|\mu|\leq N_0} |d_{n,\mu,j}^{(0)}||g_j(x+\mu h_n)| + \sum_{|\mu|\leq N_1} d_{n,\mu,j}^{(1)}|f_j(x+\mu h_n)|$$
$$\leq \Big(\sum_{0\neq|\mu|\leq N_0} |d_{n,\mu,j}^{(0)}|\Big)\sup_{y\in G_n}|g_j(y)| + \Big(\sum_{|\mu|\leq N_1} d_{n,\mu,j}^{(1)}\Big)\sup_{y\in G_n}|f_j(y)|,$$

$$\Big\{d_{n,0,j}^{(0)} - \sum_{0\neq|\mu|\leq N_0} |d_{n,\mu,j}^{(0)}|\Big\}\sup_{x\in G_n}|g_j(x)| \leq \Big(\sum_{|\mu|\leq N_1} d_{n,\mu,j}^{(1)}\Big)\sup_{x\in G_n}|f_j(x)|.$$
$$j = 1,\ldots,\iota, \quad (x,t) \in G_n \times [0,T]_n', \quad n \geq n_0,$$

and

$$\sup_{x\in G_n}|g_j(x)| \leq (1+C\tau_n)\sup_{x\in G_n}|f_j(x)|, \quad j = 1,\ldots,\iota, \quad n \geq n_1 \ (\geq n_0),$$

which, taken together, prove

$$||\hat{C}_n(t)|| \leq (1+C\tau_n), \quad t \in [0,T]_n', \quad n \geq n_1. \tag{10}$$

In order to derive the stability inequalities 11.(42a,b), we observe that

$$C_n^{(\ell)}(t)g = \sum_{|\mu|\leq N_\ell} M_n D_{n,\mu}^{(\ell)}(\cdot,t)M_n^{-1}T_n^{\mu}g = M_n\hat{C}_n^{(\ell)}(t)M_n^{-1}g, \quad g \in C(G_n),$$
$$\ell = 0,1, \quad t \in [0,T]_n', \quad n \in I,$$

$$C_n^{(0)}(t)^{-1}f = M_n\hat{C}_n^{(0)}(t)^{-1}M_n^{-1}f, \quad f \in C(G_n), \quad t \in [0,T]_n', \quad n \in I,$$

and

$$\prod_{\nu=1}^{k} C_n(t_\nu)g = M_n \prod_{\nu=1}^{k} \hat{C}_n(t_\nu)M_n^{-1}g,$$

$$\prod_{\nu=m+1}^{k} C_n(t_\nu)C_n^{(0)}(t_m)^{-1}g = M_n \prod_{\nu=m+1}^{k} \hat{C}_n(t_\nu)\hat{C}_n^{(0)}(t_m)^{-1}M_n^{-1}g, \quad g \in C(G_n),$$

$$1 \leq m \leq k \leq N.$$

By (10), it follows that

$$\left\|\prod_{\nu=m+1}^{k} \hat{C}_n(t_\nu)\right\| \leq (1+C\tau_n)^{k-m} \leq \exp(C(k-m)\tau_n) \leq \exp(CT),$$

$$0 \leq m \leq k \leq N, \quad n \geq n_1,$$

and hence

$$\left\|\prod_{\nu=1}^{k} C_n(t_\nu)\right\| \leq C_M^2\exp(CT),$$

as well as

$$\left\|\prod_{\nu=m+1}^{k} C_n(t_\nu)C_n^{(0)}(t_m)^{-1}\right\| \leq \frac{1}{\rho_0(1-q)}\|M_n\|_\infty\|M_n^{-1}\|_\infty\exp(CT)$$

$$\leq \frac{C_M^2}{\rho_0(1-q)}\exp(CT), \quad 1 \leq m \leq k \leq N, \quad n \geq n_1. \qquad \square$$

In the treatise by Meis & Marcowitz (1981), Sec. I.8, finite-difference methods of the following form (after diagonalizing) are considered,

$$(\hat{C}_n^{(0)}g)_j(x) = \left[\sum_{\mu=-\kappa}^{\kappa} \alpha_{n,\mu,j}(x) + \sum_{\mu \neq -\kappa}^{\kappa} \beta_{n,\mu,j}(x)\right] g_j(x)$$

$$- \sum_{\mu \neq -\kappa}^{\kappa} \beta_{n,\mu,j}(x)g_j(x+\mu h_n), \tag{11}$$

$$(\hat{C}_n^{(1)}g)_j(x) = \sum_{\mu=-\kappa}^{\kappa} \alpha_{n,\mu,j}(x)g_j(x+\mu h_n), \quad j = 1,\dots,\iota,$$

$$g \in C(G_n), \quad n \in I,$$

where the coefficients $\alpha_{n,\mu,j}$, $\beta_{n,\mu,j}$ are assumed to be nonnegative and uniformly bounded, and where $\sum_\mu \alpha_{n,\mu,j} \geq 1$. For methods of this type, the following corollary of Theorem 12.1 ensures inverse stability.

<u>Theorem 12.2</u>. Finite-difference methods with uniformly diagonalizable coefficients and having the form (11) are positive and satisfy the conditions (4) and (6), provided that the $\alpha_{n,\mu,j}$ and $\beta_{n,\mu,j}$ are uniformly bounded, nonnegative and fulfill $\sum_\mu \alpha_{n,\mu,j} \geq 1$; in this case, condition (6) holds with bound 1 in place of $1 + O(\tau_n)$.

Proof: The positivity condition (5a) is satisfied because $\alpha_{n,\mu,j} \geq 0$; (5b) holds since $\beta_{n,\mu,j} \geq 0$ and

$$d^{(0)}_{n,0,j} - \sum_{0 \neq |\mu| \leq \kappa} |d^{(0)}_{n,\mu,j}| = \sum_{\mu=-\kappa}^{\kappa} \alpha_{n,\mu,j} + \sum_{\substack{\mu=-\kappa\\ \mu\neq 0}}^{\kappa} \beta_{n,\mu,j} - \sum_{\substack{\mu=-\kappa\\ \mu\neq 0}}^{\kappa} \beta_{n,\mu,j}$$

$$= \sum_{\mu=-\kappa}^{\kappa} \alpha_{n,\mu,j} \geq 1.$$

The boundedness property (4) is satisfied by assumption. Condition (6) finally follows from

$$\sum_{\mu=-\kappa}^{\kappa} d^{(1)}_{n,\mu,j} \Big/ \Big\{ d^{(0)}_{n,0,j} - \sum_{\substack{\mu=-\kappa\\ \mu\neq 0}}^{\kappa} d^{(0)}_{n,\mu,j} \Big\} = \sum_{\mu=-\kappa}^{\kappa} \alpha_{n,\mu,j} \Big/ \sum_{\mu=-\kappa}^{\kappa} \alpha_{n,\mu,j} = 1. \qquad \square$$

To conclude this section, we consider several methods and ascertain under what conditions the positivity (5a,b) and the further assumptions of Theorem 12.1 are present. In Section 4.5, we have already given their representation in the form of (1). Condition (4) is trivially satisfied for all examples studied below whenever the coefficients are constant and independent of n (and of h_n).

Method 4.(8) (Crank-Nicolson method with general Θ in $0 \leq \Theta \leq 1$):

$$B^{(1)}_{n,-1} = B^{(1)}_{n,1} = ar(1-\Theta) \geq 0, \quad B^{(1)}_{n,0} = 1 - 2ar(1-\Theta) \geq 0,$$

$$\text{provided} \quad r \leq \frac{1}{2a(1-\Theta)};$$

$$B^{(0)}_{n,0} - \sum_{0\neq|\mu|\leq 1} |B^{(0)}_{n,\mu}| = \sum_{|\mu|\leq 1} B^{(0)}_{n,\mu} = 1,$$

$$\sum_{|\mu|\leq 1} B^{(1)}_{n,\mu} \Big/ \sum_{|\mu|\leq 1} B^{(0)}_{n,\mu} = 1.$$

This method is positive and satisfies condition (6) provided the mesh ratio $r = \tau_n/h_n^2$ is restricted by $r \leq 1/(2a(1-\Theta))$. Hence, for $\Theta = 1$, i.e., for the totally implicit method 4.(6), we do not need a requirement on r. $\square$

Explicit Method 4.(13) (with variable coefficients): Under the assumption that $a(x,t) \geq \alpha_0 > 0$, for all x,t, we have

$$B^{(1)}_{n,\pm 1}(x,t) = a(x,t) > 0, \quad B^{(1)}_{n,0}(x,t) = 1 - 2ra(x,t) \geq 0$$

provided $r \leq 1/(2\alpha_0)$. Then (6) is also satisfied, since

$$\sum_{|\mu|\leq 1} B^{(1)}_{n,\mu}(x,t)/B^{(0)}_{n,0}(x,t) = 1. \quad \square$$

ADI Method 4.(16) of Peaceman-Rachford: We check the positivity properties of each single factor of this product method. For the first one, we get

$$C_n^{(\ell),1}:\ B^{(1),1}_{n,\pm 1} = \tfrac{1}{2}a_2 \geq 0,\quad B^{(1),1}_{n,0} = 1-a_2 \geq 0, \text{ provided } a_2 \leq 1;$$

$$B^{(0),1}_{n,0} - \sum_{0\neq|\mu|\leq 1} |B^{(0),1}_{n,\mu}| = \sum_{|\mu|\leq 1} B^{(0),1}_{n,\mu} = 1;$$

$$\sum_{|\mu|\leq 1} B^{(1),1}_{n,\mu} / \sum_{|\mu|\leq 1} B^{(0),1}_{n,} = 1, \text{ provided } a_2 \leq 1.$$

The method defined by the first equation in 4.(16) is positive and satisfies condition (6) in case $a_2 \leq 1$, i.e., $r_2 \leq 1/a$. For the finite-difference method defined by the second equation in 4.(16), we can analogously guarantee conditions (5) and (6) under the restriction $a_1 \leq 1$, i.e., $r_1 \leq 1/a$. Altogether, we get $||C_n|| \leq 1$ for the difference operator associated with the ADI method 4.(16) provided $r_\nu \leq 1/a$, $\nu = 1,2$. The latter represents a restriction on the mesh ratio in each spatial direction analogous to that required for the one-dimensional Crank-Nicolson method. $\square$

Crank-Nicolson-Galerkin Method 4.(22):

$$B^{(1)}_{n,-1} = B^{(1)}_{n,1} = \tfrac{1}{6} + \tfrac{1}{2}ar \geq 0,\quad B^{(1)}_{n,0} = \tfrac{2}{3} - ar \geq 0, \text{ provided } r \leq \tfrac{2}{3a};$$

$$B^{(0)}_{n,\mu} = \tfrac{1}{6} - \tfrac{1}{2}ar \leq 0,\quad \mu = -1,1, \text{ provided } r \geq 1/(3a);$$

$$B^{(0)}_{n,\mu} - \sum_{0\neq|\mu|\leq 1} |B^{(0)}_{n,\mu}| = \sum_{|\mu|\leq 1} B^{(0)}_{n,\mu} = 1, \text{ provided } r \geq 1/(3a);$$

$$\sum_{|\mu|\leq 1} B^{(1)}_{n,\mu}/(B^{(0)}_{n,0} - \sum_{0\neq|\mu|\leq 1} |B^{(0)}_{n,\mu}|) = 1, \text{ provided } r \geq 1/(3a).$$

The method 4.(22) is positive, if $r \leq 2/(3a)$, and satisfies condition (6) in case $r \geq 1/(3a)$. $\square$

Friedrichs' Method 4.(33): $\iota = 2$, $N_0 = 0$, $N_1 = 1$,

$$A = c\begin{pmatrix} 0 & 1 \\ 1 & 0 \end{pmatrix},\quad PAP^{-1} = D = c\begin{pmatrix} 1 & 0 \\ 0 & -1 \end{pmatrix},\quad M_n = P^{-1},$$

$$D^{(0)}_{n,0} = I,\quad D^{(1)}_{n,-1} = \tfrac{1}{2}(I-\lambda D),\quad D^{(1)}_{n,0} = 0,\quad D^{(1)}_{n,1} = \tfrac{1}{2}(I+\lambda D).$$

$$d^{(1)}_{n,\mu,j} \geq 0, \quad |\mu| \leq 1, \quad j = 1,2, \text{ provided } 1-\lambda c \geq 0, \text{ i.e., } \lambda \leq 1/c.$$

$$\sum_{|\mu|\leq 1} d^{(1)}_{n,\mu,j}/d^{(0)}_{n,0,j} = 1, \quad j = 1,2.$$

Friedrichs' method 4.(33) is positive and satisfies condition (6) whenever the mesh ratio $\lambda = \tau_n/h_n$ is restricted by $\lambda \leq 1/c$; this restriction is called the *Courant-Friedrichs-Lewy condition*. □

Courant-Isaacson-Rees Method 4.(34): $\iota = 2$, $N_0 = 0$, $N_1 = 1$,

$$D^{(0)}_{n,0} = I, \quad D^{(1)}_{n,-1} = -\lambda D^-, \quad D^{(1)}_{n,0} = (1-\lambda|c|)I, \quad D^{(1)}_{n,1} = \lambda D^+,$$

$$D^+ = c\begin{pmatrix}1 & 0\\ 0 & 0\end{pmatrix}, \quad D^- = c\begin{pmatrix}0 & 0\\ 0 & -1\end{pmatrix} \quad (\text{since } c > 0).$$

$$d^{(1)}_{n,\mu,j} \geq 0, \quad \mu = -1,1, \quad j = 1,2; \quad d^{(1)}_{n,0,j} \geq 0, \quad j = 1,2,$$

$$\text{provided } \lambda \leq 1/c;$$

$$\sum_{|\mu|\leq 1} d^{(1)}_{n,\mu,j}/d^{(0)}_{n,0,j} = 1, \quad j = 1,2.$$

The Courant-Isaacson-Rees method 4.(34) is positive and satisfies (6) again under the Courant-Friedrichs-Lewy condition $\lambda \leq 1/c$. □

Lax-Wendroff Method 4.(35): $\iota = 2$, $N_0 = 0$, $N_1 = 1$. Because of $D^2 = c^2 I$, we have $D^{(0)}_{n,0} = I$, $D^{(1)}_{n,0} = I - \lambda^2 D^2 = (1 - \lambda^2 c^2)I$,

$$D^{(1)}_{n,-1} = -\tfrac{1}{2}\lambda D(I-\lambda D) = \tfrac{1}{2}\lambda c\begin{pmatrix}\lambda c-1 & 0\\ 0 & \lambda c+1\end{pmatrix},$$

$$D^{(1)}_{n,1} = \tfrac{1}{2}\lambda c\begin{pmatrix}\lambda c+1 & 0\\ 0 & \lambda c-1\end{pmatrix}$$

$$d^{(1)}_{n,\mu,j} \geq 0, \quad \mu = -1,1, \quad j = 1,2, \text{ provided } \lambda \geq 1/c;$$

$$d^{(1)}_{n,0,j} \geq 0, \quad j = 1,2, \text{ provided } \lambda \leq 1/c; \text{ and}$$

$$\sum_{|\mu|\leq 1} d^{(1)}_{n,\mu,j}/d^{(0)}_{n,0,j} = 1, \quad j = 1,2.$$

The Lax-Wendroff method 4.(35) is positive and satisfies (6) whenever $\lambda = 1/c$. □

For all methods considered up to now, we were able to guarantee inverse stability (with respect to the supremum norms) for the case of uniform meshes $G_n = \{x_j = jh_n : j = 0,\pm 1,\pm 2,\ldots\}$ in $\mathbb{R}$. The statements of Theorem 12.1 also provide analogous stability estimates in case of a finite spatial interval [0,1]. It is easy to see that, with Dirichlet boundary conditions, the approximate solutions obtained via the above methods satisfy the estimates (again, $h_nJ = 1$)

$$\max_{0\le j\le J} |v_j^k| \le C\Big\{\max_{0\le j\le J} |u_{0,j}| + \max_{1\le\nu\le k} (|\gamma_0^\nu|,|\gamma_1^\nu|) + \tau_n \sum_{\nu=1}^{k} \max_{1\le j\le J-1} |w_n(x_j,t_\nu)|\Big\}, \quad k = 1,\dots,N,\ n \in I, \tag{12}$$

if the aforementioned restrictions on the mesh ratio for each of the above methods are satisfied. The particular inhomogeneous right-hand sides are given in Section 4.5 for each of the methods and the γ_0^k and γ_1^k denote given boundary values at $x_0 = 0$ and $x_J = 1$, respectively. For approximate methods for the wave equations, the corresponding quantities in (12) are vectors of two components.

Theorem 12.1, moreover, provides stability estimates for methods which include approximations of derivatives in the boundary conditions. We present such a result for Example 4.(10) in case Neumann boundary conditions occur (cf. the representation of this method in Section 4.5).

Method 4.(10) (with $\alpha_0 = \beta_0 = 0$, $\alpha_1 = \beta_1 = 1$): Under the assumption $r \le 1/[2a(1-\Theta)]$ (i.e., the positivity assumption for method 4.(8)), the solutions of 4.(10) satisfy the estimates

$$\begin{aligned}\max_{0\le j\le J} |v_j^k| &\le C\Big\{\max_{0\le j\le J} |u_{0,j}| + \tau_n \sum_{\nu=1}^{k} \max_{0\le j\le J} |w_n(x_j,t_\nu)|\Big\}\\ &\le C\Big\{\max_{0\le j\le J} |u_{j,0}| + \tau_n\Big[\sum_{\nu=0}^{k} \max_{0\le j\le J} |s_j| + \frac{2a}{h_n}\max(|\gamma_0^\nu|,|\gamma_1^\nu|)\Big]\Big\},\end{aligned} \tag{13}$$

$$k = 0,\dots,N, \quad n \in I.$$

These follow immediately from (12) with the representation of the inhomogeneous terms w_n given in Section 4.5. We observe that the boundary data do appear in the stability estimate with a factor of magnitude $O(\tau_n^{-1/2})$ (since $1/h_n = (r/\tau_n)^{1/2}$). □

The aforementioned sufficient criteria for inverse stability do not apply for three-level methods which (considered as a two-level system) are not diagonalizable. For the example of the Du Fort-Frankel method, we show, however, how positivity properties can also be utilized for this case.

Du Fort-Frankel Method 4.(11): The approximate solutions $v_j^k = u_n(x_j,t_k)$, $k = 0,\dots,N$, obtained by the Du Fort-Frankel method clearly satisfy the following estimates

$$|v_j^{k+1}| \le \frac{1}{1+2ar}\{|1-2ar||v_j^{k-1}| + 2ar(|v_{j+1}^k| + |v_{j-1}^k|) + 2\tau_n|s_j^k|\},$$

$$j = 0,\pm 1,\dots, \quad k = 1,\dots,N-1.$$

If we set (analogously to the splitting 4.(11c)) $w_j^k = v_j^{k-1}$ and $\underline{v}_j^k = (v_j^k,w_j^k)$, $k = 1,\dots,N$, then it follows that

$$|v_j^{k+1}| \le \frac{|1-2ar|+2ar}{1+2ar}|\underline{v}^k|_{0,\infty} + \frac{2\tau_n}{1+2ar}|s^k|_{0,\infty}, \quad j = 0,\pm 1,\dots,$$

where the supremum norm of the vector-valued mesh functions is also denoted by

$$|\underline{v}^k|_{0,\infty} \equiv \max(\sup_j|v_j^k|, \sup_j|w_j^k|).$$

We observe that

$$|1 - 2ar| + 2ar = \begin{cases} 1 & \text{if } 0 < ar \le 1/2, \\ 4ar-1, & \text{if } ar \ge 1/2. \end{cases}$$

If we restrict r in the second case by $1 \ge ar$ ($\ge 1/2$), then $4ar-1 < 1 + 2ar$; in any case we get

$$|\underline{v}^{k+1}|_{0,\infty} \le |\underline{v}^k|_{0,\infty} + 2\tau_n|s^k|_{0,\infty} \le |\underline{v}^1|_{0,\infty} + 2\tau_n \sum_{\nu=1}^{k} |s^\nu|_{0,\infty} \tag{14}$$

$$k = 1,\dots,N-1.$$

The restriction $r \le 1/a$ is the same as that for the Crank-Nicolson method 4.(8) (with $\Theta = 1/2$). □

For all these methods - with the exception of the totally implicit method 4.(6) - we have imposed a restriction on the mesh ratio in order to satisfy conditions (5) and (6); for the methods 4.(22) and 4.(35) we additionally need the mesh ratio to be bounded from below. We call an approximation method for an initial value problem *conditionally stable* (or more precisely: *conditionally inversely stable*) if it is inversely stable only for a restricted ratio of the mesh widths in the x- and t-directions. Otherwise it is called *unconditionally (inversely) stable*. This property of being conditionally or unconditionally stable depends on the underlying norms (in this section: $|\cdot|_{E_n} = |\cdot|_{F_n} = |\cdot|_{0,\infty}$) which will be made more clear in the following sections.

12.2. THE VON NEUMANN CONDITION

In this section, we show how the results of the classical Lax-Richtmyer theory fit into our framework. It turns out that the classical stability condition, namely the uniform boundedness of the powers of the amplification matrices, is equivalent to our inverse stability inequality when the underlying spaces are discrete L^2-spaces and the coefficients are constant. In this special setting, the well-known von Neumann condition is a necessary condition for inverse stability and is, in some cases, also sufficient (cf. Theorems 12.5 to 12.7). We shall shorten our presentation at some places of the text where we cite results without proof and refer to the detailed treatment of Meis & Marcowitz (1981), Part I. Let us emphasize here that our underlying norms are the *discrete* L^2-norms and that we use the *discrete* Fourier transforms.

We now study finite-difference methods for approximating solutions of linear, pure initial value problems in which the linear difference operators $C_n^{(\ell)}$, $\ell = 0,1$, $n \in I$, are independent of t. According to Theorem 11.12, we need only investigate the semihomogeneous approximating equations

$$u_n(0) = u_{n,0}, \quad C_n^{(0)} u_n(t) = C_n^{(1)} u_n(t'), \quad t \in [0,T]_n', \quad n \in I. \tag{15}$$

We consider such methods in the spaces $C(G_n)$, $n \in I$, of bounded functions with values in $\mathbb{K}^\iota$, $\mathbb{K} = \mathbb{R}$ or $\mathbb{K} = \mathbb{C}$, defined on sets of mesh points G_n in $\mathbb{R}^d$ (with uniform mesh widths $h_n^{(1)},\dots,h_n^{(d)}$ in the corresponding spatial variables). Suppose the difference operators are of the form

$$C_n^{(\ell)} = \sum_{|\mu| \leq N_\ell} B_{n,\mu}^{(\ell)} T_n^\mu, \quad \ell = 0,1, \quad n \in I, \tag{16}$$

where $B_{n,\mu}^{(\ell)}$ are $\iota \times \iota$-matrices; $\mu = (\mu_1,\dots,\mu_d)$ are multi-indices (with modulus $|\mu| = |\mu_1| + \dots + |\mu_d|$); and T_n^μ denote the (d-dimensional) shift operators

$$(T_n^\mu f)(x) = f(\xi_1 + \mu_1 h_n^{(1)},\dots,\xi_d + \mu_d h_n^{(d)}),$$
$$x = (\xi_1,\dots,\xi_d) \in G_n, \quad n \in I. \tag{17}$$

The matrices $B_{n,\mu}^{(\ell)}$ are the coefficients of the method (15). The above representation (16) implicitly assumes that the coefficients are also independent of the spatial variable. An explicit method results when $N_0 = 0$.

It is our first aim to represent the solution of (15) by a "discrete Fourier analysis" whenever the mesh functions - and hence the solutions also - are periodic with period 2π in all spatial variables. For the time being, we restrict our treatment to the one-dimensional case (i.e., $d = 1$). A periodic mesh function defined on equidistant mesh points $x_j \equiv 2\pi j/J$, $j = 0,\pm1,\pm2,\ldots$, is determined by its value at the points in

$$[0,2\pi]'_n \equiv \{x_j = 2\pi j/J, \quad j = 0,\ldots,J-1\}.$$

J depends on $n \in I$ which we shall not explicitly indicate. The space of all bounded, periodic mesh functions can therefore be identified with $E_n = C([0,2\pi]'_n)$. Let us define the *discrete* L^2*-scalar product* and the *discrete* L^2*-norm* by

$$(f,g)_{0,n} \equiv h_n \sum_{j=0}^{J-1} (f(x_j),g(x_j)) \quad \text{and} \quad |f|_{0,n} \equiv (f,f)_{0,n}^{1/2}, \quad f,g \in E_n,$$

respectively, where $h_n = 2\pi/J$ and $(.,.)$ represents the Euclidean scalar product (with $|\cdot|$ the Euclidean norm) on $\mathbb{K}^l$. (It would be more precise to write $|\cdot|_{0,2,n}$ instead of $|\cdot|_{0,n}$, but we omit the index 2 since no other p-norm appears in this section.) By means of the exponential functions

$$v_n^{(m)}(x) = \frac{1}{\sqrt{2\pi}} \exp(imx), \quad m = 0,\ldots,J-1,$$

every $f \in E_n$ can be represented in the form

$$f(x_j) = \sum_{m=0}^{J-1} A_m v_n^{(m)}(x_j), \quad j = 0,\ldots,J-1, \tag{18}$$

with the *discrete Fourier coefficients*

$$A_m = (h_n/\sqrt{2\pi}) \sum_{\nu=0}^{J-1} f(x_\nu)\exp(-imx_\nu), \quad m = 0,\ldots,J-1.$$

Note that these are vectors in $\mathbb{K}^l$. The discrete L^2-norm of each $f \in E_n$ can be represented in terms of its discrete Fourier coefficients via the *Parseval relation* for mesh functions,

$$|f|_{0,n}^2 = \sum_{m=0}^{J-1} |A_m|^2, \quad f \in E_n. \tag{19}$$

We now associate with the difference operators $C_n^{(\ell)}$, $\ell = 0,1$, the *characteristic matrices*

$$S_n^{(\ell)}(y) = \sum_{|\mu|\le N_\ell} \exp(i\mu y) B_{n,\mu}^{(\ell)}, \quad y \in \mathbb{R}, \quad \ell = 0,1, \quad n \in I, \tag{20}$$

and obtain the following result.

Lemma 12.3. The discrete Fourier coefficients of $C_n^{(\ell)} f$ are given in terms of the discrete Fourier coefficients of $f \in E_n$ by

$$S_n^{(\ell)}(mh_n)A_m, \quad m = 0,\dots,J-1, \quad \ell = 0,1, \quad n \in I. \tag{21}$$

The mappings $C_n^{(0)}$ are continuously invertible if, and only if, the matrices $S_n^{(0)}(mh_n)$, $m = 0,\dots,J-1$, are nonsingular.

Proof: The proof of (21) follows easily by inserting f from (18) into the representation (16) of $C_n^{(\ell)}$. The solvability of the equation $C_n^{(0)} f = g$ is tantamount to determining the discrete Fourier coefficients of f from

$$\sum_{m=0}^{J-1} (S_n^{(0)}(mh_n)A_m - \tilde{A}_m) v_n^{(m)}(x) = 0, \quad x \in [0,2\pi]_n',$$

where $\tilde{A}_m$ are those of g. Since the $v_n^{(m)}$, $m = 0,\dots,J-1$, are linearly independent, the second assertion of the lemma is immediate. □

In the following, let us assume that the $C_n^{(0)}$ are invertible or, equivalently, that the $S_n^{(0)}(mh_n)$ are nonsingular. Then the solutions of (15) exist and are expressed as

$$u_n(0) = u_{n,0}, \quad u_n(t_k) = C_n u_n(t_{k-1}) = C_n^k u_{n,0}, \quad k = 1,\dots,N,$$

where $C_n \equiv C_n^{(0)-1} C_n^{(1)}$, $n \in I$. The solution $u_n(t_k)$ is thus determined for the discrete time level t_k when its associated Fourier coefficients $A_m^{(k)}$ are determined from the discrete Fourier coefficients $A_m^{(k-1)}$ of the approximation $u_n(t_{k-1})$ on the previous time level by

$$S_n^{(0)}(mh_n)A_m^{(k)} = S_n^{(1)}(mh_n)A_m^{(k-1)}, \quad m = 0,\dots,J-1. \tag{22}$$

The solutions of (22) are given by

$$A_m^{(k)} = S_{n,m} A_m^{(k-1)}, \quad m = 0,\dots,J-1,$$

where the matrices

$$S_{n,m} \equiv S_n^{(0)}(mh_n)^{-1} S_n^{(1)}(mh_n), \quad m = 0,\dots,J-1, \tag{23}$$

are called *amplification matrices*. By induction, we can easily convince ourselves that the amplification matrices associated with powers of C_n are the corresponding powers of the amplification matrices $S_{n,m}$. We therefore obtain the following representation for the solutions $u_n(t)$ of (15) by means of the discrete Fourier coefficients of the initial function.

Lemma 12.4. Let $A_m^{(0)}$, $m = 0,\dots,J-1$, be the discrete Fourier coefficients of $u_{n,0} \in E_n$. Then the solution $u_n(\cdot,t)$ $(= u_n(t))$ of (15) can be represented in the following manner,

$$u_n(x,t_k) = \frac{1}{\sqrt{2\pi}} \sum_{m=0}^{J-1} \exp(imx) S_{n,m}^k A_m^{(0)}$$

$$= \sum_{\nu=0}^{J-1} G_n(x,t_k,x_\nu) u_{n,0}(x_\nu), \quad k = 0,\dots,N, \quad x \in [0,2\pi]_n',$$

where

$$G_n(x,t_k,y) \equiv \frac{h_n}{2\pi} \sum_{m=0}^{J-1} \exp(im(x-y)) S_{n,m}^k, \quad x,y \in G_n, \quad k = 0,\dots,N. \quad \square$$

For mesh functions with values in $\mathbb{K}^1$ (i.e., $\iota = 1$), the $S_n^{(m)}(\cdot)$ are also called *characteristic functions* and $S_{n,m}$, the *amplification factors*. The latter can be expressed as

$$S_{n,m} = \frac{\sum_{|\mu| \le N_1} B_{n,\mu}^{(1)} \exp(i\mu m h_n)}{\sum_{|\mu| \le N_0} B_{n,\mu}^{(0)} \exp(i\mu m h_n)}, \quad m = 0,\dots,J-1.$$

We now present the amplification matrices and factors for the finite-difference methods introduced in Chapter 4.

Crank-Nicolson Method 4.(8) (with general $0 \le \Theta \le 1$):

$$S_n^{(0)}(y) = 1 + 2ar\Theta(1 - \cos(y))$$

$$= 1 + 4ar\Theta \sin^2(y/2),$$

$$S_n^{(1)}(y) = 1 - 2ar(1-\Theta)(1-\cos(y))$$

$$= 1 - 4ar(1-\Theta)\sin^2(y/2),$$

$$S_{n,m} = \frac{1-4ar(1-\Theta)\sin^2(mh_n/2)}{1+4ar\Theta \sin^2(mh_n/2)}, \quad m = 0,\dots,J-1;$$

Du Fort-Frankel Method 4.(11) (as a two-level system):

$$S_{n,m} = \left(\begin{array}{c|c} \frac{4ar}{1+2ar}\cos(mh_n) & \frac{1-2ar}{1+2ar} \\ 1 & 0 \end{array}\right);$$

Crank-Nicolson-Galerkin Method 4.(22):

$$S_{n,m} = \frac{1-2(ar+1/3)\sin^2(mh_n/2)}{1+2(ar-1/3)\sin^2(mh_n/2)};$$

Friedrichs' Method 4.(33):

$$S_{n,m} = \frac{1}{2}\exp(-imh_n)(I-\lambda A) + \frac{1}{2}\exp(imh_n)(I+\lambda A)$$

$$= \cos(mh_n)I + i\lambda\sin(mh_n)A;$$

Courant-Isaacson-Rees Method 4.(34):

$$S_{n,m} = -\lambda\exp(-imh_n)A^- + I - \lambda(A^+-A^-) + \lambda\exp(imh_n)A^+;$$

Lax-Wendroff Method 4.(35):

$$S_{n,m} = I + i\lambda\sin(mh_n)A + \lambda^2(\cos(mh_n)-1)A^2;$$

Discrete-Time Galerkin Method 4.(40) (for the wave equation):

$$S_n^{(0)}(y) = \left(\begin{array}{c|c} 1-(2/3-\lambda^2c^2)\sin^2(y/2) & 0 \\ 0 & 1 \end{array}\right),$$

$$S_n^{(1)}(y) = \left(\begin{array}{c|c} 2[1-(2/3+\lambda^2c^2)\sin^2(y/2)] & -[1-(2/3-\lambda^2c^2)\sin^2(y/2)] \\ 1 & 0 \end{array}\right);$$

Discrete-Time Galerkin Method 4.(46) (for the scalar hyperbolic IVP 4.(42)):

$$S_n^{(0)}(y) = 1 - \frac{2}{3}\sin^2(y/2) - i\,\frac{1}{2}\lambda c\sin(y),$$

$$S_n^{(1)}(y) = 1 - \frac{2}{3}\sin^2(y/2) + i\,\frac{1}{2}\lambda c\sin(y).$$

We are now in a position to establish necessary and sufficient conditions for the inverse stability inequality 11.(43) (to be valid in the space of periodic mesh functions equipped with discrete L^2-norms) by using properties of the amplification matrices. For this purpose, we additionally need a result which expresses the (operator) norms of the C_n in terms of those of the associated amplification matrices. Let

$$||A||_S \equiv \sqrt{\rho(A^*A)}$$

denote the *spectral norm* of a matrix A with $\rho(\cdot)$, the *spectral radius* (i.e., the maximal modulus of the eigenvalues) and $A^* = \bar{A}^T$, the associated *conjugate transpose*. We then have the relation,

$$||C_n|| \equiv \sup_{0 \neq g \in E_n} \frac{|C_n g|_{0,n}}{|g|_{0,n}} = \max_{0 \leq m \leq J-1} ||S_{n,m}||_S, \tag{24}$$

and can thereby prove the main result of this section.

Theorem 12.5. Let $||S_n^{(0)}(mh_n)^{-1}||_S$, $m = 0,\ldots,J-1$, $n \in I$, be uniformly bounded. Then the inverse stability inequality 11.(43) with respect to the discrete L^2-norms, i.e.,

$$|C_n^k g_n|_{0,n} \leq \gamma |g_n|_{0,n}, \quad g_n \in E_n, \quad k = 0,\ldots,N, \quad n \in I, \tag{25a}$$

is equivalent to

$$\max_{0 \leq m \leq J-1} ||S_{n,m}^k||_S \leq \gamma, \quad k = 0,\ldots,N, \quad n \in I. \tag{25b}$$

A sufficient condition for (25a) and (25b) is that

$$\max_{0 \leq m \leq J-1} ||S_{n,m}||_S \leq 1 + O(\tau_n) \quad (n \in I). \tag{25c}$$

The "von Neumann condition",

$$\max_{0 \leq m \leq J-1} \rho(S_{n,m}) \leq 1 + O(\tau_n) \quad (n \in I) \tag{25d}$$

is necessary for (25a) and (25b).

Proof: The discrete Fourier coefficients of $f = C_n^{(0)-1} g$ are given by $S_n^{(0)}(mh_n)^{-1}\tilde{A}_m$, with $\tilde{A}_m$ the discrete Fourier coefficients of $g \in E_n$. As a consequence of the Parseval relation, the uniform boundedness of the $C_n^{(0)-1}$, $n \in I$, (with respect to the discrete L^2-norms) is equivalent to that of $||S_n^{(0)}(mh_n)^{-1}||_S$, $m = 0,\ldots,J-1$, $n \in I$, which has been assumed at the outset. It therefore suffices to characterize the uniform boundedness of the C_n^k, $0 \leq k \leq N$, $n \in I$, in order to obtain equivalent conditions for the inverse stability inequality 11.(43). We notice that (25a) is indeed equivalent to 11.(43) since, due to the finite dimensions of the E_n, the mappings are clearly continuous and we need not distinguish between the cases $n \geq \nu_0$ and $n < \nu_0$.

(25a) $\Longleftrightarrow$ (25b). The $S^k_{n,m}$, $m = 0,\ldots,J-1$, represent the amplification matrices for C^k_n; and together with (24), we know that

$$||C^k_n|| = \max_{0\leq m\leq J-1} ||S^k_{n,m}||_S, \quad k = 0,\ldots,N, \quad n \in I,$$

Condition (25a) therefore immediately yields (25b). The converse is seen by the estimates,

$$|C^k_n g_n|_{0,n} \leq \max_{0\leq m\leq J-1} ||S^k_{n,m}||_S |g_n|_{0,n} \leq \gamma |g_n|_{0,n}, \quad g_n \in E_n,$$
$$k = 0,\ldots,N, \quad n \in I.$$

provided that (25b) holds.

(25c) $\Rightarrow$ (25b). With certain $C \geq 0$, $\nu_1 \in I$, we obtain

$$||S^k_{n,m}||_S \leq ||S_{n,m}||^k_S \leq (1+C\tau_n)^k \leq \exp(Ck\tau_n) \leq \exp(CT),$$
$$m = 0,\ldots,J-1, \quad k = 0,\ldots,N, \quad n \geq \nu_1,$$

where the third inequality is an immediate consequence of the power series expansion of the exponential function. For $n < \nu_1$, the $S^k_{n,m}$ are obviously bounded, and the proof of (25b) is complete.

(25b) $\Rightarrow$ (25d). Without loss of generality, let $\gamma \geq 1$ in (25b). A simple analysis shows that $\gamma^x \leq 1 + \gamma x$ for $0 \leq x \leq 1$. Furthermore, $\rho(S_{n,m})^k = \rho(S^k_{n,m})$ and $\rho(A) \leq ||A||_S$, and thus (25b) implies

$$\max_{0\leq m\leq J-1} \rho(S_{n,m}) \leq \gamma^{1/k}, \quad 0 \leq k \leq N(= T/\tau_n), \quad n \in I.$$

For the case $k = N$, we obtain the von Neumann condition since

$$\max_{0\leq m\leq J-1} \rho(S_{n,m}) \leq \gamma^{\tau_n/T} \leq 1 + \gamma\frac{\tau_n}{T}, \quad n \in I. \qquad \square$$

It is of great interest to know in what special situations the von Neumann condition is not only necessary but also sufficient for the inverse stability inequality (25a). The following theorem addresses this question.

Theorem 12.6. The von Neumann condition (25d) is equivalent to (25a) and (25b) in case the amplification matrices $S_{n,m}$, $m = 0,\ldots,J-1$, $n \in I$, are normal, i.e., $S_{n,m}S^*_{n,m} = S^*_{n,m}S_{n,m}$. If the coefficients $B^{(0)}_{n,0}$,

$B^{(1)}_{n,\mu}$, $|\mu| \le N_1$, of an explicit method are independent of n; if the amplification matrices are normal; and if $\tau_n \to 0$, $h_n \to 0$ $(n \in I)$, then the von Neumann condition is further equivalent to

$$||S(y)||_S \le 1, \quad y \in \mathbb{R}, \tag{25e}$$

where

$$S(y) \equiv \sum_{|\mu|\le N_1} \exp(i\mu y) B_0^{(0)-1} B_\mu^{(1)}, \quad B_\mu^{(\ell)} = B_{n,\mu}^{(\ell)}.$$

Proof: The proof relies on the fact that normal matrices satisfy $||S_{n,m}||_S = \rho(S_{n,m})$ (cf. Varga (1962), Sec. 1.3). Hence (25d) implies (25c) and (25b). As a preparatory remark for the proof of the equivalence of (25d) and (25e), we notice that, for every $y \in [0,2\pi]$, $n \in I$ there exists an $m_n \in \{0,\ldots,J-1\}$ such that

$$|y - m_n h_n| \le h_n.$$

In the present case of an explicit method, we have $S_{n,m} = S(mh_n)$, and the continuity of $S(\cdot)$ in connection with the convergence $h_n \to 0$ $(n \in I)$, implies

$$\lim_{n\in I} ||S_{n,m_n}||_S = \lim_{n\in I} ||S(m_n h_n)||_S = ||S(y)||_S.$$

If we now impose the von Neumann condition (25d), then

$$||S(y)||_S = \lim_{n\in I} ||S_{n,m_n}||_S = \lim_{n\in I} \rho(S_{n,m_n}) \le \lim_{n\in I} (1+C\tau_n) = 1,$$
$$y \in [0,2\pi],$$

for normal amplification matrices, thereby proving (25e) via the periodicity of $S(\cdot)$. Conversely, (25e) together with $S_{n,m} = S(mh_n)$, $m = 0,\ldots,J-1$, $n \in I$, immediately implies the von Neumann condition (25d). □

As a consequence of the previous theorem, we know that all conditions (25a) to (25d) are equivalent in the scalar case (i.e., $\iota = 1$). The von Neumann condition can be then rewritten in terms of the amplification factors as

$$\max_{0\le m\le J-1} |S_{n,m}| \le 1 + O(\tau_n) \quad (n \in I).$$

For general $\iota \in \mathbb{N}$, the von Neumann condition is moreover equivalent to (25a) and (25b) in case similarity transformations convert the amplification matrices into normal matrices $\Lambda_{n,m} = M^{-1}_{n,m} S_{n,m} M_{n,m}$ with $M_{n,m}$, $M^{-1}_{n,m}$ uniformly bounded. This result is an obvious generalization of the first statement in Theorem 12.6, since similarity transformations leave the spectral radius invariant.

In view of the specific methods we want to analyze, the following theorem provides another statement guaranteeing the sufficiency of the von Neumann condition for inverse stability. Its proof is in Richtmyer & Morton (1967), Sec. 4.11, where still other results of this type can be found.

Theorem 12.7. If the entries of the amplification matrices $S_{n,m}$ are uniformly bounded for n,m, and if the eigenvalues λ_s of $S_{n,m}$, with the possible exception of λ_1, satisfy

$$|\lambda_s| \leq \gamma < 1, \quad s = 2,\dots,\iota, \quad m = 0,\dots,J-1, \quad n \in I,$$

then the von Neumann condition is sufficient as well as necessary for the inverse stability inequality (25a). □

We now check the (inverse) stability properties with respect to the discrete L^2-norms of the specific finite-difference methods for which we have already calculated the amplification matrices and factors (after Lemma 12.4). At this particular juncture, we would like to mention that the following proofs of conditional (inverse) stability include showing that inverse stability is violated for certain mesh ratios. We always assume that $\tau_n \to 0$ $(n \in I)$ for the following examples. For fixed mesh ratios $r = \tau_n/h_n^2$ or $\lambda = \tau_n/h_n$, then also $h_n \to 0$ $(n \in I)$.

Proposition 12.8. The method 4.(8) is unconditionally stable for Θ in $1/2 \leq \Theta \leq 1$, which means that each of the equivalent conditions (25a-d) holds for an arbitrary $r > 0$. For Θ in $0 \leq \Theta < 1/2$, 4.(8) is inversely stable if, and only if,

$$r \leq \frac{1}{2a(1-2\Theta)}.$$

Proof: Obviously, $S_n^{(0)}(y)^{-1} \leq 1$, thereby satisfying the hypothesis of the Stability Theorem 12.5 and guaranteeing the existence and uniform boundedness of the $C_n^{(0)-1}$. The amplification factors are bounded from above since

$$S_n(y) \equiv S_n^{(1)}(y)/S_n^{(0)}(y) \leq 1, \quad y \in \mathbb{R}, \quad n \in I.$$

In order to prove and estimate from below, we set $\alpha \equiv 4ar\sin^2(y/2)$ and notice that

$$S_n(y) = \frac{1-(1-\Theta)\alpha}{1+\Theta\alpha} \lesseqgtr -1 \quad \text{in case} \quad \alpha(1-2\Theta) \gtreqless 2.$$

For Θ in $1/2 \le \Theta \le 1$, we have

$$-1 \le 1 - 2\Theta \le 0 \quad \text{and} \quad \alpha(1-2\Theta) \le 0 < 2,$$

so that $-1 \le S_n(y) \le 1$ for all $n \in I$, $y \in \mathbb{R}$. For Θ in $0 \le \Theta < 1/2$, we have $1-2\Theta > 0$, and we can conclude that

$$\alpha(1-2\Theta) = 4ar(1-2\Theta)\sin^2(y/2) \le 2 \sin^2(y/2) \le 2,$$

provided $r \le 1/[2a(1-2\Theta)]$. This again proves $|S_n(y)| \le 1$ for the alternating case.

We now let $r > 1/[2a(1-2\Theta)]$ and claim that $\{S_{n,m}\}$ tends to infinity for special choices of n,m. We notice that every sequence $m_n \in \{0,\dots,J-1\}$, $n \in I$, with $m_n h_m \to \pi$ $(n \in I)$ satisfies

$$S_{n,m_n} = S_n(m_n h_n) \to \frac{1-4ar(1-\Theta)}{1+4ar\,\Theta} \quad (n \in I).$$

As above, we have

$$\frac{1-4ar(1-\Theta)}{1+4ar\,\Theta} < -1, \quad \text{provided that} \quad 4ar(1-2\Theta) > 2.$$

For such an r, there are thus an $n_0 \in I$ and a $\rho_0 > 0$ such that

$$|S_{n,m_n}| \ge 1 + \rho_0, \quad n \ge n_0.$$

Passing to the limit $m_n \to \infty$ $(n \in I)$, we see that

$$|S_{n,m_n}|^{m_n} \to \infty \quad (n \in I)$$

results, thereby showing that inverse stability with respect to the discrete L^2-norms is violated. □

The last result states, in particular, that the Crank-Nicolson method is unconditionally stable with respect to discrete L^2-norms, whereas, in the preceding section, we needed a restriction on the mesh ratio to show inverse stability with respect to supremum norms. A similar statement holds for the Du Fort-Frankel method.

Proposition 12.9. The Du Fort-Frankel method is inversely stable (with respect to discrete L^2-norms) for all $r > 0$.

Proof: The eigenvalues of the matrix $S(y) = B_0^{(0)-1}S_n^{(1)}(y)$ are the roots of

$$\det(S(y)-\mu I) = \mu^2 - \frac{4ar}{1+2ar}\cos(y)\mu - \frac{1-2ar}{1+2ar}$$

which are given by

$$\mu_{1,2} = \frac{1}{1+2ar}\left\{2ar\cos(y) \pm \sqrt{1-4a^2r^2\sin^2(y)}\right\}.$$

For brevity, we set $\omega = 2ar$ and obtain

$$|\mu_{1,2}|^2 = \frac{\omega^2(\cos^2(y)+\sin^2(y))-1}{(1+\omega)^2} = \frac{\omega^2-1}{(\omega+1)^2} = \frac{\omega-1}{\omega+1} < 1$$

provided $1 < \omega^2\sin^2(y)$. In case $1 \geq \omega^2\sin^2(y)$, we have

$$|\mu_{1,2}|^2 \leq \frac{1}{1+\omega}(\omega+1) = 1.$$

In any case, the von Neumann condition is fulfilled. For the second case, moreover, we see that

$$|\mu_1\mu_2| = |\omega-1|/(\omega+1) < 1$$

and thus there is a root $\mu \in \{\mu_1,\mu_2\}$ satisfying $|\mu| \leq \sqrt{|\mu_1\mu_2|} < 1$. Otherwise both roots would satisfy

$$|\mu_s| > \sqrt{|\mu_1\mu_2|}, \quad s = 1,2, \text{ and } \quad |\mu_1\mu_2| = |\mu_1|\,|\mu_2| > |\mu_1\mu_2|,$$

a contradiction. Application of Theorem 12.7 finally yields the desired inverse stability inequality (25a). □

Unconditional stability with respect to discrete L^2-norms is also present for the Crank-Nicolson-Galerkin method (with continuous, piecewise linear trial and test functions).

Proposition 12.10. The Crank-Nicolson-Galerkin method 4.(22) is inversely stable (with respect to discrete L^2-norms) for all $r > 0$.

Proof: We see that

$$S_n^{(0)}(y) = 1 + 2(ar - 1/3)\sin^2(y/2) \geq 1/3, \quad y \in \mathbb{R}, \quad n \in I,$$

and hence $S_n^{(0)}(y)^{-1}$ is uniformly bounded. In order to estimate $S_{n,m}$, we set $\alpha \equiv 2\sin^2(y/2)$ and obtain

$$S_n(y) = \frac{S_n^{(1)}(y)}{S_n^{(0)}(y)} = \frac{1-\alpha/3-a\alpha r}{1-\alpha/3+a\alpha r} = 1 - \frac{2a\alpha r}{1-\alpha/3+a\alpha r}.$$

This immediately implies

$$-1 \leq S_n(y) \leq 1, \quad y \in \mathbb{R}, \quad n \in I,$$

thereby ensuring the von Neumann condition (25d). □

We present the corresponding results for the Friedrichs, Courant-Isaacson-Rees and Lax-Wendroff methods used to approximate first order hyperbolic systems having a constant, real diagonalizable matrix A (in the differential equation 4.(31)). For the proof, we refer to the standard literature.

Proposition 12.11. For a constant, real diagonalizable matrix A, Friedrichs' method (cf. 4.(33a)), the Courant-Isaacson-Rees method (cf. 4.(34a-c)), and the Lax-Wendroff method (cf. 4.(35a)) are inversely stable with respect to discrete L^2-norms if, and only if, $\lambda\rho(A) \leq 1$. □

We thus have the same restriction on the mesh ratio λ for all three methods which establishes a necessary and sufficient condition for inverse stability with respect to discrete L^2-norms. For the wave equation, this restriction directly means the Courant-Friedrichs-Lewy condition $\lambda \leq 1/c$. A comparison with the stability properties with respect to the supremum norms shows that, for the Friedrichs and Courant-Isaacson-Rees method, the same restriction has to be required as for the discrete L^2-norms, whereas the Lax-Wendroff method satisfies the positivity properties of Section 12.1 only in case of $\lambda = 1/c$. For all our methods (including those for parabolic IVP's), we observe that inverse stability with respect to discrete L^2-norms is present under weaker - or at least the same - requirements than with respect to the supremum norms.

In this context, we additionally study the stability properties of the discrete-time Galerkin method 4.(40) for the wave equation. These are interesting from the point of view that the von Neumann condition is satisfied but that inverse stability with respect to discrete L^2-norms is not present. We thereby confirm the character of the von Neumann condition as a, in general, only necessary condition for inverse stability.

Proposition 12.12. The finite-difference method 4.(40) satisfies the von Neumann condition for all $\lambda > 0$. For $\lambda = 1/c$, however, it is not inversely stable with respect to discrete L^2-norms.

Proof: (i) Setting

$$\Lambda(y) \equiv \frac{1-(2/3 + \lambda^2c^2)\sin^2(y/2)}{1-(2/3 - \lambda^2c^2)\sin^2(y/2)}, \quad y \in \mathbb{R},$$

we get

$$S_n(y) = S_n^{(0)}(y)^{-1}S_n^{(1)}(y) = \begin{pmatrix} 2\Lambda(y) & -1 \\ 1 & 0 \end{pmatrix}.$$

The eigenvalues of $S_n(y)$ are determined by

$$\mu^2 - 2\Lambda(y)\mu + 1 = 0$$

which yield $\mu_{1,2} = \Lambda(y) \pm \sqrt{\Lambda^2(y)-1}$, $y \in \mathbb{R}$. Since $\Lambda(y) \leq 1$, we can write

$$\mu_{1,2} = \Lambda(y) \pm i\sqrt{1 - \Lambda^2(y)}, \quad y \in \mathbb{R},$$

and observe that

$$|\mu_{1,2}|^2 = \Lambda(y)^2 + 1 - \Lambda(y)^2 = 1,$$

which ensures the von Neumann condition.

(ii) In case $\lambda = 1/c$, we have

$$\Lambda(y) = \frac{1+5\cos(y)}{7 - \cos(y)}, \quad y \in \mathbb{R}.$$

We shall show that there exists a sequence $m_n \in \{0,\dots,J-1\}$, $n \in I$, such that

$$||S_{n,m_n}^{\nu(n)}||_S \to \infty \quad (n \in I)$$

for every sequence $\nu(n) \to \infty$ $(n \in I)$. If we choose $m_n = 1$, $n \in I$, then $m_n h_n \to 0$ $(n \in I)$ and $\lambda_n \equiv \Lambda(m_n h_n) \to 1$ $(n \in I)$, where always $\lambda_n \leq 1$. Using the derivative

$$\Lambda'(y) = -\frac{36\sin(y)}{(7 - \cos(y))^2}$$

and $\sin(y) \leq y$ for $0 \leq y \leq h_n$, we obtain, via the Mean Value Theorem, that

$$|\Lambda(0) - \Lambda(h_n)| = 1 - \lambda_n \leq h_n \sup_{0 \leq y \leq h_n} |\Lambda'(y)| \leq h_n^2$$

for sufficiently large n. Thus, for

$$M_n \equiv S_{n,1} = \begin{pmatrix} 2\lambda_n & -1 \\ 1 & 0 \end{pmatrix} \quad \text{and} \quad M \equiv \begin{pmatrix} 2 & -1 \\ 1 & 0 \end{pmatrix}$$

we have the relation $M_n = M(1 + O(h_n^2))$, and we conclude by induction that

$$M^\nu = -(\nu-1)I + \nu M, \quad \nu = 1,2,\ldots .$$

As a consequence, we can easily show that the spectral norms of M^ν tend to infinity. Indeed, according to the property that the spectral norm is the natural matrix norm induced by the Euclidean vector norm, for the vector $\hat{z} = (1,0)^T$ we have

$$\begin{aligned} ||M^\nu||_S &= \sup_{0 \neq z \in \mathbb{C}^2} \frac{(M^\nu z, M^\nu z)^{1/2}}{(z,z)^{1/2}} \geq \frac{(M^\nu \hat{z}, M^\nu \hat{z})^{1/2}}{(\hat{z},\hat{z})^{1/2}} \\ &= \{(\nu-1)^2(\hat{z},\hat{z}) - 2\nu(\nu-1)(M\hat{z},\hat{z}) + \nu^2(M\hat{z},M\hat{z})\}^{1/2} \\ &= \{2\nu^2 + 2\nu + 1\} \to \infty \quad (\nu \to \infty). \end{aligned}$$

The eigenvalues are continuously dependent on the coefficients of a matrix, and this dependence, together with the above relation between M and M_n, yields

$$||M_n^{\nu(n)}||_S \to \infty \quad (n \in I)$$

for each sequence $\nu(n) \to \infty$ $(n \in I)$. □

In connection with the method just analyzed, let us mention that the same negative stability behavior can be observed for the particular method which approximates the second (spatial) derivative in the wave equation by the central difference quotient of second order (cf. Richtmyer & Morton (1967), Sec. 4.8 & 4.11, Forsythe & Wasow (1967), Sec. 4).

We proceed to define analogous concepts for the case of several spatial variables and to outline the corresponding results. Let us again consider finite-difference methods of the form (15) where the difference operators are given as in (16). The spaces of all mesh functions which are periodic in each spatial variable (with period 2π) are again identified with $E_n = C([0,2\pi]_n^{\prime})$ where, now, $h_n^{(s)} \equiv 2\pi/J_s$ and

$$[0,2\pi]_n^{\prime} \equiv \{x_j = (\xi_{j_1}^{(1)},\ldots,\xi_{j_d}^{(d)}),\ \xi_{j_s}^{(s)} \equiv j_s h_n^{(s)},$$
$$0 \leq j_s \leq J_s - 1,\ \ 1 \leq s \leq d\}.$$

Each $f \in E_n$ can be represented as

$$f(x_j) = \sqrt{2\pi}^{-d} \sum_{0 \leq m \leq J-1} A_m \exp(i(m,x_j)),$$

where j,m,J denote multi-indices, $(m,x_j) = \Sigma_s m_s \xi_{j_s}^{(s)}$ is the Euclidean scalar product in $\mathbb{R}^d$, and

$$\sum_{0 \le m \le J-1} = \sum_{0 \le m_1 \le J_1 - 1} \cdots \sum_{0 \le m_d \le J_d - 1}$$

are the corresponding d-fold sums. The *discrete Fourier coefficients* are given in the multidimensional case by

$$A_m = \sqrt{2\pi}^{\,-d}\, h_n^{(1)} \cdots h_n^{(d)} \sum_{0 \le \nu \le J-1} f(x_\nu) \exp(-i(m,x_\nu)),$$

$$m = (m_1, \ldots, m_d), \quad 0 \le m_s \le J_s - 1;$$

the *characteristic matrices* by

$$S_n^{(\ell)}(y) = \sum_{|\mu| \le N_\ell} \exp(-i(\mu,y)) B_{n,\mu}^{(\ell)}, \quad y \in \mathbb{R}^d, \quad \ell = 0,1;$$

and the *amplification matrices* by

$$S_{n,m} = \left(\sum_{|\mu| \le N_0} \exp\left(i \sum_{s=1}^{d} \mu_s m_s h_n^{(s)}\right) B_{\mu,n}^{(0)} \right)^{-1} \sum_{|\mu| \le N_1} \exp\left(i \sum_{s=1}^{d} \mu_s m_s h_n^{(s)}\right) B_{\mu,n}^{(1)},$$

$$0 \le m_s \le J_s - 1, \quad n \in I.$$

In order to guarantee the existence of the latter matrices, we assume that the $S_n^{(0)}(x_m)$ are nonsingular. The solutions of the approximating finite-difference equations (15) can be given in explicit form (cf. Lemma 12.4),

$$u_n(x,t_k) = \sqrt{2\pi}^{\,-d} \sum_{0 \le m \le J-1} \exp(i(m,x)) S_{n,m}^k A_m^{(0)}$$

$$= \sum_{0 \le \nu \le J-1} G_n(x,t_k,x_\nu) u_{n,0}(x_\nu), \quad k = 0,\ldots,N, \quad x \in [0,2\pi]_n',$$

where

$$G_n(x,t_k,y) = (2\pi)^{-d} h_n^{(1)} \cdots h_n^{(d)} \sum_{0 \le m \le J-1} \exp(i(m,x-y)) S_{n,m}^k.$$

The sums above are again to be understood as d-fold ones. With the underlying discrete L^2-scalar products

$$(f,g)_{0,n} = h_n^{(1)} \cdots h_n^{(d)} \sum_{0 \le j \le J-1} (f(x_j), g(x_j))$$

and the associated discrete L^2-norms $|\cdot|_{0,n}$, we again obtain the representation (24) for the norms of $C_n = C_n^{(0)-1}C_n^{(1)}$ (but now with multi-indices). The central Theorem 12.5 concerning inverse stability extends verbatim, and Theorem 12.6 and 12.7 also carry over in essence.

At this points, we would like to mention the investigations concerning inverse stability of "product methods" for several spatial dimensions described in Meis & Marcowitz (1981), Sec. I.10. Such methods result if the multi-dimensional difference operators are given as factors of (in most cases) one-dimensional operators. The amplification matrices are therefore exactly the products of the corresponding single amplification matrices. According to this observation, stability results for one-dimensional methods can be applied to multi-dimensional problems. We do not present a general investigation of such product methods, here, but immediately turn to the best-known classical *example* of a product method, namely the *ADI-method* of *Peaceman-Rachford* (cf. 4.(16)).

In this example the associated finite-difference operators are given by

$$C_n = (C_n^{(0),2})^{-1}C_n^{(1),2}(C_n^{(0),1})^{-1}C_n^{(1),1}, \quad n \in I,$$

where the factors $C_n^{(\ell),\nu}$, $\ell = 0,1$, $\nu = 1,2$, have been given in Sections 4.1 and 4.5. The corresponding amplification factors can be easily calculated from

$$S^1(y) = \frac{1 - 2a_2\sin^2(y/2)}{1 + 2a_1\sin^2(y/2)}, \quad \text{for} \quad \nu = 1,$$

$$S^2(y) = \frac{1 - 2a_1\sin^2(y/2)}{1 + 2a_2\sin^2(y/2)}, \quad \text{for} \quad \nu = 2.$$

As in Proposition 12.8, we can easily convince ourselves that the von Neumann condition is satisfied if, and only if, $r_2 \leq 1/a$ and $r_1 \leq 1/a$, respectively. For $\nu = 1$, this means that r_2 has to be restricted and that r_1 can be arbitrarily chosen; the opposite is true for $\nu = 2$. Since we are considering a product method, we can deduce that the total amplification factor is given by the product of the single ones. As a result, unconditional stability can be deduced for the present method.

Proposition 12.13. The ADI method 4.(16) is unconditionally inversely stable with respect to discrete L^2-norms.

Proof: We compute the amplification factors and get

$$S_{n,m_1,m_2} = \frac{\sum\limits_{|\mu_1|+|\mu_2|\le 2} \exp(i \sum\limits_{s=1}^{2} \mu_s m_s h_n^{(s)}) B^{(1)}_{\mu_1,\mu_2,n}}{\sum\limits_{|\mu_1|+|\mu_2|\le 2} \exp(i \sum\limits_{s=1}^{2} \mu_s m_s h_n^{(s)}) B^{(0)}_{\mu_1,\mu_2,n}} .$$

More specifically, the denominator is (cf. Section 4.1)

$$\begin{aligned}(1+a_1)(1+a_2) &- a_2(1+a_1)\cos(m_2h_n^{(2)}) - a_1(1+a_2)\cos(m_1h_n^{(1)}) \\ &+ a_1a_2\cos(m_1h_n^{(1)})\cos(m_2h_n^{(2)}) \\ &= [1 + a_2(1-\cos(m_2h_n^{(2)}))][1 + a_1(1-\cos(m_1h_n^{(1)}))] \\ &= [1 + 2a_2\sin^2(m_2h_n^{(2)}/2)][1 + 2a_1\sin^2(m_1h_n^{(1)}/2)].\end{aligned}$$

For the numerator, similar calculations yield

$$[1 - 2a_1\sin^2(m_1h_n^{(1)}/2)][1 - 2a_2\sin^2(m_2h_n^{(2)}/2].$$

Since $|1-\alpha|/|1+\alpha| \le 1$ for any nonnegative number α, we have that $|S_{n,m_1,m_2}| \le 1$. □

In general, product methods are inversely stable provided every factor has this property. (This is true for any underlying norm.) The present method shows, however, that the inverse stability factors multiply each other, unconditional stability can be present, whereas the single factors may be only conditionally stable.

12.3. INVERSE STABILITY OF GALERKIN METHODS

The analysis of the preceding sections, when applied to the Crank-Nicolson-Galerkin method (CNG method) 4.(22), has shown that this particular method is conditionally inversely stable with respect to supremum norms and unconditionally inversely stable with respect to discrete L^2-norms. These statements are valid for the CNG method when based on continuous, piecewise linear trial and test functions. (In all honesty, we have not proved conditional stability with respect to supremum norms but merely ensured inverse stability under a certain restriction on the mesh ratio.)

In this section, we prove unconditional stability for the general CNG method 4.(21) with respect to appropriate Sobolev space norms. Corresponding stability properties will be shown for the discrete-time

Galerkin methods 4.(39) and 4.(45) which approximate the generalized wave equation 4.(36) and the scalar hyperbolic IVP 4.(42), respectively. Note that we have already analyzed the inverse stability of the Galerkin method 4.(39) for the case of continuous, piecewise linear basis functions, and with the discrete L^2-norms as the underlying norms (cf. Proposition 12.12).

In this section, we again let τ_n, $n \in I$, be a null sequence of positive step widths (in the t-direction), and E_n, $n \in I$, be finite-dimensional subspaces of suitable function spaces. For simplicity, we additionally suppose that all functions are real-valued.

We first study the *CNG method* 4.(21),

$$(D_\tau^+ v_n^k, \phi_n)_0 + a(v_n^{k+1/2}, \phi_n) = (S_n^{k+1/2}, \phi_n)_0,$$
$$\phi_n \in E_n \quad (\subset H_0^1(0,1)), \quad k = 0,1,\dots,N-1,$$

where S_n^ν denotes the L^2-projection of $s^\nu = s(.,t_\nu)$ in E_n, $\nu = k,k+1$, and

$$D_\tau^+ v_n^k = \frac{1}{\tau_n}(v_n^{k+1} - v_n^k), \quad a(\psi,\phi) = (a\psi',\phi')_0.$$

The initial approximations need not yet be specified when stability properties are studied. The coefficient a can be a function of x which is assumed to be continuous and positive on $[0,1]$.

Besides the usual norm $||\cdot||_1$ in $H_0^1(0,1)$ (cf. Chapter 2), we again use the seminorm $|\phi|_1 = |\phi'|_0$ which also represents a norm on $H_0^1(0,1)$ due to the Poincaré-Friedrichs inequality. As before, we denote by $(.,.)_0$ and $|\cdot|_0$ the L^2-scalar product and the L^2-norm, respectively. With this notation, we prove the following inverse stability inequality for the solutions of the CNG method.

Theorem 12.14. Let $a \in C[0,1]$, $0 < \alpha_0 \leq a(x) \leq \alpha_1$, $x \in [0,1]$. Then the solutions $v_n^k \in E_n$, $k = 0,\dots,N$, of the CNG method 4.(21) satisfy the following estimates

$$\max_{0\leq k\leq N} |v_n^k|_1 \leq \gamma\left\{|v_n^0|_1 + \left(\tau_n \sum_{k=0}^{N-1} |S_n^{k+1/2}|_0^2\right)^{1/2}\right\}, \quad n \in I, \tag{26}$$

where $\gamma = [\max(\alpha_1,1/2)/\alpha_0]^{1/2}$.

Proof: We observe first that

$$(\rho D_\tau^+\phi^k, \phi^{k+1/2})_0 = \frac{1}{2\tau_n}[(\rho\phi^{k+1}, \phi^{k+1})_0 - (\rho\phi^k, \phi^k)_0], \tag{27}$$

for arbitrary functions $\phi^\nu \in L^2(0,1)$, $\nu = k,k+1$, and any positive, continuous function ρ. Indeed, we have

$$(\rho\phi^{k+1}, \phi^{k+1/2})_0 = \tfrac{1}{2}[(\rho\phi^{k+1}, \phi^k)_0 + (\rho\phi^{k+1}, \phi^{k+1})_0],$$

$$(\rho\phi^k, \phi^{k+1/2})_0 = \tfrac{1}{2}[(\rho\phi^k, \phi^k)_0 + (\rho\phi^k, \phi^{k+1})_0],$$

and subtraction along with division by τ_n yields the desired relation (27). Inserting $D_\tau^+ v_n^k$ in place of ϕ_n into the equations of the CNG method, we obtain

$$(D_\tau^+ v_n^k, D_\tau^+ v_n^k)_0 + a(v_n^{k+1/2}, D_\tau^+ v_n^k) = (S_n^{k+1/2}, D_\tau^+ v_n^k)_0, \quad k = 0,\ldots,N-1.$$

Rewriting $a(\psi,\phi) = (a^{1/2}\psi', a^{1/2}\phi')$ and setting $\phi^\nu = (v_n^\nu)'$, $\rho = a$ in (27), we see that

$$|D_\tau^+ v_n^k|_0^2 + \frac{1}{2\tau_n}(|a^{1/2}(v_n^{k+1})'|_0^2 - |a^{1/2}(v_n^k)'|_0^2) = (S_n^{k+1}, D_\tau^+ v_n^k)_0,$$
$$k = 0,\ldots,N-1.$$

Schwarz's inequality and the fact that $\alpha\beta \le \varepsilon\alpha^2 + \beta^2/(4\varepsilon)$, for every $\alpha,\beta \ge 0$, $\varepsilon > 0$, with $\varepsilon = 1/4$, $\alpha = |S_n^{k+1/2}|_0$, $\beta = |D_\tau^+ v_n^k|_0$ imply

$$|D_\tau^+ v_n^k|_0^2 + \frac{1}{2\tau_n}(|a^{1/2}(v_n^{k+1})'|_0^2 - |a^{1/2}(v_n^k)'|_0^2)$$
$$\le |S_n^{k+1/2}|_0 |D_\tau^+ v_n^k|_0 \le \tfrac{1}{4}|S_n^{k+1/2}|_0^2 + |D_\tau^+ v_n^k|_0^2, \quad k = 0,\ldots,N-1,$$

and thus

$$|a^{1/2}(v_n^{k+1})'|_0^2 - |a^{1/2}(v_n^k)'|_0^2 \le \tfrac{1}{2}\tau_n |S_n^{k+1/2}|_0^2, \quad k = 0,\ldots,N-1.$$

We sum up both sides (from $k = 0$ until $k = m-1$) and obtain

$$|a^{1/2}(v_n^m)'|_0^2 \le |a^{1/2}(v_n^0)'|_0^2 + \tfrac{1}{2}\tau_n \sum_{k=0}^{m-1} |S_n^{k+1/2}|_0^2, \quad m = 0,\ldots,N-1.$$

The asserted inequality (26) results from applying the estimates

$$\alpha_0 |v_n^k|_1^2 \le |a^{1/2}(v_n^k)'|_0^2 \le \alpha_1 |v^k|_1^2. \qquad \square$$

As we have shown in Section 4.5, the equations of the CNG method are expressible in the form

$$C_n^{(0)} u_n(t_{k+1}) - C_n^{(1)} u_n(t_k) = \tau_n w_n(t_{k+1}), \quad k = 0,\ldots,N-1.$$

For every element $g_n \in E_n$, there exists a uniquely determined $\hat{g}_n \in E_n$ defined by

$$(\hat{g}_n,\phi_n)_0 = (g_n,\phi_n)_1, \quad \phi_n \in E_n.$$

The underlying scalar product $(.,.)_1$ can be chosen to be

$$(\psi,\phi)_1 = (\psi',\phi')_0 \quad \text{or} \quad (\psi,\phi)_1 = (\psi,\phi)_0 + (\psi',\phi')_0, \quad \psi,\phi \in H_0^1(0,1).$$

For each $n \in I$, we define

$$[g_n]_n \equiv |g_n|_0, \quad g_n \in E_n \quad (= F_n), \tag{28a}$$

which can also be expressed in terms of the function g_n itself via

$$[g_n]_n = \sup_{0\neq\phi_n\in E_n} |(g_n,\phi_n)_1|/|\phi_n|_0. \tag{28b}$$

We view $C_n^{(\ell)}$, $\ell = 0,1$, as mappings from E_n into F_n $(= E_n)$ equipped with the norms

$$|g_n|_{E_n} = |g_n|_1, \quad |f_n|_{F_n} = [f_n]_n.$$

The Poincaré-Friedrichs inequality shows us that $|g_n|_1$ can be uniformly (in n) bounded in terms of $[g_n]_n$.

The equations of the CNG method can be equivalently written as

$$\frac{1}{\tau_n}(C_n^{(0)}u_n(t_{k+1}) - C_n^{(1)}u_n(t_k),\phi_n)_1 = (\hat{w}_n(t_{k+1}),\phi_n)_0,$$
$$\phi_n \in E_n, \quad k = 0,\dots,N-1.$$

By comparing the last relation with 4.(21), we see that $\hat{w}_n(t_{k+1})$ is chosen to be the L^2-projection of $s^{k+1/2}$ in E_n. The inequality of the last theorem directly yields the inverse stability of the associated mappings T_n, $n \in I$, (cf. 11.(3)) relative to the underlying norms,

$$||v_n||_{\infty,n} = \max_{t\in[0,T]_n} |v_n(t)|_1, \quad v_n \in X_n = C([0,T]_n,E_n), \tag{29a}$$

$$||y_n||_{2,n} = |y_n^0|_1 + \left(\tau_n \sum_{t\in[0,T]_n'} [y_n^1(t)]_n^2\right)^{1/2},$$
$$y_n \in Y_n = E_n \times C([0,T]_n',F_n). \tag{29b}$$

The unique solvability of the equations of the CNG method has already been guaranteed in Section 4.2. We know that it also follows from (26).

We proceed to analyze the stability properties of the *discrete-time*

Galerkin method 4.(39),

$$(D_\tau^2 v_n^k,\phi_n)_0 + a(v_n^{k,1/4},\phi_n) = (S_n^{k,1/4},\phi_n)_0,$$
$$\phi_n \in E_n \ (\subset H_0^1(0,1)), \quad k = 1,\ldots,N-1,$$

which approximates the generalized wave equation 4.(36). Again, S_n^ν denotes the L^2-projection of $s^\nu = s(\cdot,t_\nu)$, $\nu = k,k\pm1$. By similar techniques as above, we are able to derive the following result.

Theorem 12.15. Let $a \in C[0,1]$, $0 < \alpha_0 \le a(x) \le \alpha_1$, $x \in [0,1]$. Then the solutions $v_n^k \in E_n$, $k = 0,\ldots,N$, of the Galerkin method 4.(39) satisfy the estimates

$$\max_{0\le k\le N-1} (|D_\tau^+ v_n^k|_0^2 + |v_n^{k+1/2}|_1^2) \le \gamma^2\Big\{|D_\tau^+ v_n^0|_0^2 + |v_n^{1/2}|_1^2 + \tau_n \sum_{k=1}^{N-1} |S_n^{k,1/4}|_0^2\Big\}, \quad n \ge \nu_0, \tag{30}$$

for some $\nu_0 \in I$ and $\gamma^2 = 2\exp(4T)\max(1,\alpha_1)/\min(1,\alpha_0)$.

Proof: We observe that

$$D_\tau\phi^k \equiv \frac{1}{2\tau_n}(\phi^{k+1} - \phi^{k-1}) = D_\tau^+\phi^{k-1/2} \tag{31a}$$

for arbitrary $\phi^\nu \in L^2(0,1)$, $\nu = k\pm1$. With the definition of $\phi^{k,\Theta} \equiv (1-2\Theta)\phi^k + \Theta(\phi^{k-1} + \phi^{k+1})$, elementary calculations show that

$$(a\phi^{k,1/4},D_\tau\phi^k)_0 \doteq \frac{1}{2\tau_n}(|a^{1/2}\phi^{k+1/2}|_0^2 - |a^{1/2}\phi^{k-1/2}|_0^2),$$
$$\phi^\nu \in L^2(0,1), \quad \nu = k,k\pm1. \tag{31b}$$

We additionally need the relation

$$(D_\tau^2\phi^k,D_\tau\phi^k)_0 = \frac{1}{2\tau_n}(|D_\tau^+\phi^k|_0^2 - |D_\tau^+\phi^{k-1}|_0^2),$$
$$\phi^\nu \in L^2(0,1), \quad \nu = k,k\pm1, \tag{31c}$$

which easily follows from (31a). We now insert $\phi = \phi_n = D_\tau v_n^k$ into the equations 4.(39a) and obtain

$$(D_\tau^2 v_n^k, D_\tau v_n^k)_0 + (a(v_n^{k,1/4})', (D_\tau v_n^k)')_0$$
$$= \frac{1}{2\tau_n}\{(|D_\tau^+ v_n^k|_0^2 + |a^{1/2}(v_n^{k+1/2})'|_0^2) -$$
$$- (|D_\tau^+ v^{k-1}|_0^2 + |a^{1/2}(v_n^{k-1/2})'|_0^2\}$$

$$= (S_n^{k,1/4}, D_\tau v_n^k)_0 = (S_n^{k,1/4}, \tfrac{1}{2}(D_\tau^+ v_n^k + D_\tau^+ v_n^{k-1}))_0, \quad k = 1,\ldots,N-1.$$

Schwarz's inequality yields the estimates

$$\frac{1}{2\tau_n}\{|D_\tau^+ v_n^k|_0^2 + |a^{1/2}(v_n^{k+1/2})'|_0^2 - (|D_\tau^+ v_n^{k-1}|_0^2 + |a^{1/2}(v_n^{k-1/2})'|_0^2)\}$$

$$\leq \tfrac{1}{2}|S_n^{k,1/4}|_0(|D_\tau^+ v_n^k|_0 + |D_\tau^+ v_n^{k-1}|_0)$$

$$\leq \tfrac{1}{4}|S_n^{k,1/4}|_0^2 + \tfrac{1}{2}(|D_\tau^+ v_n^k|_0^2 + |D_\tau^+ v_n^{k-1}|_0^2), \quad k = 1,\ldots,N-1.$$

Multiplying by $2\tau_n$ and summing from $k = 1$ until $k = m$, we get

$$|D_\tau^+ v_n^m|_0^2 + |a^{1/2}(v_n^{m+1/2})'|_0^2$$

$$\leq |D_\tau^+ v_n^0|_0^2 + |a^{1/2}(v_n^{1/2})'|_0^2 + \tau_n \sum_{k=1}^{m} (|D_\tau^+ v_n^k|_0^2 + |D_\tau^+ v_n^{k-1}|_0^2)$$

$$+ \frac{1}{2}\tau_n \sum_{k=0}^{m} |S_n^{k,1/4}|_0^2, \quad m = 1,\ldots,N-1.$$

We now select a $\nu_0 \in I$ such that $\tau_n \leq 1/2$, $n \geq \nu_0$, and subtract $\tau_n|D_\tau^+ v_n^m|_0^2$ from both sides to obtain

$$|D_\tau^+ v_n^m|_0^2 + |a^{1/2}(v_n^{m+1/2})'|_0^2$$

$$\leq 2\{|D_\tau^+ v_n^0|_0^2 + |a^{1/2}(v_n^{1/2})'|_0^2\} + 4\tau_n \sum_{k=1}^{m-1} |D_\tau^+ v_n^k|_0^2$$

$$+ \tau_n \sum_{k=1}^{m} |S_n^{k,1/4}|_0^2, \quad m = 0,\ldots,N-1.$$

Application of the discrete Gronwell Lemma (cf., e.g. Fairweather (1978), Sec. 4.5) yields

$$|D_\tau^+ v_n^m|_0^2 + |a^{1/2}(v_n^{m+1/2})'|_0^2$$

$$\leq \exp(4m\tau_n)\{2(|D_\tau^+ v_n^0|_0^2 + |a^{1/2}(v_n^{1/2})'|_0^2)$$

$$+ \tau_n \sum_{k=1}^{m} |S_n^{k,1/4}|_0^2\}, \quad m = 0,\ldots,N-1.$$

The inequality $0 \leq \alpha_0 \leq a(x) \leq \alpha_1$ finally proves the desired estimate (30). □

If we take the splitting $u_n^k = v_n^{k+1/2}$, $w_n^k = D_\tau^+ v_n^k$ of the present 3-level method into a 2-level system into consideration (cf. 4.(39c)), we note that we have just proved that

$$\max_{0\le k\le N-1} (|u_n^k|_1^2 + |w_n^k|_0^2) \le \gamma^2\Big\{|u_n^0|_1^2 + |w_n^0|_0^2 + \tau_n \sum_{k=1}^{N-1} |S_n^{k,1/4}|_0^2\Big\}, \quad n \ge \nu_0.$$

This is exactly the inverse stability of the associated linear mappings T_n, $n \in I$, (cf. 11.(37)) if we equip the vector-valued functions $\underline{u}_n^k = (u_n^k, w_n^k)$ in $E_n \times E_n$ with the norm $(|u_n^k|_1^2 + |w_n^k|_0^2)^{1/2}$ and, $F_n = E_n \times \{0\}$ with the norm $[\cdot]_n$ (defined in (28a)), and if we set $p = 2$ for the norms in Y_n, $n \in I$.

In an analogous manner, we shall finally prove the inverse stability of the *discrete-time Galerkin method* 4.(45a),

$$(D_\tau^+ v_n^k - c(v_n^{k+1/2})', \phi_n)_0 = (S_n^{k+1/2}, \phi_n)_0, \quad \phi_n \in E_n \ (\subset H), \ k = 0,\dots,N-1,$$

which approximates the scalar hyperbolic IVP 4.(42). In the exact as well as in the approximating equations, we seek periodic solutions with period 1. As above, S_n^ν denotes the L^2-projection of $s^\nu = s(\cdot,t_\nu)$ in E_n, $\nu = k,k+1$.

Theorem 12.16. Let $c \in C_\#^1[0,1]$. Then there exist a $\gamma \ge 0$ and a $\nu_0 \in I$ such that solutions to 4.(45a,b) satisfy the following estimates,

$$\max_{0\le k\le N} |v_n^k|_0 \le \gamma\Big\{|v_n^0|_0 + \Big(\tau_n \sum_{k=0}^{N-1} |S_n^{k+1/2}|_0^2\Big)^{1/2}\Big\}, \quad n \ge \nu_0. \tag{32}$$

Proof: Using the identity

$$(c\psi',\psi)_0 = -\tfrac{1}{2}(c'\psi,\psi)_0, \quad \psi \in H,$$

and inserting $\phi = v_n^{k+1/2}$ into 4.(45a), we obtain

$$(D_\tau^+ v_n^k, v_n^{k+1/2})_0 + \tfrac{1}{2}(c' v_n^{k+1/2}, v_n^{k+1/2})_0 = (S_n^{k+1/2}, v_n^{k+1/2})_0, \quad k = 0,\dots,N-1.$$

Schwarz's inequality and relation (27) yield the following estimates

$$\begin{aligned}|v_n^{k+1}|_0^2 - |v_n^k|_0^2 &\le 2\tau_n |S_n^{k+1/2}|_0 |v_n^{k+1/2}|_0 - \tau_n (c' v_n^{k+1/2}, v_n^{k+1/2})_0 \\ &\le \tau_n[|S_n^{k+1/2}|_0^2 + \tfrac{1}{2}(|v_n^{k+1}|_0^2 + |v_n^k|_0^2)(1 + |c'|_{0,\infty})].\end{aligned}$$

Summing, we get

$$\begin{aligned}|v_n^m|_0^2 \le |v_n^0|_0^2 &+ \tfrac{1}{2}\tau_n(1 + |c'|_{0,\infty}) \sum_{k=0}^{m-1} (|v_n^{k+1}|_0^2 + |v_n^k|_0^2) \\ &+ \tau_n \sum_{k=0}^{m-1} |S_n^{k+1/2}|_0^2, \quad m = 0,\dots,N-1.\end{aligned}$$

For some $\nu_0 \in I$, we know that $\tau_n(1 + |c'|_{0,\infty}) \leq 1$, $n \geq \nu_0$, and thus the last estimate yields

$$\frac{1}{2}|v_n^m|_0^2 \leq |v_n^0|_0^2 + \tau_n(1 + |c'|_{0,\infty}) \sum_{k=0}^{m-1} |v_n^k|_0^2 + \tau_n \sum_{k=0}^{m-1} |s_n^{k+1/2}|_0^2, \quad n \geq \nu_0.$$

Finally, the application of the discrete version of Gronwell's Lemma yields the desired estimate with

$$\gamma^2 = 2\exp(2(1 + |c'|_{0,\infty})T). \qquad \square$$

Let us remark that, instead of $c \in C_\#^1[0,1]$, it suffices to require c to be continuous, bounded and periodic, along with c' (semi)bounded from below.

12.4. INVERSE STABILITY OF NONLINEAR METHODS

In the final section of this chapter, we investigate the stability behavior of those methods introduced in Section 4.4 for approximating nonlinear initial value problems. According to the perturbation theory developed in Part II and to the results in the previous sections of this chapter, it becomes rather clear how the analysis of this section proceeds. We namely give the linearized approximating equations - which are determined by the Fréchet-derivatives given for a series of examples in Section 12.1 - and then proceed to investigate the inverse stability properties of these (now, linear) methods by means of the tools available from the previous sections. The latter are, in particular, the positivity criteria from Section 12.1 and the stability inequalities proved in 12.3 for Galerkin methods. At this particular point, we would like to emphasize that inverse stability must be shown relative to those norms for which the differentiability conditions have been verified in Section 11.1; a corresponding requirement has to be fulfilled as far as consistency is concerned (cf. Chapter 13, below).

For our methods, we again let τ_n, $n \in I$, be a null sequence of positive step widths in the t-direction. For the case of a fixed mesh ratio ($r = h_n^2/\tau_n$ or $\lambda = h_n/\tau_n$) appearing in finite-difference methods, we then have the convergence $h_n \to 0$ $(n \in I)$. We again assume, for simplicity, that all functions considered are real-valued.

As a *first* class of *examples*, we study finite-difference approximations to the following type of *semilinear parabolic IVP's* (cf. 11.(9)),

$$u_t = a(x,t)u_{xx} + F(x,t,u,u_x), \quad (x,t) \in [0,1] \times [0,T],$$
$$u(x,0) = u_0(x), \quad u(0,t) = u(1,t) = 0.$$

We suppose that $u_0(0) = u_0(1) = 0$ and that the conditions 11.(10a,b) hold. We adopt the notation of Section 11.1 and study the stability properties of the *explicit method* 11.(11) (cf. also 4.(48a,b))

$$v_j^{k+1} = (1-2ra_j^k)v_j^k + ra_j^k(v_{j-1}^k + v_{j+1}^k)$$
$$+ \tau_n F(x_j,t_k,v_j^k,D_h v_j^k), \quad j = 1,\ldots,J-1, \quad k = 0,\ldots,N-1,$$

as well as those of the *totally implicit method* 11.(12),

$$(1 + 2ra_j^{k+1})v_j^{k+1} - ra_j^{k+1}(v_{j-1}^{k+1} + v_{j+1}^{k+1})$$
$$- \tau_n F(x_j,t_{k+1},v_j^{k+1},D_h v_j^{k+1}) = v_j^k, \quad j = 1,\ldots,J-1, \quad k = 0,\ldots,N-1,$$

with appropriately given initial and boundary conditions. According to Lemma 11.2, the differentiability conditions 11.(8a,b) are met for both methods if (u_n^0) is taken to be the sequence of restrictions of the solution u^0 of the given IVP and if the maximum norms multiplied by the reciprocals of the step sizes τ_n are the underlying norms (cf. 11.(15)),

$$|g_n|_{E_n} = \frac{1}{\tau_n} \max_{0\le j\le J} |g_n(x_j)|, \quad g_n \in E_n, \quad n \in I,$$

where $E_n = \{g_n \in C(G_n): g_n(x_0) = g_n(x_J) = 0\}$. This choice of norms stems from the dependence of F on u_x; in case F is independent of u_x, the conditions 11.(8a,b) are satisfied with respect to the usual maximum norms.

We now turn to the associated linearized equations. For the explicit method 11.(11), they are given by (cf. 11.(16))

$$v_j^{k+1} = (1 - 2ra_j^k)v_j^k + ra_j^k(v_{j-1}^k + v_{j+1}^k)$$
$$+ \tau_n(F_y v_j^k + F_z D_h v_j^k + w_j^{k+1}), \quad j = 1,\ldots,J-1, \quad k = 0,\ldots,N-1, \tag{33}$$

where $w_j^{k+1} = w_n(x_j,t_{k+1})$ is an arbitrary inhomogeneous term and the partial derivatives F_y, F_z are evaluated at

$$(x_j,t_k,u^0(x_j,t_k), \frac{1}{2h_n}(u^0(x_{j+1},t_k) - u^0(x_{j-1},t_k))).$$

For the totally implicit method 11.(12), the linearized equations are given by (cf. 11.(17))

$$\begin{aligned}(1+2ra_j^{k+1})v_j^{k+1} - ra_j^{k+1}(v_{j-1}^{k+1} + v_{j+1}^{k+1}) - \tau_n(F_y v_j^{k+1} + F_z D_h v_j^{k+1}) \\ = v_j^k + \tau_n w_j^{k+1}, \quad j = 1,\dots,J-1, \quad k = 0,\dots,N-1.\end{aligned} \tag{34}$$

The arguments of F_y and F_z are

$$(x_j, t_{k+1}, u^0(x_j, t_{k+1}), \frac{1}{2h_n}(u^0(x_{j+1}, t_{k+1}) - u^0(x_{j-1}, t_{k+1}))).$$

The following theorem provides conditions under which both linearized finite-difference methods (33) and (34) satisfy the positivity criteria (5a,b) and condition (6) from Section 12.1.

Theorem 12.16. Let the assumptions 11.(10a,b) be satisfied. Then the finite-difference methods (33) and (34) satisfy the inverse stability inequalities 11.(42a,b) with respect to the maximum norms in E_n, $n \in I$, if the following respective conditions hold,

$$1 - 2r\alpha_0 \geq 0, \quad h_n|F_z| \leq 2\alpha_0 \quad \text{in } U, \quad n \geq \nu_0 \quad \text{(for (33))}, \tag{35}$$

$$\tau_n F_y \leq 1/2 \quad \text{and} \quad h_n|F_z| \leq 2\alpha_0 \quad \text{in } U, \quad n \geq \nu_0 \quad \text{(for (34))}. \tag{36}$$

Proof: (i). First, we study the method which arises from (33) if we omit the term including F_y,

$$\begin{aligned}v_j^{k+1} &= (1 - 2ra_j^k)v_j^k + (ra_j^k - \frac{\tau_n}{2h_n} F_z)v_{j-1}^k \\ &\quad + (ra_j^k + \frac{\tau_n}{2h_n} F_z)v_{j+1}^k, \quad j = 1,\dots,J-1, \quad k = 0,\dots,N-1.\end{aligned}$$

Using the relation $\tau_n/h_n = rh_n$, we see that the coefficients of the above method will obviously satisfy the positivity condition (5a) provided (35) holds. Condition (5b) is trivially satisfied for the explicit method we consider (with $\rho_0 = 1$), and condition (6) holds, since

$$\sum_{|\mu|\leq 1} d_{n,\mu}^{(1)}(x,t) = 1.$$

Thus, this method is inversely stable with respect to the maximum norm and satisfies the inequalities of 11.(42). The given linear finite-difference method (33) is exactly the one for which the have just proved inverse stability plus a difference operator of magnitude $O(\tau_n)$. For such cases, Lemma 11.7 is available which we shall apply with $Q_n^{(1)}(t_k)$:

$E_n \to E_n$ defined by

$$(Q_n^{(1)}(t_{k+1})g_n)(x_j) = F_y g_n(x_j), \quad j = 1,\ldots,J-1, \quad k = 0,\ldots,N-1.$$

Lemma 11.7 ensures the asserted inverse stability of (33) when we take the uniform boundedness of F_y in U into account.

(ii). Using the notation of Section 12.1, we see that the coefficients of the method (34) are given by

$$d_{n,0}^{(1)} = 1, \quad (N_1 = 0, \quad N_0 = 1)$$

$$d_{n,0}^{(0)}(x_j,t_{k+1}) = 1 + 2ra_j^{k+1} - \tau_n F_y, \quad d_{n,\pm 1}^{(0)}(x_j,t_{k+1}) = -r(a_j^{k+1} \pm \tfrac{1}{2}h_n F_z).$$

Condition (4) is satisfied because of the assumptions in 11.(10a,b), and (5a) is clearly valid. In order to check (5b), we observe that

$$d_{n,\pm 1}^{(0)}(x_j,t_{k+1}) \leq 0, \quad n \geq \nu_0,$$

since $h_n|F_z| \leq 2\alpha_0$. Since, moreover, $\tau_n F_y \leq 1/2$, $n \geq \nu_0$, we see that

$$\sum_{|\mu|\leq 1} d_{n,\mu}^{(0)}(x_j,t_{k+1}) = 1 - \tau_n F_y \geq 1/2, \quad n \geq \nu_0,$$

thereby proving (5b). Condition (6) finally follows from $(1-\tau_n F_y)^{-1} = 1 + O(\tau_n)$ $(n \in I)$. □

As mentioned before, the underlying norms have to be chosen such that inverse stability *and* the differentiability requirements 11.(8a&b) hold with respect to the same norms. We easily see that the maximum norms weighted with τ_n^{-1} (cf. 11.(15)) can be chosen here since such a factor simply multiplies both sides of the inequalities in 11.(42a,b). Thus, inverse stability of the associated linearized methods with respect to the norms defined in 11.(15) is also guaranteed by Theorem 12.16.

It is remarkable that the assumptions of the last theorem are not stronger than - or coincide with - those which we have imposed in Section 4.4 in order to guarantee the solvability of the associated nonlinear systems of equations. Moreover, we see that the implicit method (34) is unconditionally inversely stable whereas, for the explicit method (33), we have required a restriction on the mesh ratio - namely $r \leq 1/(2\alpha_0)$ - for proving inverse stability. The latter restriction agrees with that required for the most simple explicit method approximating the heat equation (cf. the examples in Section 12.1) with a constant conductivity coefficient.

The last theorem, together with Theorem 11.5, ensures the inverse stability of the associated sequence of nonlinear difference operators itself. However, according to the factor τ_n^{-1} appearing in the norm of E_n - and hence in those of X_n and Y_n - inverse stability is proved only in a weak form. Namely, in the general condition of inverse stability (cf. 6.(11)), the admissible perturbations have to converge with a rate faster than $O(\tau_n)$ relative to the maximum norms. Such a requirement is caused by the weight τ_n^{-1} appearing in the norms underlying our analysis. Concerning the convergence of the finite-difference methods 11.(11) and 11.(12) carried out in the following chapter, this observation has its analogy in a weak convergence property which is discussed in the remarks following Theorem 13.5.

We consider, as a *second example*, the *quasilinear parabolic IVP* 11.(18),

$$u_t = (a(x,u)u_x)_x + F(x,t,u), \quad (x,t) \in [0,1] \times [0,T],$$
$$u(x,0) = u_0(x), \quad u(0,t) = u(1,t) = 0,$$

and assume that $u_0(0) = u_0(1) = 0$; that a classical solution u^0 exists; and that the requirements of 11.(19a,b) are satisfied. We wish to study the inverse stability of the *nonlinear Crank-Nicolson-Galerkin method* 11.(20) (cf. also 4.(55)),

$$(D_\tau^+ v_n^k, \phi_n)_0 + \tfrac{1}{2}(\alpha(v_n^{k+1})(v_n^{k+1})' + \alpha(v_n^k)(v_n^k)', \phi_n')_0$$
$$= \tfrac{1}{2}(\Phi(t_{k+1}, v_n^{k+1}) + \Phi(t_k, v_n^k), \phi_n)_0, \quad \phi_n \in E_n, \quad k = 0,\dots,N-1.$$

The mappings $C_n^{(\ell)}(t)$ and their Frechet-derivatives associated with the approximation method 11.(20) have already been presented in Section 11.1, Example 2. The corresponding linearized equations can therefore be written as

$$\begin{aligned}(D_\tau^+ v_n^k, \phi_n)_0 &+ \tfrac{1}{2}(\alpha(U_n^{k+1})(v_n^{k+1})' + \alpha(U_n^k)(v_n^k)', \phi_n')_0\\ &+ \tfrac{1}{2}(\alpha_y(U_n^{k+1})(U_n^{k+1})' v_n^{k+1} + \alpha_y(U_n^k)(U_n^k)' v_n^k, \phi_n')_0\\ &+ \tfrac{1}{2}(F_y(\cdot,t_{k+1},U_n^{k+1})v_n^{k+1} + F_y(\cdot,t_k,U_n^k)v_n^k, \phi_n)_0 = (w_n^{k+1}, \phi_n)_0,\end{aligned} \tag{37}$$
$$\phi_n \in E_n, \quad k = 0,\dots,N-1, \quad n \in I.$$

where w_n^{k+1}, $k = 0,\dots,N-1$, represent arbitrary inhomogeneous terms from E_n and where $U_n^k \equiv P_n u^0(t_k)$, $k = 0,\dots,N$, with the orthogonal projections

$P_n: E \to E_n$, $n \in I$ (with respect to $(\psi,\phi)_1 = (\psi',\phi')_0$). The α, α_y are defined by $\alpha(v) \equiv a(\cdot,v)$, $\alpha_y(v) \equiv a_y(\cdot,v)$.

In order to derive a uniform estimate for the solutions of (37), we now apply the same techniques as used in the proof of inverse stability for the CNG method in the previous section (cf. Theorem 12.14). Besides requiring $\tau_n \to 0$ $(n \in I)$, we assume that the finite-dimensional subspaces E_n, $n \in I$, of $E = H_0^1(0,1)$ satisfy the following approximability property,

$$|P_n g - g|_1 = o(\tau_n^{1/2}) \quad (n \in I). \tag{38}$$

Here, we make use of the *Landau symbol* $o(\cdot)$ defined by

$$\alpha_n = o(\beta_n) \Longleftrightarrow |\alpha_n|/|\beta_n| \to 0 \quad (n \in I).$$

(Note that under (38) I must eventually be replaced by a subset $\{n \geq \nu_0\}$, which we do not explicitly express in the notation of the following discussion.) We recall that, for the analysis of this example in Section 11.1, we have considered the mappings $C_n^{(\ell)}(t)$ as mappings from E_n into F_n $(= E_n)$ equipped with the norms (cf. 11.(25a))

$$|g_n|_{E_n} = \tau_n^{-1/2}|g_n|_1, \quad |f_n|_{F_n} = \tau_n^{-1/2} \sup_{0 \neq \phi_n \in E_n} \frac{|(f_n,\phi_n)_1|}{|\phi_n|_0} \; (= \frac{[f_n]_n}{\sqrt{\tau_n}}).$$

Additionally, we need the *inverse assumption*

$$|\phi_n|_1 \leq C_0 \tau_n^{-1/2} |\phi_n|_0, \quad \phi_n \in E_n, \quad n \in I, \tag{39}$$

already required in 11.(25b). As we have mentioned in Section 11.1, the latter requirement is met for piecewise polynomial functions on quasi-uniform meshes when the mesh ratio $r = \tau_n/h_n^2$ remains constant. In that case, (38) is satisfied whenever $|P_n g - g|_1 = O(h_n^{\mu})$ for some $\mu > 1$, e.g., for piecewise quadratics or piecewise cubics. The following analysis will show that assumption (39) is not needed in case $a(x,y)$ does not depend on y.

<u>Theorem 12.17</u>. Let the assumptions in 11.(19a,b) hold, and let the subspaces E_n, $n \in I$, satisfy the approximability property (38) along with the inverse assumption (39). Then there exist numbers $\gamma > 0$, $\nu_0 \in I$ such that the solutions $v_n^k \in E_n$, $k = 0,\dots,N$, of the equations in (37) satisfy the following estimates,

$$\max_{0\le k\le N} |v_n^k|_1 \le \gamma\Big\{|v_n^0|_1 + \Big(\tau_n \sum_{m=1}^{N} |w_n^m|_0^2\Big)^{1/2}\Big\}, \quad n \ge \nu_0. \tag{40}$$

Proof: Instead of method (37), we first consider the linearized equations at the solution u^0 itself, i.e., we replace $U_n^k = P_n u^0(t_k)$ in (37) by $\tilde{u}^k \equiv u^0(t_k)$. We observe that

$$\frac{1}{2}(\alpha(\tilde{u}^{k+1})(v_n^{k+1})' + \alpha(\tilde{u}^k)(v_n^k)', \phi_n')_0$$

$$= \frac{1}{2}([\alpha(\tilde{u}^{k+1}) + \alpha(\tilde{u}^k)](v^{k+1/2})', \phi_n')_0 + \frac{1}{4}([\alpha(\tilde{u}^{k+1}) - \alpha(\tilde{u}^k)](v_n^{k+1} - v_n^k)', \phi_n')_0.$$

For the solution u^0 of the given IVP, the difference quotients $D_\tau^+\tilde{u}^k$ remain uniformly bounded, and imply that

$$|\alpha(\tilde{u}^{k+1}) - \alpha(\tilde{u}^k)|_{0,\infty} = O(\tau_n) \quad (n \in I).$$

Setting $\rho^{k+1} \equiv [\alpha(\tilde{u}^{k+1}) + \alpha(\tilde{u}^k)]/2$, for brevity, we have that

$$\rho^{k+1} = \rho^k + \frac{1}{2}[\alpha(\tilde{u}^{k+1}) - \alpha(\tilde{u}^{k-1})] = \rho^k(1 + O(\tau_n)) \quad (n \in I)$$

where we have used the assumption that $0 < \alpha_0 \le a(x,y) \le \alpha_1$. If we now insert $\phi_n = D_\tau^+ v_n^k$ into the equations of the modified linearized method, set $\psi_n^k \equiv (v_n^k)'$, $u_x^k = (u^k)'$, $k = 0,\dots,N$, and use relation (27) (with $\rho = \rho^{k+1}$, $\phi_n^\nu = \psi_n^\nu$, $\nu = k,k+1$), we obtain

$$|D_\tau^+ v_n^k|_0^2 + \frac{1}{2\tau_n}[(\rho^{k+1}\psi_n^{k+1}, \psi_n^{k+1})_0 - (\rho^{k+1}\psi_n^k, \psi_n^k)_0]$$

$$+ \frac{1}{4}\tau_n([\alpha(\tilde{u}^{k+1}) - \alpha(\tilde{u}^k)]D_\tau^+\psi_n^k, D_\tau^+\psi_n^k)_0$$

$$+ \frac{1}{2}(\alpha_y(\tilde{u}^{k+1})\tilde{u}_x^{k+1}v_n^{k+1} + \alpha_y(\tilde{u}^k)\tilde{u}_x^k v_n^k, D_\tau^+\psi_n^k)_0$$

$$+ \frac{1}{2}(F_y(\cdot,t_{k+1},\tilde{u}^{k+1})v_n^{k+1} + F_y(\cdot,t_k,\tilde{u}^k)v_n^k,\ D_\tau^+ v_n^k)_0$$

$$= (w_n^{k+1}, D_\tau^+ v_n^k)_0, \quad k = 0,\dots,N-1.$$

By assumption, $\frac{\partial}{\partial x}(\alpha_y(\tilde{u}^\nu)\tilde{u}_x^\nu)$, $\nu = 0,\dots,N$, exist and are bounded continuous functions. Integration by parts yields the relations

$$-(\alpha_y(\tilde{u}^\nu)\tilde{u}_x^\nu v_n^\nu, D_\tau^+\psi_n^k)_0 = (\frac{\partial}{\partial x}(\alpha_y(\tilde{u}^\nu)\tilde{u}_x^\nu)v_n^\nu, D_\tau^+ v_n^k)_0$$

$$+ (\alpha_y(\tilde{u}^\nu)\tilde{u}_x^\nu\psi_n^\nu, D_\tau^+ v^k)_0, \quad \nu = k,k+1, \quad k = 0,\dots,N-1.$$

The solutions v_n^k to the linearized equations (modified from (37) as above) therefore satisfy

$$|D_\tau^+ v_n^k|_0^2 + \frac{1}{2\tau_n}[(\rho^{k+1}\psi_n^{k+1},\psi_n^{k+1})_0 - (\rho^{k+1}\psi_n^k,\psi_n^k)_0]$$
$$+ \frac{1}{4}\tau_n([\alpha(\tilde{u}^{k+1}) - \alpha(\tilde{u}^k)]D_\tau^+\psi_n^k, D_\tau^+\psi_n^k)_0$$
$$= \frac{1}{2}(\frac{\partial}{\partial x}(\alpha_y(\tilde{u}^{k+1})\tilde{u}_x^{k+1})v_n^{k+1} + \frac{\partial}{\partial x}(\alpha_y(\tilde{u}^k)\tilde{u}_x^k)v_n^k, D_\tau^+ v_n^k)_0$$
$$+ \frac{1}{2}(\alpha_y(\tilde{u}^{k+1})\tilde{u}_x^{k+1}\psi_n^{k+1} + \alpha_y(\tilde{u}^k)\tilde{u}_x^k\psi_n^k, D_\tau^+ v_n^k)_0$$
$$- \frac{1}{2}(F_y(.,t_{k+1},\tilde{u}^{k+1})v_n^{k+1} + F_y(.,t_k,\tilde{u}^k)v_n^k, D_\tau^+ v_n^k)_0$$
$$+ (w_n^{k+1}, D_\tau^+ v_n^k)_0, \quad k = 0,\ldots,N-1.$$

The absolute value of the last term on the left-hand side (of the equal sign) will be estimated and subtracted by using the inverse inequality (39), and the right-hand side can be estimated by using the equivalence of the norms $||\cdot||_1$ and $|\cdot|_1$ in $H_0^1(0,1)$. With constants $C_1,C_2 \geq 0$ (independent of n,k), we obtain

$$(1-C_1\tau_n)|D_\tau^+ v_n^k|_0^2 + \frac{1}{2\tau_n}[(\rho^{k+1}(v_n^{k+1})',(v_n^{k+1})')_0 - (\rho^{k+1}(v_n^k)',(v_n^k)')_0]$$
$$\leq C_2\{|v_n^k|_1 + |v_n^{k+1}|_1 + |w_n^{k+1}|_0\}|D_\tau^+ v_n^k|_0, \quad n \geq \nu_1.$$

Here, ν_1 is chosen such that $C_1\tau_n \leq 1/2$, $n \geq \nu_1$. We apply the inequality $\alpha\beta \leq \varepsilon\alpha^2 + \beta^2/(4\varepsilon)$ to the right-hand side with

$$\alpha = |D_\tau^+ v_n^k|_0, \quad \beta = C_2\{\text{——}\}, \quad \varepsilon = 1 - C_1\tau_n$$

and observe that the term including $|D_\tau^+ v_n^k|_0^2$ disappears. After multiplication by $2\tau_n$, we obtain

$$(\rho^{k+1}(v_n^{k+1})',(v_n^{k+1})')_0 \leq (\rho^{k+1}(v_n^k)',(v_n^k)')_0$$
$$+ C_3\tau_n\{|v_n^{k+1}|_1^2 + |v_n^k|_1^2 + |w_n^{k+1}|_0^2\}, \quad k = 0,\ldots,N-1, \quad n \geq \nu_1,$$

where C_3 denotes a constant independent of n. Using the estimates

$$\rho^{k+1} = \rho^k(1+O(\tau_n)), \quad \alpha_0|v_n^\nu|_1^2 \leq (\rho^\nu(v_n^\nu)',(v_n^\nu)')_0 \leq \alpha_1|v_n|_1^2,$$
$$\nu = k,k+1,$$

we finally come to

$$(1 - C_4\tau_n)(\rho^{k+1}(v_n^{k+1})',(v_n^{k+1})')_0$$
$$\le (1 + C_5\tau_n)(\rho^k(v_n^k)',(v_n^k)')_0 + C_3\tau_n|w_n^{k+1}|_0^2,$$
$$k = 0,\ldots,N-1, \quad n \ge \nu_2,$$

which in turn imply

$$(\rho^{k+1}(v_n^{k+1})',(v_n^{k+1})')_0 \le (1 + C_6\tau_n)^{k+1}(\rho^0(v_n^0)',(v_n^0)')_0$$
$$+ C_6\tau_n \sum_{m=0}^{k} (1 + C_6\tau_n)^{k-m}|w_n^{m+1}|_0^2, \quad k = 0,\ldots,N-1, \quad n \ge \nu_2,$$

for a sufficiently large $\nu_2 \in I$ and constants C_i, $i = 4,5,6$. Observing $0 < \alpha_0 \le \rho \le \alpha_1$ and $(1 + C_6\tau_n)^\nu \le \exp(C_6T)$, we have proved the estimate (40) for the solutions v_n^k of the (with $\tilde{u}^k$ in place of U_n^k) modified linearized equations (37).

The Fréchet-derivatives of the associated mappings T_n, $n \in I$, (cf. 11.(3), 11.(7), and 11.(13)) thus satisfy, with some $\gamma > 0$,

$$\frac{1}{\gamma} \max_{t\in[0,T]_n} |v_n(t)|_1 \le |v_n(0)|_1 + \left(\tau_n \sum_{m=1}^{N} |w_n^m|_0^2\right)^{1/2}, \quad v_n \in X_n, \quad n \ge \nu_2.$$

Here, $w_n^m \in E_n$ can be characterized by

$$(w_n^m,\phi_n)_0 = ((T_n'(u^0|[0,T]_n)v_n)(t_m),\phi_n)_1, \quad \phi_n \in E_n,$$

with

$$|w_n^m|_0 = [(T_n'(u^0|[0,T]_n)v_n)(t_m)]_n$$

where $[\cdot]_n$ is defined in 12.(28a). Lemma 11.3, together with Theorem 11.1, ensures that T_n', $n \in I$, is equicontinuous at $u^0|[0,T]_n$, $n \in I$, with respect to the norms $||\cdot||_{\infty,n}$ in X_n and $||\cdot||_{2,n}$ in Y_n (cf. 11.(5a,b)) which in turn are specified by the norms $|\cdot|_{E_n}$ and $|\cdot|_{F_n}$ given above. Thus, for $\varepsilon = 1/(2\gamma)$, there is a $\delta > 0$ such that

$$\left(\tau_n \sum_{t\in[0,T]_n'} |(([T_n'(u^0|[0,T]_n) - T_n'(z_n(t))]v_n)(t)|_{F_n}^2\right)^{1/2}$$
$$\le \frac{1}{2\gamma} \max_{t\in[0,T]_n} |v_n(t)|_{E_n}, \quad v_n \in X_n, \quad n \in I,$$

whenever

$$\max_{t\in[0,T]_n} |u^0(t) - z_n(t)|_{E_n} \leq \delta.$$

Taking into consideration the factor $\tau_n^{-1/2}$ in the definition of the norms $|\cdot|_{E_n}$ and $|\cdot|_{F_n}$, we see that the latter condition is met for $z_n(t_k) = (R_n^X u^0)(t_k) = U_n^k = P_n u^0(t)$ and $n \geq \nu_3$ ($\geq \nu_2$) since the approximability property (38) is required. Altogether we obtain

$$\frac{1}{2\gamma} \max_{t\in[0,T]_n} |v_n(t)|_{E_n} \leq |v_n(0)|_{E_n}$$
$$+ \left(\tau_n \sum_{t\in[0,T]_n'} |(T_n'(R_n^X u^0)v_n)(t)|_{F_n}^2\right)^{1/2}, \quad n \geq \nu_3,$$

which, after multiplication by $\tau_n^{1/2}$, yields the desired estimate (40) with 2γ and ν_3 in place of γ and ν_0, respectively. □

Estimate (40) ensures the uniform boundedness of the inverses of $T_n'(R_n^X u^0)$, $n \geq \nu_0$, relative to the spaces and norms which we have already chosen for the Crank-Nicolson-Galerkin method 4.(21) approximating the heat equation (here, $(R_n^X u^0)(t_k) = P_n u^0(t_k) = U_n^k$, $k = 0,\ldots,N$). This statement also holds with respect to the norms $|\cdot|_{E_n}$ and $|\cdot|_{F_n}$ including $\tau_n^{-1/2}$ since this factor multiplies simply both sides of (40). The negative powers of τ_n in the norms, however, have the consequence that the stronger approximability property (38) has to be required. Moreover, and more importantly, the inverse stability of the nonlinear mappings T_n, $n \in I$, themselves (guaranteed by Theorem 11.5) holds in a weak form allowing only perturbations with a stronger convergence property (cf. also the discussion following Theorem 12.16).

We conclude this section by investigating the stability properties of a *third* class of *examples*, namely finite-difference methods approximating *scalar, quasilinear hyperbolic IVP's* of the form 11.(26),

$$u_t - c(x,t,u)u_x = s(x,t,u), \quad t \in [0,T], \quad x \in \mathbb{R},$$
$$u(x,0) = u_0(x), \quad x \in \mathbb{R}.$$

Suppose that all functions considered are (at least) bounded and continuous and that c and s have bounded, continuous partial derivatives with respect to the third argument. Moreover, let u^0 be a solution of the given IVP. We could restrict the domains of definition of c, s and their partial derivatives to a compact neighborhood of the solution which, however, will not be specified in detail.

The finite-difference methods presented in Example 3 of Section 11.1 (the Friedrichs, Courant-Isaacson-Rees and Lax-Wendroff methods) can be expressed in the following manner,

$$v_j^{k+1} = \sum_{|\mu|\le 1} b_{\mu,j}^k(v_j^k)v_{j+\mu}^k + \tau_n s(x_j,t_k,v_j^k), \quad j = 0,\pm 1,\ldots .$$

Here, $G_n = \{x_j \equiv jh_n,\ j = 0,\pm 1,\pm 2,\ldots\}$, $n \in I$, denote uniform meshes in $\mathbb{R}$ with mesh widths h_n, $n \in I$, and the functions

$$b_{\mu,j}^k(y) = b_\mu(x_j,t_k,y), \qquad |\mu| \le 1,$$

are defined by $b_\mu(x,t,y)$ given in terms of a function $b(x,t,y)$ as in 11.(28).

For linear problems, the b_μ are independent of y, and we can directly apply the positivity criteria of Section 12.1 in order to verify the inverse stability with respect to the supremum norm in each $E_n = C(G_n)$, $n \in I$. As a result, the requirement $b_\mu \ge 0$, $|\mu| \le 1$, is sufficient for inverse stability. Indeed, then (5a) is satisfied; (5b) is trivially true for the explicit methods considered, and (6) results from $\Sigma b_\mu = 1$ (cf. 11.(29)). For each single method, this positivity requirement is only satisfied when the mesh ratio λ is restricted by the Courant-Friedrichs-Lewy condition $\lambda \le 1/|c|$. For the Lax-Wendroff method we must additionally require that $\lambda|c| = 1$ which shows that, as far as positivity properties are concerned, this method is applicable only for a constant function c (see also the discussion in Meis & Marcowitz (1981) on the Lax-Wendroff method and its generalization, the Lax-Wendroff-Richtmyer method).

Törnig & Ziegler (1966) have shown that, for nonlinear methods of the above form, the positivity condition $b_\mu \ge 0$, $|\mu| \le 1$, together with a consistency requirement, ensures the local convergence of the solutions with respect to supremum norms (cf. also Törnig (1979), 17.3.4). On the basis of our stability theory, these results are at first glance somewhat surprising because the nonlinearity is, to a certain extent, ignored when the requirements on linear methods are carried over to nonlinear ones verbatim. The following theorem, however, shows that, also for the nonlinear case, the condition $b_\mu \ge 0$ yields the inverse stability of the associated linearized equations via criteria from Section 12.1. For the above methods, the associated linearized equations are of the form (cf. 11.(31))

$$v_j^{k+1} = \sum_{|\mu|\le 1} b_\mu v_{j+\mu}^k + \tau_n\left\{\frac{1}{\lambda}\left[\frac{\partial b_1}{\partial y} D_h^+U_j^k - \frac{\partial b_{-1}}{\partial y} D_h^-U_j^k\right] + \frac{\partial s}{\partial y}\right\}v_j^k \tag{41}$$

$$+ \tau_n w_j^{k+1}, \quad j = 0,\pm1,\dots, \quad k = 0,\dots,N-1.$$

The argument of b_μ, $\partial b_\mu/\partial y$ and $\partial s/\partial y$ is (x_j,t_k,U_j^k) where $U_j^k = u^0(x_j,t_k)$. From this representation of the linearized method, we can easily deduce the following result.

Theorem 12.18. Let c, b and s be bounded, continuous, and differentiable with respect to their third arguments with resulting partial derivatives bounded and continuous. Then, in case of $b_\mu \ge 0$, $|\mu| \le 1$, the finite-difference method (41) is inversely stable with respect to the supremum norms in $E_n = C(G_n)$, $n \in I$.

Proof: The method

$$v_j^{k+1} = \sum_{|\mu|\le 1} b_\mu v_{j+\mu}^k + \tau_n w_j^{k+1}, \quad j = 0,\pm1,\dots, \quad k = 0,\dots,N-1,$$

satisfies conditions (5a,b) and (6) of Section 12.1 due to the hypotheses and the relation $\Sigma b_\mu = 1$. Therefore this method is inversely stable with respect to the supremum norms. A comparison shows that a term of magnitude $O(\tau_n)$ is added in (41), because the difference quotients $D_h^{\pm}U_j^k$ remain uniformly bounded for the solution u^0 of the given IVP. Application of Lemma 11.7 with

$$(Q_n^{(1)}(t_{k+1})g_n)(x_j) = \left\{\frac{1}{\lambda}\left[\frac{\partial b_1}{\partial y} D_h^+U_j^k - \frac{\partial b_{-1}}{\partial y} D_h^-U_j^k\right] + \frac{\partial s}{\partial y}\right\}g_n(x_j),$$

$$j = 0,\pm1,\dots, \quad k = 0,\dots,N-1, \quad n \in I,$$

yields the assertion. □

REFERENCES (cf. also References in Chapters 4 and 11)

Ciarlet (1978), Fairweather (1978), Forsythe & Wasow (1967), Meis & Marcowitz (1981), Richtmyer & Morton (1967), Törnig (1979), Törnig & Ziegler (1966)*, Varga (1962).

*Article.

Chapter 13
Convergence Analysis of Special Methods

In this chapter, we analyze the behavior of the truncation errors for various methods approximating solutions of linear and nonlinear initial value problems. Based on this analysis, and on results on inverse stability from Chapter 12, convergence results for these methods are then obtained. It is appropriate at this point to emphasize that the investigation of the truncation errors and the resulting convergence analysis must be carried out with respect to those norms for which inverse stability is assured for the method being considered, and - for nonlinear problems - for which the uniform differentiability conditions in Section 11.1 are satisfied. For finite-difference methods, the truncation errors will be examined by using Taylor series expansions; then convergence along with error estimates will be shown with respect to both the discrete supremum norms and the discrete L^2-norms. For Galerkin methods, we aim to produce quasi-optimal error estimates with respect to the approximations in the spatial variable. Attaining such estimates requires extensive investigations for rather simple problems so that we shall necessarily restrict the scope of our analysis of Galerkin methods to an example for the heat and the wave equations, respectively

In Section 13.1, we obtain results on the consistency and convergence of the Crank-Nicolson method, the Du Fort-Frankel method, and a finite-difference method approximating a nonlinear parabolic IVP; moreover, we study well-known methods for approximating hyperbolic, linear and scalar quasilinear problems. In an example where we use the Crank-Nicholson method, we show, moreover, how the approximation of Neumann boundary conditions can affect the behavior of the truncation errors and the discritization errors themselves. The discrete supremum norms and the discrete

L^2-norms are quite natural as underlying norms, and we have previously shown inverse stability with respect to these norms in Chapter 12. For the discrete L^2-norms, however, we have shown stability only for linear methods with constant coefficients. Since the discrete L^2-norm is weaker than the supremum norm, uniform estimates of the truncation errors can be used to derive error estimates also in this norm.

In Section 13.2, the convergence analysis will be carried out for two discrete-time Galerkin methods used to approximate the solutions, respectively, of the generalized inhomogeneous heat and wave equations. In our treatment, it becomes clear what the concepts of truncation error and consistency mean for these methods; to our knowledge, these concepts are not properly explained for discrete-time Galerkin methods in the literature. A careful estimate of the truncation errors by using interpolation estimates, inverse inequalities, and the so-called Aubin-Nitsche-trick yields quasi-optimal convergence rates for approximating the spatial dependence. In both examples, the time step sizes appear with magnitude of $O(\tau_n^2)$ in the error estimates; such behavior is typical when the time derivative u_t is approximated by the type of averaging used in the Crank-Nicolson method. The analysis of both examples relies on the treatment in Fairweather (1978), Sec. 4.5, and in Dupont (1973); at several places in the discussion, however, we use other norms and other initial approximations than those discussed in these works.

13.1. CONSISTENCY AND CONVERGENCE OF FINITE-DIFFERENCE APPROXIMATIONS

In this section, we study the behavior of the truncation errors for several finite-difference methods used to solve numerically both linear and nonlinear, parabolic and hyperbolic initial value problems. In particular, we shall restrict our analysis to a few typical examples of such problems and obtain, via the results on inverse stability from Chapter 12, convergence results along with error estimates in both the supremum norms and the discrete L^2-norms. Furthermore, we make clear in our analysis how the approximation of the boundary conditions (e.g., the approximation of Neumann boundary conditions for the heat equation) can influence the consistency and the convergence (with accompanying error estimates).

We shall always estimate the truncation errors with respect to the supremum norm in $C(G_n)$, where G_n, $n \in I$, denotes a finite or infinite set of mesh points in $\mathbb{R}$. We can hereby obtain for periodic mesh functions corresponding estimates in discrete L^2-norms which were basic for the analysis in Section 12.2. Indeed these latter norms are of course weaker than the supremum norms and can be estimated in terms of these.

The step sizes τ_n are always assumed to converge to zero. Whenever fixed mesh ratios $r = \tau_n/h_n^2$ or $\lambda = \tau_n/h_n$ are constant in n, the mesh widths h_n of the grids G_n must also necessarily converge to zero. Occasionally, we shall write $v(t)(x) = v(x,t)$ and $v_j^k = v(x_j,t_k)$.

We now begin to investigate finite-difference methods for approximating the solution of the *heat equation* 4.(1a,b),

$$u(x,0) = u_0(x), \quad u_t(x,t) = au_{xx}(x,t) + s(x,t), \quad (x,t) \in [0,1] \times [0,T],$$

where $a > 0$ is constant. We first consider Dirichlet boundary conditions, and obtain the following results on the consistency and convergence properties of the *Crank-Nicolson method* (cf. 4.(8) and 4.(9)).

Theorem 13.1. Suppose that the solution to the inhomogeneous heat equation 4.(1a,b) with Dirichlet boundary conditions is sufficiently smooth. Then the truncation errors of the Crank-Nicolson method 4.(9) ($\Theta = 1/2$) have the following representations,

$$d_n^1(u)(x_j,t_k) = s(x_j,t_{k-1/2}) - s(x_j,t_k) + O(\tau_n^2+h_n^2), \qquad j = 1,\dots,J-1, \quad k = 1,\dots,N, \tag{1}$$

uniformly in $n \in I$. If $r = \tau_n/h_n^2$ is bounded by $r \leq 1/a$, the errors of the solution $u_n(x_j,t_k) = v_j^k$ to 4.(9) are then estimated by

$$\max_{\substack{0\leq j\leq J\\ 0\leq k\leq N}} |u_n(x_j,t_k) - u(x_j,t_k)| = O(\tau_n) \quad (n \in I). \tag{2}$$

With respect to the discrete L^2-norms, these errors can be estimated by

$$\max_{0\leq k\leq N} \left(h_n \sum_{j=0}^{J} |u_n(x_j,t_k) - u(x_j,t_k)|^2\right)^{1/2} = O(\tau_n) \quad (n \in I) \tag{3}$$

without any restriction on the mesh ratio r.

Proof: The truncation errors of method 4.(9) (with $\Theta = 1/2$) have the form

$$d_n^1(u)(x_j,t_k) = \frac{1}{\tau_n}(u(x_j,t_k) - u(x_j,t_{k-1})) - \frac{1}{2}a((D_h^2u)(x_j,t_k) + (D_h^2u)(x_j,t_{k-1})) - s(x_j,t_k), \quad 1 \leq j \leq J-1.$$

From Taylor's formula, we conclude

$$\tau_n^{-1}(u(x_j,t_k) - u(x_j,t_{k-1})) = u_t(x_j,t_{k-1/2}) + O(\tau_n^2),$$

if the partial derivative u_{ttt} exists and is continuous. We get for the finite-difference quotients with respect to x,

$$\tfrac{1}{2}((D_h^2u)(x_j,t_k) + (D_h^2u)(x_j,t_{k-1})) = u_{xx}(x_j,t_{k-1/2}) + O(h_n^2 + \tau_n^2)$$

whenever the partial derivatives u_{xxxx} and u_{xxtt} exist and are continuous. The above relation (1) for the truncation errors then easily follows. When $r \leq 1/a$, we know from Section 12.1 that the Crank-Nicolson method satisfies an inverse stability inequality of the form 12.(12). Inserting the discretization error $e_n(x,t) = u_n(x,t) - u(x,t)$ into this inequality, we obtain the estimate,

$$\max_{\substack{0\leq j\leq J\\ 0\leq k\leq N}} |e_n(x_j,t_k)| \leq \gamma\tau_n \sum_{k=1}^{N} \max_{1\leq j\leq J-1} |\tfrac{1}{2}(s(x_j,t_k) + s(x_j,t_{k-1})) - s(x_j,t_{k-1/2})| + O(\tau_n^2 + h_n^2).$$

In order to see this, we observe that the errors themselves vanish at the boundary and otherwise satisfy the equations of the Crank-Nicolson method with right-hand sides given by (after simple calculations)

$$w_n(x_j,t_k) - s(x_j,t_k) - d_n^1(u)(x_j,t_k)$$
$$= \tfrac{1}{2}(s(x_j,t_k) + s(x_j,t_{k-1})) - s(x_j,t_{k-1/2}) + O(\tau_n^2 + h_n^2),$$
$$1 \leq j \leq J-1, \quad 1 \leq k \leq N.$$

For sufficiently smooth s, the partial derivatives s_t and s_{tt} will exist, and hence

$$\tfrac{1}{2}(s(x_j,t_k) + s(x_j,t_{k-1}))-s(x_j,t_{k-1/2}) = \tfrac{1}{8}\,\tau_n^2 s_{tt}(x_j,\tilde{t}_k) = O(\tau_n^2)$$

where $\tilde{t}_k \in (t_{k-1},t_k)$. If we now use the fact that $h_n^2 = \tau_n/r$, we obtain the desired estimate with respect to the maximum norm. If we extend the errors as odd, periodic grid functions (with period 2), then application of the results from Section 12.2 yields the second error estimate, if we observe that the truncation errors with respect to the discrete L^2-norms can be estimated by those with respect to maximum norms. □

From the representation of the truncation error for the Crank-Nicolson method, we see that we could have used, in 4.(8) and 4.(9) with

$\Theta = 1/2$, $s(x_j, t_{k+1/2})$ instead of the arithmetic mean of the values of t_k and t_{k+1}, which would also yield the same order of convergence (cf. also Marchuk (1975), Sec. 4.1). We moreover observe that, because $h_n^2 = O(\tau_n)$, the partial derivative with respect to t is better approximated (to within $O(\tau_n^2)$) than that with respect to x (to within $O(h_n^2)$). In order to approximate both partial derivatives to within $O(\tau_n^2)$ - or even to within $O(\tau_n^3)$ - we must use other methods which are, for example, discussed in Mitchell and Griffiths (1980), Chapter 2.

We now proceed to discuss the case of Neumann boundary conditions. For the Crank-Nicolson methods, we have already obtained in 12.(13) the appropriate stability inequality with respect to the maximum norms which shows that the boundary data appear with a factor of magnitude $O(\tau_n^{-1/2})$. As a consequence, the order of the convergence is lowered by 1/2.

Theorem 13.2. Suppose the solution u of the inhomogeneous heat equation 4.(1a,b) with Neumann boundary conditions can be extended beyond $x = 0$ and $x = 1$ for all $t \in [0,T]$ and that this extension is sufficiently smooth. If $r \leq 1/a$, then the errors in the solution $u_n(x_j,t_k) = v_j^k$ of the Crank-Nicolson method 4.(10) (with $\Theta = 1/2$) can be estimated by

$$\max_{\substack{0\leq j\leq J \\ 0\leq k\leq N}} |u_n(x_j,t_k) - u(x_j,t_k)| = O(\tau_n^{1/2}) \quad (n \in I). \tag{4}$$

Proof: Since the solution u of the given initial-boundary-value problem can be extended sufficiently smoothly beyond $x = 0$ and $x = 1$, we see that (with $u_j^k = u(x_j,t_k)$)

$$-\frac{u_1^k - u_{-1}^k}{2h_n} = \gamma_0^k + O(h_n^2), \quad \frac{u_{J+1}^k - u_{J-1}^k}{2h_n} = \gamma_1^k + O(h_n^2),$$

or, equivalently, that

$$u_{-1}^k = u_1^k + 2h_n\gamma_0^k + O(h_n^3), \quad u_{J+1}^k = u_{J-1}^k + 2h_n\gamma_1^k + O(h_n^3), \quad k = 0,\dots,N.$$

The errors $e_j^k = e_n(x_j,t_k) = u_n(x_j,t_k) - u(x_j,t_k)$ satisfy the following equations when $x = x_0$:

$$\begin{aligned}(1+ar)e_0^k &- are_1^k - (1-ar)e_0^{k-1} - are_1^{k-1} \\ &= arh(\gamma_0^k + \gamma_0^{k-1}) + \frac{1}{2}\tau_n(s_0^k + s_0^{k-1}) \\ &\quad - (1 + ar)u_0^k + aru_1^k + (1 - ar)u_0^{k-1} + aru_1^{k-1}\end{aligned}$$

$$= -(1 + ar)u_0^k + \frac{1}{2}\, aru_1^k + \frac{1}{2}\, ar(u_1^k + 2h_n\gamma_0^k)$$

$$+ (1 - ar)u_0^{k-1} + \frac{1}{2}\, aru_1^{k-1} + \frac{1}{2}\, ar(u_1^{k-1} + 2h_n\gamma_0^{k-1})$$

$$+ \frac{1}{2}\,\tau_n(s_0^k + s_0^{k-1})$$

$$= -(u_0^k - u_0^{k-1}) + \frac{1}{2}\, a\tau_n(D_h^2 u_0^k + D_h^2 u_0^{k-1})$$

$$+ \frac{1}{2}\,\tau_n(s_0^k + s_0^{k-1}) + O(h_n^3)$$

$$= \tau_n\{-u_t(x_0,t_{k-1/2}) + au_{xx}(x_0,t_{k-1/2})\}$$

$$+ \frac{1}{2}\,\tau_n(s_0^k + s_0^{k-1}) + O(\tau_n^3 + \tau_n h_n^2 + h_n^3),$$

$$= \tau_n[-s(x_0,t_{k-1/2}) + \tfrac{1}{2}(s_0^k + s_0^{k-1})] + O(\tau_n^3 + \tau_n h_n^2 + h_n^3),$$

$$k = 1,2,\ldots,N.$$

At $x = x_J$, we have

$$(1+ar)e_J^k - are_{J-1}^k - (1-ar)e_J^{k-1} - are_{J-1}^{k-1}$$

$$= \tau_n[-s(x_J,t_{k-1/2}) + \tfrac{1}{2}(s_J^k + s_J^{k-1})] + O(\tau_n^3 + \tau_n^2 h_n^2 + h_n^3),$$

$$k = 1,\ldots,N.$$

The grid points $x_1,\ldots,x_{J-1}$ are to be treated as in the proof of the previous theorem. If we make use of

$$\tfrac{1}{2}(s_j^k + s_j^{k-1}) - s(x_j,t_{k-1/2}) = O(\tau_n^2), \quad j = 0,\ldots,J,$$

and observe that $h_n^2 = O(\tau_n)$, then

$$\tau_n^{-1}[(C_n^{(0)}e_n)(x_j,t_k) - (C_n^{(1)}e_n)(x_j,t_{k-1})] = \begin{cases} O(h_n), & j = 0,J, \\ O(\tau_n), & j = 1,\ldots,J-1. \end{cases}$$

Applying 12.(13) for $v_j^k = e_j^k$, we obtain the asserted error estimate (4). □

Our final remark concerning method 4.(8) is that we can obtain the same order of convergence with respect to the maximum norm for arbitrary Θ in $[0,1]$, as was stated in Theorems 13.1 and 13.2, if we assume that the mesh ratio is bounded by $r \leq 1/(2a(1-\Theta))$.

We now turn to investigating the consistency and convergence of the *Du Fort-Frankel method* 4.(11) for the case of Dirichlet boundary conditions. Corresponding to its formulation as a two-level system, the truncation errors of this scheme are vectors with two components,

$$\begin{aligned} d_n^1(\underline{u})_1(x,t) &= \frac{1}{2\tau_n}(u(x,t) - u(x,t-2\tau_n)) \\ &\quad - \frac{a}{h_n^2}(u(x+h_n,t-\tau_n) - u(x,t) - u(x,t-2\tau_n) + u(x-h_n,t-\tau_n))-s(x,t), \\ d_n^1(\underline{u})_2(x,t) &= 0, \end{aligned} \tag{5}$$

where $\underline{u}(x,t) = (u(x,t),u(x,t-\tau_n))$. Having rewritten in a straightforward manner this method as a two-level system, we observe that the second component of the truncation errors vanishes. We express the first component $d_n^1(\underline{u})_1$ as $d_n^1(u)$ for simplicity. The following interesting results shows that the Du Fort-Frankel method is consistent with the heat equation when $r = \tau_n/h_n^2$ is constant, but with a hyperbolic IVP whenever $\lambda = \tau_n/h_n$ is constant.

Lemma 13.3. If u is sufficiently smooth, then the truncation errors of the Du Fort-Frankel method have the form

$$\begin{aligned} d_n^1(u)(x_j,t_{k+1}) &= (u_t)_j^k - a(u_{xx})_j^k - s_j^{k+1} \\ &\quad + a\,\frac{\tau_n^2}{h_n^2}(u_{tt})_j^k - a\,\frac{h_n^2}{12}(u_{xxxx})_j^k + O(\tau_n^2 + \tau_n^3/h_n^2 + h_n^4), \end{aligned} \tag{6}$$

$$1 \le j \le J-1, \quad k = 0,\dots,N-1.$$

If $r = \tau_n/h_n^2$ is constant for all n, then

$$d_n^1(u)(x_j,t_{k+1}) = s_j^k - s_j^{k+1} + O(\tau_n) \ (n \in I), \quad 1 \le j \le J-1,\ 0 \le k \le N-1,$$

where u is the solution of the heat equation 4.(1). If, however, $\lambda = \tau_n/h_n$ is constant for all n, then

$$d_n^1(u)(x_j,t_{k+1}) = s_j^k - s_j^{k+1} + O(\tau_n^2) \ (n \in I), \quad 1 \le j \le J-1,\ 0 \le k \le N-1,$$

where u is a solution of the hyperbolic partial differential equation,

$$u_t + a\lambda^2 u_{tt} = au_{xx} + s.$$

Proof: Using Taylor series expansions, we see that

$$u(x_j,t_{k+1}) + u(x_j,t_{k-1}) = 2u(x_j,t_k) + \tau_n^2(u_{tt})_j^k + O(\tau_n^3),$$

$$u(x_{j+1},t_k) + u(x_{j-1},t_k) = 2u(x_j,t_k) + h_n^2(u_{xx})_j^k + \frac{1}{12}h_n^4(u_{xxxx})_j^k + O(h_n^6).$$

The truncation errors therefore become

$$d_n^1(u)(x_j,t_{k+1}) = (u_t)_j^k + a\frac{\tau_n^2}{h_n^2}(u_{tt})_j^k - a(u_{xx})_j^k$$
$$- \frac{a}{12}h_n^2(u_{xxxx})_j^k - s_j^{k+1} + O(\tau_n^2 + \frac{\tau_n^3}{h_n^2} + h_n^4),$$
$$1 \le j \le J-1, \quad 0 \le k \le N-1,$$

and the desired representation follows. With constant $r = \tau_n/h_n^2$, $n \in I$, we have, for the solution of the inhomogeneous heat equation, that

$$d_n^1(u)(x_j,t_{k+1}) = s_j^k - s_j^{k+1} + O(\tau_n),$$

where $s_j^k - s_j^{k+1} = O(\tau_n)$ for sufficiently smooth s. If $\lambda = \tau_n/h_n$ is constant for all $n \in I$, we have the other consistency result of the theorem. □

To prove convergence of the Du Fort-Frankel method, we combine its inverse stability inequality 12.(14) with the representation of the truncation errors in the previous lemma. We recall that an initial approximation is needed for $t_1 = \tau_n$ which can be obtained by a rather simple explicit method (cf. 4.(12)).

Theorem 13.4. If u denotes a sufficiently smooth solution of the inhomogeneous heat equation 4.(1) with Dirichlet boundary conditions, and $r = \tau_n/h_n^2$ is constant for all n and bounded by $r \le 1/a$, then the errors in the solutions $u_n(x_j,t_k) = v_j^k$ of the Du Fort-Frankel method 4.(12) are estimated by

$$\max_{\substack{0\le j\le J\\0\le k\le N}} |u_n(x_j,t_k) - u(x_j,t_k)| \le \max_{0\le j\le J} |u_n(x_j,t_1) - u(x_j,t_1)|$$
$$+ T \max_{\substack{1\le j\le J-1\\1\le k\le N}} |s_j^{k-1} - s_j^k - d_n^1(u)(x_j,t_k)| = O(\tau_n) \quad (n \in I). \tag{7}$$

Proof: The errors $e_j^k = u_n(x_j,t_k) - u(x_j,t_k)$ satisfy the following equations (cf. Lemma 13.3)

$$\frac{1}{2\tau_n}(e_j^{k+1} - e_j^{k-1}) - \frac{a}{h_n^2}(e_{j+1}^k - e_j^{k+1} - e_j^{k-1} + e_{j-1}^k)$$
$$= s_j^k - \{d_n^1(u)(x_j,t_{k+1}) + s_j^{k+1}\} = O(\tau_n), \quad 1 \le j \le J-1, \quad 1 \le k \le N-1.$$

If we now set $\underline{v}^{k+1} = (e_n(t_{k+1}), e_n(t_k))$ in 12.(14) and observe that $\underline{v}^1 = (e_n(t_1), e_n(t_0)) = (e_n(t_1), 0)$, then

$$\max_{0\le j\le J} |e_n(x_j,t_{k+1})| \le \max_{0\le j\le J} |e_n(x_j,t_1)|$$

$$+ 2T \max_{\substack{1\le j\le J-1 \\ 1\le \nu\le k}} |s_j^\nu - s_j^{\nu+1} - d_n^1(u)(x_j,t_{\nu+1})|, \quad k = 1,\ldots,N-1.$$

The last term on the right-hand side is of accuracy $O(\tau_n)$ by Lemma 13.3. For the error at time t_1, we have (cf. 4.(12))

$$\begin{aligned} e_n(x_j,t_1) &= ar(u_0(x_{j+1}) - 2u_0(x_j) + u_0(x_{j-1})) \\ &\qquad - (u(x_j,t_1) - u_0(x_j)) + \tau_n s_j^0 \\ &= \tau_n\{au_{xx}(x_j,t_0) - u_t(x_j,t_0)\} + \tau_n s_j^0 + O(\tau_n h_n^2 + \tau_n^2), \quad j = 1,\ldots,J-1, \end{aligned}$$

thereby yielding (7). □

As an example of a *nonlinear parabolic IVP* we consider

$$u(0) = u_0, \quad u_t = a(x,t)u_{xx} + F(x,t,u,u_x), \quad (x,t) \in [0,1] \times [0,T],$$

with homogeneous Dirichlet boundary conditions (cf. 4.(47) and 11.(9)). We recall that the differentiability and stability properties of both the explicit method given by 11.(11) and the totally implicit method given by 11.(12) have been studied in a supremum norm setting (cf. Sections 11.1 and 12.4). It should be also recalled that, due to the dependence of F on u_x, the underlying norms in $E_n = \{g_n \in C(G_n): g_n(0) = g_n(1) = 0\}$ are maximum norms multiplied by τ_n^{-1} (cf. Lemma 11.2). With regard to stability properties, the implicit method 11.(12) is preferable (cf. Theorem 12.16) where we now consider arbitrary inhomogeneous terms w_j^k,

$$\begin{aligned} &\frac{1}{\tau_n}(v_j^{k+1} - v_j^k) = a_j^{k+1} D_h^2 v_j^{k+1} + F(x_j, t_{k+1}, v_j^{k+1}, D_h v_j^{k+1}) + w_j^{k+1}, \quad 1 \le j \le J-1, \\ &v_j^0 = u_0(x_j), \quad j = 0,\ldots,J; \quad v_0^{k+1} = v_J^{k+1} = 0, \quad k = 0,\ldots,N-1. \end{aligned} \tag{8}$$

In the following theorem, we shall analyze the truncation errors for this method and obtain convergence by using the results from Chapter 11. Also, we shall again make use of the Landau symbol $o(\cdot)$ in our discussion in addition to the symbol $O(\cdot)$.

<u>Theorem 13.5</u>. Suppose that the solution u^0 of the IVP 11.(9) exists and is sufficiently smooth, and that the hypotheses of Theorem 12.16 together with 12.(36) are satisfied. Then the truncation errors of the method defined by (8) are estimated by

$$\max_{\substack{1\le j\le J-1\\ 1\le k\le N}} |d_n^1(u^0)(x_j,t_k)| = O(\tau_n + h_n^2) \quad (n \in I). \tag{9}$$

The errors in the solutions $u_n(x_j,t_k) = v_j^k$ of the approximating equations (8) are then estimated by

$$\max_{\substack{0\le j\le J\\ 0\le k\le N}} |u_n(x_j,t_k) - u^0(x_j,t_k)| \le C \max_{\substack{1\le j\le J-1\\ 1\le k\le N}} (|d_n^1(u^0)(x_j,t_k)|+|w_j^k|), \quad n \ge \nu_0, \tag{10}$$

if, additionally, $d_n^1(u^0)(x_j,t_k) = o(\tau_n)$ and $w_j^k = o(\tau_n)$ $(n \in I)$ uniformly for all j,k.

<u>Proof</u>: The truncation errors can be expressed as

$$\begin{aligned} d_n^1(u^0)(x_j,t_{k+1}) &= D_\tau^+U_j^k - a_j^{k+1}D_h^2U_j^{k+1} - F(x_j,t_{k+1},U_j^{k+1},D_hU_j^{k+1}) \\ &= D_\tau^+U_j^k - (u_t^0)_j^{k+1} - a_j^{k+1}[D_h^2U_j^{k+1} - (u_{xx}^0)_j^{k+1}] \\ &\quad - [F(x_j,t_{k+1},U_j^{k+1},D_hU_j^{k+1}) - F(x_j,t_{k+1},U_j^{k+1},(u_x^0)_j^{k+1})] \\ &= O(\tau_n + \alpha_1 h_n^2 + |F_z|h_n^2) = O(\tau_n+h_n^2) \quad (n \in I),\ j = 1,\dots,J-1, \end{aligned}$$

where $U_j^k = u^0(x_j,t_k)$ and $D_\tau^+U_j^k = (U_j^{k+1} - U_j^k)/\tau_n$. Applying Theorem 11.9 with $p = 1$, with $w^0 = 0$, and with each of

$$E_n = F_n = \{g \in C(G_n)\colon g(x_0) = g(x_J) = 0\},$$

$$G_n = \{x_j = jh_n,\ 0 \le j \le J\},\quad n \in I,$$

equipped with the maximum norms multiplied by τ_n^{-1} (cf. 11.(15)), we can show that the approximating equations (8) are uniquely solvable whenever

$$\sum_{k=1}^{N} \max_{1\le j\le J-1} |w_j^k| \le \tau_n\sigma/2.$$

This inequality is satisfied for all $n \ge \nu_0 \in \mathbb{N}$, provided that $w_j^k = o(\tau_n)$ $(n \in I)$. The general error estimate 11.(53) finally yields the error estimate of (10). □

The result of this theorem allows us to conclude that convergence of the approximations is assured whenever the truncation errors and the inhomogeneous terms approach zero at a rate faster than that in (9). This is the case for the test problem

$$u(0) = 0, \quad u_t = u_{xx} + (u_x)^2$$

where the solution is identically zero and the truncation errors therefore vanish at the solution. Theorem 13.5 states that the solutions of the associated inhomogeneous finite-difference equations (8) converge in case $w_j^k = o(\tau_n)$. Those methods, whose solutions converge only under stronger approximability requirements on the inhomogeneous right-hand sides, are denoted as *stably convergent* in Ansorge and Hass (1970), §6. (This notion is due to Dahlquist (1956).) Whenever F does not depend on u_x, we can use the maximum norm for the grid functions, and (10) then is valid without further restrictions on $d_n^1(u^0)$ and w_j^k; in this case the solutions converge as $O(\tau_n)$, whenever $w_j^k = O(\tau_n)$ $(n \in I)$.

For a general F and the particular method considered in the last theorem, the assured stable convergence is not fully satisfactorily. We conjecture, however, that a convergence result can be stated - not containing any restriction in the sense of stable convergence - when stronger norms are chosen. To verify such a statement, it should be noted that the whole analysis including differentiability properties, inverse stability, and consistency has to be carried out with respect to the same underlying norms. (Concerning other norms, we refer to the discussion following the proof of Lemma 11.2.) We like to emphasize once more that any convergence statement is only relative to the chosen norms; in the above Theorem 13.5, the maximum norms.

We now examine the consistency properties of Friedrichs' method and of the Lax-Wendroff method for approximating solutions to the *wave equation*. We recall that the wave equation can be expressed in the form 4.(31) as a first order system of equations,

$$u(0) = u_0, \quad u_t = Au_x, \quad -\infty < x < \infty, \quad t \in [0,T].$$

The truncation errors are then vectors of two components. We have the following result on the consistency of the methods 4.(33) and 4.(35), where the maximum norm of a vector in $\mathbb{R}^2$ is simply denoted by $|\cdot|$.

Theorem 13.6. Suppose that the solution $u = (u_1, u_2)$ of the wave equation 4.(31) is sufficiently smooth and that the mesh ratio $\lambda = \tau_n/h_n$ is constant in n. Then the truncated errors of Friedrichs' method are estimated by

$$\sup_{\substack{x\in G_n \\ t\in[0,T]_n'}} |d_n^1(u)(x,t)| = O(\tau_n) \quad (n \in I), \tag{11}$$

whereas those of the Lax-Wendroff method are estimated by

$$\sup_{\substack{x\in G_n \\ t\in[0,T]_n'}} |d_n^1(u)(x,t)| = O(\tau_n^2) \quad (n \in I). \tag{12}$$

Proof: (i) (Friedrichs' method). By definition,

$$\begin{aligned} d_n^1(u)(x,t) &= \frac{1}{\tau_n}\left(u(x,t) - \tfrac{1}{2}(u(x+h_n,t') + u(x-h_n,t'))\right) \\ &\quad - \frac{1}{2h_n} A(u(x+h_n,t') - u(x-h_n,t')) \\ &= \frac{1}{\tau_n}(u(x,t) - u(x,t')) + O(h_n^2/\tau_n) - Au_x(x,t') + O(h_n) \\ &= u_t(x,t') - Au_x(x,t') + O(\tau_n + h_n). \end{aligned}$$

For these relations, it suffices to require the continuity and boundedness of u_{tt} and u_{xx}.

(ii) (Lax-Wendroff method). If we use the representation 4.(35a), we obtain for the truncation errors,

$$\begin{aligned} d_n^1(u)(x,t) &= \frac{1}{\tau_n}(u(x,t) - u(x,t')) - \frac{1}{2h_n} A(u(x+h_n,t') - u(x-h_n,t')) \\ &\quad - \frac{\tau_n}{2h_n^2} A^2(u(x+h_n,t') - 2u(x,t') + u(x-h_n,t')) \\ &= u_t(x,t') + \frac{\tau_n}{2} u_{tt}(x,t') - Au_x(x,t') \\ &\quad - \frac{\tau_n}{2} A^2 u_{xx}(x,t') + O(\tau_n^2 + h_n^2 + \tau_n h_n). \end{aligned}$$

Here, we have assumed the existence of continuous and bounded derivatives u_{ttt} and u_{xxx}. Since $u_{tt} = A^2 u_{xx}$, the asserted estimate (12) for the truncation errors follows. □

We can easily verify that the Courant-Isaacson-Rees method possesses the same order of consistency as Friedrichs' method. These results on consistency for both Friedrichs' and the Courant-Isaacson-Rees method are also valid for hyperbolic systems of first order with coefficients depending on x. Our consistency analysis for the Lax-Wendroff method is restricted to the constant coefficient case; note that inverse stability with respect to the supremum norm has been shown in Chapter 12 only for this case. For a discussion on the applicability of this method and on its generalization to treat problems with variable coefficients (i.e., the Lax-Wendroff-Richtmyer method), the reader is referred to Meis & Marcowitz (1981).

We conclude this section by considering the *quasilinear hyperbolic IVP*

$$u_t - c(x,t,u)u_x = s(x,t,u), \quad x \in \mathbb{R}, \quad t \in [0,T]$$

(cf. 4.(58) and 11.(26)). Extending Friedrichs' method and the Courant-Isaacson-Rees method to this class of nonlinear problems had been done in Section 4.4, and the investigations of the differentiability properties and inverse stability with respect to the supremum norm were carried out in Sections 11.1 and 12.4, respectively. We do not consider here the Lax-Wendroff method because of the reasons listed above (concerning, in particular, inverse stability).

Our methods can be expressed as

$$v_j^{k+1} = \sum_{\mu=-1}^{1} b_\mu(x_j,t_k,v_j^k)v_{j+\mu}^k + \tau_n s(x_j,t_k,v_j^k),$$

$$j = 0,\pm1,\ldots, \quad k = 0,1,\ldots,N-1,$$

where the b_μ are given in terms of a function b by 11.(28), i.e.,

$$b_0 = 1 - \lambda b, \quad b_{\pm1} = \tfrac{1}{2}\lambda(\pm c+b).$$

The last relations yield directly the consistency of each method.

Theorem 13.7. Suppose that the function c is continuous and bounded and that a solution u^0 of the nonlinear IVP 11.(26) exists and possesses continuous, bounded partial derivatives up to second order. Suppose, moreover, that $\lambda = \tau_n/h_n$ is constant for all $n \in I$. Then the truncation errors of both Friedrichs' method and the Courant-Isaacson-Rees method can be estimated by

$$\sup_{\substack{x\in G_n \\ t\in[0,T]_n'}} |d_n^1(u^0)(x,t)| = O(\tau_n). \tag{13}$$

Proof: In the course of the proof, we shall not explicitly display the dependence of b, b_μ, and s on the arguments (x_j,t_k,U_j^k), $U_j^k = u^0(x_j,t_k)$. From 11.(28), we obtain the following representation for the truncation errors,

$$\begin{aligned}
\tau_n d_n^1(u^0)(x_j,t_{k+1}) &= U_j^{k+1} - \sum_{\mu=-1}^{1} b_\mu U_{j+\mu}^k - \tau_n s \\
&= U_j^{k+1} - U_j^k + (1-b_0)U_j^k - \sum_{0\neq|\mu|\leq 1} b_\mu U_{j+\mu}^k - \tau_n s \\
&= \tau_n D_\tau^+ U_j^k + \lambda b U_j^k - \frac{\lambda}{2}\{(-c+b)U_{j-1}^k + (c+b)U_{j+1}^k\} - \tau_n s \\
&= \tau_n D_\tau^+ U_j^k - \frac{\lambda}{2} c(U_{j+1}^k - U_{j-1}^k) \\
&\quad - \frac{\lambda}{2} b(U_{j+1}^k - 2U_j^k + U_{j-1}^k) - \tau_n s, \quad j = 0,\pm 1,\ldots, \ k = 0,\ldots,N.
\end{aligned}$$

The regularity assumptions on u^0 imply that, for some $\hat{x}_j \in (x_{j-1},x_{j+1})$,

$$d_n^1(u^0)(x_j,t_{k+1}) = -bh_n u_{xx}^0(\hat{x}_j,t_k) + O(\tau_n+h_n).$$

The function b is equal to $1/\lambda$ for Friedrichs' method and to $1-\lambda|c|$ for the Courant-Issacson-Rees method. Thus b is bounded; and (13) is shown, since $h_n = \tau_n/\lambda$. □

Inverse stability with respect to the supremum norm is present for both Friedrichs' method and the Courant-Issacson-Rees method in case the hypotheses of Theorem 12.18 are met. In addition to basic regularity assumptions on c and s, this means essentially that the Courant-Friedrichs-Lewy condition $\lambda \leq 1/|c|$ must be satisfied. In order to apply the conclusions of Theorem 12.18 to the Courant-Isaacson-Rees method, we need c to be of constant sign. Under the assumptions of this theorem, we then obtain the convergence of the approximations as $O(\tau_n)$ whenever the initial approximates are chosen as the restriction of the initial function to the grid - or as an $O(\tau_n)$ approximation thereof.

13.2. CONSISTENCY AND ERROR ANALYSIS OF DISCRETE-TIME GALERKIN METHODS

In this section, we study both the Crank-Nicolson-Galerkin method 4.(21) and the (discrete-time) Galerkin method 4.(39) for approximating solutions of heat and wave equations, respectively. In our discussion, we will make clear the meaning of consistency for these methods. It is useful to recall that these methods are inversely stable with respect to appropriately chosen norms (see Section 12.3). Now, if consistency with respect to such norms is also present, convergence of the approximations to the exact solution is assured whenever the initial approximations are suitably chosen.

For our convergence analysis, we shall require the existence of discrete approximations, which is assured via interpolation estimates on spaces of piecewise polynomial functions. Moreover, we shall need convergence results for approximations to variational equations (cf. Chapter 9), since suitable Ritz-Galerkin approximations are chosen as restrictions (in the sense of Section 5.2) in the finite-dimensional subspaces where the approximations are sought.

We conclude the introduction of this section by some remarks on nonlinear Galerkin methods. The corresponding convergence proofs proceed principally in the same manner as in the linear case. Indeed, we construct the appropriate error equations, insert the error itself into an inverse stability inequality we have at our disposal (cf., e.g., 12.(40) for method 11.(20)), and then proceed to estimate the resulting terms. Concerning the latter task, we could also make use of Taylor series expansions; however, this does not in general lead to optimal orders of convergence and requires stringent regularity properties. The analysis of this and the previous chapters thus provides the tools for treating nonlinear Galerkin methods; however, such a treatment lies beyond the scope of this book, and we must restrict ourselves to the rather lengthy analysis of two sample linear Galerkin methods.

In Sections 4.5 and 12.3, we have seen that the Crank-Nicolson-Galerkin (CNG) method can be expressed in the general form 11.(2), which means here that

$$\frac{1}{\tau_n}(C_n^{(0)}u_n(t_{k+1}) - C_n^{(1)}u_n(t_k),\phi_n)_1 = (S_n^{k+1/2},\phi_n)_0, \quad \phi_n \in E_n,$$

$$k = 0,1,\dots,N-1,$$

where $S_n^\nu = P_n^0 s^\nu$, $\nu = k,k+1$, denotes the L^2-projection in E_n of $s^\nu = s(\cdot,t_\nu)$, and τ_n, $n \in I$, is a null sequence of step sizes. The associated mappings $C_n^{(\ell)}: E_n \to F_n$, $n \in I$, $\ell = 0,1$, are defined by

$$(C_n^{(0)} g_n,\phi_n)_1 = (g_n,\phi_n)_0 + \frac{\tau_n}{2} a(g_n,\phi_n),$$

$$(C_n^{(1)} g_n,\phi_n)_1 = (g_n,\phi_n)_0 - \frac{\tau_n}{2} a(g_n,\phi_n), \quad \phi_n \in E_n, \quad n \in I.$$

We use the scalar product $(\psi,\phi)_1 = (\psi',\phi')_0$, which is definite on $H_0^1(0,1)$ (via the Poincaré inequality), and defines a norm $|\cdot|_1$ equivalent to the usual norm $||\cdot||_1 = (|\cdot|_0^2 + |\cdot|_1^2)^{1/2}$. Also $a(\psi,\phi) \equiv (a\psi',\phi')_0$, and $a(\cdot)$ is assumed to lie in $C[0,1]$ and to satisfy the estimates $0 < \alpha_0 \le a(x) \le \alpha_1$, $x \in [0,1]$. Then $a(.,.)$ represents a scalar product on $H_0^1(0,1)$, whose associated norm, given for arbitrary $\phi \in H_0^1(0,1)$ by $|\phi|_a \equiv a(\phi,\phi)^{1/2}$, is equivalent to $|\cdot|_1$ and $||\cdot||_1$. The E_n, $n \in I$, are taken to be finite-dimensional subspaces of $H_0^1(0,1)$ and are equipped with norm $|\cdot|_1$; the F_n, $n \in I$, are equal to E_n with norms given by

$$[g_n]_n = \sup_{0 \neq \phi_n \in E_n} \frac{|(g_n,\phi_n)_1|}{|\phi_n|_0}, \quad g_n \in F_n, \quad n \in I \tag{14}$$

(cf. the discussion in Section 12.3 after Theorem 12.14).

We can express the given IVP 4.(18a,b) as

$$u(0) = u_0, \quad u_t(t) - Au(t) = s(t), \quad t \in [0,T],$$

and we shall henceforth consider this differential-operational equation in $L^2(0,1)$. More precisely, A is the operator obtained by restricting the operator $\hat{A}: H_0^1(0,1) \to H^{-1}(0,1)$ (defined via the bilinear form $\langle\phi,\hat{A}g\rangle = a(g,\phi)$, $g,\phi \in H_0^1(0,1)$) to the domain of definition $D(A) \equiv \{g \in H_0^1(0,1): \hat{A}g \in L^2(0,1)\}$. The unique solvability of the given IVP with its solution in $X = C([0,T], H_0^1(0,1)) \cap C^1([0,T],L^2(0,1))$ is guaranteed by well-known results on semigroups of operators. Indeed, the above defined operator $\hat{A}$ is elliptic on $E = H_0^1(0,1)$ and thus its restriction $A: D(A) \to F(= L^2(0,1))$ generates a semigroup of contractions on F (cf. Aubin (1979), Chapter 13 and 14).

In order to define the truncation errors of the CNG method, we must specify the sequence of restrictions $R_n^E: E \to E_n$, $R_n^F: F \to F_n$, $n \in I$. Among the several possibilities, we choose

$$R_n^E g \in E_n : a(R_n^E g, \phi_n) = a(g, \phi_n), \quad \phi_n \in E_n, \tag{15a}$$

$$R_n^F f \in F_n : (R_n^F f, \phi_n)_1 = (f, \phi_n)_0, \quad \phi_n \in E_n, \quad n \in I. \tag{15b}$$

Using the norms given by (14) and the L^2-projections P_n^0, we get directly the relation $[R_n^F f]_n = |P_n^0 f|_0$, $n \in I$. Note that $R_n^E g$ is nothing other than the Ritz-Galerkin approximation of g in E_n.

We have the existence of discrete approximations $A(E, \pi E_n, \lim^E)$ and $A(F, \pi F_n, \lim^F)$ in case

$$|R_n^E g - g|_a \to 0 \quad (n \in I), \quad g \in E, \tag{16a}$$

$$|P_n^0 f - f|_0 \to 0 \quad (n \in I), \quad f \in F, \tag{16b}$$

respectively. That (R_n^E) has the properties of a restriction sequence is clear; that (R_n^F) has such properties is due to the fact that $[R_n^F f]_n \to 0$ implies $f = 0$ in the L^2-sense, because of (16b) and $[R_n^F f]_n = |P_n^0 f|_0$. Clearly, it suffices to have (16a) and (16b) on a dense subspace of E and of F, respectively, in order to have discrete approximations.

Suppose that the E_n are spaces of piecewise polynomial functions associated with a sequence of uniform meshes in $[0,1]$ (cf. Fairweather (1978), Section 2.3),

$$E_n = \overset{\bullet}{M}{}^s_\ell = \{v \in C^\ell[0,1]: \ v|I_j \in P_s(I_j), \ j = 1,\dots,J\} \cap H_0^1(0,1),$$
$$0 \le \ell \le s,$$

with $h_n \to 0$. Then the existence of discrete approximations follows from the interpolation estimates,

$$|g - \pi_n g|_m \le C h_n^{\mu+1-m} |g|_{\mu+1}, \quad g \in H^{\mu+1}(0,1), \tag{17}$$

where $0 \le m \le \ell+1$, $m-1 \le \mu \le s$, and $\pi_n g$ is the E_n-interpolant of g (cf., e.g., Ciarlet (1978), Section 3.2). In Section 9.2, we have already shown quasi-optimal convergence in $|\cdot|_1$ for the Ritz-Galerkin approximations $R_n^E g$ in E_n which implies, for $E_n = \overset{\bullet}{M}{}^s_0$, $0 \le \mu \le s$, that

$$|R_n^E g - g|_1 \le C h_n^\mu |g|_{\mu+1}, \quad g \in E \cap H^{\mu+1}(0,1). \tag{18a}$$

Using the Aubin-Nitsche trick, we get for the L^2-norms and for $1 \le \mu \le s$ that

$$|R_n^E g - g|_0 \leq Ch_n^{\mu+1}|g|_{\mu+1}, \quad g \in E \cap H^{\mu+1}(0,1). \tag{18b}$$

We are now in a position to estimate the truncation errors for the CNG method. We recall from the general definition in 11.(46) that

$$d_n^1(u)(t) = \frac{1}{\tau_n}[C_n^{(0)} R_n^E u(t) - C_n^{(1)} R_n^E u(t')] - R_n^F(u_t(t) - Au(t)), \quad u \in X.$$

With $s^k \equiv u_t(t_k) - Au(t_k)$, the above expression is equivalent to

$$(d_n^1(u)(t_k), \phi_n)_1 = (D_\tau^+ U_n^{k-1}, \phi_n)_0 + a(U_n^{k-1/2}, \phi_n) - (S_n^k, \phi_n)_0,$$
$$\phi_n \in E_n, \quad k = 1, \ldots, N, \tag{19}$$

where

$$U_n^k = R_n^E u(t_k) \quad \text{and} \quad S_n^k = P_n^0 s^k, \quad k = 0, \ldots, N, \tag{20}$$

since $(R_n^F s^k, \phi_n)_1 = (s^k, \phi_n)_0 = (S_n^k, \phi_n)_0$, $\phi_n \in E_n$. In order to better represent $d_n^1(u)$ for the purposes of obtaining error estimates, we note that

$$v^{k-1/2} + \frac{1}{2}\tau_n D_\tau^+ v^{k-1} = v^k, \quad k = 1, \ldots, N, \tag{21}$$

where v^{k-1}, v^k are any arbitrary functions in $L^2(0,1)$. The following lemma gives a more practical representation of $d_n^1(u)$.

<u>Lemma 13.8</u>. The truncation errors satisfy the following relations,

$$(d_n^1(u)(t_k), \phi_n)_1 = (\rho^{k-1}, \phi_n)_0 + (D_\tau^+ \eta^{k-1}, \phi_n)_0 - \frac{1}{2}\tau_n (D_\tau^+ S_n^{k-1}, \phi_n)_0, \tag{22}$$
$$\phi_n \in E_n, \quad k = 1, \ldots, N, \; n \in I,$$

where

$$\rho^k = D_\tau^+ u^k - \left(\frac{\partial u}{\partial t}\right)^{k+1/2}, \quad \eta^k = R_n^E u(t_k) - u^k, \quad u^k = u(t_k), \quad k = 0, \ldots, N.$$

<u>Proof</u>: Using (19)-(21), we get that

$$\begin{aligned}(d_n^1(u)(t_k), \phi_n)_1 &= (D_\tau^+ U_n^{k-1}, \phi_n)_0 + a(U_n^{k-1/2}, \phi_n) - (S_n^k, \phi_n)_0 \\ &= (D_\tau^+ U_n^{k-1}, \phi_n)_0 + a(U_n^{k-1/2}, \phi_n) - (S_n^{k-1/2}, \phi_n)_0 \\ &\quad - \frac{1}{2}\tau_n (D_\tau^+ S_n^{k-1}, \phi_n)_0, \quad \phi_n \in E_n, \quad k = 1, \ldots, N, \quad n \in I.\end{aligned}$$

By definition of $s^k = u_t(t_k) - Au(t_k)$,

$$(u_t^{k-1/2},\phi_n)_0 + a(u^{k-1/2},\phi_n) = (s^{k-1/2},\phi_n)_0 = (S_n^{k-1/2},\phi_n)_0,$$
$$\phi_n \in E_n, \quad n \in I,$$

so that, with $u^k = u(t_k)$, $u_t^k = u_t(\cdot,t_k)$ and because of $a(U_n^{k-1/2},\phi_n) = a(u^{k-1/2},\phi_n)$, we have

$$\begin{aligned}(d_n^1(u)(t_k),\phi_n)_1 &= (D_\tau^+ U_n^{k-1},\phi_n)_0 + a(U_n^{k-1/2},\phi_n)\\ &\quad - (u_t^{k-1/2},\phi_n)_0 - a(u^{k-1/2},\phi_n) - \frac{1}{2}\tau_n(D_\tau^+ S_n^{k-1},\phi_n)_0\\ &= (D_\tau^+ U_n^{k-1},\phi_n)_0 - (u_t^{k-1/2},\phi_n)_0 - \frac{1}{2}\tau_n(D^+ S_n^{k-1},\phi_n)_0\\ &= (D_\tau^+\eta^{k-1},\phi_n)_0 + (\rho^{k-1},\phi_n)_0 - \frac{1}{2}\tau_n(D_\tau^+ S_n^{k-1},\phi_n)_0,\end{aligned}$$
$$\phi_n \in E_n, \quad k = 1,\dots,N. \qquad \square$$

In order to prove consistency, we must estimate the three terms on the right-hand side of (22). We first give an integral representation for ρ^k (cf. Fairweather (1978), Section 4.5) which is more appropriate for error estimates. The proof relies on the Peano Kernel Theorem (cf. Davis (1963)). Suppose that $u_{ttt} \in L^2([0,T],L^2(0,1))$, i.e.,

$$||u_{ttt}||_{L^2(L^2)} \equiv \left(\int_0^T\int_0^1 |u_{ttt}(x,t)|^2 dxdt\right)^{1/2} < \infty,$$

where the above integrals are Lebesgue integrals. Then

$$\rho^k(x) = \frac{1}{2\tau_n}\int_k^{t_{k+1}} (t-t_k)(t-t_{k+1})u_{ttt}(x,t)dt. \tag{23}$$

We can now estimate the norm of ρ^k.

<u>Lemma 13.9</u>. Suppose $u_{ttt} \in L^2([0,T],L^2(0,1))$. Then

$$\left(\tau_n \sum_{k=0}^{N-1} |\rho^k|_0^2\right)^{1/2} \le \frac{\tau_n^2}{\sqrt{20}}||u_{ttt}||_{L^2(L^2)}. \tag{24}$$

<u>Proof</u>: Using (23) and Schwarz's inequality, we get

$$\begin{aligned}|\rho^k(x)|^2 &\le \frac{1}{4\tau_n^2}\left(\int_{t_k}^{t_{k+1}}|(t-t_k)(t-t_{k+1})u_{ttt}(x,t)(x,t)|dt\right)^2\\ &\le \frac{1}{4\tau_n^2}\int_{t_k}^{t_{k+1}}(t-t_k)^2(t-t_{k+1})^2 dt \int_{t_k}^{t_{k+1}}|u_{ttt}(x,t)|^2 dt\end{aligned}$$

$$\leq \frac{1}{4\tau_n^2} \underbrace{\left(\int_{t_k}^{t_{k+1}} (t-t_k)^4 dt\right)^{1/2}}_{\frac{1}{5}\tau_n^5} \underbrace{\left(\int_{t_k}^{t_{k+1}} (t-t_{k+1})^4 dt\right)^{1/2}}_{\frac{1}{5}\tau_n^5} \int_{t_k}^{t_{k+1}} |u_{ttt}(x,t)|^2 dt.$$

Integration yields

$$\int_0^1 |\rho^k(x)|^2 dx \leq \frac{1}{20} \tau_n^3 \int_0^1 \int_{t_k}^{t_{k+1}} |u_{ttt}(x,t)|^2 dt dx,$$

and summation over k gives

$$\tau_n \sum_{k=0}^{N-1} |\rho^k|_0^2 \leq \frac{1}{20} \tau_n^4 \int_0^T \int_0^1 |u_{ttt}(x,t)|^2 dx dt. \quad \square$$

We let each $L^2([0,T],H^i(0,1))$, $i = 0,1,\ldots$, be the space of functions $v(x,t)$ for which

$$||v||_{L^2(H^i)} \equiv \left(\int_0^T \int_0^1 |\frac{\partial^i v}{\partial x^i}(x,t)|^2 dx dt\right)^{1/2} < \infty.$$

For brevity, we label each respective space as $L^2(H^i)$. For estimating the second term on the right-hand side of (22), we need the following result.

Lemma 13.10. If $v_t \in L^2(H^i)$ for some $i \geq 0$, then for $v^k = v(\cdot,t_k)$, $k = 0,\ldots,N$,

$$\tau_n \sum_{k=0}^{N-1} |D_\tau^+ v^k|_i^2 \leq ||v_t||_{L^2(H^i)}^2, \tag{25}$$

where $|g|_i \equiv |g^{(i)}|_0$ is the seminorm for functions in $H^i(0,1)$.

Proof: We write $v^{(i)}(t) = \partial^i v/\partial x^i(\cdot,t)$ and apply the identity

$$[v^{(i)}(t_{k+1}) - v^{(i)}(t_k)](x) = \int_{t_k}^{t_{k+1}} \frac{\partial v^{(i)}}{\partial t}(x,t) dt.$$

Schwarz's inequality yields

$$|D_\tau^+ v^k|_i^2 = \frac{1}{\tau_n^2} \int_0^1 |[v^{(i)}(t_{k+1}) - v^{(i)}(t_k)](x)|^2 dx$$

$$= \frac{1}{\tau_n^2} \int_0^1 \left|\int_{t_k}^{t_{k+1}} \frac{\partial v^{(i)}}{\partial t}(x,t) dt\right|^2 dx$$

$$\leq \frac{1}{\tau_n^2}\int_0^1 \left[\tau_n^{1/2}\left\{\int_{t_k}^{t_{k+1}} |\frac{\partial v^{(i)}}{\partial t}(x,t)|^2 dt\right\}^{1/2}\right]^2 dx$$

$$= \frac{1}{\tau_n}\int_{t_k}^{t_{k+1}} |\frac{\partial v}{\partial t}(.,t)|_i^2 dt, \quad k = 0,\ldots,N-1.$$

The desired estimate follows by summation. □

From (20), we conclude that

$$D_\tau^+ U_n^k = R_n^E D_\tau^+ u^k \quad \text{and} \quad D_\tau^+ S_n^k = P_n^0 D_\tau^+ s^k, \quad k = 1,\ldots,N.$$

The projections R_n^E defined by (15a) and the L^2-projections P_n^0 have their respective (operator) norms bounded by unity for every n, i.e.

$$|R_n^E g|_a \leq |g|_a \quad \text{and} \quad |P_n^0 f|_0 \leq |f|_0, \quad n \in I.$$

Using (25), we can now estimate the third term in the right-hand side of (22) by

$$\tau_n \sum_{k=0}^{N-1} |D_\tau^+ S_n^k|_0^2 \leq \tau_n \sum_{k=0}^{N-1} |D_\tau^+ s^k|_0^2 \leq ||s_t||^2_{L^2(L^2)}, \tag{26}$$

if $s_t \in L^2(L^2)$.

The estimate for the remaining second term, $D_\tau^+\eta^{k-1}$, in (22) depends on how well $u^k = u(.,t_k)$ can be approximated by the projections U_n^k. This, in turn, depends on the approximability properties of the subspaces E_n. For special subspaces - say, of continuous piecewise polynomial functions, $E_n = \overset{\bullet}{M}{}_0^s$ - we have the estimates (18a,b) which are a consequence of the interpolation estimates of (17). Application of (18b) to $g = D_\tau^+ u^k$ yields

$$|D_\tau^+\eta^k|_0 = |D_\tau^+ U_n^k - D_\tau^+ u^k|_0 \leq Ch_n^{\mu+1}|D_\tau^+ u^k|_{\mu+1}, \quad k = 0,\ldots,N-1,$$

and thus,

$$\tau_n \sum_{k=0}^{N-1} |D_\tau^+\eta^k|_0^2 \leq C^2 h_n^{2(\mu+1)}\tau_n \sum_{k=0}^{N-1} |D_\tau^+ u^k|_{\mu+1}^2 \leq C^2 h_n^{2(\mu+1)} ||u_t||^2_{L^2(H^{\mu+1})} \tag{27}$$

follows from Lemma 13.10, if $u_t \in L^2(H^{\mu+1})$, $1 \leq \mu \leq s$.

With the inequalities (24), (26) and (27), we have completely estimated the truncation errors. If we now insert $u_n(t_k) - U_n^k$ into the inverse stability inequality 12.(26), we obtain the convergence of the

Crank-Nicolson-Galerkin method with accompanying error estimates. The following theorem summarizes our results on the consistency and convergence of the CNG method in case of piecewise polynomial subspaces.

Theorem 13.11. Suppose the solution $u \in X$ of the heat equation 4.(18a,b) has partial derivatives $u_t \in L^2(H^{\mu+1})$, and $u_{ttt} \in L^2(L^2)$, and, moreover, suppose $s_t \in L^2(L^2)$. Then the truncation errors of the Crank-Nicolson-Galerkin method can be estimated by

$$\left(\tau_n \sum_{k=1}^{N} [d_n^1(u)(t_n)]_n^2\right)^{1/2} \le C\{\tau_n^2 ||u_{ttt}||_{L^2(L^2)} + h_n^{\mu+1}||u_t||_{L^2(H^{\mu+1})} + \tau_n||s_t||_{L^2(L^2)}\}, \quad n \in I, \tag{28}$$

where $E_n = \dot{M}_0^s$, $1 \le \mu \le s$. The errors in the approximations can be estimated by

$$\max_{0\le k\le N} |u_n(t_k) - U_n^k|_1 \le C\{|u_n(0) - U_n^0|_1 + \tau_n^2||u_{ttt}||_{L^2(L^2)} + h_n^{\mu+1}||u_t||_{L^2(H^{\mu+1})}\}, \quad n \in I. \tag{29}$$

If $\tau_n \to 0$, $h_n \to 0$, and the approximation at $t = 0$ is given by $u_n(0) = U_n^0$, $n \in I$, then

$$\max_{0\le k\le N} |u_n(t_k) - u(t_k)|_1 = O(\tau_n^2 + h_n^{\mu}) \quad (n \in I), \tag{30}$$

whenever $u(t) \in H^{\mu+1}(0,1)$, $t \in [0,T]$.

Proof: Schwarz's inequality applied to (22) yields

$$|(d_n^1(u)(t_k),\phi_n)_1| \le (|\rho^{k-1}|_0 + |D_\tau^+\eta^{k-1}|_0 + \frac{1}{2}\tau_n|D_\tau^+s_n^{k-1}|_0)|\phi_n|_0, \quad \phi_n \in E_n, \quad n \in I.$$

Now,

$$[d_n^1(u)(t_k)]_n^2 \le \frac{5}{2}(|\rho^{k-1}|_0^2 + |D_\tau^+\eta^{k-1}|_0^2 + \frac{\tau_n}{2}|D_\tau^+s_n^{k-1}|_0^2), \quad k = 1,\ldots,N,$$

where $[\cdot]_n$ is the norm defined in (14). We have twice used the following inequality

$$(\alpha + \beta)^2 \le (1 + 2\varepsilon)\alpha^2 + (1 + \frac{1}{2\varepsilon})\beta^2, \quad \alpha,\beta \ge 0, \quad \varepsilon > 0,$$

with suitably chosen quantities. Summing over k, and using (24), (26) and (27), we get the desired estimate (28).

The errors $e_n^k = u_n(t_k) - U_n^k$, $k = 0,\ldots,N$, satisfy (cf. (19) and (22))

$$\begin{aligned}(D_\tau^+ e_n^k, \phi_n)_0 &+ a(e_n^{k+1/2}, \phi_n) \\ &= (S_n^{k+1/2}, \phi_n)_0 - (d_n^1(u)(t_{k+1}), \phi_n)_1 - (S_n^{k+1}, \phi_n)_0 \\ &= -(\rho^k + D_\tau^+ \eta^k, \phi_n)_0, \quad \phi_n \in E_n, \quad k = 0,\ldots,N-1, \quad n \in I,\end{aligned}$$

since

$$S_n^{k+1/2} - S_n^{k+1} = -\frac{1}{2}\tau_n D_\tau^+ S_n^k$$

by (21). Inserting e_n^k into the stability inequality 12.(26), and applying (24) and (27), we get (29). For the case $u_n(0) = U_n^0$, we obtain the convergence relation

$$\max_{0\leq k\leq N} |u_n(t_k) - U_n^k|_1 = O(\tau_n^2 + h_n^{\mu+1}) \quad (n \in I).$$

If, moreover, $u \in C([0,T], H^{\mu+1}(0,1))$, the interpolation estimate (17) along with (18a) yields

$$\max_{0\leq k\leq N} |U_n^k - u(t_k)|_1 = O(h_n^\mu) \quad (n \in I),$$

and thus (30) is proved. □

At this point, we would like to make a few remarks about the preceding theorem. If we choose for the approximation at $t = 0$ the L^2-projection of u_0 (as suggested in 4.(21)) instead of the Ritz-Galerkin approximation U_n^0, then a term $O(h_n^\mu)$ appears in (29). To see this, we must use the fact that an inverse inequality (cf. Ciarlet (1978), Theorem 3.2.6) is available for spaces E_n of polynomial functions satisfying an inverse assumption. This inequality, together with (17) and (18b), yields

$$|P_n^0 u_0 - U_n^0|_1 \leq Ch_n^{-1}|P_n^0 u_0 - U_n^0|_0 \leq Ch_n^{-1}(|P_n^0 u_0 - u_0|_0 + |u_0 - U_n^0|_0) = O(h_n^\mu),$$

since $u_0 \in H^{\mu+1}$ by hypothesis. The estimate given by (30) remains the same in this case.

A comparison of (29) with (28) (when $u_n(0) = U_n^0$) shows that the approximations are accurate to within $O(\tau_n^2 + h_n^{\mu+1})$, whereas the truncation errors converge to zero as $O(\tau_n + h_n^{\mu+1})$. This is due to the additional term $D_\tau^+ S_n^{k-1}$ occurring in (22) which would not have been present if R_n^F were defined differently. In the error equations, this term does not appear.

We also note that if u is sufficiently regular and if E_n consists of piecewise polynomials of correspondingly high degree, the powers $\mu+1$ and μ of h in (29) and (30), respectively, can be increased. However, the power 2 of τ_n cannot be improved. If the mesh ratio $r = \tau_n/h_n^2$ is constant for all n, then we must select our finite element functions to be at least piecewise cubic polynomials in order to obtain altogether an error of $O(\tau_n^2)$ in (30).

We now examine the consistency and convergence properties of the discrete-time *Galerkin method* 4.(39) for approximating the generalized wave equation 4.(37), The inverse stability of this method has already been shown in Theorem 12.15. We express this method as a two-level system 4.(39c), i.e., we seek vectors $u_n(t_k) = (u_{n,1}^k, u_{n,2}^k) \in E_n \times E_n$ as solutions of

$$(D_\tau^+ u_{n,2}^{k-1}, \phi_n)_0 + a(u_{n,1}^{k-1/2}, \phi_n) = (s^{k,1/4}, \phi_n)_0, \quad \phi_n \in E_n,$$

$$D_\tau^+ u_{n,1}^{k-1} - u_{n,2}^{k-1/2} = 0, \quad k = 1,\ldots,N-1,$$

where again E_n, $n \in I$, are finite-dimensional subspaces of $H_0^1(0,1)$, $a(\psi,\phi) = (a\psi',\phi')_0$, $s^k = s(.,t_k)$, and $v^{k,1/4} = (v^{k+1} + v^{k-1})/4 + v^k/2$. We assume that $a \in C[0,1]$ satisfies $0 < \alpha_0 \le a(x) \le \alpha_1$, $x \in [0,1]$. We first show that the functions $u_{n,1}^k$ and $u_{n,2}^k$ are approximations of $u^{k+1/2}$ and $D_\tau^+ u^k$, respectively, where $u^k = u(.,t_k)$, and u is the solution of 4.(37). Toward this end, we note that the method can be also expressed in the general form

$$C_n^{(0)} u_n(t_k) = C_n^{(1)} u_n(t_{k-1}) + \tau_n w_n(t_k), \quad k = 1,\ldots,N-1,$$

where each $C_n^{(\ell)}$ maps $E_n \times E_n$ into itself and has components determined by

$$((C_n^{(0)} g_n)_1, \phi_n)_1 = (g_{n,2}, \phi_n)_0 + \frac{1}{2}\tau_n a(g_{n,1}, \phi_n), \quad \phi_n \in E_n,$$

$$(C_n^{(0)} g_n)_2 = g_{n,1} - \frac{1}{2}\tau_n g_{n,2},$$

$$((C_n^{(1)} g_n)_1, \phi_n)_1 = (g_{n,2}, \phi_n)_0 - \frac{1}{2}\tau_n a(g_{n,1}, \phi_n), \quad \phi_n \in E_n,$$

$$(C_n^{(1)} g_n)_2 = g_{n,1} + \frac{1}{2}\tau_n g_{n,2}, \quad g_n = (g_{n,1}, g_{n,2}) \in E_n \times E_n.$$

The inhomogeneous terms are given by

$$(w_n(t_k)_1,\phi_n)_1 = (s^{k,1/4},\phi_n)_0, \quad \phi_n \in E_n, \quad w_n(t_k)_2 = 0.$$

We let $(\psi,\phi)_1 = (\psi',\phi')_0$ be the underlying scalar product in $H_0^1(0,1)$ and equip $E_n \times E_n$ and $F_n = E_n \times \{0\}$ with the respective norms

$$|g_n|_{E_n} \equiv (|g_{n,1}|_1^2 + |g_{n,2}|_0^2)^{1/2}, \quad g_n \in E_n \times E_n, \tag{31a}$$

$$|f_n|_{F_n} \equiv [f_{n,1}]_n \quad \text{with} \quad [f_{n,1}]_n \equiv \sup_{0 \neq \phi_n \in E_n} \frac{|(f_{n,1},\phi_n)_1|}{|\phi_n|_0}, \qquad f_n = (f_{n,1},0) \in F_n, \quad n \in I. \tag{31b}$$

As before, $P_n^0\colon L^2(0,1) \to E_n$ denote the L^2-projections onto E_n, $n \in I$, and $P_n^1\colon E \to E_n$, $E = H_0^1(0,1)$, $n \in I$, are given by

$$G_n = P_n^1 g \in E_n\colon (G_n,\phi_n)_0 + a(G_n,\phi_n) = (g,\phi_n)_0 + a(g,\phi_n), \qquad g \in E, \quad \phi_n \in E_n, \quad n \in I. \tag{32}$$

More consisely, $P_n^1 g$ is the Ritz-Galerkin approximation of the solution g to

$$-(ag')' + g = w \quad \text{in} \quad [0,1], \quad g(0) = g(1) = 0.$$

For the spaces $E_n = \overset{\bullet}{M}{}_\ell^s$ of polynomial functions, we have the interpolation estimates at our disposal for estimating the errors in the projections P_n^0 and P_n^1, and, moreover, we have the estimates (18a,b) for P_n^1 used in lieu of R_n^E. Concerning the latter result, we note that the bilinear form in (32) is elliptic on $H_0^1(0,1)$ (with respect to either $|\cdot|_1$ or $||\cdot||_1$), and hence the convergence results of Section 9.2 are applicable.

In order to estimate the truncation errors, we define the sequence of restriction operators

$$R_n^X\colon C([0,T],E) \to C([0,T]_n, E_n \times E_n) \quad (E = H_0^1(0,1)),$$

and

$$R_n^Y\colon E \times C([0,T],F) \to E_n \times E_n \times C([0,T]_n',F_n) \quad (F = L^2(0,1)),$$

in terms of restrictions $R_n^E\colon E \times E \to E_n \times E_n$ and $R_n^F\colon F \times \{0\} \to F_n \times \{0\}$. The latter are defined by

$$R_n^E g \equiv \left(\frac{1}{2} P_n^1(g_1+g_2), \frac{1}{\tau_n} P_n^1(g_1-g_2)\right), \quad g = (g_1,g_2) \in E \times E, \tag{33a}$$

and

$$R_n^F f \equiv (f_{n,1},0) \quad \text{where} \quad f_{n,1} \in F_n\colon (f_{n,1},\phi_n)_1 = (f_1,\phi_n)_0, \quad \phi_n \in E_n, \quad f = (f_1,0), \quad f_1 \in F. \tag{33b}$$

To every $u \in C([0,T],E)$, we associate a two component vector function $\underline{u}(t) = (u(t-\tau_n),u(t))$, $t \in [0,T-\tau_n]$. Then,

$$(R_n^X u)(t_k) = R_n^E \underline{u}(t_k) = (U_n^{k+1/2}, W_n^k), \quad k = 0,\ldots,N-1,$$

where $U_n^k = P_n^1 u(t_k)$, $W_n^k = D_\tau^+ U_n^k$. We observe that we must omit the last mesh point $t_N = T$ in the definition of the meshes $[0,T]_n$ and $[0,T]_n'$ because of the two component functions.

The truncation errors now possess two components of which the second vanishes, since

$$d_n^1(u)(t_k)_2 = D_\tau^+ U_n^{k-1/2} - W_n^{k-1/2} = 0.$$

The remaining component is expressed simply as $d_n^1(u)(t_k) \equiv d_n^1(u)(t_k)_1$, and

$$\begin{aligned}(d_n^1(u)(t_k),\phi_n)_1 &= \tau_n^{-1}(W_n^k - W_n^{k-1},\phi_n)_0 \\ &\quad + a\left(\tfrac{1}{2}(U_n^{k+1/2} + U_n^{k-1/2}),\phi_n\right) - (s^k,\phi_n)_0 \\ &= (D_\tau^2 U_n^k,\phi_n)_0 + a(U_n^{k,1/4},\phi_n) - (S_n^k,\phi_n)_0, \quad \phi_n \in E_n.\end{aligned}$$

Here $s(t)(x) = u_{tt}(x,t) - (au_x)_x(x,t)$, $s^k = s(t_k)$, $S_n^k = P_n^0 s^k$, and $U_n^{k,1/4}$ is defined as above. It is easy to see that with

$$\eta^k \equiv U_n^k - u^k, \quad \rho^k \equiv D_\tau^2 u^k - u_{tt}^{k,1/4}, \quad k = 1,\ldots,N-1, \tag{34}$$

the truncation errors satisfy

$$(d_n^1(u)(t_k),\phi_n)_1 = (D_\tau^2 \eta^k - \eta^{k,1/4} + \rho^k,\phi_n)_0 + \frac{1}{4}\tau_n^2(D_\tau^2 s^k,\phi_n)_0, \quad \phi_n \in E_n, \quad k = 1,\ldots,N-1. \tag{35}$$

For ρ^k, we have

$$\rho^k = \frac{1}{12}\int_{-\tau_n}^{\tau_n} (\tau_n - |t|)(3-2[1-|t|/\tau]^2)\frac{\partial^4 u}{\partial t^4}(\cdot,t_k+t)\,dt, \tag{36}$$

if $\partial^4u/\partial t^4 \in L^2(L^2)$ (cf. Dupont (1973), Sec. 3). We can now estimate the individual terms on the right-hand side of (35).

Lemma 13.12. Assume that $\partial^4u/\partial t^4 \in L^2(L^2)$. Then there exists a $C > 0$ such that

$$\tau_n \sum_{k=1}^{N-1} |\rho^k|_0^2 \leq C\tau_n^4 \left\|\frac{\partial^4 u}{\partial t^4}\right\|_{L^2(L^2)}^2 . \tag{37}$$

Proof: Two applications of Schwarz's inequality yield

$$|\rho^k(x)|^2 \leq \frac{1}{144}\left(\int_{-\tau_n}^{\tau_n} (\tau_n-|t|)^4 dt\right)^{1/2}\left(\int_{-\tau_n}^{\tau_n} (3-2[1-\frac{|t|}{\tau_n}]^2)^4 dt\right)^{1/4} \int_{-\tau_n}^{\tau_n} \left|\frac{\partial^4 u}{\partial t^4}(x,t_k+t)\right|^2 dt.$$

Here,

$$\int_{-\tau_n}^{\tau_n} (\tau_n-|t|)^4 dt = 2\int_0^{\tau_n} s^4 ds = \frac{2}{5}\tau_n^5,$$

$$\int_{-\tau_n}^{\tau_n} (3-2[1-\frac{|t|}{\tau_n}]^2)^4 dt = 2\tau_n \int_0^1 (3-2s^2)^4 ds \equiv 2C_0\tau_n,$$

and hence

$$|\rho^k(x)|^2 \leq \frac{1}{\sqrt{5}\cdot 72}\sqrt{C_0}\,\tau_n^3 \int_{-\tau_n}^{\tau_n} \left|\frac{\partial^4 u}{\partial t^4}(x,t_k+t)\right|^2 dt, \quad k = 1,\ldots,N-1.$$

Integration and summation over k gives (with $C_1 \equiv \sqrt{C_0}/(\sqrt{5}\cdot 72)$)

$$\tau_n \sum_{k=1}^{N-1} |\rho^k|_0^2 \leq C_1\tau_n^4 \sum_{k=1}^{N-1} \int_{t_{k-1}}^{t_{k+1}} \int_0^1 \left|\frac{\partial^4 u}{\partial t^4}(x,t)\right|^2 dxdt$$

$$\leq 2C_1\tau_n^4 \left\|\frac{\partial^4 u}{\partial t^4}\right\|_{L^2(L^2)} . \quad \square$$

For estimating $D_\tau^2\eta^k$, we make use of the following representation,

$$D_\tau^2 v^k = \tau_n^{-2}\int_{-\tau_n}^{\tau_n} (\tau_n-|t|)\frac{\partial^2 v}{\partial t^2}(\cdot,t_k+t)dt, \quad k = 1,\ldots,N-1, \tag{38}$$

which is valid for every v such that $v_{tt} \in L^2(L^2)$. Our estimate is now given by the following lemma.

Lemma 13.13. Let $u_{tt} \in L^2(H^{\mu+1})$ and $E_n = \overset{\bullet}{M}{}_0^s$ for some μ in $1 \leq \mu \leq s$. Then there exists a $C > 0$, such that

$$\begin{aligned}\tau_n \sum_{k-1}^{N-1} |D_\tau^2 \eta^k|_0^2 &\leq C\tau_n h_n^{2(\mu+1)} \sum_{k=1}^{N-1} |D_\tau^2 u^k|_{\mu+1}^2 \\ &\leq Ch_n^{2(\mu+1)} ||u_{tt}||_{L^2(H^{\mu+1})}^2 .\end{aligned} \tag{39}$$

Proof: Now $\eta_1^k = P_n^1 u^k - u^k$, and application of (18b) (with R_n^E and g replaced by P_n^1 and $D_\tau^2 \eta^k$, respectively) yields

$$|D_\tau^2 \eta^k|_0 \leq Ch_n^{\mu+1} |D_\tau^2 u^k|_{\mu+1}, \quad 1 \leq \mu \leq s.$$

Schwarz's inequality applied to (38) (with $v = u^{(\mu+1)}$) now gives

$$|D_\tau^2 u^{(\mu+1)}(x,t_k)|^2 \leq \tau_n^{-4} \int_{-\tau_n}^{\tau_n} (\tau_n - |t|)^2 dt \int_{-\tau_n}^{\tau_n} |u_{tt}^{(\mu+1)}(x,t_k+t)|^2 dt$$

$$= \frac{2}{3} \tau_n^{-1} \int_{-\tau_n}^{\tau_n} |u_{tt}^{(\mu+1)}(x,t_k+t)|^2 dt,$$

$$|D_\tau^2 u^k|_{\mu+1}^2 \leq \frac{2}{3} \tau_n^{-1} \int_{t_{k-1}}^{t_{k+1}} |u_{tt}(t)|_{\mu+1}^2 dt, \quad k = 1,\ldots,N-1,$$

and

$$\tau_n \sum_{k=1}^{N-1} |D_\tau^2 u^k|_{\mu+1}^2 \leq \frac{4}{3} ||u_{tt}||_{L^2(H^{\mu+1})}^2 .$$

The asserted estimate (39) then follows from the first estimate in the proof. □

In order to estimate $\eta^{k,1/4}$ in (35), we again apply (18b) for the case when $E_n = \overset{\bullet}{M}{}_0^s$ and obtain

$$\tau_n \sum_{k=1}^{N-1} |\eta^{k,1/4}|_0^2 \leq T \max_{0\leq k\leq N} |\eta^k|_0^2 \leq Ch_n^{\mu+1} \max_{0\leq k\leq N} |u^k|_{\mu+1}^2, \tag{40}$$

if $u(t) \in H^{\mu+1}(0,1)$ for all $t \in [0,T]$. For the remaining term in (35), we get

$$\tau_n \sum_{k=1}^{N-1} |D_\tau^2 s^k|_0^2 \leq \frac{4}{3} ||s_{tt}||_{L^2(L^2)}^2 \tag{41}$$

whenever $s_{tt} \in L^2(L^2)$. We can use the representation of $D_\tau^2 s^k$ in (38) to prove (41) by a procedure analogous to that used in the previous lemma. We omit the details.

All the terms of the truncation errors have now been estimated, and, inserting the error into the inverse stability inequality 12.(30), we get convergence along with accompanying error estimates. In the following theorem, we summarize these results and specify a suitable approximation of the initial data.

Theorem 13.14. Suppose that

$$\frac{\partial^2 s}{\partial t^2} \in L^2(L^2), \quad \frac{\partial^2 u}{\partial t^2} \in L^2(H^{\mu+1}), \quad \frac{\partial^4 u}{\partial t^4} \in L^2(L^2), \quad \text{and}$$

$$u(t), u_t(t) \in H^{\mu+1}(0,1), \quad t \in [0,T].$$

Then, for $E_n = \overset{\bullet}{M}{}^s_0$, the truncation errors of the Galerkin method 4.(39) are estimated by

$$\left(\tau_n \sum_{k=1}^{N-1} [d_n^1(u)(t_k)]_n^2\right)^{1/2} \le C\left\{\tau_n^2\left(\left\|\frac{\partial^4 u}{\partial t^4}\right\|_{L^2(L^2)} + \left\|\frac{\partial^2 s}{\partial t^2}\right\|_{L^2(L^2)}\right)\right. \tag{42}$$

$$\left. + h_n^{\mu+1}\left(\|u_{tt}\|_{L^2(H^{\mu+1})} + \|u\|_{L^\infty(H^{\mu+1})}\right)\right\},$$

$$n \in I.$$

If we choose

$$u^0_{n,1} = U_n^{1/2} \; (= \tfrac{1}{2} P_n^1(u^0+u^1)), \quad u^0_{n,2} = W_n^0 \; (= D_\tau^+ P_n^1 u^0), \quad n \in I,$$

as our approximations to the initial data, then the errors can be estimated by

$$\max_{0\le k\le N-1} (|u^k_{n,1} - U_n^{k+1/2}|_1^2 + |u^k_{n,2} - D_\tau^+ U_n^k|_0^2)^{1/2} = O(\tau_n^2 + h_n^{\mu+1}) \; (n \in I). \tag{43}$$

If $E_n = \overset{\bullet}{M}{}^s_0$ satisfies the usual inverse inequality and $\lambda = \tau_n/h_n$ is constant in n, then the errors in the approximations $u_n^{k+1} \equiv u^k_{n,1} + \frac{1}{2}\tau_n u^k_{n,2}$ to U_n^{k+1} itself are estimated by

$$\max_{0\le k\le N-1} |u_n^{k+1} - U_n^{k+1}|_1 = O(\tau_n^2 + h_n^{\mu+1}) \quad (n \in I). \tag{44}$$

Proof: Schwarz's inequality applied to (35) gives

$$|(d_n^1(u)(t_k), \phi_n)_1| \le (|D_\tau^2 \eta^k|_0 + |\eta^{k,1/4}|_0 + |\rho^k|_0$$

$$+ \tfrac{1}{4}\tau_n^2 |D_\tau^2 s^k|_0)|\phi_n|_0, \quad \phi_n \in E_n, \quad n \in I,$$

and, using the definition of the norm $[\cdot]_n$ (cf. 31b)), we get

$$\tau_n \sum_{k=1}^{N-1} [d_n^1(u)(t_k)]_n^2 \le 4\tau_n \sum_{k=1}^{N-1} (|D_\tau^2\eta^k|_0^2 + |\eta^{k,1/4}|_0^2$$
$$+ |\rho^k|_0^2 + \frac{1}{16}\tau_n^2|D_\tau^2 s^k|_0^2).$$

The estimate (42) follows from the estimates in (37), (39), (40), and (41).

(ii) The errors $e_{n,1}^k = u_{n,1}^k - U_n^{k+1/2}$, $e_{n,2}^k = u_{n,2}^k - W_n^k$ satisfy the equations

$$(D_\tau^+ e_{n,2}^k, \phi_n)_0 + a(e_{n,1}^{k-1/2}, \phi_n)_0$$
$$= (s^{k,1/4}, \phi_n)_0 - (d_n^1(u)(t_k), \phi_n)_1 - (s^k, \phi_n)_0$$
$$= -(D_\tau^2\eta^k - \eta^{k,1/4} + \rho^k, \phi_n)_0 + (s^{k,1/4} - s^k - \frac{1}{4}\tau_n D_\tau^2 s^k, \phi_n)_0,$$

where $s^{k,1/4} - s^k$ is precisely $\frac{1}{4}\tau_n D_\tau^2 s^k$, $k = 1,\dots,N-1$. Now, inserting $e_{n,1}^k$ and $e_{n,2}^k$ (in place of $v_n^{k+1/2}$ and $D_\tau^+ v_n^k$, respectively) into the inverse stability inequality 12.(30), we obtain

$$\max_{0\le k\le N-1} (|e_{n,1}^k|_1^2 + |e_{n,2}^k|_0^2) \le \gamma^2\Big(|e_{n,1}^0|_1^2 + |e_{n,2}^0|_0^2$$
$$+ \tau_n \sum_{k=1}^{N-1} |D_\tau^2\eta^k - \eta^{k,1/4} + \rho^k|_0^2\Big).$$

Our error estimate (43) now follows from the estimates (37), (39), and (40) if we choose $u_{n,1}^0 = U_n^{1/2}$, $u_{n,2}^0 = W_n^0$.

(iii) For $u_n^{k+1} \equiv u_{n,1}^k + \frac{1}{2}\tau_n u_{n,2}^k$, we have the following relation by virtue of (21),

$$u_n^{k+1} - U_n^{k+1} = u_{n,1}^k - U_n^{k+1/2} + \frac{1}{2}\tau_n(u_{n,2}^k - D_\tau^+ U_n^k).$$

Thus,

$$|u_n^{k+1} - U_n^{k+1}|_1 \le |u_{n,1}^k - U_n^{k+1/2}|_1 + \frac{1}{2}C\lambda|u_{n,2}^k - D_\tau^+ U_n^k|_0$$
$$= O(\tau_n^2 + h_n^{\mu+1}) \quad (n \in I)$$

results from (43) and the inverse inequality. □

This theorem points out that $u_{n,1}^k$ and $u_{n,2}^k$ constitute approximations to $U_n^{k+1/2}$ and $D_\tau^+ U_n^k$, respectively, and hence also to $u^{k+1/2}$ and $D_\tau^+ u^k$, respectively, where $u^k = u(t_k)$, $U_n^k = P_n^1 u^k$ and u is the solution of the given IVP. Moreover, we conclude from the error estimate (44) that u_n^{k+1} is an approximation to u^{k+1} itself to within the same order of accuracy. We actually compute u_n^{k+1}, $k = 0,\ldots,N-1$ (denoted as v^{k+1} in Section 4.3) by this scheme and these are related to the $u_{n,1}^k$, $u_{n,2}^k$ by $u_{n,1}^k = u_n^{k+1/2}$, $u_{n,2}^k = D_\tau^+ u_n^k$ (cf. 4.(39a-c)).

REFERENCES (cf. also References in Chapters 4, 11 and 12)

Ansorge & Hass (1970), Aubin (1979), Bramble & Sammon (1980)*, Ciarlet (1978), Dahlquist (1956)*, Davis (1963), Dupont (1973)*, Fairweather (1978), Mitchell & Griffiths (1980), Törnig (1963)*, Wahlbin (1978)*.

*Article

Bibliography

BOOKS AND MONOGRAPHS

Agmon, S. (1965): Elliptic boundary value problems. Van Nostrand, New York.

Ames, W. F. (1977): Numerical methods for partial differential equations. Academic Press-Nelson, New York-London.

Anselone, P. M. (1971): Collectively compact operator approximation theory. Prentice-Hall, Englewood Cliffs.

Ansorge, R. (1978): Differenzenapproximation partieller Anfangswertaufgaben. Teubner, Stuttgart.

Ansorge, R., Hass, R. (1970): Konvergenz von Differenzenverfahren für lineare und nichtlineare Anfangswertaufgaben. Lecture Notes in Math., Vol. 159. Springer, Berlin.

Atkinson, K. E. (1976): A survey of numerical methods for the solution of Fredholm integral equations of the second kind. SIAM, Philadelphia.

Aubin, J.-P. (1972): Approximation of elliptic boundary-value problems. Wiley, New York.

Aubin, J.-P. (1979): Applied functional analysis. Wiley, New York.

Babuska, I., Prager, M., Vitasek, E. (1966): Numerical processes in differential equations. Interscience, New York.

Bachman, G., Narici, L. (1969): Functional Analysis. Academic Press, New York.

Baker, C. T. H. (1977): The numerical treatment of integral equations, Clarendon Press, Oxford.

Böhmer, K. (1974): Spline Funktionen. Teubner, Stuttgart.

Bohl, E. (1981): Finite Modelle gewohnlicher Randwertaufgaben. Teubner, Stuttgart.

de Boor, C. (1978): A practical guide to splines. Applied Math. Sciences, Vol. 27. Springer, New York.

Bourbaki, N. (1965): Intégration, Ch. 1-4. Hermann, Paris.

Braun, M. (1983): Differential equations and their applications. Applied Math. Sciences, Vol. 15. Springer, New York.

Ciarlet, P. G. (1978): The finite element method for elliptic problems. Studies in Math. and its Applications, Vol. 4. North-Holland, Amsterdam.

Coddington, E. A., Levinson, N. (1955): Theory of ordinary differential equations. McGraw-Hill, New York.

Collatz, L. (1966): The numerical treatment of differential equations (3rd edition). Springer, Berlin.

Courant, H., Hilbert, R. (1966): Methods of mathematical physics, Vol. I & II. Interscience, New York.

Davis, P. J. (1963): Interpolation and approximation. Blaisdell, Waltham, Mass.

Dieudonné, J. (1969): Foundations of modern analysis. Academic Press, New York.

Dunford, N., Schwartz, J. T. (1966): Linear Operators I. Interscience, New York.

Edwards, K. E. (1965): Functional analysis. Theory and applications. Holt, Rinehart and Winston, New York.

Fairweather, G. (1978): Finite Element Galerkin methods for differential equations. Lecture Notes in Pure and Appl. Math. Vol. 34. Dekker, New York-Basel.

Forsythe, G. E., Wasow, W. R. (1960): Finite-difference methods for partial differential equations. Wiley, New York.

Gallagher, R. H. (1975): Finite element analysis. Fundamentals. Prentice Hall, Englewood Cliffs.

Garabedian, P. R. (1964): Partial differential equations. Wiley, New York.

Gladwell, I., Wait, R. (eds.) (1979): A survey of numerical methods for partial differential equations. Clarendon Press, Oxford.

Gottlieb, D., Orszag, S. A. (1977): Numerical analysis of spectral methods: Theory and applications. Regional Conference Series in Applied Mathematics, Vol. 26. SIAM, Philadelphia.

Grigorieff, R. D. (1973b): Theorie der Näherungsverfahren. Skriptum, Fachbereich Mathematik, TU Berlin, SS 1973.

Hartman, P. (1964): Ordinary differential equations. Wiley, New York.

Hille, E., Phillips, R. S. (1957): Functional analysis and semi-groups. AMS Coll. Publ., Vol. 31. AMS, Providence.

von der Houwen, P. J. (1968): Finite difference methods for solving partial differential equations. Math. Centre Tracts, Vol. 20. Math. Centrum, Amsterdam.

van der Houwen, P. J. (1977): Construction of integration formulas for initial value problems. North-Holland, Amsterdam.

Isaacson, E., Keller, H. B. (1966): Analysis of numerical methods. Wiley, New York.

John, F. (1967): Lectures on advanced numerical analysis. Notes on Math. and its Applications. Gordon & Breach, New York.

John, F. (1982): Partial differential equations. Applied Math. Sciences, Vol. 1. Springer, New York.

Kantorovich, L., Akilov, G. (1964): Functional analysis in normed spaces. Pergamon Press, Oxford.

Kato, T. (1966): Perturbation theory for linear operators. Springer, Berlin.

Keller, H. B. (1968): Numerical methods for two-point boundary-value problems. Blaisdell, Waltham, Mass.

Keller, H. B. (1976): Numerical solution of two-point boundary value problems. Regional Conference Series in Applied Math., Vol. 24. SIAM Philadelphia.

Krasnoselskii, M. A. (1964): Topological methods in the theory of nonlinear integral equations. Pergamon Press, Oxford.

Krasnoselskii, M. A., Vainikko, G. M., Zabreiko, P. P., Rutitskii, Y. B., Stetsenko, V. Y. (1972): Approximate solution of operator equations. Wolters-Noordhoff, Groningen.

Kuratowski, K. (1966): Topology, Vol. I. Academic Press, New York.

Linz, P. (1979): Theoretical numerical analysis. An introduction to advanced techniques. Wiley-Interscience, New York.

Lions, J. L., Magenes, E. (1972): Nonhomogeneous boundary value problems and applications. I, II, III. Springer, Berlin, 1972-73.

Luenberger, D. G. (1969): Optimization by vector space methods. Wiley, New York.

Marchuk, G. I. (1975): Methods of numerical mathematics. Applications of Mathematics, Vol. 2. Springer, New York.

Meis, T., Marcowitz, U. (1981): Numerical solution of partial differential equations. Applied Math. Sciences, Vol. 32. Springer, New York.

Michlin, S. G. (1969): Numerische Realisierung von Variationsmethoden. Akademie-Verlag, Berlin.

Mikhlin, S. G. (1970): Mathematical physics, an advanced course. North-Holland, Amsterdam.

Mikhlin, S. G., Smolitskiy, K. L. (1967): Approximate methods for solution of differential and integral equations. American Elsevier, New York.

Mitchell, A. R. (1969): Computational methods in partial differential equations. Wiley, London.

Mitchell, A. R., Griffiths, D. F. (1980): The finite difference method in partial differential equations. Wiley, Chichester.

Mitchell, A. R., Wait, R. (1977): The finite element method in partial differential equations. Wiley-Interscience, London.

Oden, J. T., Reddy, J. N. (1976): An introduction to the mathematical theory of finite elements. Wiley, New York.

Ortega, J. M., Poole, W. G. (1981): An introduction to numerical methods for differential equations. Pitman, London.

Ortega, J. M., Rheinboldt, W. C. (1970): Iterative solution of nonlinear equations in several variables. Academic Press, New York.

Rektorys, K. (1980): Variational methods in mathematics, science and engineering. D. Reidel, Dodrecht.

Richtmyer, R. D., Morton, K. W. (1967): Difference methods for initial-value problems. Interscience, New York.

Riesz, F., Sz.-Nagy, B. (1968): Vorlesungen über Funktionalanalysis. Verlag Deutscher Wissenschaften, Berlin.

Rinow, W. (1961): Die innere Geometrie der metrischen Räume. Springer, Berlin.

Rjabenki, V. S., Filippow, A. F. (1960): Über die Stabilitat von Differenzengleichungen. Deutscher Verlag der Wissenschaften, Berlin.

Rudin, W. (1966): Real and complex analysis. Mc-Graw-Hill, New York.

Smirnow, W. I. (1964): Lehrgang der höheren Mathematik. Teil II. Deutscher Verlag der Wissenschaften, Berlin.

Smirnow, W. I. (1971): Lehrgang der höheren Mathematik. Teil V. Deutscher Verlag der Wissenschaften, Berlin.

Smith, G. D. (1978): Numerical solution of partial differential equations. Finite difference methods. Oxford Applied Mathematics and Computing Science Series. Clarendon Press, Oxford.

Smithies, F. (1958): Integral Equations. Cambridge Univ. Press, Cambridge.

Stetter, H. J. (1973): Analysis of discretization methods for ordinary differential equations. Springer Tracts in Nat. Phil., Vol. 23. Springer, Berlin.

Stoer, J. (1979): Einführung in die Numerische Mathematik I. Heidelberger Taschenbücher, Bd. 105. Springer, Berlin.

Stoer, J. Bulirsch, R. (1978): Einführung in die Numerische Mathematik II. Heidelberger Taschenbücher, Bd. 114. Springer, Berlin.

Strang, G., Fix, G. J. (1973): An analysis of the finite element method. Prentice-Hall, Englewood Cliffs.

Stummel, F. (1973b): Approximation methods in analysis. Aarhus Univ. Lecture Notes Series, Vol. 35.

Stummel, F., Hainer, K. (1982): Praktische Mathematik. Teubner, Stuttgart. (engl.: Introduction to numerical analysis, Scottish Academic Press, Edinburgh, 1980).

Törnig, W. (1979): Numerische Mathematik fur Ingenieure und Physiker. Band 1 and 2. Springer, Berlin.

Vainberg, M. M. (1964): Variational methods for the study of nonlinear operators. Holden-Day, San Francisco.

Vainikko, G. (1976): Funktionalanalysis der Diskretisierungsmethoden. Teubner-Verlag, Leipzig.

Varga, R. S. (1962): Matrix iterative analysis. Prentice-Hall, Englewood Cliffs.

Varga, R. S. (1971): Functional analysis and approximation theory in numerical analysis. Regional Conf. Series in Applied Math., Vol. 3. SIAM, Philadelphia.

Wachspress, E. L. (1966): Iterative solution of elliptic systems. Prentice-Hall, Englewood Cliffs.

Walter, W. (1976): Gewöhnliche Differentialgleichungen. Eine Einführung. Heidelberger Taschenbücher, Bd. 110. Springer, Berlin.

Yanenko, N. N. (1971): The method of fractional steps. Springer, Berlin.

Yosida, K. (1968): Functional analysis. Springer, Berlin.

ARTICLES

Anselone, P. M. (1965): Convergence and error bounds for approximate solutions of integral and operator equations. In: Error in digital computation, Vol. 2 (Proc. Symp., Madison, Wisc., 1965), 231-252. Wiley, New York.

Anselone, P. M. Ansorge, R. (1979): Compactness principles in nonlinear operator approximation theory. Numer. Funct. Anal. Optim. 1, 589-618.

Anselone, P. M., Ansorge, R. (1981): A unified framework for the discretization of nonlinear operator equations. Numer. Funct. Anal. Optim. 4, 61-99.

Anselone, P. M., Moore, R. H. (1964): Approximate solutions of integral and operator equations. J. Math. Anal. Appl. 10, 268-277.

Anselone, P. M., Palmer, T. W. (1968): Collectively compact sets of linear operators. Pac. J. Math. 25, 417-422.

Atkinson, K. E. (1967): The numerical solution of Fredholm integral equations of the second kind. SIAM J. Numer. Anal. 4, 337-348.

Aubin, J. P. (1967a): Approximation des espaces de distribution et des opérateurs différentiels. Bull. Soc. Math. France Mém. 12, 1-139.

Aubin, J. P. (1967b): Behaviour of the error of the approximate solutions of boundary value problems for linear elliptic operators by Galerkin's and finite difference methods. Ann. di Pisa 21, 599-637.

Babuska, I., Aziz, A. K. (1972): Survey lectures on the mathematical foundations of the finite element method. In: The mathematical foundations of the finite element method with applications to partial differential equations (Proc. Conf. Baltimore, 1972), 5-359, Academic Press, New York.

Baker, G. A. (1976): Error estimates for finite element methods for second order hyperbolic equations. SIAM J. Numer. Anal. 13, 564-576.

Brakhage, H. (1960): Über die numerische Behandlung von Integralgleichungen nach der Quadraturformelmethode. Numer. Math. 2, 183-196.

Bramble, J. H., Sammon, P. H. (1980): Efficient higher order single step methods for parabolic problems: Part I. Math. Comp. 35, 655-677.

Browder, F. E. (1967): Approximation-solvability of nonlinear functional equations in normed linear spaces. Arch. Rat. Mech. Anal. 26, 33-42.

Chartres, B., Stepleman, R. (1972): A general theory of convergence for numerical methods. SIAM J. Numer. Anal. 9, 476-492.

Ciarlet, P. G. (1970): Discrete maximum principle for finite-difference operators. Aequationes Math. 4, 338-352.

Ciarlet, P. G., Schultz, M. H., Varga, R. S. (1967): Numerical methods of high-order accuracy for nonlinear boundary value problems. I. One dimensional problem. Numer. Math. 9, 394-430.

Dahlquist, G. (1956): Convergence and stability in the numerical integration of ordinary differential equations. Math. Scand. 4, 33-53.

von Dein, H. (1976): Konvergenzbedingungen bei der numerischen Lösung nichtlinearer Anfangswertaufgaben mittels Differenzenverfahren. In: Numerische Behandlung von Differentialgleichungen, insbesondere mit der Methode der finiten Elemente (Proc. Conf. Oberwolfach, 1975), ISNM, Vol. 31. Birkhäuser, Basel.

Dorr, F. W. (1970): The numerical solution of singular perturbations of boundary value problems. SIAM J. Numer. Anal. 281-313.

Douglas, J., Dupont, J. (1974): Galerkin approximations for the two-point boundary problem using continuous, piecewise polynomial spaces. Numer. Math. 22, 99-109.

Dupont, T. (1973): L^2-estimates for Galerkin methods for second order hyperbolic equations, SIAM J. Numer. Anal. 10, 880-889.

Graham, I. G., Sloan, I. H. (1982): On the compactness of certain integral operators. J. Math. Anal. Appl., to appear.

Grigorieff, R. D. (1969): Approximation von Eigenwertproblemen und Gleichungen zweiter Art in Hilbertschen Räumen. Math. Ann. 183, 45-77.

Grigorieff, R. D. (1972): Über die Fredholm-Alternative bei linearen approximationsregulären Operatoren. Applicable Anal. 2, 217-227.

Grigorieff, R. D. (1973a): Zur Theorie linearer approximationsregulärer Operatoren. I. Math. Nachr. 55, 233-249, II. Math. Nachr. 55, 251-263.

Grigorieff, R. D. (1975): Über diskrete Approximationen nichtlinearer Gleichungen 1. Art. Math. Nachr. 69, 253-272.

Ikebe, Y. (1972): The Galerkin method for the numerical solution of Fredholm integral equations of the second kind. SIAM Rev. 14, 465-491.

John, F. (1968): On quasi-isometric mappings, I. Comm. Pure Appl. Math. 21, 77-110.

Kreth, H. (1975): Der Nachweis der Existenz verallgemeinerter Lösungen quasilinearer Anfangswertaufgaben mittels Differenzenverfahren. Computing 15, 251-261.

Nelson, P., Jr., Victory, H. D., Jr. (1979): Theoretical properties of one-dimensional discrete ordinates. SIAM J. Numer. Anal. 16, 270-283.

Parter, S. V. (1980): On the roles of "stability" and "convergence" in semidiscrete projection methods for initial-value problems. Math. Comp. 34, 127-154.

Pereyra, V. (1967): Iterated deferred corrections for nonlinear operator equations. Numer. Math. 10, 316-323.

Petryshyn, W. V. (1967a): Projection methods in nonlinear numerical functional analysis. J. Math. Mech. 17, 353-372.

Petryshyn, W. V. (1967b): Remarks on the approximation-solvability of nonlinear functional equations. Arch. Rat. Mech. Anal. 26, 43-49.

Petryshyn, W. V. (1968a): On the approximation solvability of nonlinear equations. Math. Ann. 177, 156-164.

Petryshyn, W. V. (1968b): On projectional-solvability and the Fredholm alternative for equations involving linear a-proper operators. Arch. Rat. Mech. Anal. 30, 270-284.

Radon, J. (1919): Über lineare Funktionaltransformationen und Funktionalgleichungen. Sitzsber. Akad. Wiss. Wien 128, 1083-1121.

Reinhardt, H.-J. (1975a): Nonlinear mappings in metric discrete limit spaces and their topological properties. Collect. Math. 26, 173-204.

Reinhardt, H.-J. (1975b): On the existence of generalized solutions and the convergence of difference methods for nonlinear initial-value problems. Manuscripta Math. 17, 151-170.

Reinhardt, H.-J. (1977): Differentiable difference approximations for nonlinear initial-value problems. I. Applicable Anal. 6 (1976/77), no. 4, 281-298, II. Applicable Anal. 7 (1977/78), no. 1, 81-96.

Sloan, I. H. (1980a): A review of numerical methods for Fredholm equations of second kind. In: The Application and Numerical Solution of Integral Equations (Proc. Sem., Australian Nat. Univ., Canberra, 1978), 51-74. Sijthoff & Noordhoff, Alphen aan den Rijn.

Sloan, I. H. (1980b): On choosing the points in product integration. J. Math. Phys. 21, 1032-1039.

Sloan, I. H. (1981): Analysis of general quadrature methods for integral equations of the second kind. Numer. Math. 38, 263-278.

Sloan, I. H., Noussair, E., Burn, B. J. (1979): Projection methods for equations of the second kind. J. Math. Anal. Appl. 69, 84-103.

Stetter, H. J. (1965a): Asymptotic expansions for the error of discretization algorithms. J. SIAM Numer. Anal. 2, Ser. B., 265-280.

Stetter, H. J. (1966): Stability of nonlinear discretization algorithms. In: Numerical Solution of Partial Differential Equations (Proc. Symp., College Park, 1965), 111-123. Academic Press, New York.

Stummel, F. (1970): Diskrete Konvergenz linearer Operatoren. I. Math. Ann. 190 (1970), 45-92. II. Math. Z. 120 (1971), 231-264. III. In: Linear Operators and Approximation (Proc. Conf. Oberwolfach, 1971), 196-215. Birkhäuser, Basel, 1972.

Stummel, F. (1972): Singular perturbations of elliptic sesquilinear forms. In: Differential equations (Proc. Conf. Dundee, 1972), 155-180. Lect. Notes in Math., Vol. 280. Springer, Berlin.

Stummel, F. (1973a): Discrete convergence of mappings. In: Topics in numerical analysis (Proc. Royal Irish Academy Conf., Dublin 1972), 285-310. Academic Press, London.

Stummel, F. (1974a): Perturbation theory for Sobolev spaces. Proc. Royal Soc. Edinburgh 73A (1974/75), 1-49.

Stummel, F. (1974b): Perturbations of non-linear integral operators. Proc. Royal Soc. Edinburgh 74A (1974/75), 55-70.

Stummel, F. (1975): Discretely uniform approximation of continuous functions. J. Approx. Theory 13, 178-191.

Stummel, F. (1976a): Perturbation of domains in elliptic boundary value problems. In: Application of methods of functional analysis to problems of mechanics (Proc. Conf., Marseille, 1975), 110-136. Lect. Notes in Math. Vol. 503. Springer, Berlin.

Stummel, F. (1976b): Stability and discrete convergence of differentiable mappings. Rev. Roumaine Math. Pures Appl. 21, 63-96.

Stummel, F. (1977): Approximation methods for eigenvalue problems in elliptic differential equations. In: Numerik und Anwendungen von Eigenwertaufgaben und Verzweigungsproblemen (Proc. Conf. Oberwolfach, 1976), 133-165. ISNM, Vol. 38. Birkhäuser, Basel.

Stummel, F., Reinhardt, H.-J. (1973): Discrete convergence of continuous mappings in metric spaces. In: Numerische, insbesondere approximationstheoretische Behandlung von Funkionalgleichungen (Proc. Conf. Oberwolfach, 1972), 218-242. Lecture Notes in Math., Vol. 333. Springer, Berlin.

Thomée, V., Wahlbin, L. (1975): On Galerkin methods in semilinear parabolic problems. SIAM J. Numer. Anal. 12, 378-389.

Törnig, W. (1963): Über Differenzenverfahren in Rechteckgittern zur numerischen Lösung quasilinearer hyperbolischer Differentialgleichungen. Numer. Math. 5, 353-370.

Törnig, W., Ziegler, M. (1966): Bemerkungen zur Konvergenz von Differenzenapproximationen für quasilinear hyperbolische Anfangswertprobleme in zwei unabhängingen Veränderlichen. Z. Angew. Math. Mech. 46, 201-210.

Trotter, H. F. (1958): Approximation of semi-groups of operators. Pac. J. Math. 8, 887-919.

Vainikko, G. M. (1967): Galerkin's perturbation method and the general theory of approximate methods for nonlinear equations. U.S.S.R. Comput. Math. Math. Phys. 7/4, 1-41.

Vainikko, G. M. (1969): The compact approximation principle in the theory of approximation methods. U.S.S.R. Comput. Math. Phys. 9/4, 1-32.

Wahlbin, L. B. (1978): Maximum norm error estimates in the finite element method with isoparametric quadratic elements and numerical integration. RAIRO Anal. Numér. 12, no. 2, 173-202.

Witsch, K. (1978): Projektive Newton-Verfahren und Anwendungen auf nichtlinearer Randwertaufgaben. Numer. Math. 31, 209-230.

Wolf, R. (1974): Über linear approximationsreguläre Operatoren. Math. Nachr. 59, 325-341.

Atkinson, K., Graham, I., Sloan, I. (1983): Piecewise continuous collocation for integral equations, SIAM J. Numer. Anal., 20, 172-186.

Glossary of Symbols

$B(E,E)$ 112, $B(E,F)$ 153

$\dot{B}_\rho(v)$, $B_\rho(u)$ 70, 153

$C(G)$ 52, $C^r[a,b]$ 5, $C(I_h)$, $C(I_h')$ 6, $C_0^\infty(a,b)$ 28

$D_h^\pm v_h$, D_h, D_h^2 7, D_τ^+ 336

$H_0^1(a,b)$ 28, $H^2(0,1)$ 89

$L^2(a,b)$ 28, $L^p(G)$, L^p 140, $L_0^p(G)$ 140

Lim sup G_n, Lim inf G_n, Lim G_n 133

$\ell^\infty(E_n)$ 161

meas 141

$\mu(u)$ 145

$N(L)$ 153

$\nu_n \rightharpoonup \nu$ 143

$O(\cdot)$ 7, $o(\cdot)$ 347

$W^{1,2}(a,b)$ 28, $(.,.)_o$ 22, $(.,.)_1$ 22, $(.,.)_2$ 23, $\langle w,\Psi\rangle$ 30, $[u,v]_1$ 24

$||\cdot||_0$, $||\cdot||_1$ 22, $||\cdot||_p$ 271, $||\cdot||_{o,p}$, $||\cdot||_{o,p,G}$ 140, 148

$||\cdot||_{o,\infty}$ 247, $||\cdot||_\infty$ 227, 271, 309, $|\cdot|_{o,\infty}$ 310

$|\cdot|_{o,n}$ 320, $||\cdot||_{p,n}$ 272, $|||\cdot|||_{p,n}$ 179, 290

$|u,E_n|$ 125

∇u 24

$\ni$: such that

$[0,T]_n$, $[0,T]_n'$ 113

Subject Index

W

Applied Mathematical Sciences

cont. from page ii

39. Piccinini/Stampacchia/Vidossich: **Ordinary Differential Equations in R^n.**
40. Naylor/Sell: **Linear Operator Theory in Engineering and Science.**
41. Sparrow: **The Lorenz Equations: Bifurcations, Chaos, and Strange Attractors.**
42. Guckenheimer/Holmes: **Nonlinear Oscillations, Dynamical Systems and Bifurcations of Vector Fields.**
43. Ockendon/Tayler: **Inviscid Fluid Flows.**
44. Pazy: **Semigroups of Linear Operators and Applications to Partial Differential Equations.**
45. Glashoff/Gustafson: **Linear Optimization and Approximation: An Introduction to the Theoretical Analysis and Numerical Treatment of Semi-Infinite Programs.**
46. Wilcox: **Scattering Theory for Diffraction Gratings.**
47. Hale et al.: **An Introduction to Infinite Dimensional Dynamical Systems — Geometric Theory.**
48. Murray: **Asymptotic Analysis.**
49. Ladyzhenskaya: **The Boundary-Value Problems of Mathematical Physics.**
50. Wilcox: **Sound Propagation in Stratified Fluids.**
51. Golubitsky/Schaeffer: **Bifurcation and Groups in Bifurcation Theory, Vol. I.**
52. Chipot: **Variational Inequalities and Flow in Porous Media.**
53. Majda: **Compressible Fluid Flow and Systems of Conservation Laws in Several Space Variables.**
54. Wasow: **Linear Turning Point Theory.**
55. Yosida: **Operational Calculus: A Theory of Hyperfunctions.**
56. Chang/Howes: **Nonlinear Singular Perturbation Phenomena: Theory and Applications.**
57. Reinhardt: **Analysis of Approximation Methods for Differential and Integral Equations.**
58. Dwoyer/Hussaini/Voigt (eds.): **Theoretical Approaches to Turbulence.**

MIX
Papier aus verantwortungsvollen Quellen
Paper from responsible sources
FSC® C105338

If you have any concerns about our products,
you can contact us on
ProductSafety@springernature.com

In case Publisher is established outside the EU,
the EU authorized representative is:
Springer Nature Customer Service Center GmbH
Europaplatz 3, 69115 Heidelberg, Germany

Printed by Libri Plureos GmbH
in Hamburg, Germany